The Art of Natural History

38 De la uolubelle. Capitolo. xxxviij.

Dixe diasconde che de questa uolubelle sene truoua piu spetie. una
spetia sie chiama cussus co e elera. e questa sie la uolubelle ma
ore e la saluega. e questa herba sie simele a la uolubelle menore
ma le piu dura cha la menore. E questa uolubele maore ha tre spe
cie luna e biancha. e chiamase biancha. per la bianchecça del so fruto.
La segonda spetia e chiama negra. per la negreça del so fruto. La
terça sie chiama chacus lupolo. e no ha fruto o uer somença e ha i
rami menu. e le foie menu. e uerde pine de canton amare e grose
Ancora sene truoua una altra spetia che uen chiama acinn. i
menore cha le soura dite spetie. la quale ha le foie simele a le foie
de la elera. un pucho menore e ha li rami longi i quale se inuolçe
a ogna cosa e a ogno arbore che li se apresso. e nasse in le uigne e in li
formenti e in li campi. Ancora se ne truoua una altra spetia i
che se chiama athin ———— e questa e la uollubelle menore
e a someiase in le foie a la uolubelle de soura. noma che le foie de i
questa e menore e piu ironde. e soura questa e pili. e ha li rami me
nu de longeça de cinque palmi o uer de sye che uen fura de una ray
xe e sie piene de foie e le foie suo ha sauore pontico e nasce entro el
formento e in gy luogi desmestigi. De la uertu del cussus. Halie

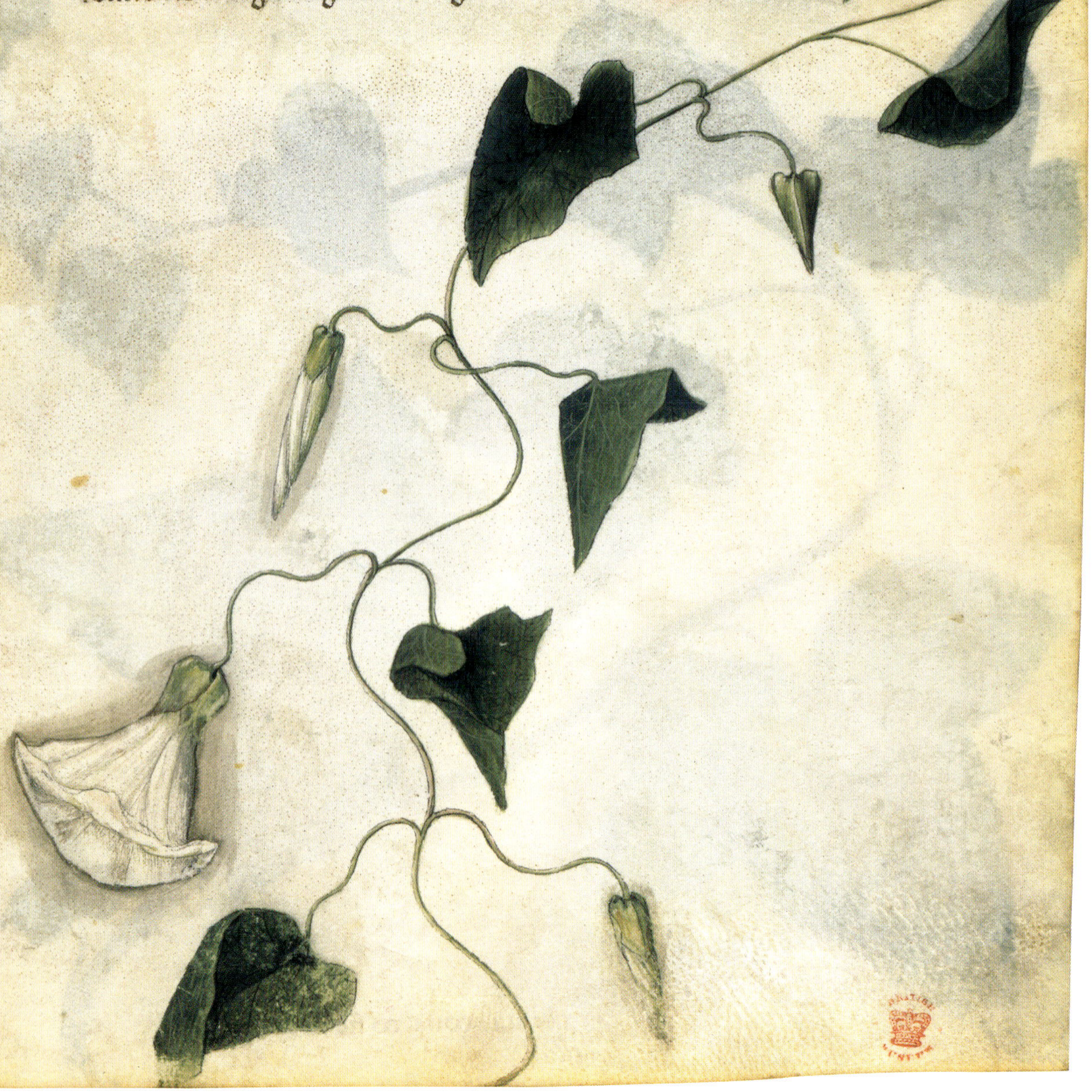

STUDIES IN THE HISTORY OF ART • 69 •
Center for Advanced Study in the Visual Arts
Symposium Papers XLVI

The Art of Natural History: Illustrated Treatises and Botanical Paintings, 1400–1850

Edited by Therese O'Malley and Amy R. W. Meyers

National Gallery of Art, Washington
Distributed by Yale University Press
New Haven and London

This volume was produced by the Center for Advanced Study in the Visual Arts and the Publishing Office, National Gallery of Art, Washington
www.nga.gov

Editorial Board
JOHN OLIVER HAND, *chairman*
SUSAN M. ARENSBERG
SARAH FISHER
THERESE O'MALLEY

Editor in Chief
JUDY METRO

Series Editor
THERESE O'MALLEY

Program Assistants
COLLEEN KELLY HOWARD
ANNE NELLIS
JESSICA EVANS

Managing Editor
CYNTHIA WARE

Production Manager
CHRIS VOGEL

Design Manager
WENDY SCHLEICHER

Editor
MONICA S. RUMSEY

Editorial Assistant
MAGDA NAKASSIS

Prepress production by Marquand Books, Inc., Seattle
The type is Trump Mediaeval, set by Maggie Lee.
Layout by Tina Kim
Color separations by iocolor, Seattle

The text paper is Stora Enso Matt.
Printed and bound in China by Four Colour Imports Ltd., Louisville, Kentucky

Distributed by Yale University Press, New Haven and London
www.yalebooks.com

Abstracted and indexed in BHA (Bibliography of the History of Art) and Art Index

Proceedings of the symposium "The Art and History of Botanical Painting and Natural History Treatises," organized by the Center for Advanced Study in the Visual Arts and sponsored by the Samuel H. Kress Foundation in honor of Franklin D. Murphy. The symposium was held 3–4 May 2002 in Washington.

The Center for Advanced Study in the Visual Arts was founded in 1979 at the National Gallery of Art to foster the study of the history, theory, and criticism of art, architecture, urbanism, and photographic media, through programs of meetings, research, publication, and fellowships, and through the formation of a community of scholars.

Library of Congress Cataloging-in-Publication Data

The art of natural history : illustrated treatises and botanical paintings, 1400–1850 / edited by Therese O'Malley and Amy R. W. Meyers.
p. cm.—(Studies in the history of art ; 69) (Symposium papers / Center for Advanced Study in the Visual Arts ; 46)
Includes bibliographical references and index.
"Proceedings of the symposium 'The Art and History of Botanical Painting and Natural History Treatises,' sponsored by the Samuel H. Kress Foundation in honor of Franklin D. Murphy. The symposium was held 3–4 May 2002 in Washington."
ISBN 978-0-300-12158-2 (alk. paper)
1. Natural history illustration—Congresses. 2. Art and science—Congresses. I. Meyers, Amy R. W. II. O'Malley, Therese. III. Center for Advanced Study in the Visual Arts (U.S.)
QH46.5.A78 2008
760'.0443—dc22

2007034083

ISBN 978-0-300-12158-2
ISSN 0091-7338

Frontispiece: Anonymous, *Convolvulus* (bindweed), from the Carrara Herbal, c. 1400, woodcut. The British Library, Egerton MS 2020, folio 94r

Contents

Preface

The papers gathered in this volume were presented at the symposium "The Art and History of Botanical Painting and Natural History Treatises," held at the National Gallery of Art on 3–4 May 2002. The symposium was organized by the Center for Advanced Study in the Visual Arts and supported, as is this publication, by an endowment generously established by the Samuel H. Kress Foundation in memory of Franklin D. Murphy. The proceedings of the first Kress-Murphy Symposium were published in 2003 as *The Treatise on Perspective: Published and Unpublished,* edited by Lyle Massey. Therese O'Malley, associate dean of CASVA, and Amy R. W. Meyers, director of the Yale Center for British Art, have served as expert scholarly editors for this volume, and we are deeply grateful to them for preparing the manuscript for publication. Cynthia Ware, managing editor of the series, has played an essential role throughout.

The present volume focuses on illustration as an expression of visual thinking, based on experience, about natural history in the period 1400–1850. The significance of relationships between word and image is now a staple of scholarship: but even more telling in this context, and often not nearly as well understood, are the rhetoric and signifying power of such images and their performance in diverse cultures, especially across linguistic boundaries in the New World and the Old.

The prominence given to the image in the symposium was also a response to the display of remarkable botanical illustrations then on exhibition at the National Gallery of Art. *The Flowering of Florence: Botanical Art for the Medici,* organized by Lucia Tongiorgi Tomasi and Gretchen A. Hirschauer, included, in addition to paintings and drawings, works in pietra dura and even embroidered silk. This second Kress-Murphy symposium was further inspired by the legacy of Franklin D. Murphy. Paolo Manuzio, whose emphasis on "making knowledge visible" is cited by the editors, was the son of the great printer Aldus Manutius. Franklin D. Murphy was an avid collector of Aldine volumes, and the topic under examination continues in the Aldine tradition of the dissemination of knowledge in a visually persuasive way.

Among the outcomes of the new archaeology of knowledge advanced by Michel Foucault in the 1960s was the recognition that representing, speaking, classifying, and exchanging all belong to the fundamental codes of the early modern culture of human sciences. Natural history, furthermore, was central to Foucault's thinking because of its privileged place in the new culture of observation and experience. In the vanguard of a redefinition of history itself, "natural history" was reconceived in the seventeenth century as the examination of experience and knowledge, not the compilation and reconciliation of authorities. The essays that follow are indebted to this new understanding of the structure of natural history. They are also, however, founded on the more intense analysis of specific historical cases and the attentive and critical approach to the question of the styles and traditions of representation that now characterize the field.

ELIZABETH CROPPER
Dean, Center for Advanced Study in the Visual Arts

PICTORES OPERIS,

Heinricus Füllmaurer. Albertus Meyer.

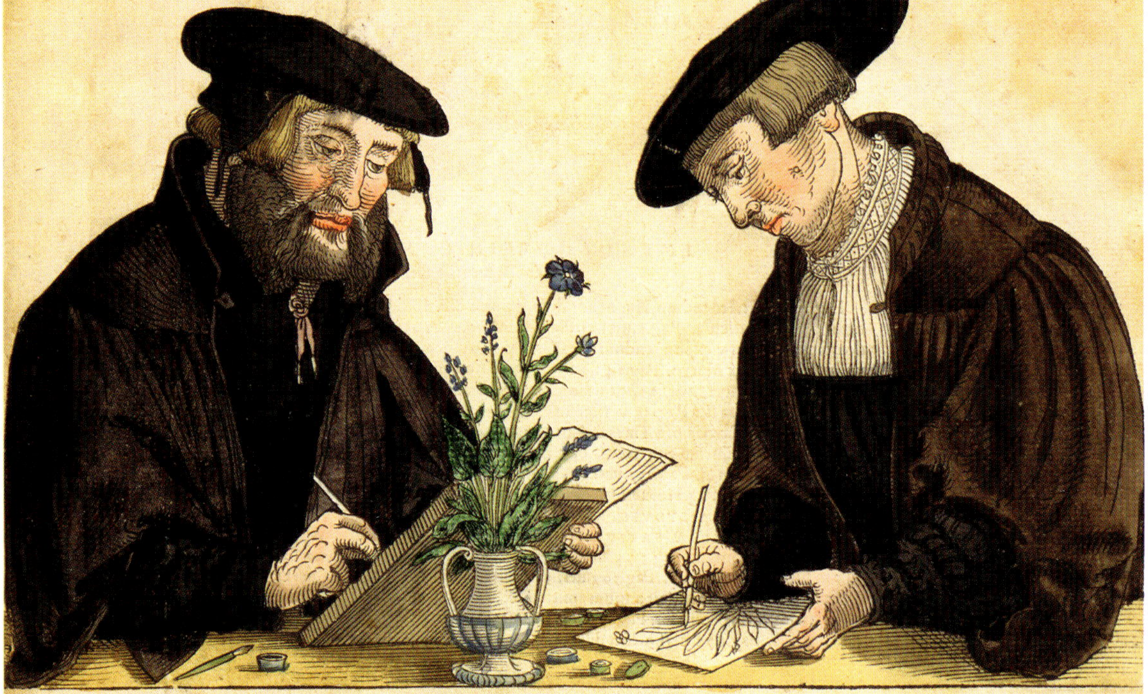

SCVLPTOR

Vitus Rodolph. Speckle.

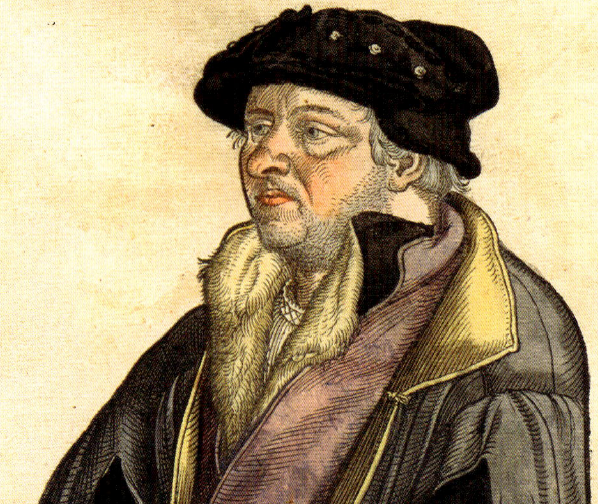

THERESE O'MALLEY
National Gallery of Art, Center for Advanced Study in the Visual Arts

AMY R. W. MEYERS
Yale Center for British Art

Introduction

> . . . I here present to the World my imperfect Indeavours; which though they shall prove no other way considerable, yet, I hope, they may be in some measure useful to the main Design of a *reformation* in Philosophy, if it be only by shewing, that there is not so much requir'd towards it, any strength of *imagination*, or exactness of *Method*; or depth of *Contemplation* . . . as a *sincere Hand*; and a *faithful Eye*, to examine, and to record, the things themselves as they appear.
>
> —Robert Hooke, from the preface to *Micrographia* (1665)

De historia stirpium commentarii insignes, published in 1542 by Leonhard Fuchs (1501–1566), is celebrated for its approximately five hundred fifty plates, including one remarkable page (shown opposite) on which the three artists who produced the illustrations are portrayed, one drawing, one transferring the drawing onto a wood block, and one holding the cutting iron with which the woodcuts are made. Similarly, *Plantae selectae* (1750–1773), by Christoph Jacob Trew (1695–1769), with illustrations by Georg Dionysius Ehret (1708–1770) and engravings by Johann Jakob Haid (1704–1767) and his son, Johann Elias Haid (1739–1809), shows author, artist, and engraver in full-page portraits (figs. 1, 2). Both examples include live flowers, in vases or hand held, linking these illustrators to the artistic traditions of still life painting and manuscript illumination. The portraits exemplify issues, discussed in the following essays, in the illustration of books and manuscripts on natural history, botany, medicine, and garden design. Contrary to earlier scholarship, which considered scientific illustrations not as the work of artists but as objective, "in the same way as a camera,"[1] the contributions to this volume demonstrate a more complex role for illustrators who were both artists and either naturalists or gardeners and who brought to their work both aesthetic judgment and empirical observation. The portraits represent the importance of the illustrator in visual thinking, or, in the words of the sixteenth-century naturalist Paolo Manuzio, in "making knowledge visible."[2] The process of depiction was parallel to the process of scientific study in that it was dependent upon direct experience with objects and natural processes. Robert Hooke's recommendation for the scientific illustrator carries with it the full sense of responsibility on the part of the artist whose task it was to represent two-dimensionally the physical object and the natural knowledge that resulted from that translation.

The works under consideration in this volume, which were produced primarily in Italy, Spain, Germany, England, and Holland, represent a moment in the development of natural sciences during the early modern period when nature assumed a new kind of authority. Our authors have selected illustrated works that raise questions about

Artist unknown, portraits of Heinrich Füllmaurer (illustrator), Albert Meyer (illustrator), and Veit Rudolf Speckle (engraver), from Leonhard Fuchs, *De historia stirpium commentarii insignes* (Basel, 1542), woodcut
Dumbarton Oaks Research Library and Collection, Washington, D.C.

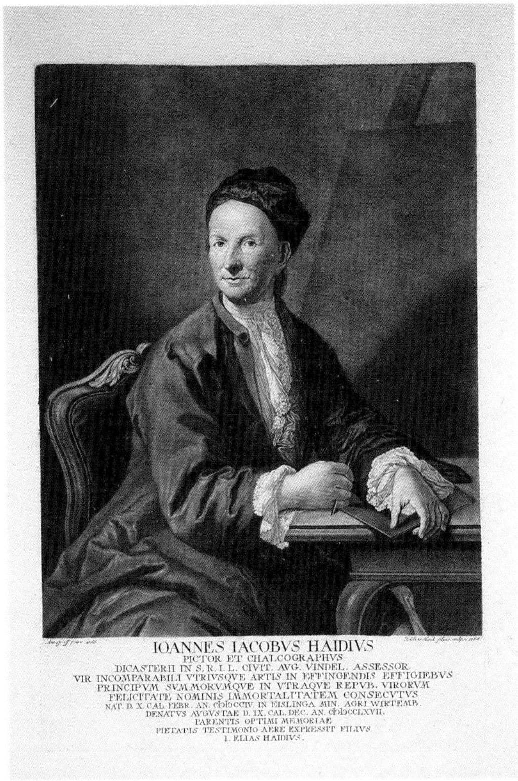

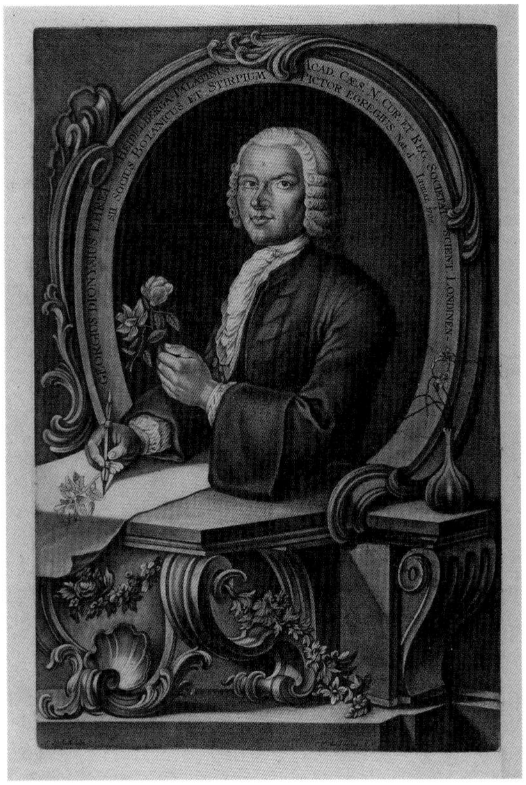

1. Anton Graf and Johann Elias Haid, portrait of Johann Jakob Haid, from Christoph Jacob Trew, *Plantae selectae* (Nuremberg, 1750–1773), engraving
Library of Congress, Washington, D.C.

2. Johann Jakob Haid, portrait of Georg Dionysius Ehret, from Christoph Jacob Trew, *Plantae selectae* (Nuremberg, 1750–1773), engraving
The LuEsther T. Mertz Library of The New York Botanical Garden, The Bronx, New York

the relationship of text and images on one hand and states of natural knowledge on the other. From a wide range of disciplinary perspectives, they consider innovation, radical disjuncture, and contradiction in the development and transmission of knowledge in illustrated manuscripts and books, as well as the patronage, readership, reception, and technologies of production of those works.

In the seventeenth century, new techniques of observation (the microscope) and of representation (improved pigments and printmaking methods) pushed naturalistic artisans to the center of the momentous shift in nature study and the evolving authority of nature. Empiricism was a touchstone for the new science, which was driven by imperial power and new technologies but was also inflected by traditions of knowing that had informed much of the history of philosophy for centuries. Drawings produced in the field were of particular importance in the transformation of the study of nature and in the emergence of empiricism. Field study took place both in situ and in laboratories and botanical gardens, where close scrutiny and the microscopic lens propelled the new sciences forward. Naturalist-artists were concerned not only with natural production but also with the effects of natural processes such as the passage of time and environmental change on plants as they were transported around the world from originating habitats to gardens and collections. Botanical gardens, the laboratories for experimentation and study of collected transported specimens, were centers for the production of many treatises on botany and natural history.

The emergence of natural science has been linked to the history of books, but this relationship is generally discussed in terms of texts. In contrast, these illustrated works, some well known, some newly discovered, are considered in this volume in terms of the images' role and relationship to text in the production of knowledge. It is our contention that visual thinking was part of the construction of knowledge gained through the processes of description.

The essays that follow reflect the international canon of fundamental works that emerged in the study of nature from the Renaissance through the Enlightenment and on through the professionalization of science in the nineteenth century. They also consider a fraction of the vast body of written and illustrated journals, notebooks, field notes, and sketches that has recently begun to be studied. The breadth of selection in this volume suggests the importance of looking beyond the best-known publications to the full range of compilations of texts and images.

In her examination of artisanal knowledge and the representation of nature in sixteenth-century Germany, Pamela H. Smith raises several issues that are picked up in subsequent essays, particularly the idea that natural knowledge derived from direct experience and from subjecting that experience to representation in words and images. She argues that the label "innovative and scientific," often applied to the volume *Herbarum vivae eicones* (1532–1536), written by Otto Brunfels (1488–1534), is warranted not by Brunfels' text but by the illustrations, the work of Hans Weiditz (c. 1495–c. 1536). She sees Weiditz in the tradition of artists who made nature studies, led by the Flemish painters, then Martin Schongauer (c. 1450–1491), and, finally, Albrecht Dürer (1471–1528). She uses the example of the sudden appearance of naturalism in the fifteenth century in Padua, where the Carrara family, patrons of Cennino Cennini (c. 1370–c. 1440) and his methods of life casting, attempted to create a theater of state through their patronage of artisanal mastery. Competition for patronage, livelihood, and fame elevated both the status of artisanal skill and the knowledge of nature. Rather than a Petrarchan, or top-down, approach to imitation, Smith argues for two-way intellectual traffic between artisans and scholars.

Alain Touwaide looks at competing scientific models in Renaissance medicine. Through an examination of the treatises of Ferrara physician Nicolo Leoniceno (1428–1524), he argues that the progression of botany and pharmacology was impeded by Leoniceno's emphasis on philology and the recovery of ancient texts. In a redirection of effort, writers such as Pietro Andrea Mattioli (1500–1577) and Andrés de Laguna (1499–1560) reunited discursive text with practical uses of plants. They followed German forerunners who, like Weiditz, challenged textual tradition with artistic innovation, and Fuchs, who relied on personal observation guided by the classical textual heritage.

Claudia Swan examines the various uses of botanical treatises in the Netherlands in the period around 1600. She places the botanical garden of Leiden University at the center of different modes of botanical studies, as reflected in various treatises produced there. In two works of Carolus Clusius (1526–1609), *Rariorum plantarum historia* (1601) and *Exoticorum libri decem* (1605), images served to complement verbal description with a morphological account of the natural world. Fransiscus Raphelengius (d. 1628), who incorporated pages from three different treatises in his codex, or manuscript volume (c. 1618), drew from the works of Rembert Dodanaeus (1517–1585), Mathias Lobelius (1538–1616), and Clusius, including original watercolors that were the basis of woodcuts for Clusius' treatise. The *Botanologicon* (1534) of Euricius Cordus (1484–1535) was based on an autopic experience of nature and on collating experience with knowledge previously acquired. Cordus wrote this text as a point of departure for an expedition to gather more information about plants en route: "their names and reputed virtues, as I may gather from the old wives I meet upon the way."

Janice Neri considers visual techniques and artistic traditions of still life painting, manuscript illumination, and natural history illustration deployed in the first representations of what was seen through the microscope in two-dimensional images published by Robert Hooke (1635–1703) in his *Micrographia* (1665). She argues that published treatises actually worked to conceal the varied circumstances of their production. She comes to this conclusion through an important new discovery of a series of sketches that she attributes to Hooke. This discovery sheds light on his early working practices and the development of visual and verbal observations. The printed images we know so well were complex translations of Hooke's early sketches. Neri's visual analysis of the sketches and printed illustrations explains Hooke's dilemmas in representation—for example, how to depict relative size or scale. This Hooke solved by adding drawings of actual size next to the magnified images. He also identified magnified images by enclosing objects in round frames. Framing foregrounded the art of looking, seeing, emphasizing virtual witnessing that referred to the artist's experience and was important for establishing his credibility. Neri argues that a community was created through the sharing of visual images, and that they therefore played a performative role.

The discovery of the natural environment of the New World is the subject of several of the treatises discussed in this volume. Jesús Carrillo Castillo's essay concerns a group of natural history treatises that were the first to disseminate, in printed form, comprehensive accounts of American natural phenomena. They represent a hybrid genre, highly ideological but at the same time descriptive and narrative, synchronic and diachronic. With imperial and ecclesiastical authority for their work, these authors recognized the power of the state to produce truth and stimulate belief. Carrillo Castillo focuses on the *Historia general y natural de las Indias,* by Gonzalo Fernández de Oviedo y Valdés (1478–1557), in which eye witnessing was related to the classical narrative tradition of autopsia, the act of seeing with one's own eyes. The main source of knowledge, Oviedo believed, was the spontaneous understanding of things derived from direct experience, which must, however, be subjected to a process of representation, through translation into words and images, that would render them comprehensible to author, artist, and reader. The representation of experience was peculiarly critical in the New World, where chaos and disorder prevailed. Carrillo Castillo argues that Oviedo's use of images must be understood as a trope referring to the exteriority of new objects with regard to language and convention. To paraphrase the chronicler, "Neither the painting of my pen nor that of my words can represent the fruit properly nor provide full satisfaction."

In a case study on the art of memory, Lina Bolzoni examines *Rhetorica christiana* (1579), the sacred oratory of Diego Valadés (1533–after 1582). The technique of mnemonic systems saw its greatest development in the same period as the birth and early diffusion of illustrated books. Bolzoni argues that Valadés' illustrations of Mexican plants and precious gems reveal an ambition that was nothing less than to provide the tool of communication and memory by which the Old World and the New World could meet on common ground.

Alessandro Tosi traces the naturalist's task, "to catch Nature in the act," between the time of Jacopo Ligozzi (1547–1627) and the modern tradition identified first with Alexander Wilson (1766–1813). Newly formed associations such as the Società Botanica Fiorentina promulgated a new definition of the naturalist's work as the description and naming of new species, a task that was inseparable from precise visual description in botanical and zoological illustration. Important works such as the five-volume illustrated treatise *Storia naturale degli uccelli* (1767–1776), produced by the Florentine society, spawned a new figurative culture marked by technical experimentation and stylistic research. Naturalistic depiction from life was the new mode that marked the rejection of the unnatural stiffness resulting from copying from stuffed skins or dried specimens; the new style was exemplified by the elegant *Pomona italiana* (1817–1839) and Charles-Lucien Bonaparte's *Iconographia della fauna italica* (1832–1841).

In her study of women artists, Lucia Tongiorgi Tomasi argues that the advent of modern science coincided with the emergence of natural history illustration, a genre in which women always played a role. She addresses the prevailing notion that women were restricted in their scientific and artistic endeavors to, on the one hand, herbal remedies and, on the other, the lower arts, such as embroidery. However, women did participate in, and received occasional acknowledgment for, a broader range of activities. Such luminaries include Maria Sibylla Merian (1647–1717), an artist-naturalist who produced treatises in which the text reinforced her exquisite drawings, and Giovanna Garzoni (1600–1670), whose luminous style was critically acclaimed during her lifetime. Tongiorgi Tomasi also presents in this context artists about whom we only have meager information, such as Isabella Cattani Parasole (active 1597–1616), Anna Maria Vaiana (act. 1627; d. c. 1655), Sister Plautilla Nelli (1523–1588), Barbara Regina Dietzsch (1706–1783), and others.

Eric A. de Jong's essay reports the rediscovery and identification of a sixteenth-century treatise on garden designs as a part of a three-volume set of albums made for Rudolf II, Holy Roman Emperor and renowned collector. The importance of this attribution is fourfold. It is very rare documentation of garden design in the Northern Renaissance. It shows the dissemination and influence of the Dutch painter and architect Hans Vredeman de Vries (1526–c. 1606). It identifies Hans Puechfeldner, Rudolf II's gardener and treatise writer. Finally, it establishes the status of garden art in Rudolf II's famous *Kunstkammer*, in which naturalia and objets d'art were collected and treasured.

Hans Hubert proposes that in the court of the Medici in the early sixteenth century, the fine arts stood in a complex relationship to an increasingly scientific approach to nature and natural collecting. Arts were active in the new processes of acquiring knowledge but passive in the face of loss of meaning and traditional iconic motifs. Cosimo de' Medici, influenced by the naturalist Francesco Redi (1626–1698), who established the intellectual premise of collecting naturalia, strove to be the ultimate Christian prince with the obligation to protect and cultivate God's creations. In his case study of the botanically labeled gallery paintings made by Bartolomeo Bimbi (1648–1730) for the Casino della Topaia, Hubert emphasizes that these portraits of fruits and vegetables are larger than life size and replete with numbering and legends, thereby showing a correspondence with scientific illustration and serving a double function of delight and usefulness.

Mark Laird looks at the impact of the environmental matrix of weather change and pollution in the history of horticulture as revealed through new forms of horticultural literature that appeared in the late 1720s. He focuses on two key volumes, *Historia plantarum rariorum*, by John Martyn, and the Society of Gardeners' *Catalogus plantarum*. These treatises represented an evolution from a scientific and medical alliance to a focus on the practice of acclimatizing plants and taxonomic standardization for decorative and commercial purposes. The burgeoning literature resulted in a new collectivity, with communities springing up in such places as coffeehouses and including individual collectors, amateurs, physicians, and nurserymen alike. One key to the reception history of these treatises is the language in which they are written. The *Catalogus*, produced by the guild of practicing nurserymen and gardeners, was in English, suggesting a wide audience. The *Historia*, in Latin, addressed a medicobotanical readership. These parallel volumes represented a limited "professional" audience for the *Historia* and a trade readership for the *Catalogus*. In an appendix to this essay, Margaret Riley has provided a history of the Temple Coffee House in London as a very active site for the exchange of plants, drawings, and information outside the botanical garden or university.

Forces behind the emergence of modern science and its theoretical and practical motivations were economic, political, and religious as well as intellectual. Exploration, classification, and the production of treatises and illustrations served commercial ambition too, a subject that is only barely addressed here. These essays, however, are models for future attempts to understand the authorial motivations and external factors that shaped illustrated works on natural history. Moreover, they demonstrate that treatises provide rich material for a critical history of ideas and visual cultures of both art and science.

NOTES

We thank Anne Nellis, research associate, and Jessica Evans, program assistant, of the Center for Advanced Study in the Visual Arts, for their help in preparing this volume for publication.

1. To say that interpretation of the role of the naturalist-artist has changed, this quotation epitomizes the long-standing attitude that this volume seeks to revise: "The unconventional term of illustrator is used circumspectly: the production of true-to-life portrayals of plants occurs almost always under the constant scrutiny of the client, as a rule a botanist. The person contracted to do the illustration has no freedom whatsoever in the work—no possibility of self-expression—and works in the same way as a camera. To describe the illustrator as an artist with the usual freedoms of self-expression seems inappropriate." H. Walter Lack, *Garden Eden: Masterpieces of Botanical Book Illustration* (New York, 2001), 15.

2. Quoted in Paul L. Rose, "The Accademia Veneziana: Science and Culture in Renaissance Venice," *Studi veneziani* 11 (1969): 231.

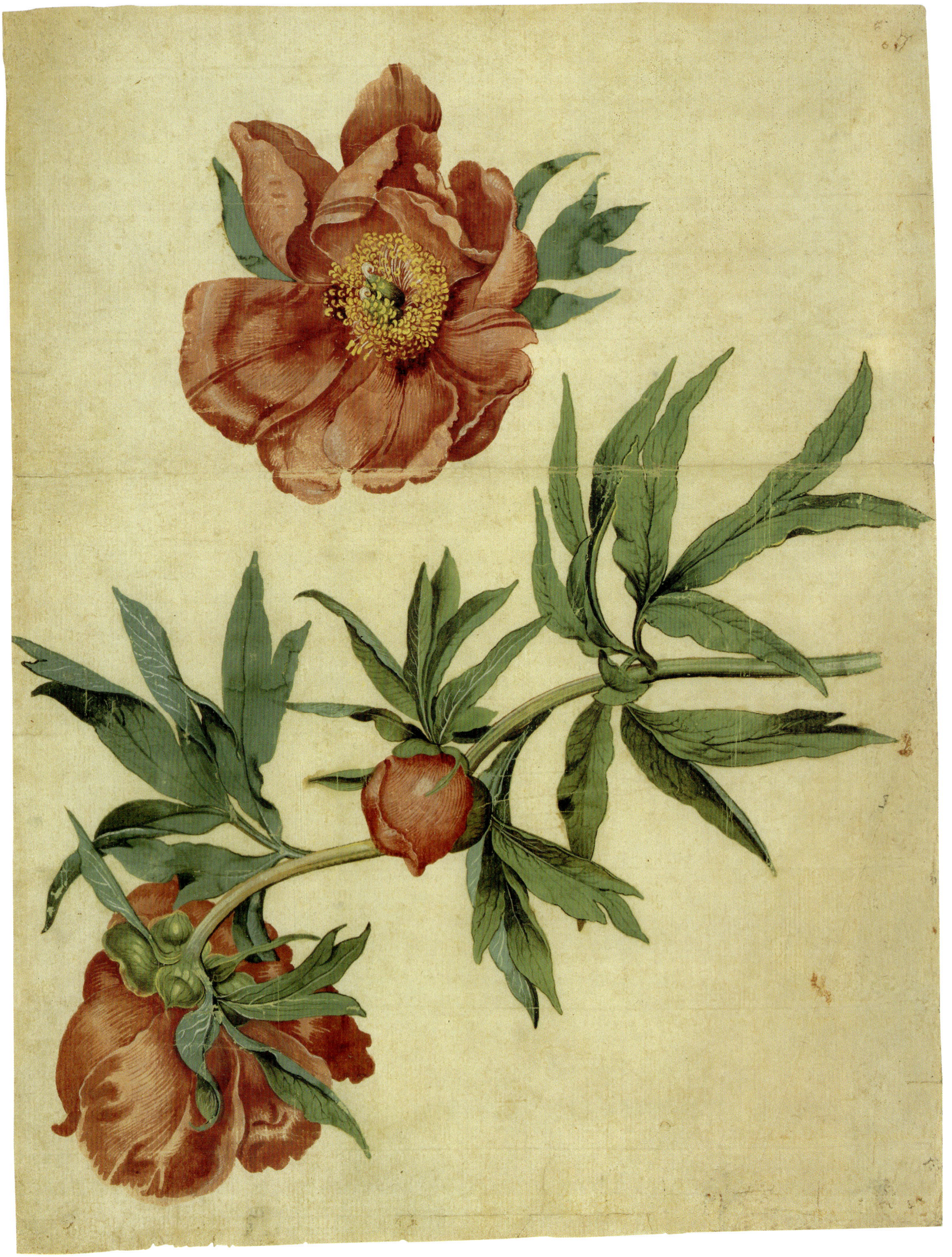

PAMELA H. SMITH
Columbia University

Artisanal Knowledge and the Representation of Nature in Sixteenth-Century Germany

It has always seemed curious that the credit for what has often been informally referred to as the first scientific printed herbal—*Herbarum vivae eicones,* published in Strasbourg in 1532–1536—is generally assigned to the author of the text, Otto Brunfels (1488–1534), when it was the naturalistic illustrations of the draftsman and block cutter Hans Weiditz (c. 1495–c. 1536) that resulted in the book being labeled "scientific" (fig. 1). Indeed, Brunfels apparently had a somewhat cranky relationship with the illustrator and the illustrations. He did not hire Weiditz; that was the idea of the book's entrepreneurial printer, Johann Schott, and Brunfels complained that he could not get Weiditz to work according to his plan. After publication, Brunfels continued to worry that the "dead lines" of the illustrations were receiving more attention than his writing.[1] Moreover, Brunfels' text is based on descriptions borrowed from classical authorities, while Weiditz's preliminary drawings for the book's illustrations were made from his own careful observation of the plant specimens. Indeed, Weiditz's original watercolor studies for the woodblock prints describe the plants in minute detail and include annotations that name the plant; note the day, month, and year the drawing was made; and describe the size, color, and texture of the plant.[2]

Weiditz's "scientific" artisanship represented at least the third generation of the "nature study," preceded immediately by the studies of the German painter and engraver Albrecht Dürer (1471–1528). Dürer himself was following the model of the two previous generations, the first of which included Martin Schongauer (c. 1450–1491), who, in turn, was probably carrying on the practices of the Flemish painters with whom he studied.[3] Much has already been written about the close observation and claims to natural verisimilitude in early printed herbals, first in the *Herbarum,* by Otto Brunfels and Hans Weiditz, then a decade later in *De historia stirpium commentarii insignes* (Basel, 1542), by Leonhard Fuchs (1501–1566). The appearance of these herbals marks an important development, one that would have been impossible without the northern workshop tradition of nature studies on which these herbals were modeled. However, in order to understand the workshop practices that lay behind the naturalism of the "scientific illustrations" of these herbals, we must reach farther back, to the sudden appearance of naturalism in a manuscript herbal that appeared a century and a half before that of Weiditz and Brunfels—not in Germany, but in early fifteenth-century Padua (figs. 2 and 3).

The plant studies in what is now known as the Carrara Herbal are highly naturalistic and provide a good point of entry into the problems raised when the history of the Weiditz and Brunfels herbal is recounted as part of a story of naturalistic "illustration" rather

Martin Schongauer, *Peonies,* c. 1472–1473, watercolor and body color on paper
The J. Paul Getty Museum

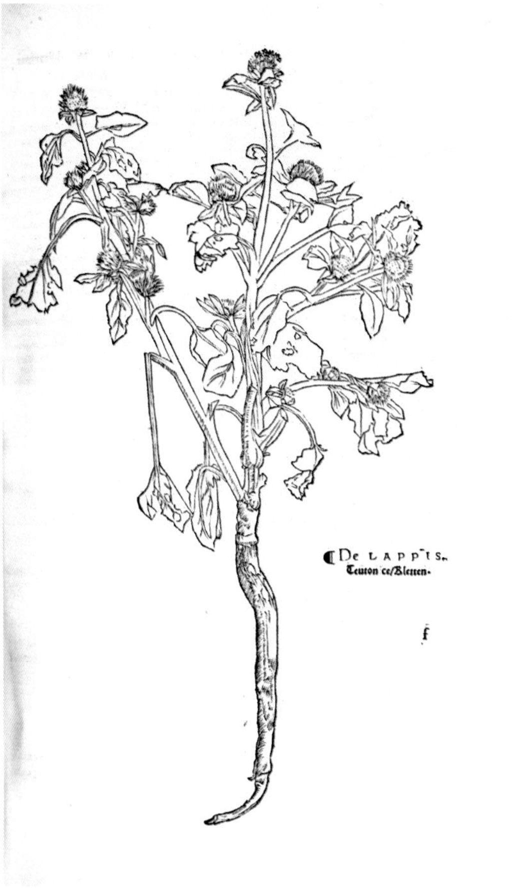

than as the "study of nature." This essay examines the significance that naturalism held for early artisans and their patrons, one that transcends the boundaries implied in the category of auxiliary "scientific illustration" and signals the need to rethink the relationship between art and science in this period. Moreover, the role of naturalism in these works indicates the far greater significance attributable to the category of "nature study."[4]

1. Hans Weiditz, *De lappis* (burdock), from Otto Brunfels and Hans Weiditz, *Herbarum vivae eicones* (Strasbourg, 1532–1536), volume 2, page 65, woodcut
The Huntington Library, San Marino, California

The Social and Political Context of Artisanal Nature Study

The Carrara Herbal, held today by the British Library (Egerton MS 2020), is a vernacular Italian translation of the *Liber aggregatus in medicinis simplicibus,* a treatise on medicinal plant simples originally written by the ninth-century Arab author Serapion the Younger (Ibn Sarabi). This translated volume was written by Jacopo Filippo in the closing years of the fourteenth century. Most of its illustrations show the mark of close observation, and some, such as the violet (fig. 4), were drawn from pressed

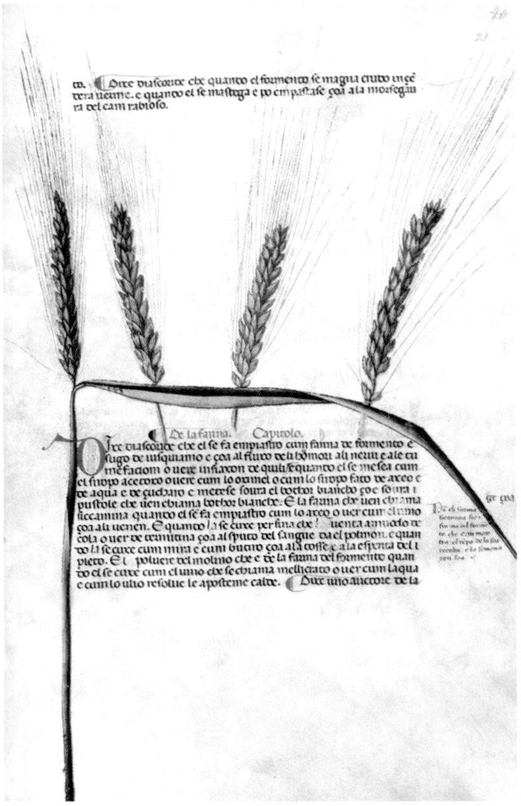

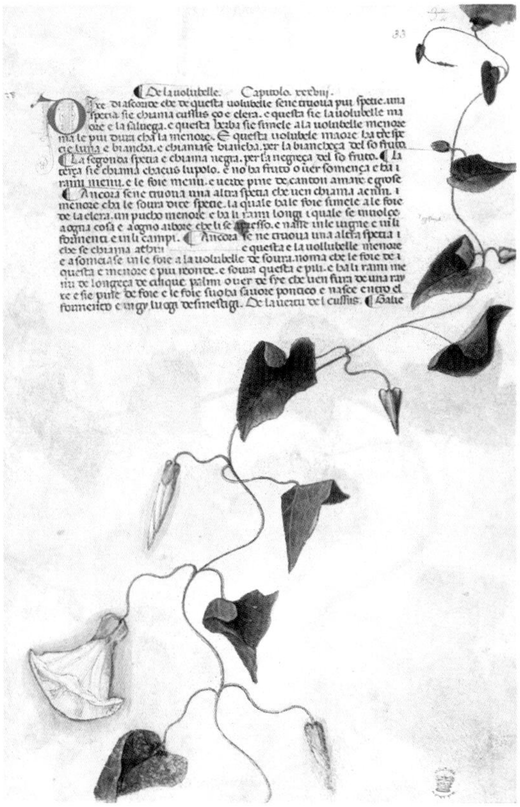

2. Anonymous, *De la fanna* (ears of wheat), from the Carrara Herbal, c. 1400, woodcut
The British Library, Egerton MS 2020, folio 21

3. Anonymous, *Convolvulus* (bindweed), from the Carrara Herbal, c. 1400, woodcut
The British Library, Egerton MS 2020, folio 94r

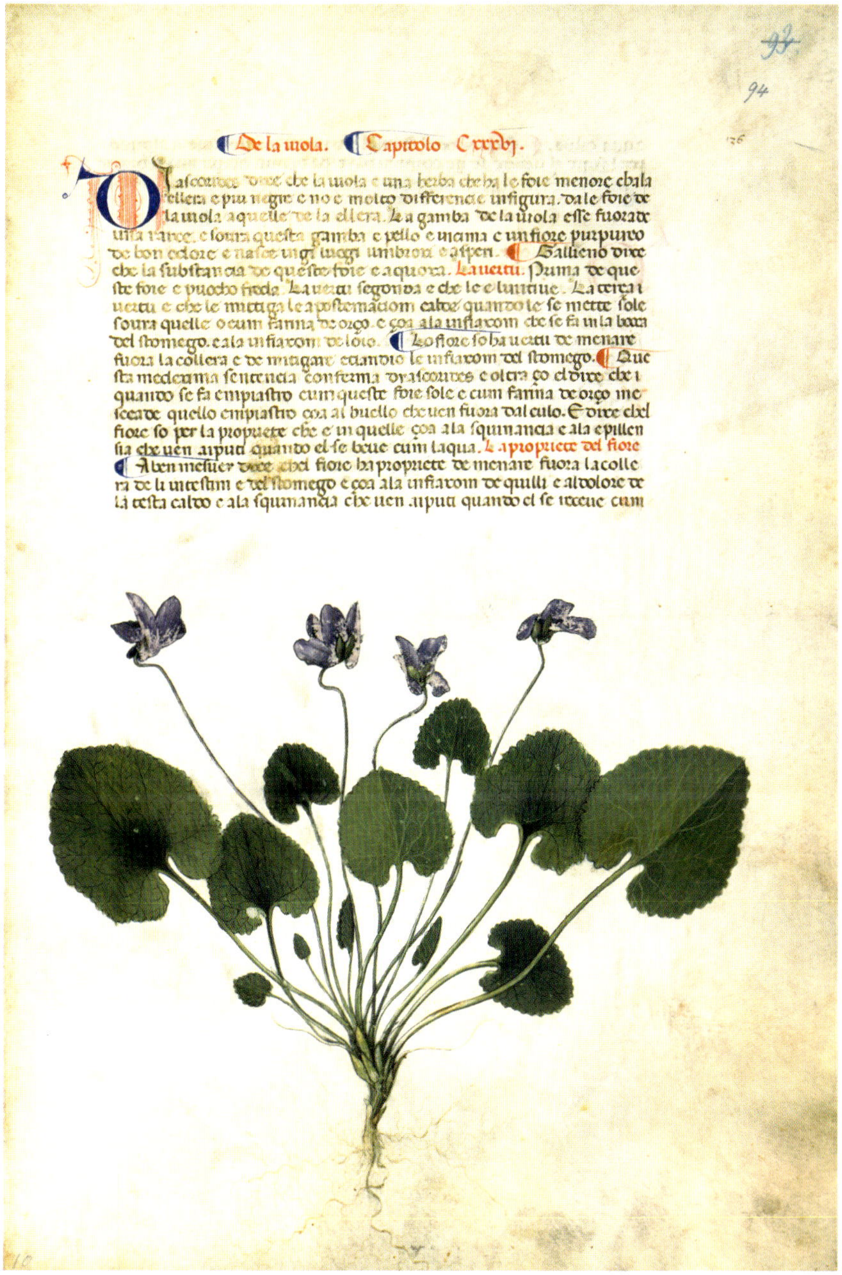

4. Anonymous, *La viola* (violet), from the Carrara Herbal, c. 1400, woodcut
The British Library, Egerton MS 2020, folio 33

specimens. These practices were probably partly associated with the medical faculty at the University of Padua, but they also reflect the abiding interest in nature of the book's patron, Francesco II, the last lord of Padua (reigned 1390–1405).[5] The Carrara family, who ruled the commune of Padua from 1318 to 1405 (with some interruptions), commissioned not only this herbal but also a number of remarkably naturalistic portraits—both painted works and funerary monuments—some of which show evidence that the artist worked from death casts of the individuals' faces and hands. Further, the Carrara ornamented their palace with paintings of lifelike animals. Perhaps most significant, Francesco II was also patron to the Florentine painter Cennino d'Andrea Cennini (c. 1370–c. 1440), the author of *Il libro dell'arte,* a manual for painters written around 1400.[6] In his book, Cennini praised the artist's struggle to imitate nature, viewing it as the "triumphal gateway" through which an apprentice must pass to emerge as an accomplished artisan.[7] His manual ends with a section on "a very useful" means of "copying and imitating things from nature . . . called casting."[8] By this he meant casting from life, in particular casting the faces of living persons in plaster.[9] He included instructions on casting a whole figure from life and "how to make a cast of your own person."[10] Casting has often been understood to be completely separate from any sort of scientific illustration, constituting instead a merely mechanical means of copying. This essay proposes to shed light on some of the ambitions and claims of artisans and shows how casting can, in fact, also be classified as nature study.

Norberto Gramaccini has argued, in an article that is mainly concerned with the life casting carried out by Andrea Riccio (1470–1532) and others in Padua from about 1500, that the Carrara family's interest in nature was related to their own precarious and "non-natural" hold on power. The family rose through the communal government of the Paduan city-state and then seized power from it, with the result that they derived their legitimacy from no accepted authority, neither that of the pope nor that of the emperor. Instead, they based their right to rule on the language of jurists at the University of Padua, scholars who claimed that government must imitate nature. Claiming the *natural* reasonableness of their rule, the Carrara family walked an uneasy course between the commune's claims that they were unnatural tyrants and the pope's assertion that they stood outside any divinely ordered hierarchy.[11] The Carrara appear to have employed nature (and artisans) in order to construct a theater of state that would bolster their claims to legitimate rule; or, at the very least, they used nature as a means of displaying their mastery. In this light, the naturalism of the Carrara Herbal can be seen as an artisan's response to the ongoing discourse on nature and as a statement declaring the artisan's own key role in constructing this theater of state. In this respect, the illustrations in this herbal are analogous to Cennini's written treatise on painting. Both derive from the elevation of the artisan and the deployment of nature within the theater of state that constituted a fundamental component of government in Padua.

Artisanal Claims to Knowledge and Skill

The Carrara Herbal and Cennini's treatise can also be put in a more general artisanal context. Before the beginning of the fifteenth century, very few texts were written by artisans; but suddenly, around 1400, artisans appeared, in the words of one historian of technology, to "resort to pen and paper in a field where their fathers had been satisfied with memory and the spoken word."[12] The result was a veritable explosion of technical treatises between 1405 and 1420. The artisans of northern Italy and southern Germany seem to have felt particularly compelled to record their modes of working, especially those in architecture, fortification building, and gunnery.[13] Historians have accounted for this sudden shift to a new medium largely by pointing to the new methods of warfare and the greater fragmentation of Europe into competing noble territories, both of which gave these practitioners new importance in society.[14] Furthermore, these author-artisans stood at the intersection of courtly and commercial forces. Freed from (or never possessing) guild membership through the rise of the courts and faced with the greater intensity of commerce, they were forced to become more self-conscious about their practices and to make explicit claims about their skill and power because they were competing for patronage, livelihood, and *fama*.

The number of practicing artisans in this period who combined craft with authorship is striking,[15] and it would appear valid to say that the artist who illuminated the Carrara Herbal can be viewed as one such artisan who began to articulate a given mode of working, not by writing but by producing a new "naturalistic" representation of nature. In so doing, this artist was making a claim to a certain elevated status, artisanal skill, and knowledge of nature.

The ways in which scholars have viewed the Carrara Herbal and Cennini's treatise resonate with considerations that relate to the attribution of the Weiditz-Brunfels treatise. In a germinal article published in 1950, Otto Pächt saw the Carrara Herbal as, among other important milestones, an expression of scientific empiricism resulting from a heroic mix of science and art. Pächt spoke of this illustrator's "courage to turn his back on all pattern-books and to look nature straight in the face."[16] In contrast to this heroic view of a new science, this naturalism might instead be seen as emerging out of a new self-consciousness on the part of the artisan. Caught between patronage and market forces similar to those that operated on fortification builders and gunners, the artisan who created the illustrations in the Carrara Herbal articulated the methods of expertise not in treatises but in paint.

Artisanal and Scientific Knowledge

In considering early naturalism and botanical illustration—what he calls "scientific illustration"—James Ackerman has asserted that "studio artists furthered the aims of science in the early Renaissance."[17] Although in his other articles on art and science Ackerman has argued that Renaissance artists realized "a new perception of the natural world . . . and of new ways of communicating that perception," his use of the phrase "scientific illustration," like that of many other historians of art and science, seems to imply that the art plays an auxiliary role to the text, the more serious business of science.[18] A related problem lies in the assumption that "science" was already a fully formed activity by the sixteenth century.[19]

Another aspect of the Weiditz-Brunfels problem can be illuminated by looking again at Cennini. Art historians have often viewed the artist-author as influenced by a learned Petrarchan cult of imitation that was current at the Paduan court, from Petrarch's last years there (1368–1374) through the end of the fourteenth and into the fifteenth century. Writing in 1366, Petrarch advised the fourteenth-century Italian novelist Giovanni Boccaccio:

> An imitator must take care to write something similar yet not identical to the original, and that similarity must not be like the image to its original in painting where the greater the similarity the greater the praise for the artist, but rather like that of a son to his father. While often very different in their individual features, they have a certain something our painters call an air [*aria*] about the face and eyes, that produces a resemblance; seeing the son's face we are reminded of the father's although if it came to measurement, the features would all be different, but there is something subtle that creates this effect. We must thus see to it that if there is something similar, there is also a great deal that is dissimilar, and that the similar be elusive and unable to be extricated except in silent meditation, for the resemblance is to be felt rather than expressed.[20]

The exchange between painters and humanists in fourteenth-century Padua was no doubt lively, but it is important to ask why the assumption persists that the influence went from top to bottom, especially when Petrarch—in laying out his model of imitation—twice cites the practices of painters when discussing degrees of verisimilitude. One might think of Cennini as *codifying* the current practices of painters in his text, rather than deriving prescriptive principles from scholars for artists to follow. No doubt the exchange went both ways, but the following discussion outlines the importance of the intellectual traffic *up*, from artisans to scholars.

Sixteenth-Century Artisanal Nature Study

Let us return to nature-study traditions in northern Europe. As Otto Pächt has noted, the attention to the details of nature apparent in the Carrara Herbal had no following in Italy, where only the artist Pisanello (Antonio Pisano, c. 1395–1455) made more frequent use of this "discovery of the non-human world."[21] But in the North, particularly in the Burgundian Netherlands, another noble family foreign to the Lowlands attempted to "naturalize" its right to rule. In this case, the art of the Burgundian Netherlands mobilized a discourse on the relationship between nature and art that reverberated throughout Europe, especially in the lands under Habsburg rule,[22] but also among the urban patriciate, whose consumption rivaled that of the noble courts. In the case of nature studies by German artists—none of which took the form of botanical treatises—these artists' ambitions were similar to those articulated by the anonymous illustrator of the Carrara Herbal. In their treatises on art and in their nature studies, these artists made claims about their own standing and their knowledge of the materials and processes of nature, as well as about the epistemological status of their knowledge.

Martin Schongauer, the great engraver and painter of Colmar, brought Flemish particularism and naturalism, as well as the practice of the nature study, to the Rhine (fig. 5).[23] Schongauer's father was an Augsburg goldsmith who settled in Colmar around 1445 and headed a workshop in which Martin and his four brothers were trained and later worked. Schongauer probably traveled in Flanders, where he became especially familiar with the work of the Flemish painter Rogier van der Weyden

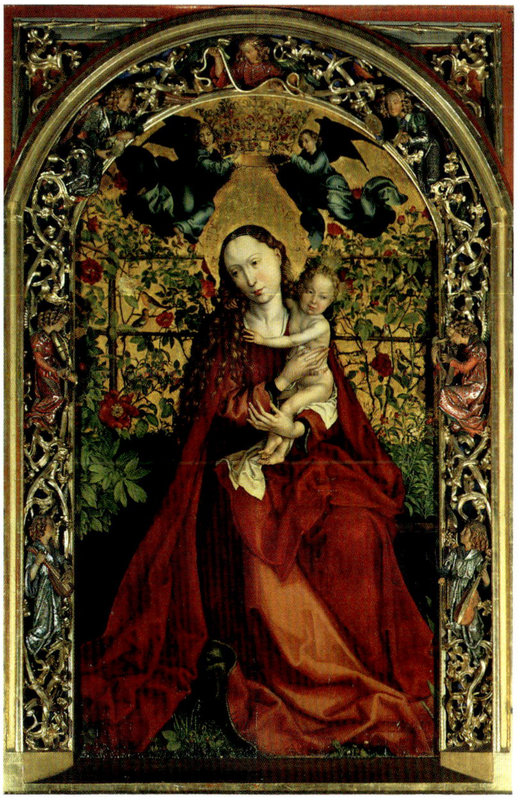

5. Martin Schongauer, *Madonna of the Rose Garden*, 1473, oil on panel
Dominican Church, Colmar; © Giraudon / Art Resource, New York

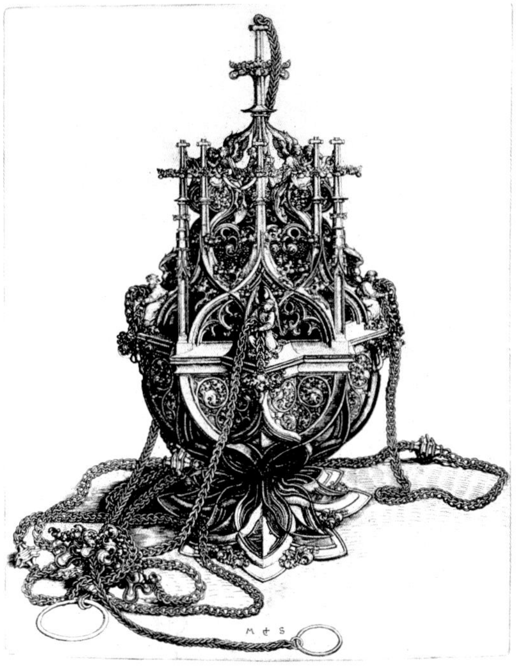

6. Martin Schongauer, *Censor*, 1470–1474, engraving
Germanisches Nationalmuseum, Nuremberg

(c. 1399–1464). Schongauer's engravings show an intense attention to detail employed in the service of a new painterly illusionism and naturalism, such as in the remarkably meticulous engraving of a church censor (fig. 6). His paintings also reveal a close and sensitive observation of nature, achieved by preparatory nature studies such as that of peonies (fig. 7). Schongauer's experience as a goldsmith probably turned his focus naturally to engraving, because the print-making technique of engraving emerged out of the goldsmith's practice of engraving precious metals. Eventually, he began to make engravings unlike any seen before. Schongauer apparently had a keen awareness of the commercial possibilities of engravings, because his plates are deeply incised, a technique which ensured that they would stand up to many printings.[24] Moreover, his prints have complex compositions, and he strove for the same high degree of naturalism that was preferred by the Flemish painters among whom he had trained. Because his prints were disseminated more widely than his paintings, they played an essential role in spreading his fame and his name, quite literally, for he included his initials on his engravings. Monograms like that of Schongauer, and Dürer after him, which emerged in the context of engravings, imitated the marks that goldsmiths had used since the Middle Ages to indicate maker and to guarantee precious metal content.[25] Schongauer was among the very first print engravers to sign his work with his initials, and in doing so he indicated a strong awareness of his own authorship.[26] In 1492 Albrecht Dürer traveled to Colmar to visit Schongauer, only to find that Schongauer had died the previous year. Still, Dürer was apparently able to obtain some of Schongauer's drawings from his brothers.

Dürer's nature studies and engravings, in turn, would become avidly sought after in his own lifetime, as well as in the later sixteenth and early seventeenth centuries. In addition, his art would come to represent a founding moment for nature studies in Europe, including the "scientific illustration" of Hans Weiditz. Dürer's nature studies, of course, are well known; among them is *The Large Turf* (fig. 8). His drawing *Madonna with a Multitude of Animals* (fig. 9) can be seen as a kind of catalogue of his (and apparently Schongauer's) nature studies, for it contains not only a copy of Schongauer's study of peonies but also Dürer's own studies of an iris, a stag beetle, a mayfly, a snail, a dog, a fox, and various birds. This drawing points, however, to a greater

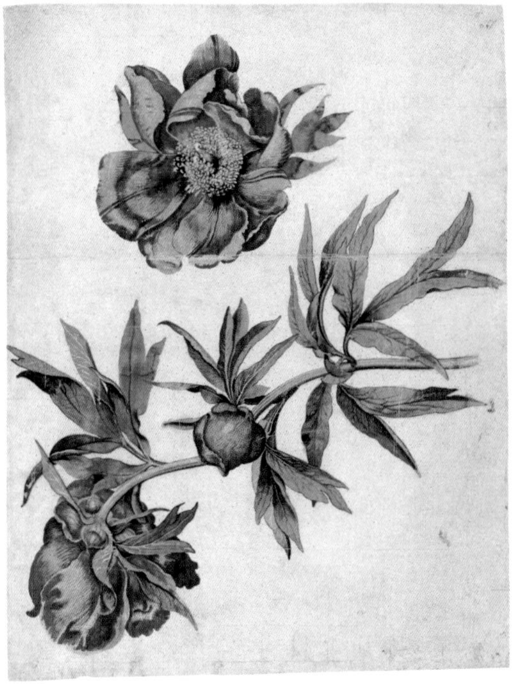

7. Martin Schongauer, *Peonies*, c. 1472–1473, watercolor and body color on paper
The J. Paul Getty Museum

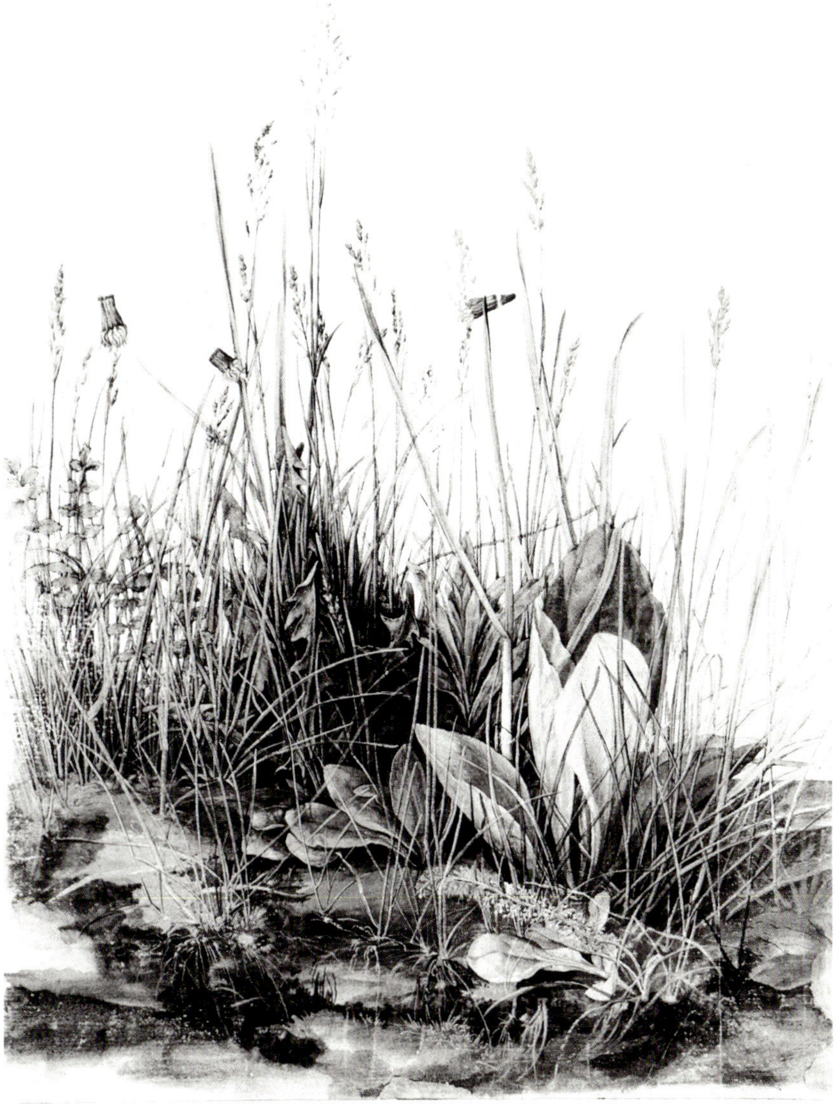

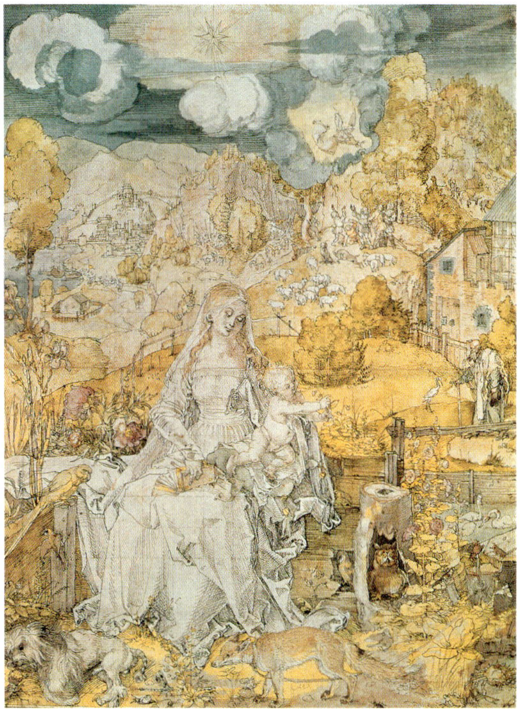

8. Albrecht Dürer, *The Large Turf*, 1503, pen and ink, watercolor, and gouache
Graphische Sammlung Albertina, Vienna

9. Albrecht Dürer, *Madonna with a Multitude of Animals*, 1503, ink and watercolor
Graphische Sammlung Albertina, Vienna; © Erich Lessing / Art Resource, New York

overall significance of nature studies. To understand this deeper meaning, it is helpful to consider a study drawing—one that was probably also in Dürer's possession, from the workshop of Martin Schongauer and his four goldsmith brothers—that gives insight into the multivalent meaning that these creatures and, more important, their accurate representation could possess (fig. 10). The drawings on this sheet—of a snail, an asp, a lizard, and a frog—were probably based on the study of dead, and possibly dried, specimens. This was likely an apprentice's study sheet, because the draftsman's observation is faulty in some instances: the snail shell, as a case in point, is a physical impossibility because the snail's body could not fit into it. The drawing appears to have been corrected for verisimilitude. For example, it was probably the master who noted corrections to the snail's antennae (although not the shell) and the lack of webbed feet on the frog, and who also admonished the apprentice to make the animals more lifelike in place of their slightly desiccated look. Finally, in the lower right corner, the master has alluded to the purpose of these images: they probably served as models for amulets that would first be cast in metal, then perhaps filled with the dried and pulverized animal and, upon completion, as the inscription reads, were "to be placed on the head of a 29-week pregnant woman." Frogs and other amphibious animals were often used during the Middle Ages as amulets in childbirth.[27] These drawings of amulet models provide us with valuable insight into a web of connections. The verisimilitude of the apprentice's drawings mattered because knowledge and efficacy resided in nature. The imitation of nature could ultimately bring natural processes into operation, processes that were bound up with the natural workings of the human body. The artisan sought to imitate nature in order to activate the powers inherent in nature.

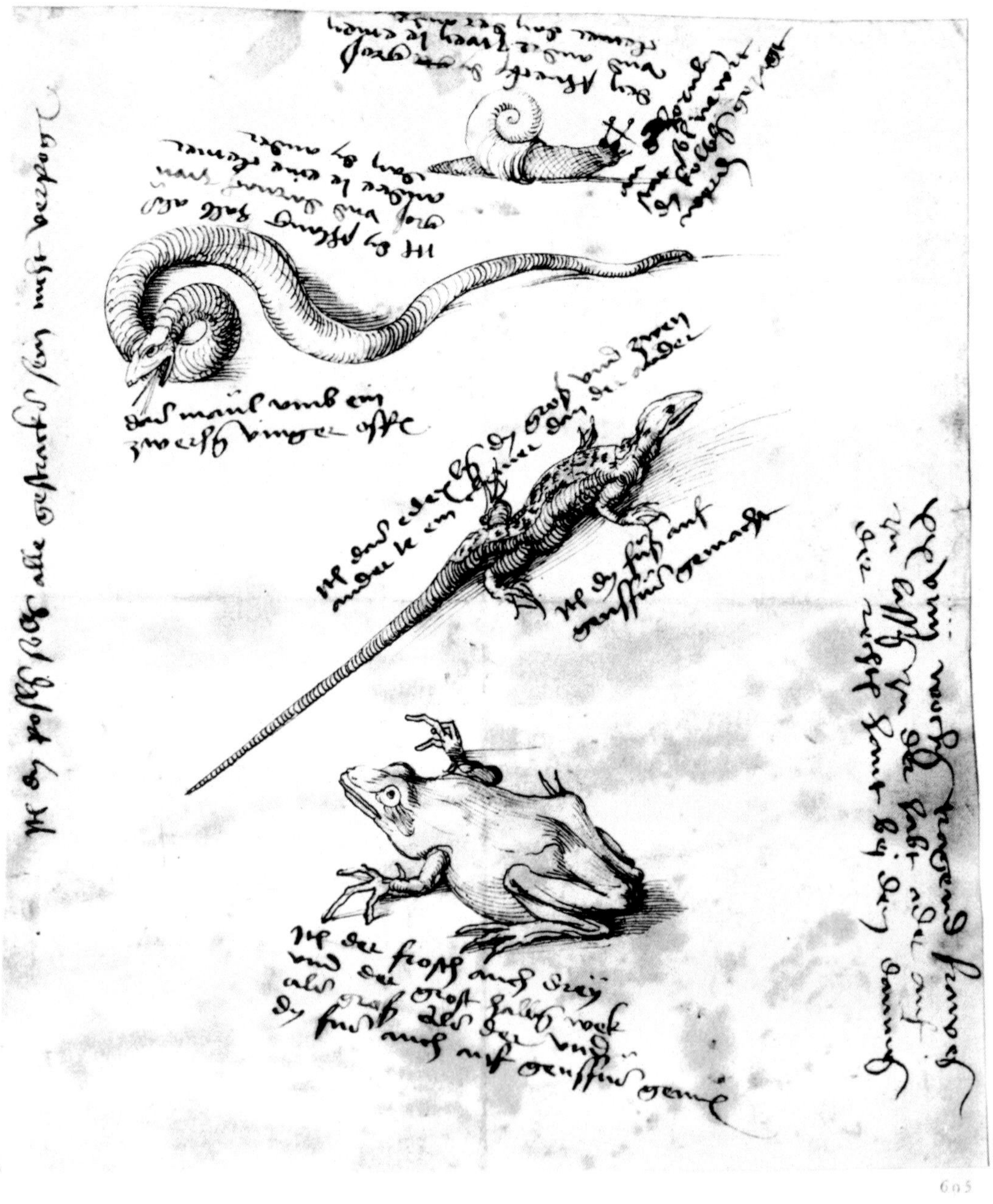

10. Martin Schongauer (workshop of), *Snail, Snake, Lizard, and Frog*, c. 1495, pen and black ink and traces of green watercolor
Kupferstichkabinett, Berlin (KdZ 695)

Parallels between Art and Science: Dürer and Paracelsus

The writings of Dürer's contemporary, the alchemist, physician, and astrologer Theophrast von Hohenheim, also called Paracelsus (1493–1541), provide additional evidence for the view that the imitation of nature could be a source of knowledge and power. The writings of Paracelsus are helpful to the understanding of artisans who worked at this time because he strove to articulate an artisanal mode of cognition and a vernacular epistemology. In his works, he constantly held up artisans and their knowledge of the material they worked as models for gaining a productive knowledge of nature. He contrasted this productive or useful knowledge, which benefited humankind, to the sterile knowledge of scholars.

One of the central components of Paracelsus' matter theory was the concept that all generation occurred through putrefaction, which is a process of decay and regeneration. According to Paracelsus, in putrefaction, particular forms of life appeared first: "serpents, toads, frogs, salamanders, spiders, bees, ants, and many worms" (fig. 11).[28] The centrality of putrefaction to the life cycle would thus indicate the importance of those types of creatures portrayed on the Schongauer study sheet.

The lizard on the study sheet appears again in Schongauer's engraving *The Flight into Egypt,* where it scurries up a dragon tree (fig. 12).[29] In a woodcut of the same subject, Dürer placed these reptiles around a spring, the source of all generation, past which the Holy Family journeys (fig. 13). These creatures, creeping forth from rotting matter and hovering around springs, were believed to be evidence of nature's generative and transformative powers. Thus Dürer's *Madonna with a Multitude of Animals* (see fig. 9), also set near a spring, pointed to the generative powers of nature, as indicated by the variety and great number of animals surrounding Mary. Moreover, as art historian Larry Silver has pointed out, the spiritual drama being played out in the background of Dürer's drawing, namely that of the angel announcing the Nativity to the shepherds, makes clear the profound correspondence and resonance between divine and natural powers.[30]

An even more significant intersection between the ideas and ambitions of Paracelsus and those of Dürer can be found in their shared epistemology. Paracelsus is often cited as a model of empiricism in the Renaissance because of his praise of experience. A close examination of his idea of experience reveals, however, that it does not resemble any modern concept of empiricism. He viewed experience not as part of a process of reasoning but as a union of the divine powers of the mind and the entire body with the divine spirit in matter. In explaining this concept, Paracelsus drew upon the scholastic terminology of theory and practice but inverted the traditional understanding of these terms. Since the time of Aristotle in the fourth century B.C.E., theory had been regarded as certain knowledge, while experience was knowledge of particulars that could never be certain. Paracelsus defined *scientia,* or certain knowledge, as the divine power contained in natural things that the

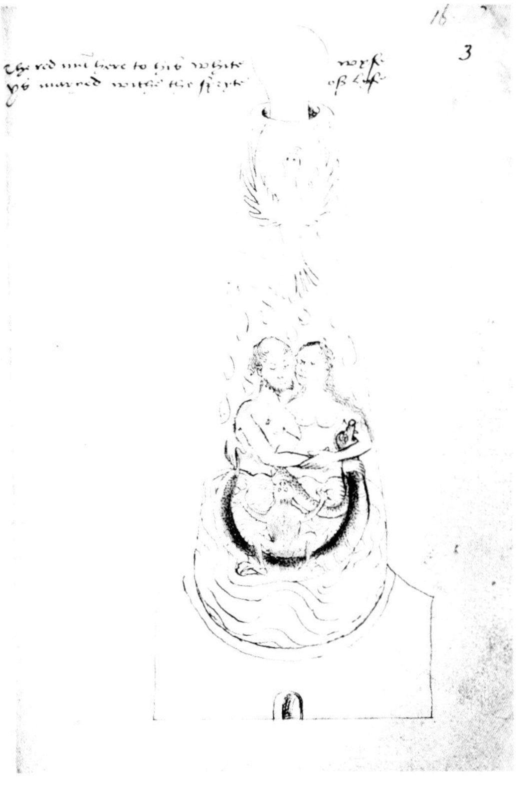

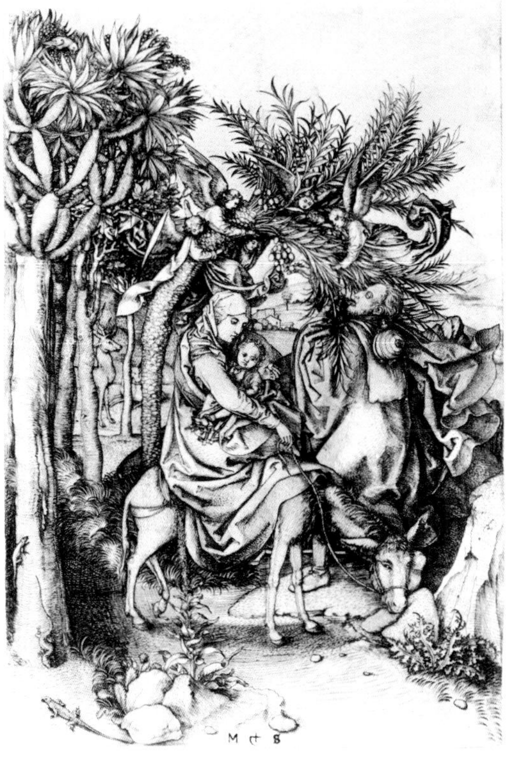

11. Anonymous, *Putrefaction,* alchemical manuscript, sixteenth century, brown ink
The British Library, Egerton MS 845, folio 16

12. Martin Schongauer, *The Flight into Egypt,* c. 1470, engraving
Kupferstichkabinett, Berlin

physician must "overhear" and with which he must achieve bodily union in order to gain knowledge of medicines.[31] For Paracelsus, certain knowledge was *in* nature. Experience was the process by which the physician united with nature and learned this science. Paracelsus writes:

Scientia is inherent in a thing. . . . For instance, the pear tree has *scientia* in itself, and we who see its works have *experientia* of its *scientia*. . . . Thus in this book, I show the way *scientia* enters into you.[32]

This is a remarkable inversion of the concepts of theory and practice, one that Paracelsus derived from artisans.

Again, there is a marked resemblance between Paracelsus' ideas and those of Dürer. Dürer also stated that certain knowledge (Paracelsus' *scientia*) lay in nature and that this certainty was expressed by naturalistic representation:

But life in nature manifests the truth of these things. Therefore observe it diligently, go by it and do not depart from nature arbitrarily, imagining to find the better by thyself, for thou wouldst be misled. For, verily, "art" [i.e., *Kunst* as opposed to *Brauch*] is embedded in nature; he who can extract it has it.[33]

For Dürer, as for Paracelsus, nature had an immediacy and a primacy; nature, not words about nature, was the certain *scientia,* which artisans got to know by directly engaging nature with the senses and the body. The practice of their art was a physical imitation of nature. This concept comes through in Paracelsus' idea of a bodily union between the natural philosopher and his object of study,[34] as well as in his statement that the creative power of humans is similar to God's but lies in the human body, rather than in the Word. What God created by the Word, Paracelsus wrote, a human being creates "with his body and his instruments [that is, sense organs]."[35] The extraction of true *scientia* from nature also involved reading the signs of natural things. Paracelsus believed that signs could be recognized by studying the proportions of individual persons, a system Dürer had attempted to lay out a decade earlier.[36]

Similarly, for Paracelsus the aim of the physician–natural philosopher was to make visible the invisible virtues that God had hidden in Creation.[37] For Dürer, art brought out what was hidden in nature.[38] Both men emphasized that the proof of an artisan's (or a physician's) successful extraction of *scientia* from nature could be seen in the tangible effects of the productive knowledge that resulted therefrom.

Paracelsus also articulated the significance of exact replication (such as that displayed in Dürer's precise nature studies) when he wrote about the place of repetition in the process of knowledge-getting. A master is someone who possesses *scientia* that he

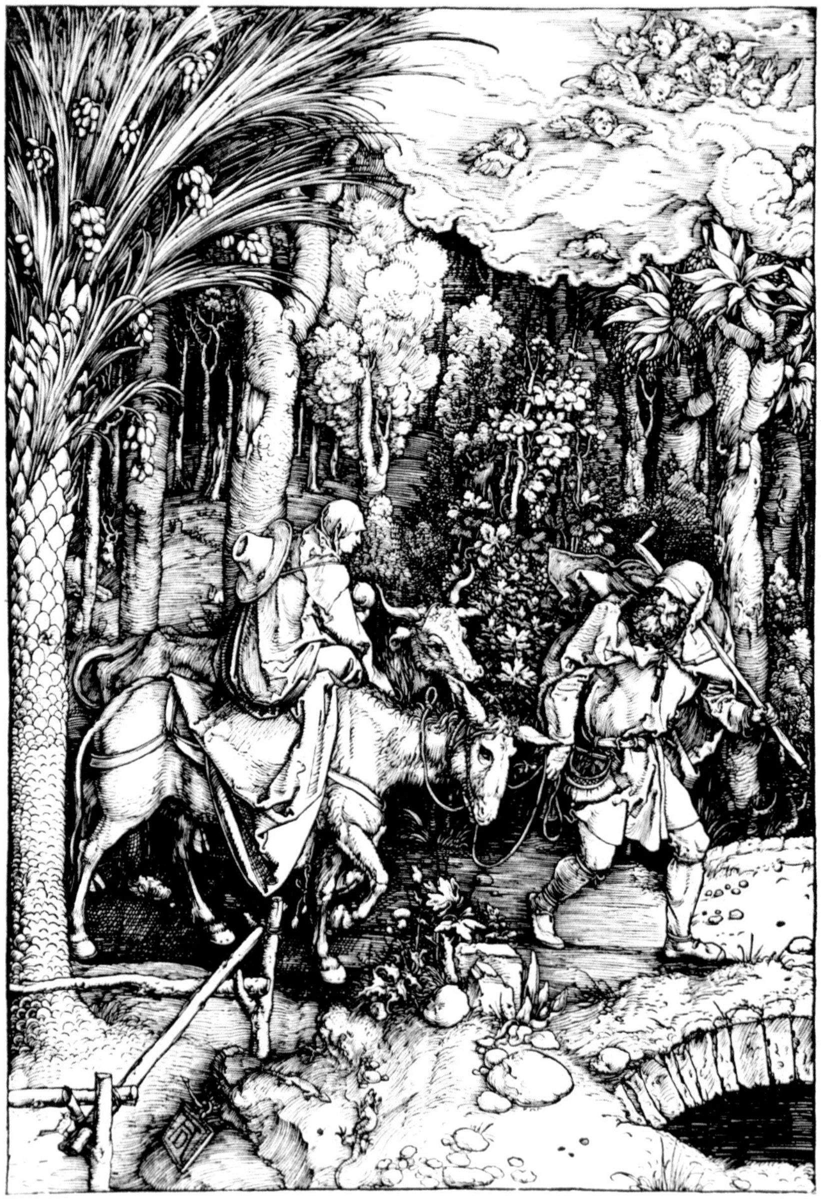

13. Albrecht Dürer, *The Flight into Egypt,* from *Life of the Virgin,* c. 1504, woodcut
Cincinnati Art Museum

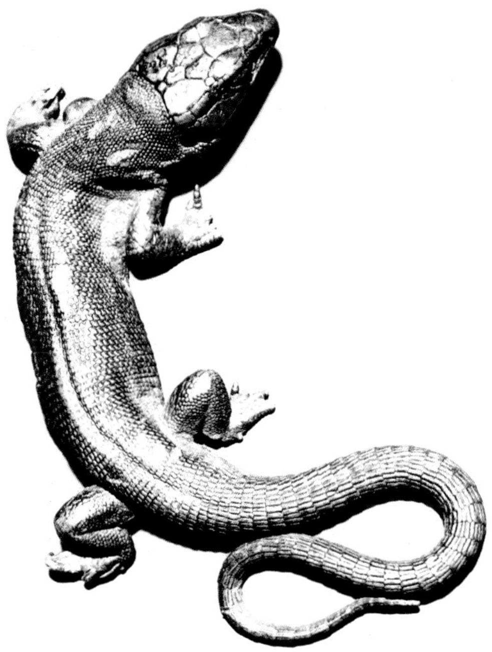

14. Wenzel Jamnitzer, *Life Cast of a Lizard*, sixteenth century, lead
Kunstgewerbemuseum, Staatliche Museen Preussischer Kulturbesitz, Berlin

has extracted from nature.[39] To learn it, a student must copy or replicate (*ablernen*, to learn through copying) nature. Dürer, too, spoke of the extensive *abmachen*, or reproducing after life, that makes a great artist.[40] Nature was prolific and copious, a creative force that might be imitated by art. To both Dürer and Paracelsus, the idea of imitating nature signified more than just producing a mirrorlike image of created nature; rather it meant imitating, reproducing, and harnessing the creative power of nature itself.

Copying Nature to Meld Science and Art

This understanding of the imitation of nature and the significance of precise replication may be easier to grasp in the operations of metalworkers. The metallurgical expertise that was involved in smelting and in casting three-dimensional objects, for example, is closer to our present view of what constitutes scientific knowledge of natural processes than the painters' work of observing nature and manipulating pigments and other materials necessary to the production of two-dimensional nature studies. To conclude, then, let us turn to the work of the Nuremberg master goldsmith Wenzel Jamnitzer (1508–1585). During his life, Jamnitzer was best known for his extreme naturalism, which he achieved by an exact replication of nature through the process of "casting from life" (fig. 14). This technique involved capturing a small animal alive, killing it by immersing it in vinegar and urine so that it was not deformed by crushing it, and then posing it in a lifelike manner by attaching it with threads to a clay base.[41] A thin solution of plaster and fine casting sand was painted over the animal, and the whole thing was then fired in a kiln to harden the plaster and burn out the organic matter. This process resulted in creating a mold that was first cleaned out with a warm oil (and probably subsequently refired) and then filled with molten metal.[42] Jamnitzer made casts of many animals, but he achieved particular fame for his ability to cast delicate flowers and grasses (fig. 15).

The most renowned work in Jamnitzer's lifetime—his life casting—might be considered by some to be his least ambitious, as it involved what could simply be called the mechanical reproduction of plant and animal forms. And yet it is clear that this famed goldsmith displayed many of the same ambitions to theorize artisanal practice that Dürer had expressed a generation before.[43] Those ambitions can be found in his extremely complex cosmic fountain, a masterpiece now known only through contemporary descriptions. Cast in metal and measuring ten feet high by five feet across (about three meters high by one and one-half meters across), the fountain was commissioned by Emperor Maximilian II in 1556 and was completed twenty-two years later, in 1578. The project was Jamnitzer's attempt to represent in a single sculptural object the entire divine, human, and political cosmos.[44] These ambitions were also expressed both in his book, *Perspectiva corporum regularium*, published in 1568, and in the perspective and metalworking instruments he invented (fig. 16).[45] A portrait by Nicolas Neufchatel offers a visual representation of Jamnitzer's strivings (fig. 17). In his left hand, Jamnitzer holds his invention for calculating the mixtures of metals (with the specific weights of the metals inscribed upon it) and in the other he holds a pair of dividers. In a niche behind him stands an ornamental vase with his signature plants cast from life, while a small gold statue of Neptune is on the table before him. A drawing of the statue on paper points to his skill in drawing and design, for which he was also well known.[46]

Finally, Jamnitzer's understanding of what it meant to imitate nature is indicated in the *Mother Earth Centerpiece* (or *Merkel Centerpiece)* of 1549, now in the Rijksmuseum, Amsterdam (fig. 18).[47] This elaborate silver-gilt centerpiece features Jamnitzer's specialty, grasses and flowers, cast from life and springing from an egg-shaped vessel at the top, while reptiles, also cast from life, creep forth from the earth around the base. The central female figure represents Mother Earth, and the whole piece symbolizes the fertility and generative powers of nature. It bears the inscription *Sum terra, Mater Omnium / Onusta caro pondere / Nascentium ex me fructuum* (I am the Earth, mother of all things / Laden with the precious burden / Of the fruits which are produced from myself).[48] While these lines refer to the fecundity of Mother Nature, they also suggest that Jamnitzer meant to display his own powers of creation in this work and to demonstrate his ability to imitate nature. His was a two-pronged imitation of nature: he demonstrated that he could produce an absolute likeness of nature, and, in creating this likeness, he also showed that he possessed a full understanding of and ability to harness the processes of nature by means of smelting and metallurgy.

Artisans as Investigators of Nature

This essay has proposed that, in the fifteenth and sixteenth centuries, some artisans were self-aware experts on the processes and transformations of nature. The highly skilled artists Martin Schongauer, Albrecht Dürer, and Wenzel Jamnitzer, for example, used their acute powers of observation and the skillful mastery of their art to transform their direct access to nature and their ability to effect productive knowledge into a form of authority as "certain" knowledge, or "*scientia.*" While they strove to establish their status as experts, they also gave voice to their knowledge. These artisans were engaged in a kind of theorizing about nature and, while so engaged, they articulated a body of claims about nature and about the bases of knowledge, which they expressed through naturalistic representation. The most important of these claims was that nature is the primary foundation of certain knowledge, that this knowledge can be extracted by engaging with nature, and that close observation and replication of the objects and processes of nature result in a facsimile of nature that both comprises

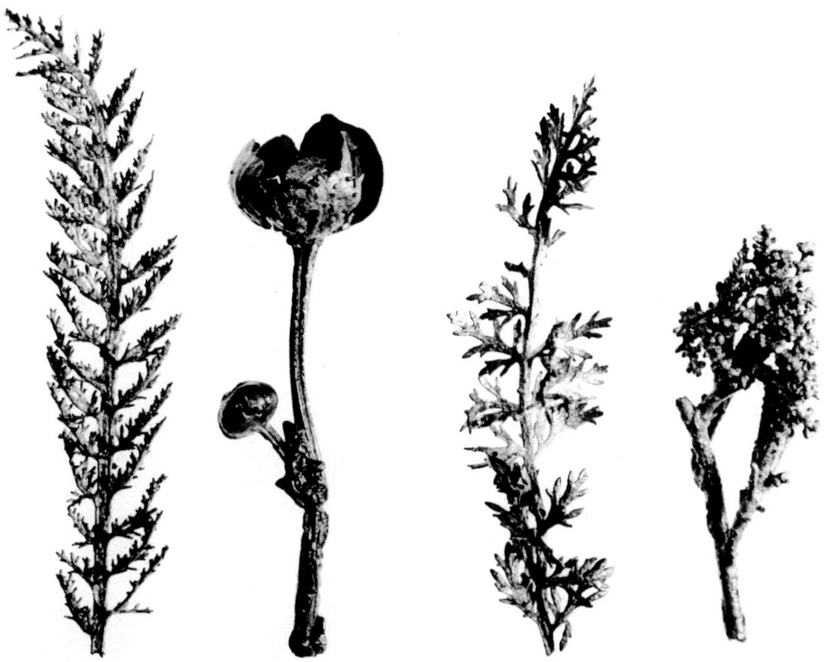

15. Wenzel Jamnitzer, *Four Natural Castings of Plants,* c. 1540, silver
Germanisches Nationalmuseum, Nuremberg

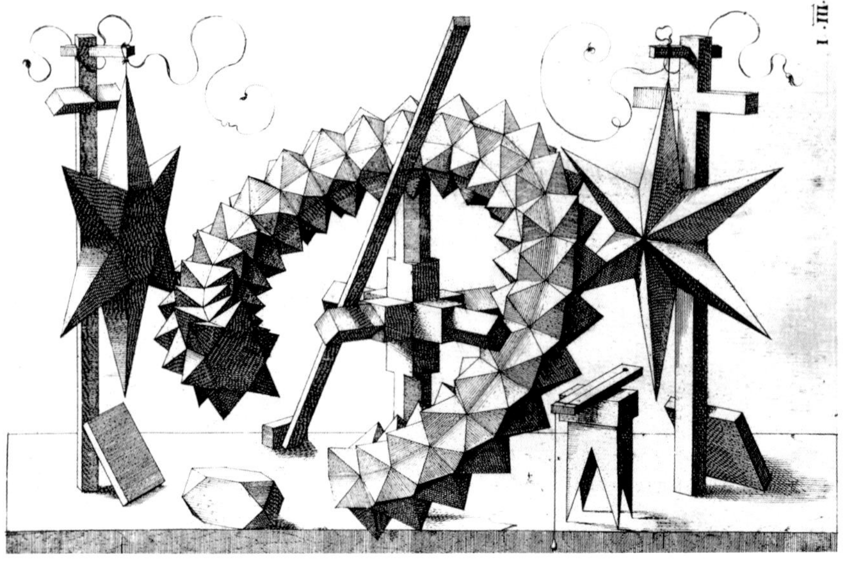

16. Jost Amman, engraver, *Forms in Perspective,* after a drawing by Wenzel Jamnitzer, from *Perspectiva corporum regularium* (Nuremberg, 1568), unpaginated, engraving
The British Library

17. Nicolas Neufchatel, *Portrait of Wenzel Jamnitzer*, c. 1562–1563, oil
Musée d'art et d'histoire, Geneva

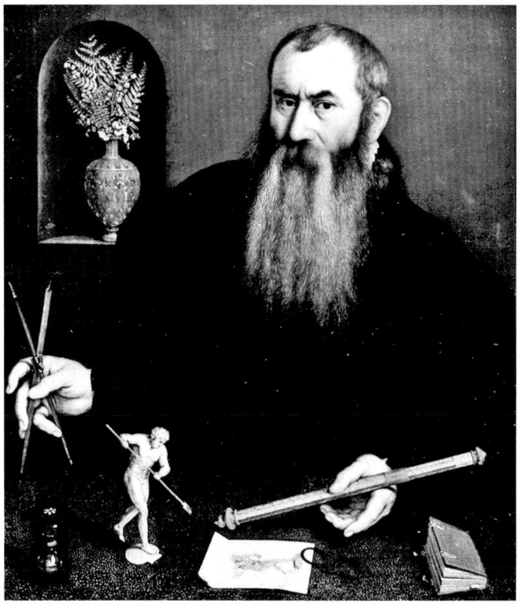

knowledge and proves the expertise of the practitioner.

In the sixteenth century, individuals who wished to know and to take possession of nature looked to art (in the active sense of *ars*, or making, not just fine art) as the medium through which to accomplish this goal. This was true not only of physicians and scholars, like Paracelsus, but also of princes and city governors, who came to regard artisans as possessing the key that would unlock the productive powers of nature. Naturalistic representation reinforced the claims of the artisanal epistemology at the same time that this epistemology was being disseminated by artisans, by their naturalistic images and objects, and by individuals such as Paracelsus (and eventually many others).[49] The fashion for naturalistic representation fostered a lively trade in natural and artificial objects and their images. The patronage and commodity value of these goods created a climate favorable to the investigation and representation of nature; it helped to raise the status of the imitators and the knowers of nature; and it initiated an extended discussion (in the media of both texts and objects) about the boundary between art and nature and the human capability of transcending this boundary.[50]

In 1953 William M. Ivins, Jr., argued that visual artists, especially in such areas as botanical illustration, furthered science by making possible a new mode of visual communication. Artists developed this powerful new tool through their naturalistic "pictorial statements" and, even more important, they made these visual statements reproducible by the development of printmaking. As Ivins wrote,

> Communication is absolutely necessary for scientific and especially technological development, and to be effective it must be accurate and exactly repeatable. . . . The trying and testing cannot be done without exact repeatability of communication. What one or two men have thought and done does not become science until it has been adequately communicated to other men.[51]

Ivins makes an important point, but as demonstrated in this essay, artisans furthered science in an even more profound way than by communicating information or making

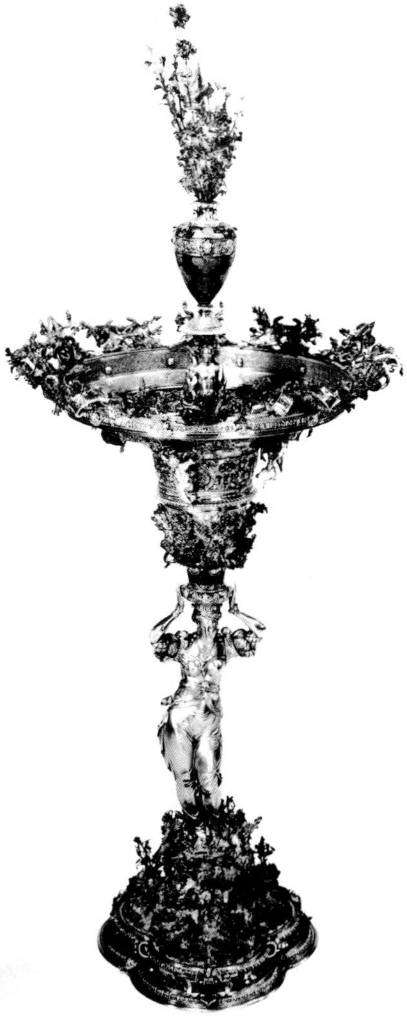

18. Wenzel Jamnitzer, *Mother Earth Centerpiece (Merkel Centerpiece)*, 1549, silver, cast, chased, stamped, etched, gilded, and painted
Rijksmuseum, Amsterdam

possible the accurate description of the objects of nature. Rather, they changed the view of what constituted positive, certain knowledge. They answered in new ways the questions of what the foundations of "scientific" knowledge are and how such knowledge is to be gained. Their techniques of observation and representation were deeply important in the development of empirical science, but perhaps more fundamental were their claims about the primacy of nature and the power and promise of natural knowledge. Artist-artisans not only helped constitute the very aims and methods of the study of nature during the fifteenth and sixteenth centuries; they also articulated a new kind of authority for nature.

By the same token, the "nature study"—broadly conceived—of Hans Weiditz has significance beyond its value as "scientific illustration." Rather, nature study epitomizes an entire approach to nature and epistemology and enables us to place naturalistic artisans—from the anonymous illustrator of the Carrara Herbal to the goldsmith Wenzel Jamnitzer—at the center of that momentous shift in attitudes toward nature and the material world that has come to be known as the Scientific Revolution.

NOTES

1. F. David Hoeniger, "How Plants and Animals Were Studied in the Mid-Sixteenth Century," in *Science and the Arts in the Renaissance,* ed. John W. Shirley and F. David Hoeniger (Washington, D.C., 1985), 130–131.

2. See the excellent discussion of Weiditz in David Landau and Peter Parshall, *The Renaissance Print: 1470–1550* (New Haven and London, 1994), 247–253.

3. Fritz Koreny, "A Coloured Flower Study by Martin Schongauer and the Development of the Depiction of Nature from van der Weyden to Dürer," *The Burlington Magazine* 133 (1991): 588–597, esp. 597.

4. Since 2002, when this essay was originally prepared, I have pursued various parts of its argument in *The Body of the Artisan: Art and Experience in the Scientific Revolution* (Chicago, 2004).

5. On the Carrara Herbal, see Felix Andreas Baumann, *Das Erbario Carrarese und die Bildtradition des Tractatus de herbis* (Bern, 1974). For discussion of the patronage of naturalism in art by Cosimo de' Medici in Tuscan Italy several centuries later, see the essay by Hans W. Hubert in this volume.

6. See Benjamin G. Kohl, *Padua under the Carrara, 1318–1405* (Baltimore, 1998). On dating Cennini's work, see Erling Skaug, "Cenniniana: Notes on Cennino Cennini and His Treatise," *Arte cristiana* 81 (1993): 15–22. For a close analysis of the innovative nature of Cennini's treatise, see Christiane Kruse, "Fleisch werden—Fleisch malen: Malerei als 'incarnazione,' Mediale Verfahren des Bildwerdens im *Libro dell'Arte* von Cennino Cennini," *Zeitschrift für Kunstgeschichte* 63 (2000): 305–325.

7. Cennino Cennini, *Il libro dell'arte,* trans. Daniel V. Thompson, Jr. (New York, 1960), 15 (hereafter cited as Cennini 1960).

8. Cennini 1960, 123.

9. Cennini noted that "if the person whom you are casting is very important, as in the case of lords, kings, popes, emperors, you mix this plaster with tepid rose water; and for other people a tepid spring or well or river water is good enough." Cennini 1960, 126.

10. Cennini 1960, 129.

11. Norberto Gramaccini, "Das genaue Abbild der Natur—Riccios Tiere und die Theorie des Naturabgusses seit Cennini," in *Natur und Antike in der Renaissance* (Frankfurt, 1985), 211–214. A more detailed study of the phenomenon of rulers seeking legitimacy by means of deploying a discourse on nature is warranted. The Hohenstaufen emperor Frederick II of Sicily, the Carrara family, and the Burgundian dukes all fostered an efflorescence of cultural production, one aspect of which appears to have been giving a new place and significance to nature.

12. Bert S. Hall, "Der Meister sol auch kennen schreiben und lesen": Writings about Technology ca. 1400–ca. 1600 A.D. and Their Cultural Implications," in *Early Technologies,* ed. Denise Schmandt-Besserat (Malibu, 1979), 48.

13. This is also discussed in Hall 1979, 53.

14. Martin Kemp, *The Science of Art: Optical Themes in Western Art from Brunelleschi to Seurat* (New Haven and London, 1990) discusses this. See also Pamela O. Long, "Power, Patronage, and the Authorship of *Ars:* From Mechanical Know-How to Mechanical Knowledge in the Last Scribal Age," *Isis* 88 (1997): 1–41; and, by the same author, "The Contribution of Architectural Writers to a 'Scientific' Outlook in the Fifteenth and Sixteenth Centuries," *Journal of Medieval and Renaissance Studies* 15 (1985): 265–298; and, more recently, *Openness, Secrecy, Authorship: Technical Arts and the Culture of Knowledge from Antiquity to the Renaissance* (Baltimore, 2001). See also Mario Carpo, *L'architettura dell'età della stampa: Oralità, scrittura, libro stampato e riproduzione meccanica dell'imagine nella storia delle teorie architettoniche* (Milan, 1998); English translation, Mario Carpo, *Architecture in the Age of Printing* (Cambridge, Mass., and London, 2001). Carpo argues that changes in organization and division of labor on the building site helped lead to the Renaissance architect's new self-conception.

15. For artisanal authorship, see Long 2001.

16. Otto Pächt, "Early Italian Nature Studies and the Early Calendar Landscape," *Journal of the Warburg and Courtauld Institutes* 13 (1950): 31. Artistic naturalism, like the scientific revolution, is often discussed in a heroic mode.

17. James S. Ackerman, "Early Renaissance 'Naturalism' and Scientific Illustration," in *The Natural Sciences and the Arts: Aspects of Interaction from the Renaissance to the Twentieth Century, An International Symposium,* ed. Allan Ellenius (Uppsala, 1985), 2.

18. James S. Ackerman, "The Involvement of Artists in Renaissance Science," in *Science and the Arts in the Renaissance,* ed. John W. Shirley and F. David Hoeniger (Washington, D.C., 1985), 126.

19. This is true of several of the essays in Shirley and Hoeniger 1985. Some of the essays also assume a clearly articulated separation between the activities of scientists and those of artists.

20. Francesco Petrarch, *Familiar Letters,* quoted in Andrea Bolland, "Art and Humanism in Early Renaissance Padua: Cennini, Vergerio and Petrarch on Imitation," *Renaissance Quarterly* 49 (1996): 481.

21. Pächt 1950, 32. About Guarino's efforts at *ekphrasis* (the description of one art form by means of another)—which he learned from his Greek tutelage in *progymnasmata* (rhetorical exercises) under Manuel Chrysoloras—and the way in which it interacted with Pisanello's descriptive painting to bring about a reinforcement of both endeavors, see Michael Baxandall, "Guarino, Pisanello and Manuel Chrysoloras," *Journal of the Warburg and Courtauld Institutes* 28 (1965): 183–204. On the function of Pisanello's drawings (and those of other Milanese artisans) as model books, see Robert W. Scheller, *Exemplum: Model-Book Drawings and the Practice of Artistic Transmission in the Middle Ages (ca. 900–ca. 1470),* trans. Michael Hoyle (Amsterdam, 1995). There seems to have been a conjunction among the humanists' use of *ekphrasis,* physicians' descriptions of particulars, and artists' precise rendering of nature. On physicians' descriptions, see Lorraine Daston and Katharine Park, *Wonders and the Order of Nature* (Cambridge, 1998), chapter 4. On scholarly attention to particulars, see also Gerhard Baader, "Anatomie, Konsilienliteratur und der neue Naturalismus in Italien im Spätmittelatler und Frühhumanismus," in *Die Kunst und das Studium der Natur,* ed. Wolfram Prinz and Andreas Beyer (Weinheim, 1987), 127–139.

22. Maximilian I, who did so much to shape the Holy Roman Empire politically and culturally, was ruler of the Netherlands before he became emperor. Charles V was brought up in Mechelen by his aunt, who was regent of the Netherlands. See Wim Blockmans and Walter Prevenier, *The Promised Lands: The Low Countries under Burgundian Rule, 1369–1530,* trans. Elizabeth Fackelman (Philadelphia, 1999), 73. See also Thomas DaCosta Kaufmann, *Mastery of Nature: Aspects of Art, Science, and Humanism in the Renaissance* (Princeton, 1993) and, by the same author, *Court, Cloister, and City: The Art and Culture of Central Europe, 1450–1800* (Chicago, 1995).

23. Koreny 1991, 591–592.

24. Landau and Parshall 1994, 50–55.

25. Joseph Leo Koerner, *The Moment of Self-Portraiture in German Renaissance Art* (Chicago, 1993), 203–205.

26. Markus Nass, "Stellung und Bedeutung des Monogramms Martin Schongauers in der Graphik des 15. Jahrhunderts," in *Martin Schongauer: Druckgraphik,* ed. Harmut Krohm and Jan Nicolaisen (Berlin, 1991), 48–62.

27. Fedja Anzelewsky et al., *From Schongauer to Holbein: Master Drawings from Basel and Berlin* [exh. cat., National Gallery of Art, Washington] (Washington, 1999), 39–42.

28. Paracelsus, *Die neun Buecher De Natura Rerum* (1537), in *Sämtliche Werke,* ed. Karl Sudhoff (Munich, Berlin, 1928), 11:314.

29. Krohm and Nicolaisen 1991, 83–84, no. 3c (Lehrs 7). See Robert Koch, "Schongauer's Dragon Tree," in "Tribute to Wolfgang Stechow," ed. Walter Strauss, special issue, *Print Review* 5 (1976): 114–119; also Götz Pochat, *Der Exotismus während des Mittelalters und der Renaissance* (Stockholm, 1970),

118–136. It may be significant that the sap from the dragon tree, which resembled blood, was used both to stop the flow of blood and as an ingredient in printing ink. See Colin T. Eisler, *Dürer's Animals* (Washington, D.C., 1991), 133.

30. Larry Silver, "Nature and Nature's God: Landscape and Cosmos of Albrecht Altdorfer," *Art Bulletin* 81 (1999): 194–214. See also Larry Silver and Pamela H. Smith, "Splendor in the Grass: The Powers of Nature and Art in the Age of Dürer," in *Merchants and Marvels: Commerce, Science, and Art in Early Modern Europe*, ed. Pamela H. Smith and Paula Findlen (New York, 2002).

31. Walter Pagel, *Paracelsus: An Introduction to Philosophical Medicine in the Era of the Renaissance* (1958; 2nd rev. ed., Basel and New York, 1982), 51.

32. See Paracelsus, *Das Buch Labyrinthus medicorum genant* (1553), in Sudhoff 1928, 11:192.

33. Albrecht Dürer, *Vier bücher von menschlicher Proportion* (1528), book 3, quoted in Erwin Panofsky, *The Life and Art of Albrecht Dürer* (Princeton, 1955), 279–280.

34. This aspect of his thought has usually been attributed to Neoplatonic influences, but two recent works on Paracelsus have suggested the vernacular nature of his ideas. See, for example, Stefan Rhein, "'Mein bart hat mer erfaren dan alle euer hohe schulen': Ein Zwischenruf zur Quellenfrage bei Paracelsus," in *Paracelsus: Das Werk, die Rezeption*, ed. Volker Zimmermann (Stuttgart, 1995), 99–104; and, in the same volume, Gundolf Keil, "Mittelalterliche Konzepte in der Medizin des Paracelsus," 173–193.

35. Paracelsus, *Astronomia Magna: oder Die gantze Philosophia sagax der grossen und kleinen Welt des von Gott hocherleuchten erfahrnen und bewerten teutschen Philosophi und Medici* (1571), in Sudhoff 1928, 12:56. For another view of the meaning of sensory immersion for Paracelsus, see J. R. R. Christie, "The Paracelsian Body," in *Paracelsus: The Man and His Reputation, His Ideas, and Their Transformation*, ed. Ole Peter Grell (Leiden, 1998).

36. Paracelsus 1571 in Sudhoff 1928, 12:92. To read the signs of a person, one studied the *chiromantia*, the outer form of the body, the hands, feet, veins, lines, folds, and so on; the *physionomia*, the face and the whole head; the *substantia*, the form of the whole body; and the *mos et usus*, the manner in which a person stands or presents his or her body (Paracelsus 1571 in Sudhoff 1928, 12:176). Dürer published *Vier bücher von menschlicher Proportion* in 1528.

37. Paracelsus 1571 in Sudhoff 1928, 12:59.

38. Panofsky 1955, 273.

39. Paracelsus 1553 in Sudhoff 1928, 11:193.

40. Paracelsus 1553 in Sudhoff 1928, 11:191. Although today *abmachen* means to undo, Panofsky translated Dürer's use as "reproducing nature from life." See Erwin Panofsky, *Idea: A Concept in Art Theory*, trans. Joseph J. S. Peake (Columbia, S.C., 1968), 121–126. Dürer's usage was probably closer to the modern *abschreiben*, to copy verbatim. Dürer also used the term *abkunterfet* to mean portraying after life or with verisimilitude.

41. A sixteenth-century goldsmith's manuscript treatise (c. 1570–1594) on the technique of casting from life survives in the Bibliothèque nationale de France, Paris. It documents these methods and is discussed by Leonard N. Amico in *Bernard Palissy: In Search of Earthly Paradise* (Paris and New York, 1996), 86–92.

42. The process is described in Gualterius Rivius, *De architectura* (Nuremberg, 1542), 95.

43. Ernst Kris, in "Der Stil 'Rustique': Die Verwendung des Naturabgusses bei Wenzel Jamnitzer und Bernard Palissy," *Jahrbuch der kunsthistorischen Sammlungen in Wien* 1 (1928): 158, noted that Jamnitzer's casting from life could be most usefully compared to Dürer's nature studies, such as *The Large Turf*.

44. This marvel of metalwork was melted down in the eighteenth century; only the Four Seasons, which formed the base of this fountain, still survive, in the Kunsthistorisches Museum, Vienna. See J. F. Hayward, "The Mannerist Goldsmiths: Wenzel Jamnitzer," *The Connoisseur* 164 (1976): 148–154; and Gerhard Bott, ed., *Wenzel Jamnitzer und die Nürnberger Goldschmiedekunst 1500–1700* [exh. cat., Germanisches Nationalmuseum, Nuremberg] (Munich, 1985). A record of this marvel from the early 1640s (from which this description is derived) states that it was "not only physics and metaphysics, but also politics, as well as many wonderful philosophical and poetical secrets displayed and proven to the eyes" (nit allein physica und metaphysica sondern auch politica mit vielen schön philosophischen und poetischen geheimnuses für die augen gestellet und fürgewiesen). The manuscript, in the Germanisches Nationalmuseum, was published in *Jahrbuch der Kunsthistorischen Sammlungen des Allerhöchsten Kaiserhauses* 7 (1888), here quoted in German from Erik Forssman, "Renaissance, Manierismus und Nürnberger Goldschmiedekunst," in Nuremberg 1985, 3. In the inventory of the *Kunstkammer* of Rudolf II (1607–1611), the fountain is described as being contained in eighteen boxes marked with a particular seal. The contents of box number 8 included a small book in which the entire meaning of the fountain was neatly written out on parchment. In box number 9 were two empty drawers in which the waterwheel was to be laid if the boxes were moved "overland." This box also conveniently contained a screwdriver, with which the waterwheel could be dismantled. These documents are reprinted in Klaus Pechstein, "Der Goldschmied Wenzel Jamnitzer," in Nuremberg 1985, 67–70.

45. For an illustration of the extraordinary range of Jamnitzer's ambition, see Nuremberg 1985. See also Kemp 1990, 63 and 173.

46. Jamnitzer's perception of himself as a designer and his skill in drawing are made clear in Klaus Pechstein, "Zeichnungen von Wenzel Jamnitzer," *Anzeiger des Germanischen Nationalmuseums* (1970): 81–95. Jamnizter was probably instrumental in formulating the requirements for the new (1572) Nuremberg goldsmiths' ordinance, which required that the candidate complete a work according to a drawn model (Pechstein 1970, 89).

47. Klaus Pechstein, "Der Merkelsche Tafelaufsatz von Wenzel Jamnitzer," *Mitteilungen des Vereins für Geschichte der Stadt Nürnberg* 61 (1974): 90–121.

48. The entire inscription is quoted in Pechstein 1974, 95–96.

49. I treat this process at length in Pamela H. Smith, *The Body of the Artisan: Art and Experience in the Scientific Revolution* (Chicago, 2004).

50. Martin Kemp, "Wrought by No Artist's Hand: The Natural, the Artificial, the Exotic, and the Scientific in some Artifacts from the Renaissance," in *Reframing the Renaissance*, ed. Claire Farago (New Haven, 1995), shows the various and interpenetrating values assigned to objects that teetered on the edge of the nature and art divide. See also Peter Parshall, "Introduction: Art and Curiosity in Northern Europe," in "Art and Curiosity," ed. Peter Parshall, special issue, *Word & Image* 11 (1995): 327–331.

51. William M. Ivins, Jr., *Prints and Visual Communication* (1953; reprint Cambridge, Mass., 1969). Erwin Panofsky also argued that art was central in the formation of science. See "Artist, Scientist, Genius: Notes on the 'Renaissance-Dämmerung,'" in *The Renaissance: A Symposium* (New York, 1953), 77–93. For more recent publications, see especially Peter Parshall, "*Imago contrafacta:* Images and Facts in the Northern Renaissance," *Art History* 16 (1993): 554–579; David Summers, *The Judgment of Sense: Renaissance Naturalism and the Rise of Aesthetics* (Cambridge, 1987); and Kemp 1990.

Drmek aneb Abrahamůw Strom. Vitex. Abrahamſbawm.

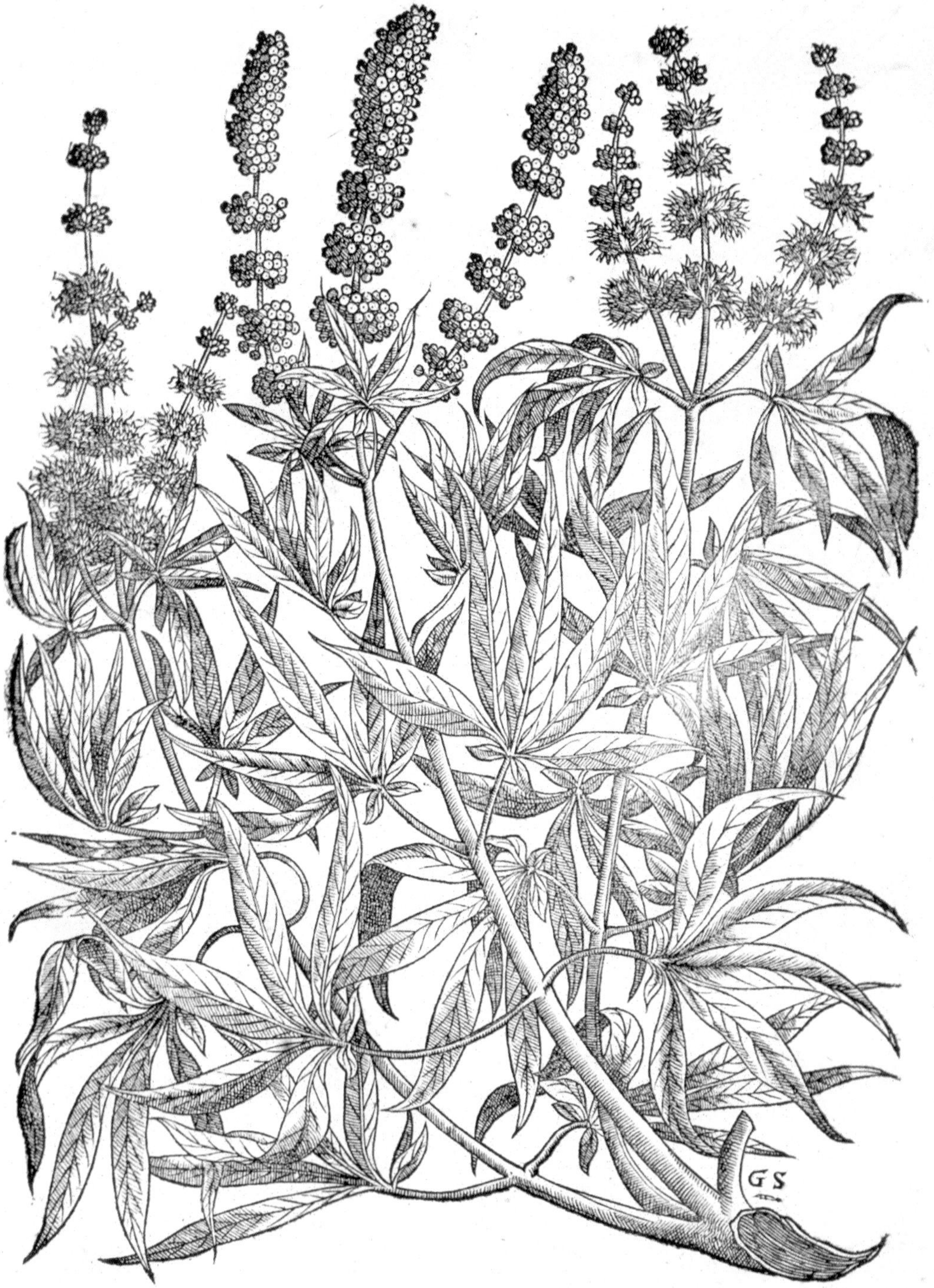

O Wrbě Kapitola LI.

Wrbij množſtwij geſt/některé gſau tak wyſoké/zwláſſť we Wlaſſých/že z nich Tyče a Kolij k Wodruom dělagij. Giné gſau nižſſý/naryſſawé a Hebké/ vžitečné k ſwazowánij Wobručuow. Giné gſau opět twrdſſý a pewněgſſý/z nichž Koſſýky a Nuoſſe dělagij. Giné gſau opět malé/kteréž obyčegně podlé Plotuow a na březých roſtau/tenké magijce prautij/k Koſſýkuom/ Nuoſſem / Koſſynám a k giným těm podobným nádo-

ALAIN TOUWAIDE
Smithsonian Institution, National Museum of Natural History

Botany and Humanism in the Renaissance: Background, Interaction, Contradictions

According to tradition, botanical knowledge in Renaissance Italy progressed in two steps: during the second half of the fifteenth century and, after an eclipse of more than forty years, from c. 1540 onward. It then culminated with translations of the Greek text of Pedanius Dioscorides' *De materia medica* (65 C.E.): into Italian (first published in 1544) by Pietro Andrea Mattioli (1500–1577) (fig. 1) and into Castilian (first published in 1555) by Andrés de Laguna (1499–1560) (fig. 2). No explanation is proposed for this progression, which is all the more surprising when one takes into account such earlier manuscript productions as the Carrara Herbal or the Codex Bellunensis.

Through an examination of this evolution in the wider context of botany and pharmacology (for the two disciplines were then closely associated), it will become evident that the development of botanical knowledge was determined by the work of the Ferrara physician Nicolao Leoniceno (1428–1524). In his attempt to establish the theoretical principles for a scientific approach to medical and scientific literature of classical antiquity, Leoniceno impeded the evolution of knowledge of botany and pharmacology. Although he did not create a new medical profile (elite physicians already specialized in textual interpretation), he did substitute new texts for those that elite physicians had traditionally analyzed. Because of the nature of the texts that he introduced (Greek classical works), he transformed the activity of these physicians, proposing a new method (philology) and a new purpose (the recovery of ancient legacy). As a result, the profile of elite physicians was transformed in the sense that they had to become erudite philologists. Leoniceno thus separated the construction of the scientific text—and, consequently, of science itself—from the practice of science, in this case botany and medicine. This is the legacy of Leoniceno and the problem that humanist physicians had to address. To this end, they had to develop new and imaginative solutions to restore the link between scientific text and scientific activity. Almost contemporaneously, Mattioli and Laguna, coming after their German forerunners, proposed new and original—though different—paradigms of botanical and medical studies.

In analyzing Leoniceno's epistemic strategy, I shall focus on its side effects, that is, the arrested development of plant science during the Renaissance. However paradoxical it may appear at first glance because of its negative consequences, Leoniceno's method will become more understandable when it is placed back in its specific context, that of medical botany and pharmacology at the end of the Middle Ages and the dawn of the Renaissance. Because Leoniceno's strategy was largely determined by the evolution of botany and therapeutics from antiquity onward, and consisted in restoring ancient

Pietro Andrea Mattioli, *Herbarz: ginak Bylinář* (Prague: G. Melantrych, 1562), chapter 51, folio 52r: *Vitex*
Biblioteca Nazionale Centrale di Roma

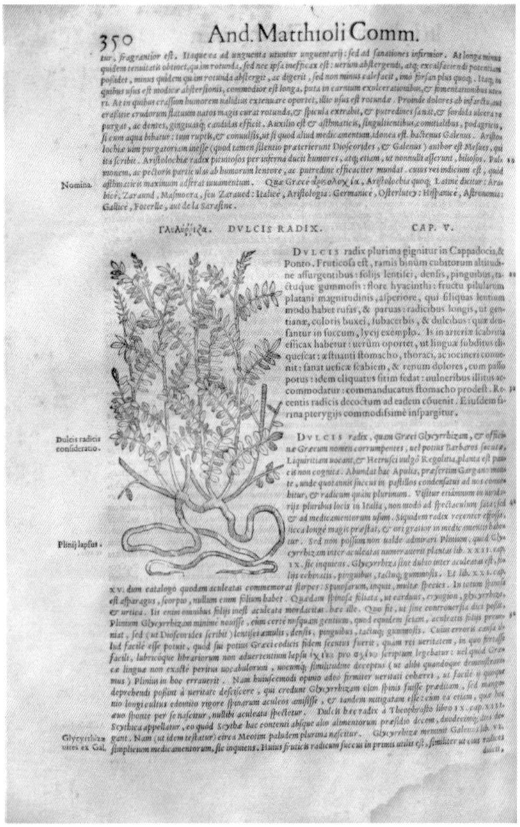

350 And. Matthioli Comm.

tur, fragrantior est. Itaque ea ad unguenta utuntur unguentarij: sed ad sanationes infirmior. At longa minus quidem tenuitatis obtinet, quàm rotunda, sed nec ipsa inefficax est: uerùm abstergendi, atq; excalfaciendi potentiam posidet, minus quidem quàm rotunda abstergit, ac digerit, sed non minus calefacit, imò forsan plus quoq;. Itaq; in quibus usus est modicæ abstersionis, commodior est longa, puta in carnium exulcerationibus, & fomentationibus uteri. At in quibus crassum humorem ualidius extenuare oportet, illic usus est rotundæ. Proinde dolores ab infarctu, aut crassitie crudorum flatuum natos magis curat rotunda, & spicula extrahit, & putredines sanat, & sordida ulcera repurgat, ac dentes, gingiuasq; candidas efficit. Auxilio est & asthmaticis, singultientibus, comitialibus, podagricis, si cum aqua bibatur: tum ruptis, & conuulsis, ut si quod aliud medicamentum idonea est. hactenus Galenus. Aristolochiæ uim purgatoriam inesse (quod tamen silentio præterierunt Dioscorides, & Galenus) author est Mesues, qui ita scribit. Aristolochiæ radix pituitosos per infrema ducit humores, atq; etiam, ut nonnulli asserunt, biliosos. Pulmonem, ac pectoris particulas ab humorum lentore, ac putredine efficaciter mundat. cuius rei indicium est, quod

Nomina. asthmaticis maximum afferat iuuamentum. Quæ Græcè ἀριστολοχία, Aristolochia quoq; Latinè dicitur: Arabicè, Zarawnd, Masmocta, seu Zaraued: Italicè, Aristologia: Germanicè, Osterlutey: Hispanicè, Astromenia: Gallicè, Foterlle, aut de la Sarasine.

ΓΛΥΚΥΡΡΙΖΑ. DVLCIS RADIX. CAP. V.

Dulcis radix plurima gignitur in Cappadocia, & Ponto. Fruticosa est, ramis binum cubitorum altitudine assurgentibus: folijs lentisci, densis, pinguibus, tactuque gummosis: flore hyacinthi: fructu pilularum platani magnitudinis, asperiore, qui siliquas lentium modo habet rufas, & paruas: radicibus longis, ut gentianæ, coloris buxei, subacerbis, & dulcibus: quæ densantur in succum, lycij exemplo. Is in arteriæ scabritia efficax habetur: uerùm oportet, ut linguæ subditus deliquescat: æstuanti stomacho, thoraci, ac iocineri conuenit: sanat uesicæ scabiem, & renum dolores, cum passo potus: idem eliquatus sitim sedat: uulneribus illitus accommodatur: commanducatus stomacho prodest. Recentis radicis decoctum ad eadem cõuenit. Eiusdem farina pterygijs commodissimè inspargitur.

Dulcis radicis consideratio. Dulcis radix, quam Græci Glycyrrhizam, & officinæ Græcum nomen corrumpentes, uel potius Barbaros secutæ, Liquiritiam uocant, & Hetrusci uulgò Regolitia, planta est paucis non cognita. Abundat hac Apulia, præsertim Gargano monte, unde quotannis succus in pastillos condensatus ad nos conuehitur, & radicum quàm plurimum. Visitur etiãmnum in uiridarijs pluribus locis in Italia, non modò ad spectaculum sata; sed & ad medicamentorum usum. Siquidem radix recenter effossa, sicca longè magis præstat, & ori gratior in medicamentis habe-

Plinij lapsus. tur. Sed non possum non ualde admirari Plinium, quòd Glycyrrhizam inter aculeatas numerauerit plantas lib. XXII. cap. IX. sic inquiens. Glycyrrhiza sine dubio inter aculeatas est, folijs echinatis, pinguibus, tactuq; gummosis. Et lib. XXI. cap. XV. dum catalogo quodam aculeatas commemorat stirpes: Spinosarum, inquit, multæ species. In totum spinosa est asparagus, scorpio, nullum enim folium habet. Quædam spinosa foliata, ut carduus, eryngion, glycyrrhiza, & urtica. Iis enim omnibus folijs inest aculeata mordacitas. hæc ille. Quo fit, ut sine controuersia dici posit, Plinium Glycyrrhizam minimè nouisse, cùm certè nusquam gentium, quod equidem sciam, aculeatis folijs proueniat, sed (ut Dioscorides scribit) lentisci æmulis, densis, pinguibus, tactuq; gummosis. Cuius erroris causa illud facilè esse potuit, quod sui potius Græci codicis fidem secutus fuerit, quàm rei ueritatem, in quo fortasse facili, lubricóque librariorum non aduertentium lapsu ἐχίνα pro σχίνῳ scriptum legebatur: uel quod Græcæ linguæ non exactè peritus uocabulorum, uocumq; similitudine deceptus (ut alibi quandoque demonstrauimus) Plinius in hoc errauerit. Nam huiuscemodi opinio adeò firmiter ueritati cohæret, ut facilè ij quoque deprehendi posint à ueritate desciscere, qui credunt Glycyrrhizam olim spinis fuisse præditam, sed mangonio longi cultus edomito rigore spinarum aculeos amisisse, & tandem mitigatam esse: cùm ea etiam, quæ hoc æuo sponte per se nascitur, nullibi aculeata spectetur. Dulcis hæc radix à Theophrasto libro IX. cap. XIII. Scythica appellatur, eo quòd Scythæ hac contenti absque alio alimentorum præsidio decem, duodecimq; dies de-

Glycyrrhizæ uires ex Gal. gant. Nam (ut idem testatur) circa Meotim paludem plurima nascitur. Glycyrrhizæ meminit Galenus lib. VI. simplicium medicamentorum, sic inquiens. Huius fruticis radicum succus in primis utilis est, similiter ut eius radices dulcis,

1. Pietro Andrea Mattioli, *Commentarii . . . in libros sex Pedanii Dioscoridis Anazarbei de materia medica* (Venice: Vincentii Valgrisii, 1558), book 3, chapter 5, page 350: *Dulcis radix*
Biblioteca Nazionale Centrale di Roma

practice, it is necessary first to look farther back in time, to the development of pharmacology in antiquity.

Ancient Botany

The Treatises of Theophrastus

Born in Eresos (Mitylini, Greece), Theophrastus (371/370–287/286 B.C.E.) moved to Athens, where he studied with Plato (427–347 B.C.E.).[1] After Plato's death, he followed Aristotle (384–322 B.C.E.) in his travels to Asia Minor and Macedonia. When both were back in Athens, Aristotle founded a school (335 B.C.E.), where Theophrastus continued his studies.[2]

A polymath author of no fewer than 225 works, according to the ancient biographer Diogenes Laertius (third century C.E.), Theophrastus is best known for his two treatises on botany: *Peri phutōn istorias* (Latin, *De historia plantarum,* in which *historia* is properly translated as *inquiry*)[3] and *Peri phutōn aitōn* (Latin, *De causis plantarum*).[4] Following Aristotelian science—especially Aristotle's *Peri zōōn moriōn* (Latin, *De partibus animalium*) and *Peri zōōn istorias* (Latin, *De animalium historia*), Theophrastus first analyzed all the parts of plants and developed a botanical lexicon *(De historia plantarum).* Then he studied the genesis of plants *(De causis plantarum).*[5] In these works, he collected all currently known data, as Aristotle did for zoology.

Pliny and the Naturalis historia

Though Pliny (23/24–79 C.E.) was not the author of a treatise specifically devoted to botany or pharmacology, his work has to be taken into consideration here, for his encyclopedia of natural sciences, *Naturalis historia,* played a major role during the Middle Ages and at the beginning of the Renaissance.[6] In this all-embracing work, he dealt with plants in books 12 through 19 (books 12 and 13, exotic trees; 14 and 15, fruit trees; 16, forest trees; 17, trees used in economic life; 18, agriculture; 19, garden plants) and with pharmacology in books 20 through 27. Though it could well be described as a "panorama of the world," Pliny's encyclopedia does not evince a specific method of scientific investigation or a particular theory of pharmacology.[7]

Dioscorides and De materia medica

Very little is known about Dioscorides, the author of a treatise on pharmacology entitled in Greek *Peri ulēs iatrikēs* (henceforth cited by its exact Latin rendering, *De materia medica*) supposedly published in 77 C.E.[8] Nevertheless, he has been credited with a medical education in Alexandria and a career as a military physician in the troops of a Roman emperor, identified variously as Claudius (r. 41–54 C.E.) or Nero (r. 54–68 C.E.).

Evidence does not support such a picture, however, making a reassessment of Dioscorides' life necessary. He was surely born in Anazarba (Cilicia)[9] at some time before Galen (129–after 216? C.E.), by whom he is cited, but his life dates cannot be stated with precision. Even the name of Areios, to whom *De materia medica* was dedicated, provides no sure chronological reference, for that person is not clearly identified. The

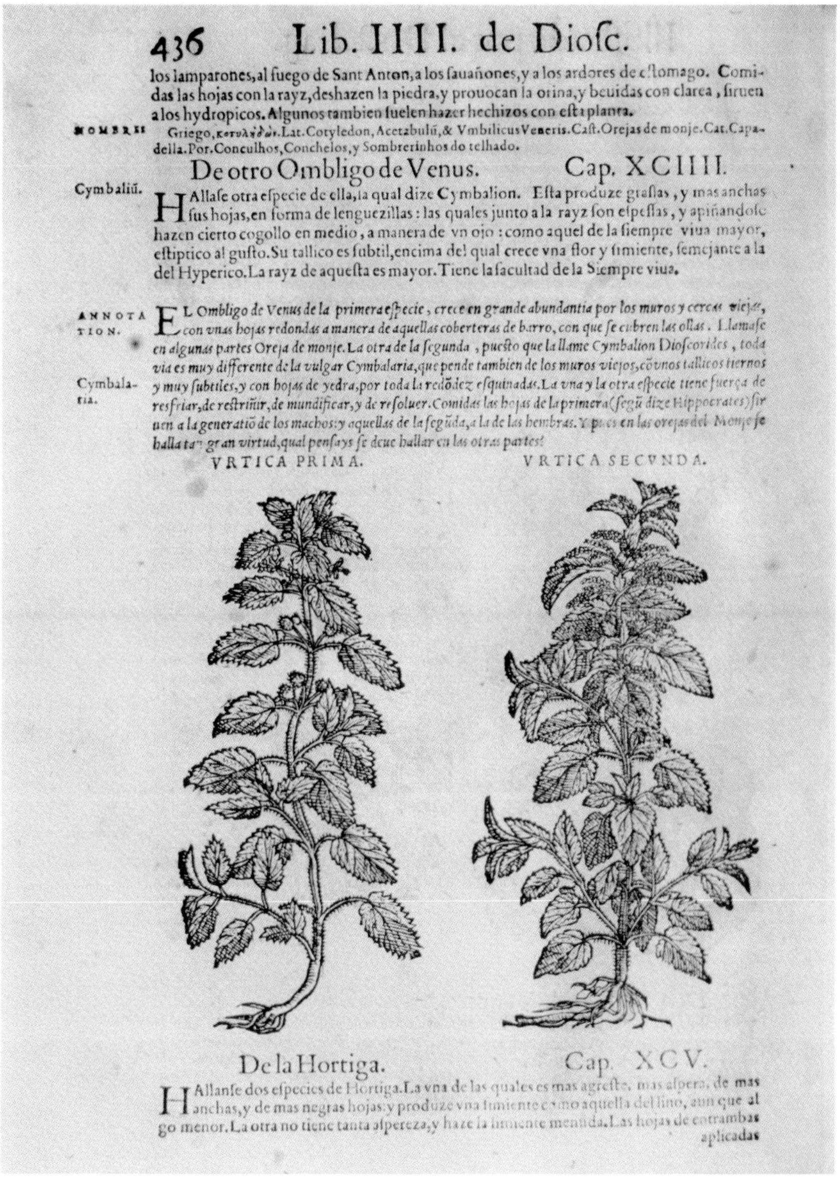

436 Lib. IIII. de Dioſc.

los lamparones, al fuego de Sant Anton, a los ſauañones, y a los ardores de eſtomago. Comidas las hojas con la rayz, deshazen la piedra, y prouocan la orina, y beuidas con clarea, ſiruen a los hydropicos. Algunos tambien ſuelen hazer hechizos con eſta planta.

NOMBRES Griego, κοτυληδών. Lat. Cotyledon, Acetabulũ, & Vmbilicus Veneris. Caſt. Orejas de monje. Cat. Capadella. Por. Conculhos, Conchelos, y Sombrerinhos do telhado.

De otro Ombligo de Venus. Cap. XCIIII.

Cymbaliũ. HAllaſe otra eſpecie de ella, la qual dize Cymbalion. Eſta produze graſſas, y mas anchas ſus hojas, en forma de lenguezillas: las quales junto a la rayz ſon eſpeſſas, y apiñandoſe hazen cierto cogollo en medio, a manera de vn ojo: como aquel de la ſiempre viua mayor, eſtiptico al guſto. Su tallico es ſubtil, encima del qual crece vna flor y ſimiente, ſemejante a la del Hyperico. La rayz de aqueſta es mayor. Tiene la facultad de la Siempre viua.

ANNOTATION. *EL Ombligo de Venus de la primera eſpecie, crece en grande abundantia por los muros y cercas viejas, con vnas hojas redondas a manera de aquellas coberteras de barro, con que ſe cubren las ollas. Llamaſe en algunas partes Oreja de monje. La otra de la ſegunda, pueſto que la llame Cymbalion Dioſcorides, toda via es muy differente de la vulgar Cymbalaria, que pende tambien de los muros viejos, cõ vnos tallicos tiernos y muy ſubtiles, y con hojas de yedra, por toda la redõdez eſquinadas. La vna y la otra eſpecie tiene fuerça de resfriar, de reſtriñir, de mundificar, y de reſoluer. Comidas las hojas de la primera (ſegũ dize Hippocrates) ſiruen a la generatiõ de los machos: y aquellas de la ſegũda, a la de las hembras. Y pues en las orejas del Monje ſe halla tan gran virtud, qual penſays ſe deue hallar en las otras partes?* Cymbalaria.

VRTICA PRIMA. VRTICA SECVNDA.

De la Hortiga. Cap. XCV.

HAllanſe dos eſpecies de Hortiga. La vna de las quales es mas agreſte, mas aſpera, de mas anchas, y de mas negras hojas: y produze vna ſimiente como aquella del lino, aun que algo menor. La otra no tiene tanta aſpereza, y haze la ſimiente menuda. Las hojas de entrambas aplicadas

2. Andrés de Laguna, *Pedacio Dioscorides Anazarbeo, Acerca de la materia medicinal, y de los venenos mortiferos* (Salamanca: Mathias Gast, 1570), book 3, chapter 95, page 436: *Urtica prima, Urtica secunda*
Biblioteca Nazionale Centrale di Roma

only evidence of Dioscorides' life in the current state of historical documentation is the mention of another person in the preface of *De materia medica*: Laekanios Bassos, qualified as *kratistos*, who can be identified with some exactness with Julius Laecanius Bassus, a Roman senator in 64 C.E. Since the Greek adjective *kratistos* is the equivalent of the Roman title of *clarus* given to ex-senators, Dioscorides' reference to him would have to be related to a period after 64 C.E. On the other hand, Laecanius Bassus died before 70/77 C.E. Along with Areios, Dioscorides mentioned Laecanius in describing the circumstances in which he wrote *De materia medica*, after spending many years studying medicinal plants. Hence it is reasonable to conclude that *De materia medica* was completed between 64 and 70/77 C.E. On this basis, it can be hypothesized that Dioscorides was born sometime during the first quarter of the first century C.E.

Though mainly devoted to plants, *De materia medica* deals with all natural substances then used for the preparation of medicines, thus including animal and mineral products as well as plant material.[10] A chapter is devoted to each substance.[11] For plants, the treatise usually presents the following information: a description of the entire plant; the part to be used in medicine (for example, roots, leaves, flowers, seeds, or gum); the state of conservation (fresh or dry); its medicinal properties, therapeutic uses, preparation, and side effects (if any); tests for authenticity (in the case of a dry drug or a product resulting from a preparation); and uses in crafts, cosmetics, or veterinary medicine.

Conserved in some fragments of papyrus[12] and Byzantine manuscripts (the earliest of which dates to c. 512 C.E.),[13] *De materia medica* is currently known in several versions.[14] In the one that is believed to be original, the treatise is divided into five "books," or entities, each of which has a well-defined and coherent content. However, this division is not necessarily original: the five books are of similar length, about twenty-five hundred lines, the maximum length of a papyrus roll, and thus of a book, during Dioscorides' time. *De materia medica* thus constituted an uninterrupted sequence of more than a thousand chapters, arranged from "iris" to "black substances used to produce ink," so as to form a *scala naturae* (ladder of nature).

In more than twenty manuscript versions,[15] the text of *De materia medica* is accompanied by representations of plants, the authenticity and origin of which are still debated.[16] Like the text, these representations appear in different versions that do not always correspond to the texts. Contrary to a thesis widely diffused in the current literature, the most ancient form of the illustrations was not necessarily realistic. The supposedly schematic pictures correspond exactly to the descriptions of the plants in the text they illustrate. Realism in the illustrations is the result of later, free reinterpretations of the earlier, schematic plant illustrations.

From *De materia medica* onward, botany was conflated with pharmacology in the late antique and medieval worlds. It was the achievement of Renaissance scientists to separate the two disciplines and to develop botany as an independent science.

Galen's Treatises on Simple and Compound Medicines

Born in 129 C.E. in Pergamon, Galen studied first in Asia Minor and then in Alexandria, at the great medical school of that time.[17] After he completed his education and began his career, he returned to Pergamon, where he practiced medicine as a physician to the gladiators, gaining formidable firsthand knowledge of human anatomy. Called to Rome by the emperor Marcus Aurelius, Galen stayed in the capital for a long time (apart from a short break and many travels), teaching and writing abundantly on all medical topics. He probably lived until 216 C.E.[18]

Although Galen did not write on botany, he did author a very long treatise on simple medicines, *De simplicium medicamentorum temperamentis et facultatibus,* in which he relied on Dioscorides' *De materia medica* for the descriptions of the plants used as medicines.[19] Nevertheless, he approached materia medica from a viewpoint that was very different from that of Dioscorides.[20] In strict accordance with his reinterpretation of Hippocratic medicine, he explained the action of therapeutic substances by their physical properties, qualities that resulted from the matter of which they were made. Like any other object of the sublunar world (natural or not), drugs were made of one (or more) of the four ultimate components of the *kosmos:* earth, water, air, and fire. By the same token, the action of medicines (cold, wet, dry, hot) directly resulted from their matter. Since the human body was also a compound of particles, an interaction between medicinal products and the body was possible. The intensity of the action of a drug on the human body could be measured with great exactitude: the effect directly depended on the quantity (and possibly also on the mixture) of the materials of which the drug was made.[21] Galen's analysis concluded with a theory of the physical structure of drugs, namely, that they are made of particles of different shapes, which he also considered to be true of bodily organs. Thus, the matching of similar shapes makes possible an interaction between drugs and the body. According to Galen, the circulation of drug particles in the human body was regulated by the weight of the particles (heavy particles fell and light ones rose).[22]

Galen also devoted a treatise to compound medicines, *De antidotis.*[23] From the second century B.C.E. onward, indeed, a new therapeutic strategy developed. Its creation is traditionally attributed to Mithridates Eupator (c. 132–63 B.C.E.), king of Pontos, a region of Asia Minor on the southern shore of the Black Sea. This strategy consisted of combining drugs so as to exert a wide spectrum of therapeutic actions. The most famous medicine resulting from this strategy was the so-called *thēriakē,* initially used to treat the effects of all kinds of animal venom.[24]

Postclassical Pharmacology

The Eastern Mediterranean World

Though Theophrastus' treatises were groundbreaking, they do not seem to have been widely disseminated, either in antiquity or, later, in the Byzantine world or in the West.[25] Neither an Arabic nor a medieval Latin translation is currently known. *De materia medica,* however, was widely read, used, and adapted in Byzantium, at least as early as the fourth century C.E. Contrary to an interpretation generally espoused, degradation was not the principle underlying the transmission of the text of *De materia medica* from one copy to another. In the original version (whatever its form may have been), the number of medicines described (1,066 entries) far exceeded what a traditional society could manage. For this reason, an initial reorganization of the text was aimed at making it possible to use it in daily practice, which entailed a reduction in the number of products. The products eliminated were those used to treat the fewest diseases, rare diseases, or diseases no longer present in the Byzantine world.

At the same time, early Byzantine physicians such as Oribasius (fourth century), Aetius (sixth century), and Paul of Egina (seventh century) associated Dioscorides' data with Galen's materialistic pharmacology of simple medicines in different ways.[26] Both the revision of *De materia medica* and its association with Galen's treatise allowed new data, resulting from physicians' personal experience, to be introduced into the two associated texts. At the same time, the complementary field of compound medicines also underwent a typical evolution, fostering the development of therapeutics.

These trends were mirrored in the Arabic world.[27] After the emergence of Arab military and political power during the seventh century and the establishment of a capital in Baghdad in the eighth century, Arab or Arabic-speaking scientists absorbed the sciences developed by earlier civilizations, including Indian, Persian, and Greek, through a systematic transfer of knowledge during the ninth century.[28] *De materia medica* and Galen's *De simplicium medicamentorum temperamentis et facultatibus* were translated by Hunayn ibn Ishaq (808–873/877), who worked alone or in collaboration with Istafan ibn Basil.[29]

Arab or Arabic-speaking scientists quickly advanced beyond this heritage. In the field of simple medicines, al-Biruni (973–after 1050) composed a large encyclopedia of simple drugs similar to *De materia medica*,[30] while Ibn Sina (980–1037), better known in the West as Avicenna, reproduced and further developed Galen's materialistic system. The Andalusian-born pharmacologist Ibn al-Baytar (c. 1188–1248) described all the medicinal herbs of southern Spain, northern Africa, and the eastern Mediterranean basin in a manner similar to that of Dioscorides.

As for compound medicines, Arab or Arabic-speaking scientists raised the question of their overall property: was it equal to the sum of the properties of each drug or was it a different and new one resulting from the interaction of the properties of the various drugs that were combined? Opting for the second, Arab or Arabic-speaking physicians developed theories to determine the properties of compound drugs. Al-Kindi (d. c. 873) created a mathematical algorithm,[31] while Avicenna took a more philosophical approach: the properties of compound medicines cannot be determined, for they depend not only on the interaction of the ingredients, but also on the individual patient's response to the treatment. Clinical trials (individually made on each patient and repeated during the treatment, since the patient's response to the treatment could vary over time) were the only way to know the actual properties of compound medicines.

Plant Knowledge in the West before Translations from Arabic

Because of the overlapping of botany and medical botany, plant knowledge in the West during the early Middle Ages mainly relied on treatises on therapeutics.[32] After the classical period, pharmacology continued previous practice. Many new treatises were written before the eleventh century, reflecting continuity with ancient science (not only Roman, as generally admitted, but also Greek) and not a break, as assumed by traditional understanding of the Dark Ages. Instead, reinforcing the continuity of classical science, new material from the late antique and early Byzantine centers of scientific production (Alexandria and Constantinople, successively) arrived in the West (for example, Ravenna). Many new works were written (in Latin); rooted in classical knowledge, they also absorbed local data coming from the field.

In contrast to Theophrastus' treatises, *De materia medica* met with exceptional good fortune in the West. Quickly diffused throughout the Mediterranean basin, its Greek text was translated into Latin in a time and place not yet identified but possibly in Africa in the sixth century. This version was described by Renaissance scholars as *Longobardus*, because its most ancient manuscript is written in a minuscule hand, known to humanists as *longobarda littera* in reference to the Lombards, who arrived in Italy in 568 C.E. Like the texts, visual representations of plants were recovered from antiquity, reproducing the ancient system and thus closely adhering to its textual information.

The West after the Assimilation of Arabic Pharmacology

A systematic transfer of knowledge by translation from Arabic into Latin took place in Salerno, in southern Italy, from the end of the eleventh century onward.[33] Supposedly initiated by Constantine the African (d. after 1085 C.E.), this enterprise may simply have been an expansion of a previous campaign of translation from Arabic into Greek.[34] The activity further extended to Toledo (Spain) and Montpellier (France), providing Western scientists with a wide range of new texts. As a result, Arabic pharmacology (including simple and compound medicines) was reproduced in the West.

In the case of simple medicines, *De materia medica* was translated anew, supposedly by Constantine the African. It is not known, however, whether the new Latin text was based on a Greek or an Arabic version. Whatever the case, the chapters of the new translation are arranged in alphabetical order by the names of the relevant substances. In addition, this version was further annotated by Pietro d'Abano (1257–c. 1315).[35]

As for compound medicines, the theoretical speculation of Arab or Arabic-speaking scientists was reproduced by Western pharmacologists. Al-Kindi's mathematical approach was developed in many works entitled in Latin *Liber de gradibus*,[36] around the same time that Avicenna's relativist theory was published.

In rendering Arabic texts into Latin, translators often did not properly translate some Arabic terms, but instead transliterated them into the Latin alphabet. This was especially the case for technical terms and plant names. Current scholarship generally considers this to have been the result of the translators' ignorance, but it may also have been intentional, at least in part. A comparison with other examples of translations (from Greek into Arabic, and conversely) shows that transliterations could directly result from the method of translation itself. For instance, technical terms were not translated in the original copy of the translation, but were simply transliterated and sometimes followed by a blank space, to be filled in later with the translation. This was especially true when the translator was a native speaker of the language being translated and not the language of the translation. On the other hand, transliteration often reproduced local usage in multicultural areas where bidirectional lexical borrowing was common. This was the case in southern Italy, where transliterated terms were integrated into translations so that readers could more easily associate the written data with their personal knowledge and experience. Once the Arabs had been expelled from southern Italy by the Norman conquest in the late eleventh century, there was a consequent tendency toward cultural uniformity, so that transliterations no longer corresponded to the usage of a multiethnic culture but instead became a source of confusion and, inevitably, of error. In the thirteenth century, Simon of Genoa (1270–1303?) provided a reference work aimed at standardizing medical terminology: the *Synonima medicinae seu Clavis sanationis*.[37] This work lists terms and variants of terms that had come into use over time, together with the most commonly accepted Latin equivalent for each. His enterprise did not succeed, however, in introducing clarity into the scientific lexicon.[38]

Both lexical imprecision and the difficulty of evaluating the properties of compound medicines combined to make pharmacology a highly uncertain science. For that reason, in 1499, the Florentine college of physicians promulgated a code listing authorized medicines and explaining their composition.[39] The so-called Ricettario Fiorentino aimed at definitively eliminating ambiguity, at the same time offering patients the guarantee of safety for medicines prepared according to the code.

Botany and Pharmacology in Fifteenth-Century Italy

Early Knowledge of De materia medica

The Greek text of *De materia medica* does not seem to have existed in the West until the fifteenth century.[40] Pietro d'Abano visited Constantinople and saw a copy of Dioscorides' text, but he probably neither reproduced it nor obtained another copy. The first manuscript of Dioscorides' Greek text that can be traced in the West is an early fourteenth-century copy made in Constantinople.[41] This copy seems to have been in Florence at the beginning of the fifteenth century, then passing to the library of the Medici, from which Angelo Poliziano (1454–1494), a Florentine classical scholar, borrowed it.[42] Bessarion (1403?–1472), a Byzantine prelate who settled in Italy after the Council of Ferrara (1438), had two copies in his personal library, which he donated to the republic of Venice: one was a fourteenth-century copy from Constantinople and the other, from the mid-fifteenth century, was written in Crete by Michael Apostoles (c. 1420–after 1474/1486).[43] About the same time, another Byzantine prelate, Cardinal Isidore of Kiev (c. 1385–1463), brought to Rome a mid-fifteenth-century copy, partially destroyed, that he himself completed.[44] Federico II Maria de Montefeltro (1422–1482), duke of Urbino, owned a mid-fifteenth-century copy from Constantinople[45] and the scholar Nicolo Leonico Tomeo (1456–1531), often confused in current bibliography with his quasi-homonym Nicolao Leoniceno, had the privilege of possessing the draft of the Byzantine recension, that is, the revision of *De materia medica* made in Constantinople around 1340.[46] Another copy circulated from Constantinople to Corfu and Venice, generating new manuscripts.[47] One of them was copied in 1482 by Demetrios Triboles (active fifteenth century). An undated copy was made in Venice by the copyist Ioannes, who is not well known but who identified himself as coming from Korónē in the Peloponnese.[48] A third was probably made in Venice; it is now in Paris.[49] This copy, in turn, was further used as a model for several others. One was made for his own collection by the humanist Giorgio Valla (1447–1500), who worked in collaboration with the Byzantine scribe Zacharias Kallierges (before 1473–after 1524).[50] Another copy, made by an unidentified scribe, belonged to the Ferrara physician Nicolao Leoniceno.[51] Finally, another was the property of Ermolao Barbaro (1453–1493),[52] the Venetian humanist who also owned another copy of *De materia medica* that is currently lost.

Early Assimilation of Greek Botany and Pharmacology

Though Greek manuscripts continued to arrive from Byzantium and were reproduced in the West in increasingly greater numbers, they did not immediately contribute to the diffusion of Greek botany, for knowledge of the Greek language was still limited.[53] Theodoros of Gaza (1400–c. 1476),[54] a Greek émigré from Byzantium to Italy, translated Theophrastus' *Historia plantarum* and *De causis plantarum* from Greek into Latin. Completed sometime in 1453 or early 1454, this translation first circulated in manuscript form. It was printed in 1483 in Treviso, after the death of its translator. As noted above, five years earlier, in 1478, the Latin version of Dioscorides' *De materia medica,* with commentary by Pietro d'Abano, had been published in Colle Val d'Elsa in Tuscany, by Johann von Medemblick. At the same time, Ermolao Barbaro completed a Latin translation, with commentary, of Dioscorides' Greek text in five books.[55] Barbaro did not have it published before his death, however, probably because he was working on a philological revision of Pliny's *Naturalis historia,* printed in two volumes in 1492–1493. Similarly, Giorgio Valla translated from Greek into Latin fragments of Dioscorides' *De materia medica* and Galen's *De simplicium medicamentorum temperamentis et facultatibus,* as early Byzantine and Arabic scientists had done. His translation constituted books 30 through 35 of his monumental encyclopedia of sciences, *De expetendis et fugiendis rebus,* first published in 1501 in Venice by Aldus Manutius (c. 1452–1515).[56]

In a second phase, the Greek texts of ancient botanical treatises were finally put into print: Theophrastus' treatise was published in 1497[57] and Dioscorides' text in 1499; both were printed in Venice by Manutius, with the collaboration of Nicolao Leoniceno, if not at his request.[58]

Though seemingly the result of a natural evolution, this shift from Latin translations to Greek texts was probably caused by another dynamic, which requires the mention of other works by Valla. In addition to his translations of the works of Dioscorides and Galen, discussed above, Valla wrote a treatise entitled *De humani corporis partibus,* which is a translation of anatomical definitions extracted from ancient (Greek) medical texts.[59] He also translated from Greek into Latin treatises on topics as different as diagnosis by means of urine analysis, dietetics, and epidemiology[60]—particularly smallpox. The last was *De pestilentia,* a Latin translation of the Greek version of a treatise on smallpox and measles by the Arabic physician ar-Razi (865–925 C.E.).[61] In the preface to his translation, Valla explicitly stated why he undertook this version: he hoped to help cure the plagues of his time. He was probably referring to syphilis, which is thought to have first appeared in the West in 1494. Because of the skin eruption it provokes, syphilis was compared to smallpox. As a result, the therapies for smallpox in ar-Razi's treatise were expected to be transferable to syphilis. In any case, Valla's work was strictly philological, excluding specific medical considerations. It was limited to an exact rendering of Greek classical texts into Latin, so as to make them useful in the daily practice of contemporary physicians.[62]

Nicolao Leoniceno: Toward a Scientific Method to Assimilate Ancient Science

At the turn of the fifteenth century, Leoniceno of Ferrara was well known as an erudite physician and teacher.[63] Not only did he own an impressive collection of Greek manuscripts of classical medical and philosophical texts, but he also actively contributed to the enterprise of editing, translating, and commenting on ancient medical texts (especially those of Hippocrates and Galen).[64]

In 1492 Leoniceno published a short pamphlet commonly titled *De Plinii et plurium aliorum in medicina erroribus.*[65] According to traditional interpretation, Leoniceno compared the newly available text of Dioscorides with Pliny's *Naturalis historia,* which was the main reference work of that time in matters of natural sciences.[66] Discovering many similarities between the two texts, he also noticed differences, particularly information in Pliny's text that seemed bizarre when compared with its equivalent in Dioscorides. Going further in his analysis, Leoniceno concluded that Pliny must have misunderstood Greek texts such as those of Dioscorides, if not Dioscorides' treatise itself. Following principles of textual criticism then in their early phases, Leoniceno proposed to replace a later epiphenomenon with its source, that is, Pliny's *Naturalis historia* with Dioscorides' *De materia medica:* in other words, a Latin work with its Greek source.

Relying on a now obsolete concept of the Renaissance,[67] this historical interpretation of Leoniceno's work focuses on only one aspect of his pamphlet of late 1492, the criticisms of Pliny. It overlooks another of its components, the denunciation of the mistakes attributed to Arab or Arabic-speaking scientists. This was nothing new at that time: polemics against Arabic science, particularly against medicine and pharmacology, were strident.[68] Furthermore, Western attitudes toward Arabic culture and Islam were aggressive: Byzantines who fled Constantinople after its conquest by the Turks in 1453 proposed to launch a new crusade,[69] and theological debates with the Muslim world—best represented by the Ottomans—were intense.[70]

The picture is more complicated than that, and must include other works by Leoniceno that are linked with the so-called Pliny controversy. These are, in the order in which they were written,

Epistola ad Hermolaum Barbarum, 1493

De epidemia quam vulgo morbum Gallicum vocant, 1497

De tiro seu vipera, c. 1498

Nicolaus Leonicenus Hieronymo Menochio, 1503

De Plinii, et plurium aliorum medicorum erroribus . . . novum opus, 1507

De dipsade et pluribus allis serpentibus, c. 1507

Apologia contra obtrectatores, 1522

In this context, *De Plinii* (1492) takes on a new dimension: it was a part of a large epistemological enterprise aimed at renovating scientific knowledge by providing it with a solid method.

Leoniceno's goal evolved over time. In its first phase, he aimed at eliminating the later versions of Greek scientific texts, that is, their classical Latin translations and their Arabic versions, together with the medieval Latin translations of the latter made in the West, from Constantine the African onward. Leoniceno's argument was simple: since the original texts were again available, they made obsolete all their later translations, rearrangements, and epiphenomena. The substitution was all the more necessary because, he believed, all the later versions of Greek scientific texts misinterpreted the originals. In this way, Leoniceno transferred the principle of textual criticism, then in its formative period, to scientific activity. Gradually discovering more and more copies of classical texts (whether in Greek or in Latin), Renaissance humanists compared them, discovering their divergences. By distinguishing between accurate and mistaken variant readings, they began to reconstruct ancient texts, defining the rules of textual criticism. By transferring these philological principles to science, Leoniceno fostered not only the use of Greek texts, but also the (re)creation of an unambiguous lexicon of plant names and technical terms. To this end, he strongly relied on Greek scientific terminology, avoiding—and even rejecting—any medieval Latin or Arabic influences.

In further applying these principles to pharmacology, Leoniceno promoted a return to Dioscorides' *De materia medica* in Greek and the use of Greek scientific designations of plants. This was something more than a philological exercise, however. Leoniceno's aim was to transform pharmacology into a reliable practice (if not a real science). Not only did he restore a therapeutic strategy of simple medicines (thus eliminating compound medicines and the speculation associated with them), but, in the matter of simples, he also fostered an unambiguous nomenclature.

Later on, Leoniceno shifted from the restoration of primary sources to an exact knowledge of the matter being dealt with. Knowledge had to be acquired not from books, but from direct contact with reality. In this new formulation, his strategy was not limited to botany and pharmacology, but was a universal epistemic method capable of being applied to any field of knowledge. Significantly enough, in its most recent formulation, philology no longer mattered. The only important point was a proper knowledge of the object of study. Hence Leoniceno's formula *Loquantur ipsi ut velint . . . modo quis serpens sit tirus . . . non ignorant* (They may speak as they want . . . provided they do not ignore what kind of snake the viper is).[71]

Though he did not explicitly target Valla's method for the assimilation of ancient science, Leoniceno's epistemology implied criticism of such an approach. The practical application of ancient texts could not rely on a purely philological translation, but had to be based on an exact knowledge of the data. This is particularly clear in Leoniceno's treatise on syphilis: he tried to show that ancient descriptions of diseases do not correspond to contemporary ones (and that ancient diseases themselves do not correspond to contemporary ones). He thus also rejected the conclusions drawn from lexical similarities between one language and another.

Overall, Leoniceno's epistemology was not simply a preference for Greek texts over Latin or Arabic translations (including their medieval Latin versions). It was a wider strategy aimed primarily at recovering the sources of science and then at reconstructing science by a proper method of acquiring knowledge. He applied this strategy to pharmacology, restoring the use of simple medicines so as to construct a reliable science of medicines, which could not be based on compound medicines because their action either was too complicated to define or was unpredictable.

Though this strategy was apparently justified, it was nonetheless misleading in its transfer to pharmacology. It amounted in fact to discarding the work of updating classical texts from antiquity onward—that is, the contributions introduced cumulatively into those texts by generation upon generation of practitioners who constantly revised the ancient heritage on the basis of their personal experience. In other words, Leoniceno's epistemology undid the progressive adjustments of ancient science over time, in a sense transporting Renaissance scientists back to antiquity. This is the real reason why Manutius published Theophrastus' and Dioscorides' Greek texts: these works not only helped Leoniceno overcome traditionalists' opposition provoked by *De Plinii* but also served as a tool for reviving the ancient practice of science.[72]

After the publication of *De Plinii* and the two Greek texts, almost no major new botanical or pharmacological work appeared in Italy until the 1540s. Exceptions included a second edition of Dioscorides' Greek text, published by the heirs and successors of Manutius in 1518;[73] two Latin translations of, and commentaries on, Dioscorides, those of Ermolao Barbaro, left unpublished by their author, and those of Marcello Virgilio Adriani (1464–1521);[74] and botanico-pharmacological works by Giovanni Manardo (1462–1536)[75] and Antonius Musa Brasavola (1500–1555).[76] The Greek edition of *De materia medica* did not contradict Leoniceno's method, so it carried on the influence of the 1499 edition after the first went out of print. Barbaro's translation and commentaries were published by the author's heirs, in reaction to the publication of the Latin translation of Dioscorides by Jean Ruelle (1474–1537) in Paris in 1516.[77] Adriani's translation was, in a certain sense, an anachronistic work of Ciceronian rhetoric on the order of early Italian literary humanism; as for Manardo and Brasavola, they were Leoniceno's pupils.

Botany and Pharmacology in Italy from 1540 Onward

The writing of new botanical treatises interrupted in Italy by Leoniceno's methodological proposal was ultimately resumed by Mattioli and Laguna. Some work was done before the publication of their major achievements, however. Illustrated herbals were published,[78] as well as a first Italian translation of *De materia medica* (1542) by Sebastiano Fausto and an illustrated version of Mattioli's Italian translation of *De materia medica* (Mantua, 1549; fig. 3). But the illustrated herbals were an extension of German production, the 1542 Italian translation had no major impact, and the 1549 illustrated Mattioli was a pirated hybrid that combined Mattioli's text with a German set of pictures. It was printed only once and was not widely distributed. Botany and pharmacology became fields of intense activity, however.[79] As evidence of this development, botanic gardens were created almost simultaneously in Pisa and Padua.[80] Nevertheless, Leoniceno's methodology continued to exert its influence for a certain time, requiring humanists to find new ways to reconcile the production of texts and the practice of science. In this pursuit, scholars have compared botanical and anatomical studies. Andreas Vesalius (1514–1564), for example, developed a new paradigm of experimentalism, writing a text, *De humani corporis fabrica* (Basel, 1543), based on his studies of actual human bodies. Though such comparisons suggest a coherent explanation of the almost simultaneous renewal of the fields of anatomy and botany, they are not justified. As recent research has demonstrated, Vesalius did not necessarily aim at constructing a new anatomy; he probably intended only to verify Galen's anatomical system, thus reactivating it.[81] In so doing, he (re)created the anatomical lexicon, in a manner not dissimilar to that fostered by Leoniceno.

In the search for a new approach,[82] Andrés de Laguna, though Spanish, has to be associated with Mattioli. Not only did Laguna freely reproduce Mattioli's plant

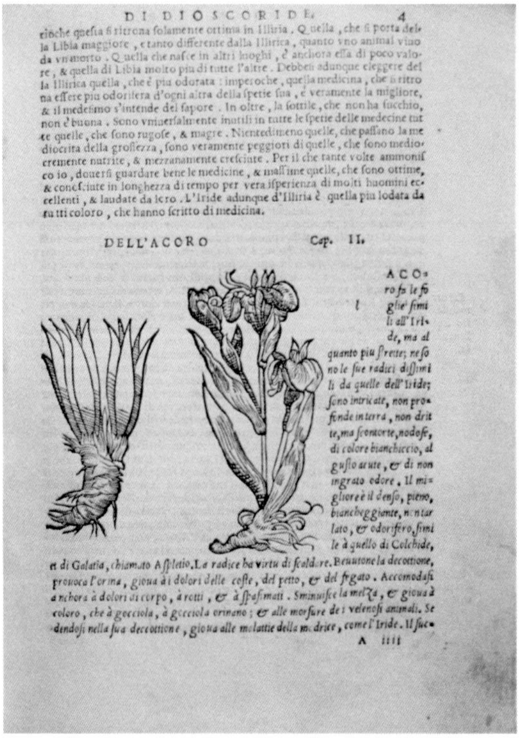
DI DIOSCORIDE. 4

cinche queſta ſi ritrona ſolamente ottima in Illiria. Quella, che ſi porta della Libia maggiore, e tanto differente dalla Illirica, quanto vno animal viuo da vn morto. Quella che naſce in altri luoghi, è anchora eſſa di poco valore, & quella di Libia molto piu di tutte l'altre. Debbeſi adunque eleggere della Illirica quella, che è piu odorata: imperoche, quella medicina, che ſi ritrona eſſere piu odorifera d'ogni altra della ſpetie ſua, è veramente la migliore, & il medeſimo s'intende del ſapore. In oltre, la ſottile, che non ha ſucchio, non è buona. Sono vniuerſalmente inutili in tutte le ſpetie delle medecine tutte quelle, che ſono rugoſe, & magre. Nientedimeno quelle, che paſſano la mediocrita della groſſezza, ſono veramente peggiori di quelle, che ſono mediocremente nutrite, & mezzanamente creſciute. Per il che tante volte ammoniſco io, douerſi guardare bene le medicine, & maſſime quelle, che ſono ottime, & conoſciute in longhezza di tempo per vera iſperienza di molti huomini eccellenti, & laudate da loro. L'Iride adunque d'Illiria è quella piu lodata da tutti coloro, che hanno ſcritto di medicina.

DELL'ACORO Cap. II.

ACORO ha le foglie ſimili all'Iride, ma alquanto piu ſtrette; ne ſono le ſue radici diſſimili da quelle dell'Iride; ſono intricate, non profonde in terra, non dritte, ma ſtorte, nodoſe, di colore bianchiccio, al guſto acute, & di non ingrato odore. Il migliore è il denſo, pieno, biancheggiante, non tarlato, & odorifero, ſimile à quello di Colchide, et di Galatia, chiamato Aſpletio. La radice ha virtu di ſcaldare. Beuutone la decottione, prouoca l'orina, gioua à i dolori delle coſte, del petto, & del fegato. Accomodaſi anchora à dolori di corpo, à rotti, & à ſpaſimati. Sminuiſce la melza, & gioua à coloro, che à gocciola, à gocciola orinano; & alle morſure de i veleno ſi animali. Sedendoſi nella ſua decottione, gioua alle malattie della madrice, come l'Iride. Il ſuc-

A iiii

3. *Il Dioscoride dell'eccellente Dottor . . . P. Andrea Matthioli* (Mantua: Iacomo Rossinello, 1549), book 1, chapter 2, page 4: *Dell'acoro*
Biblioteca Nazionale Centrale di Roma

representations; he also worked according to the methods developed by Leoniceno. Furthermore, he completed his Castilian translation of *De materia medica* in the intellectual milieu of Rome. But neither Mattioli's nor Laguna's enterprise was completely new: German scientists had already published new botanical works before them, though in a fairly different context.

The German Forerunners of Mattioli and Laguna

The production of illustrated herbals in Germany began as early as the mid-1470s with Conrad von Megenberg's *Buch der natur* (Augsburg, 1475), followed by the *Herbarius latinus* (Mainz, 1484), *Der Gart der Gesundheit* (Mainz, 1485), and the *Hortus sanitatis* (Mainz, 1491) (fig. 4). Humanistic interest in the restoration of classical scientific texts, however, did not start in Germany as early as it did in Italy (that is, around the mid-fifteenth century), for knowledge of Greek was not as widely diffused in Germany. Approximately one generation after the first Italian humanists, classical studies were fostered in Germany by such figures as Johannes Cuno (c. 1463–1513), Johannes Reuchlin (1454/1455–1522), Willibald Pirckheimer (1470–1530), and Beatus Rhenanus (1485–1547).[83] Even though some German intellectuals of that time harshly fought against Italian influence,[84] Leoniceno's work was known in Germany. Nicolas Pol (c. 1460–1532), physician to Emperor Maximilian I (1459–1519) in Innsbruck, owned a copy of *De Plinii,* which he probably acquired shortly after its publication.[85] Nevertheless, Leoniceno's method did not prevent further activity, as it did in Italy. The frontier between the practice of botany and pharmacology, on the one hand, and the production of text, on the other, was permeable, allowing a constant transfer of data from the field to book production, as was the case in ancient and medieval pharmacology.

This does not mean that botany in the German world evolved in a linear way without opposition. A scholar such as Janus Cornarius (c. 1500–1558), for example, was opposed to plant representations, asserting in the introduction to his edition of Dioscorides' Greek text (Basel, 1529) that pictures could never be the equivalent of reality and could not show all the possible variations of a single plant over time.[86] Instead, Cornarius suggested that authors only mention the names of plants (especially the vernacular names), since, according to him, they perfectly identified the plants. It was the herbals by Otto Brunfels and Leonhard Fuchs, which followed Cornarius' edition of Dioscorides' Greek text, and another herbal published by Johannes Soter the same year in Cologne, that tacitly answered Cornarius' objections.

In 1530 the Strasbourg printer Johann Schott (1477–c. 1548) published the first volume of an herbal containing text written by Otto Brunfels (c. 1488–1534) and plant representations by Hans Weiditz (c. 1500–1537): *Herbarum vivae eicones.*[87] The title page of this volume is particularly eloquent (fig. 5): it includes pictures of Apollo and Dioscorides, together with an evocation of idyllic gardens. The maxims accompanying the two portraits are no less programmatic: the one related to Apollo refers to the discovery of the therapeutic properties of

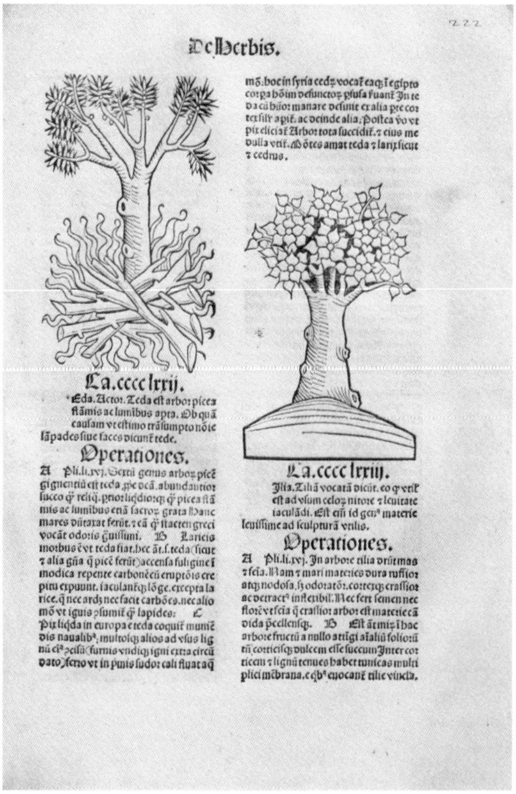

4. *Hortus sanitatis* (Mainz: I. Meydenbach, 1491), chapters 452 and 453, folio 222r: *Teda, Tilia*
Biblioteca Nazionale Centrale di Roma

plants, while the caption under Dioscorides' picture celebrates the merits of physicians and the value of health. Brunfels' purpose was to arrange medicinal plants described in ancient scientific literature (particularly that of Theophrastus and Dioscorides) in correspondence with contemporary German plant species. Such a program broke with previous practice and aimed at making ancient knowledge accessible again, according to Leoniceno's strategy.

Brunfels did not fully achieve the Leonicenian program, however. He returned neither to the original texts of ancient authors (Theophrastus and Dioscorides) nor to their Renaissance translations; instead, he mainly used medieval and Renaissance commentaries. Even though he knew Leoniceno's work, he still followed a medieval method, gathering the opinions of different authors on a topic and discussing them further. Not even in the discussion did he apply Leoniceno's analytical method. Rather than critically analyzing the data he collected, Brunfels associated them according to medieval concordism, though he did show some independence in certain cases.

The Brunfels-Weiditz herbal was the result of an uncoordinated association of textual descriptions and plant representations. In fact, the decision to include illustrations in the work did not even come from Brunfels, but rather from the publisher, Johann Schott. As a result, these plant representations constituted a sort of interference in Brunfels' approach to botany, all the more because they resulted from a different iconographical strategy. Supposedly a student of Albrecht Dürer (1471–1528), Weiditz worked directly from nature.[88] The illustrator's realism was so thorough that specimens gathered in the field were exactly reproduced, even going so far as to include their imperfections (fig. 6). The title of the herbal, which may be translated as "Lifelike Pictures of Plants," reflects Weiditz's strategy.

Rather than a work conceived in a unitary way by a single author, Brunfels' herbal is a heterogeneous association of scholarly material that was not properly assimilated with its original plant representations. More than the text, these representations correspond to Leoniceno's epistemic strategy (although such correspondence is probably coincidental). The distance between knowledge and the object of knowledge is canceled, even where the object of study is almost physically present in the book. Such an approach to nature, fostered by the Reformation and its theology, implicitly superseded the ancient and medieval system of plant representations, even though it did not immediately replace it.[89]

Problems of this kind are absent from *De historia stirpium commentarii insignes* by Leonhard Fuchs (1501–1566).[90] Fully

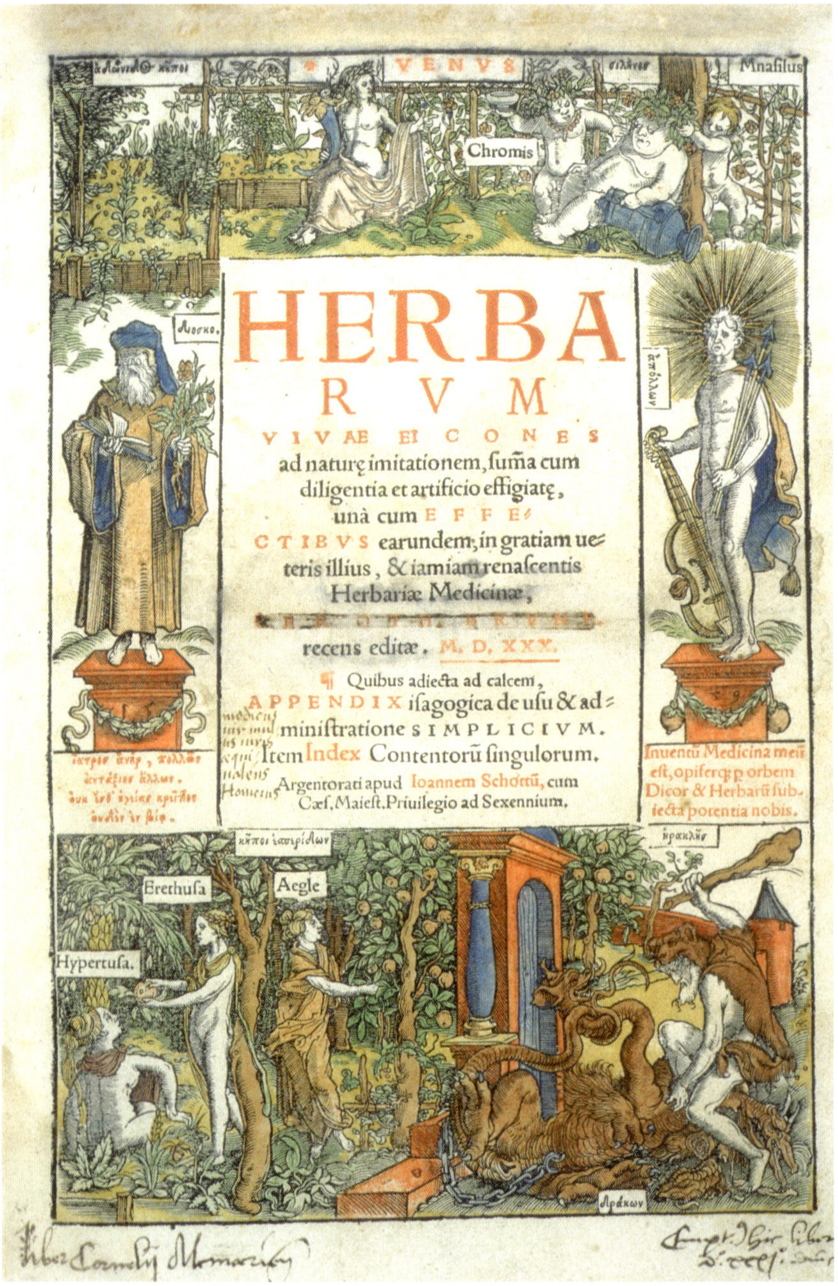

5. Otto Brunfels, *Herbarum vivae eicones* (Strasbourg: Ioannes Schottus, 1530), title page
Biblioteca Nazionale Centrale di Roma

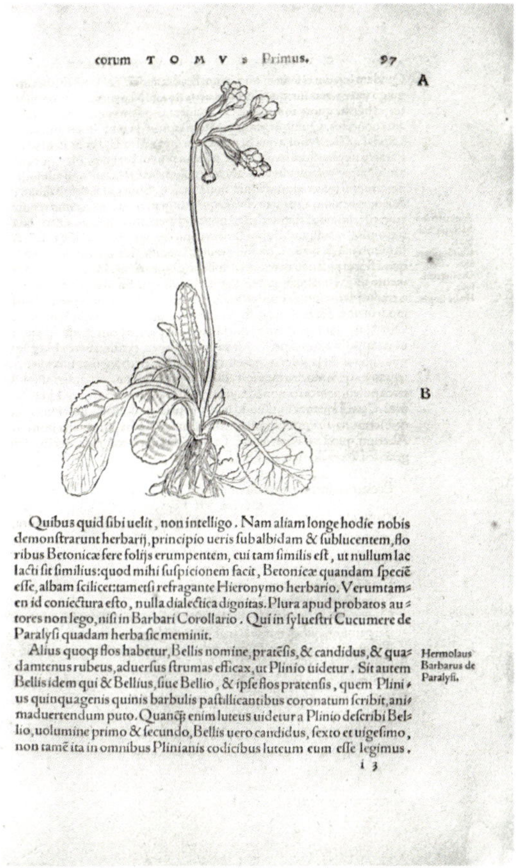

corum TOMVS Primus. 97

A

B

Quibus quid ſibi uelit, non intelligo. Nam aliam longe hodie nobis demonſtrarunt herbarij, principio ueris ſubalbidam & ſublucentem, floribus Betonicæ fere folijs erumpentem, cui tam ſimilis eſt, ut nullum lac lacti ſit ſimilius: quod mihi ſuſpicionem facit, Betonicæ quandam ſpeciē eſſe, albam ſcilicet: tametſi refragante Hieronymo herbario. Verumtamen id coniectura eſto, nulla dialectica dignitas. Plura apud probatos autores non lego, niſi in Barbari Corollario. Qui in ſyluestri Cucumere de Paralyſi quadam herba ſic meminit.

Alius quoq; flos habetur, Bellis nomine, pratēſis, & candidus, & quadamtenus rubeus, aduerſus ſtrumas efficax, ut Plinio uidetur. Sit autem Bellis idem qui & Bellius, ſiue Bellio, & ipſe flos pratenſis, quem Plinius quinquagenis quinis barbulis paſtillicantibus coronatum ſcribit, animaduertendum puto. Quanq; enim luteus uidetur a Plinio deſcribi Bellio, uolumine primo & ſecundo, Bellis uero candidus, ſexto et uigeſimo, non tamē ita in omnibus Plinianis codicibus luteum eum eſſe legimus. Hermolaus Barbarus de Paralyſi.

i 3

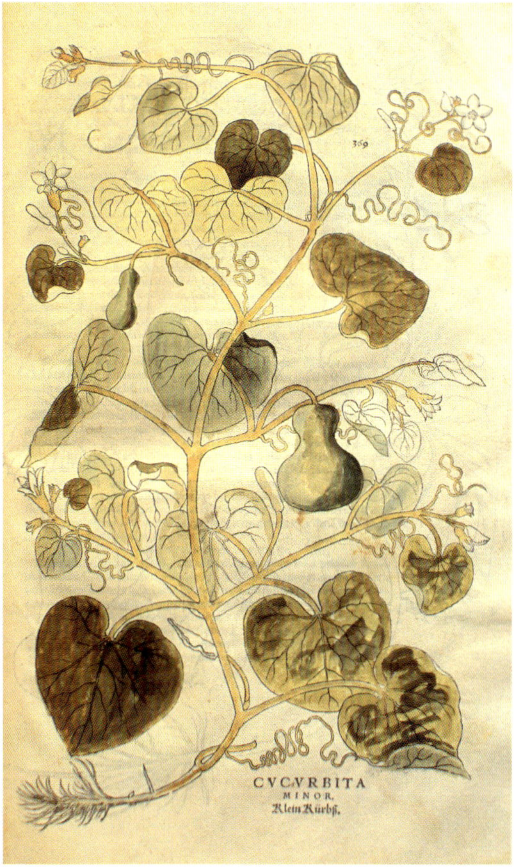

6. Otto Brunfels, *Herbarum vivae eicones* (Strasbourg: Ioannes Schottus, 1530), rhapsodia 15, folio 97: *De herba paralysi*
Biblioteca Nazionale Centrale di Roma

7. Leonhard Fuchs, *De historia stirpium commentarii insignes* (Basel: Officina Isingriniana, 1542), chapter 137, page 369: *Cucurbita minor*
Biblioteca Nazionale Centrale di Roma

mastering Leoniceno's method, Fuchs first wrote a refutation of contemporaneous physicians (1530).[91] Then he published his herbal (Basel, 1542) with plant representations by Albrecht Meyer and woodcuts by Heinrich Füllmaurer (d. 1545) and Veit Rudolf Speckle (d. 1550). A translator of—and commentator on—Hippocrates, Fuchs was interested in therapeutics, as was Brunfels.[92]

Fuchs' work, however, is different from Brunfels', especially from the perspective of visual representation.[93] Not only did Fuchs' plant descriptions result from a more personal approach, but his plant representations also represent a conceptual work in accordance with the analytical study in the text. Nature here is remodeled. Illustrations do not represent individual specimens gathered in the field (as do Weiditz's illustrations), but instead stand for their concept, capable of generating in the viewer all the possible variations to be encountered in nature, according to a Platonic definition of the idea. Perception of nature is mediated by a work of generalization: observation of a great number of specimens is followed by the abstraction of their common characteristics so as to produce a general definition, which, in turn, is graphically represented (fig. 7).

This was, in a certain sense, a full application of Leoniceno's epistemological program: knowledge is (re)built by a full mastery of its object, and the resulting knowledge is expressed by a universal and unambiguous concept. Further proof of the usefulness of Fuchs' work is provided by its many translations and re-editions, and the publication of small-size editions containing only the plant representations (fig. 8).

Pietro Andrea Mattioli

In spite of his claims, Pietro Andrea Mattioli[94] began his work in botanical illustration modestly: his 1544 Italian translation of *De materia medica*[95] was nothing more than a translation of the Latin version published in 1516 by the French herbalist Jean Ruelle

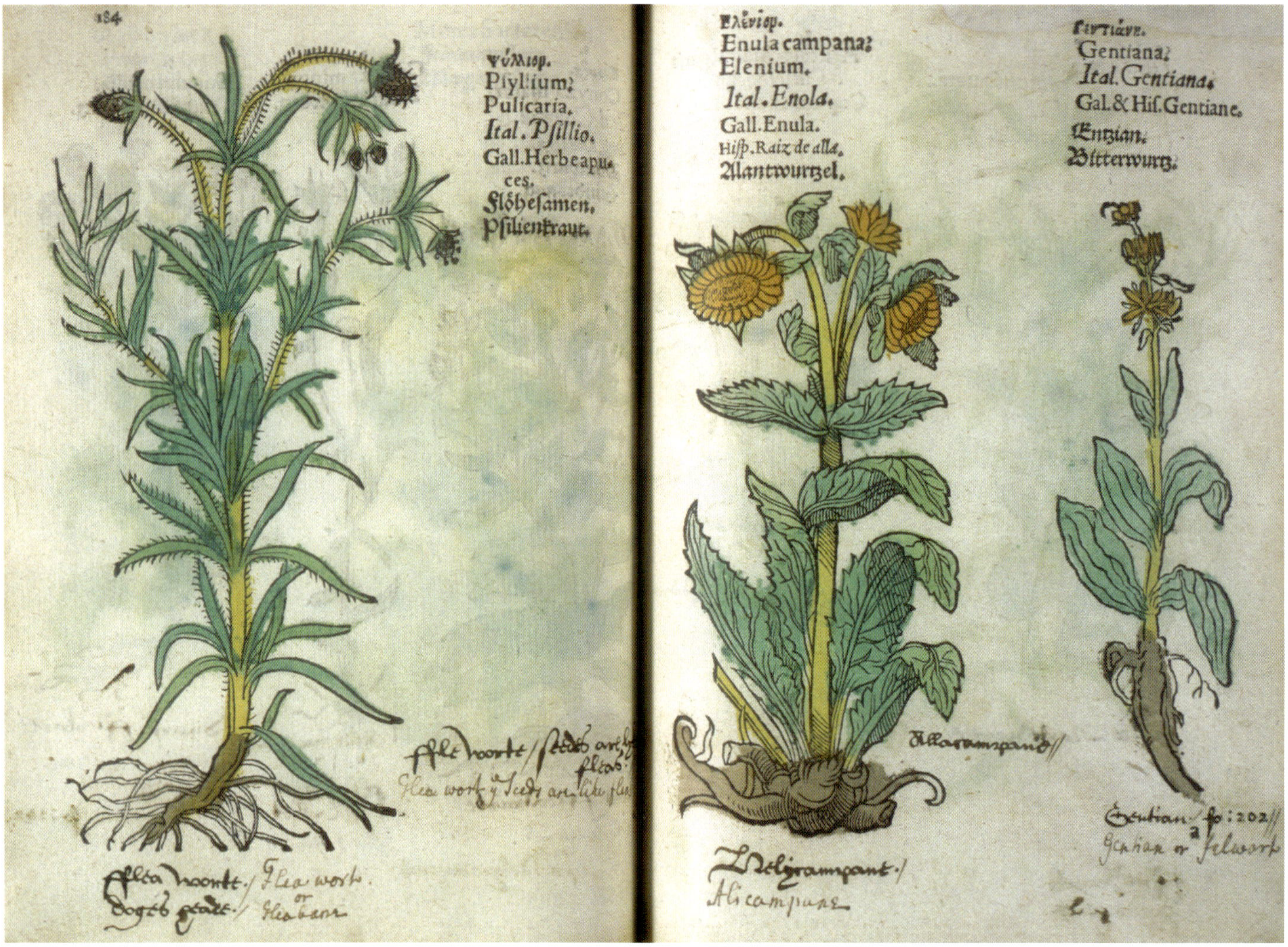

8. Leonhard Fuchs, *Primi de stirpium historia commentariorum tomi vivae imagines, in exiguam angustioremque formam contractae* (Basel: Officina Isingriniana, 1545), pages 184–185: *Psylium, Enula campana, Gentiana*
Biblioteca Nazionale Centrale di Roma

and widely reissued from that date on. This translation, in turn, relied on the edition published in 1499 in Venice by Manutius, for no manuscript was available in Paris, as Ruelle explicitly stated.

In the reissues and translations of his work,[96] Mattioli progressively developed a philological analysis of the text, gradually enlarging the number of sources consulted and consequently increasing the impact of philological analysis on his own work. This explains his programmatic and triumphant titles claiming that, by collating innumerable manuscripts, he corrected thousands of mistaken passages that had thus far gone unnoticed. Close examination reveals, however, that the number of corrections is by far smaller than that announced and that the collated sources were not necessarily manuscripts, but were often previously printed works. Be that as it may, Mattioli accumulated the contributions of his predecessors, in long chains of enumerations of the same nature as medieval commentaries, as did many of his contemporaries. To the textual data, however, Mattioli also added new information from the field. Not only did he report his personal experiences as a physician, but also those of a large number of colleagues and friends (who rarely remained such for a long time because of the Italian's irascible nature). He urged them to send specimens and seeds or, if this was not possible, information and representations of the plants. These informants were mainly botanists rather than pharmacists or practitioners, giving rise to the so-called Botanical Republic.[97]

At the same time and in the same way, Mattioli introduced into Dioscorides' text descriptions of additional plants, such as the banana tree, attaching them to the text on the basis of the differences, similarities, or other characteristics (fig. 9). By this process of agglutination, he gradually transformed his commentary on Dioscorides into an original work. But in so doing, he overloaded his translation and commentary with unrelated information, making it necessary eventually to relinquish Dioscorides' text altogether and to create new botanical works *ab novo*, independent of ancient treatises.

Andrés de Laguna

Andrés de Laguna took a different approach to *De materia medica* in his Castilian translation and commentary, *Pedacio Dioscorides Anazarbeo* (1555).[98] His work was philological in nature from its very inception. He had published his commentary on *De materia medica* in Latin in Lyon in 1554. Relying on Ruelle's translation, as had Mattioli, he aimed at improving it. To this end, Laguna consulted a Greek version of Dioscorides' manuscript, which he described as *antiquissimus* and which he had received on loan from the Spanish humanist Juan Paez de Castro (d. 1570).[99] Collating Ruelle's Latin translation and this Greek version of the manuscript, Laguna scrupulously analyzed their many divergences—thousands, according to him—from a strictly philological point of view. In so doing, he showed great philological acumen and a deep knowledge of the mechanisms of deterioration of texts, which was the unavoidable result of handwritten reproduction. As a consequence, Laguna opted for a different reading of many passages, reconstructing what he thought to be Dioscorides' original text, in perfect concordance with Leoniceno's program.

In the Castilian translation and interpretation of *De materia medica*, the Spanish scholar chose a different method from that of Mattioli. In his career, Laguna had traveled throughout Europe—from Spain to Germany and Italy, including France, England, and the Low Countries—constantly crossing boundaries of any kind, whether geopolitical, religious, or professional. In his approach to botany and therapeutics, Laguna repeatedly referred to his personal field experience as having included contacts with unschooled healers, charlatans, and others who did not belong to the elite "Botanical Republic."

Plant Representation in the Works of Mattioli and Laguna

The creation of a new botanical nomenclature and the description of plants on the basis of direct observation (firsthand or not) by means of an unambiguous botanical lexicon necessarily had an impact on plant representation and required revisions of the previous system. In *Pedacio Dioscorides Anazarbeo*, Laguna reproduced, in large part, Mattioli's representations (figs. 10 and 11). His original plant representations, on the other hand, constitute two groups: those from the East and those from the New World (including *Lignum indicum, Coccus indicus, Nux metella*, and *Milium indicum*; fig. 12). While Mattioli does not seem to have

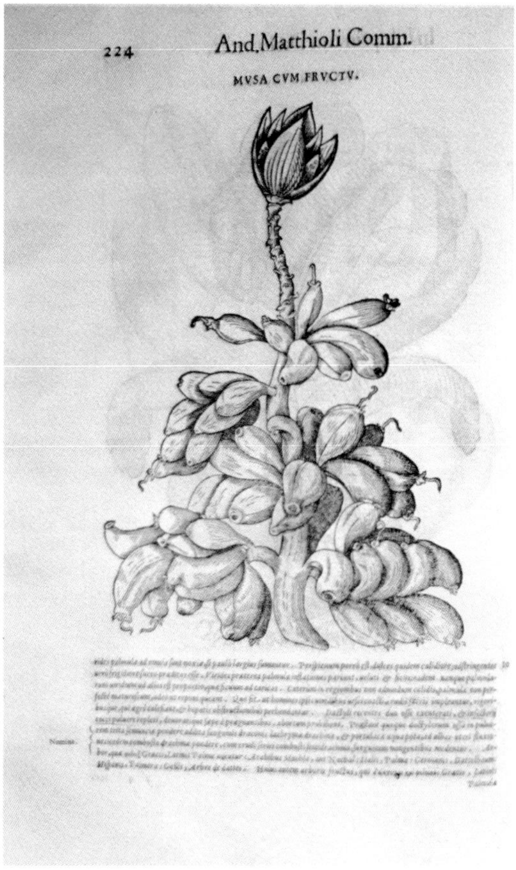

9. Pietro Andrea Mattioli, *Commentarii in sex libros Pedacii Dioscoridis Anazarbei de materia medica* (Venice: Officina Valgrisiana, 1565), page 224: *Musa cum fructu*
Biblioteca Nazionale Centrale di Roma

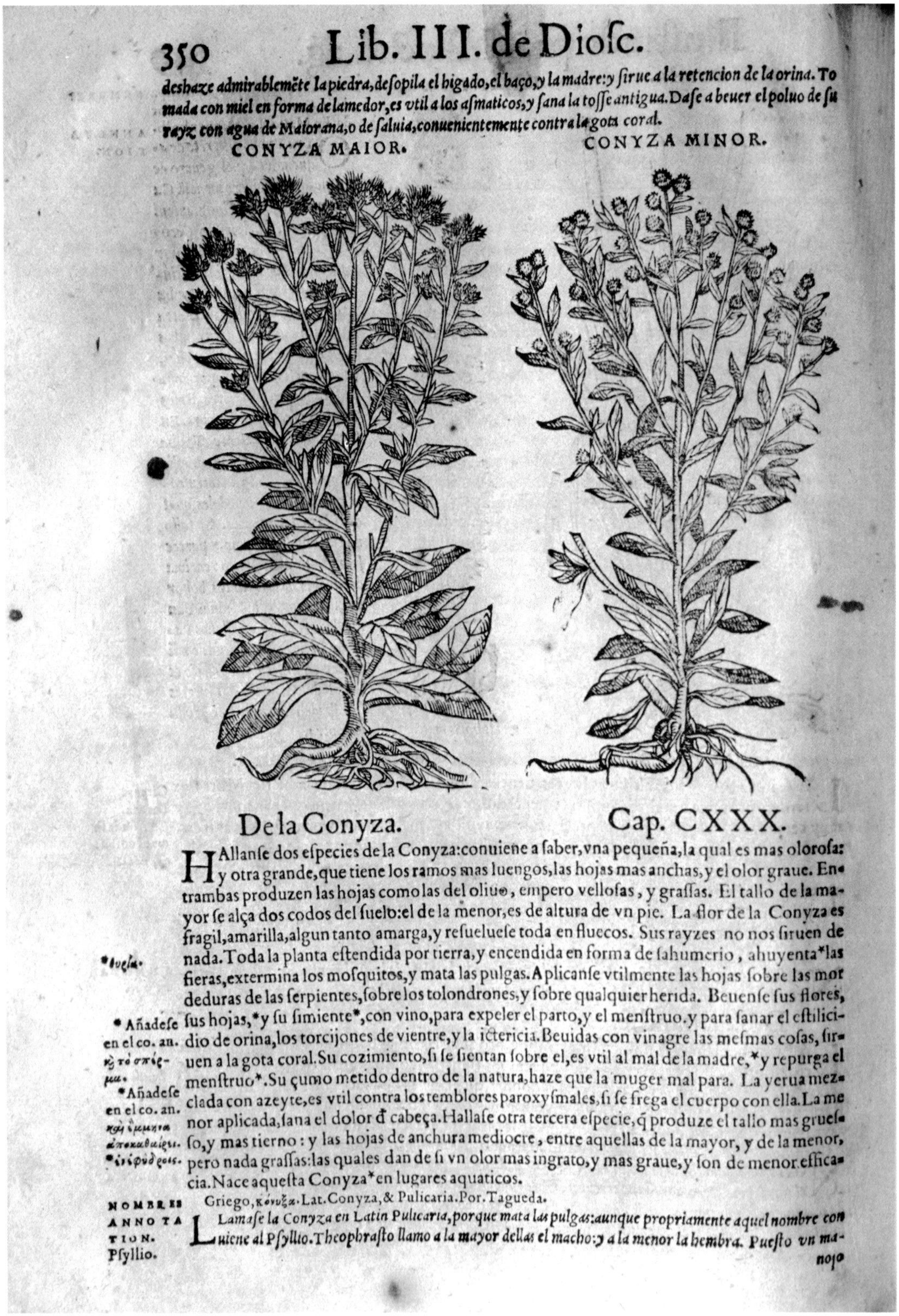

350 Lib. III. de Diosc.

desbaze admirablemẽte la piedra, desopila el higado, el baço, y la madre: y sirue a la retencion de la orina. Tomada con miel en forma de lamedor, es vtil a los asmaticos, y sana la tosse antigua. Dase a beuer el poluo de su rayz con agua de Maiorana, o de saluia, conuenientemente contra la gota coral.

CONYZA MAIOR. CONYZA MINOR.

De la Conyza. Cap. CXXX.

Hallanse dos especies de la Conyza: conuiene a saber, vna pequeña, la qual es mas olorosa: y otra grande, que tiene los ramos mas luengos, las hojas mas anchas, y el olor graue. Entrambas produzen las hojas como las del oliuo, empero vellosas, y grassas. El tallo de la mayor se alça dos codos del suelo: el de la menor, es de altura de vn pie. La flor de la Conyza es fragil, amarilla, algun tanto amarga, y resueluese toda en fluecos. Sus rayzes no nos siruen de nada. Toda la planta estendida por tierra, y encendida en forma de sahumerio, ahuyenta* las fieras, extermina los mosquitos, y mata las pulgas. Aplicanse vtilmente las hojas sobre las mordeduras de las serpientes, sobre los tolondrones, y sobre qualquier herida. Beuense sus flores, sus hojas, *y su simiente*, con vino, para expeler el parto, y el menstruo. y para sanar el estilicidio de orina, los torcijones de vientre, y la ictericia. Beuidas con vinagre las mesmas cosas, siruen a la gota coral. Su cozimiento, si se sientan sobre el, es vtil al mal de la madre, *y repurga el menstruo*. Su çumo metido dentro de la natura, haze que la muger mal para. La yerua mezclada con azeyte, es vtil contra los temblores paroxysmales, si se frega el cuerpo con ella. La menor aplicada, sana el dolor đ cabeça. Hallase otra tercera especie, q̃ produze el tallo mas gruesso, y mas tierno: y las hojas de anchura mediocre, entre aquellas de la mayor, y de la menor, pero nada grassas: las quales dan de si vn olor mas ingrato, y mas graue, y son de menor efficacia. Nace aquesta Conyza* en lugares aquaticos.

*ἰοζία. * Añadese en el co. an. καὶ τὸ σπέρμα. *Añadese en el co. an. καὶ ἔμμηνα ἀποκαθαίρει. *ἐν ὑδρηροῖς.

NOMBRES Griego, κόνυζα. Lat. Conyza, & Pulicaria. Por. Tagueda.

ANNOTATION. *Llamase la Conyza en Latin Pulicaria, porque mata las pulgas: aunque propriamente aquel nombre conuiene al Psyllio. Theophrasto llamo a la mayor dellas el macho: y a la menor la hembra. Puesto vn manojo*

Psyllio.

10. Andrés de Laguna, *Pedacio Dioscorides Anazarbeo, Acerca de la materia medicinal, y de los venenos mortiferos* (Salamanca: Mathias Gast, 1570), book 3, chapter 130, page 350: *Conyza maior, Conyza minor*
Biblioteca Nazionale Centrale di Roma

11. Pietro Andrea Mattioli, *Commentarii . . . in libros sex Pedanii Dioscoridis Anazarbei de materia medica* (Venice: Vincentii Valgrisii, 1558), book 3, chapter 19, page 452: *Conyza maior, Conyza minor, Conyza media*
Biblioteca Nazionale Centrale di Roma

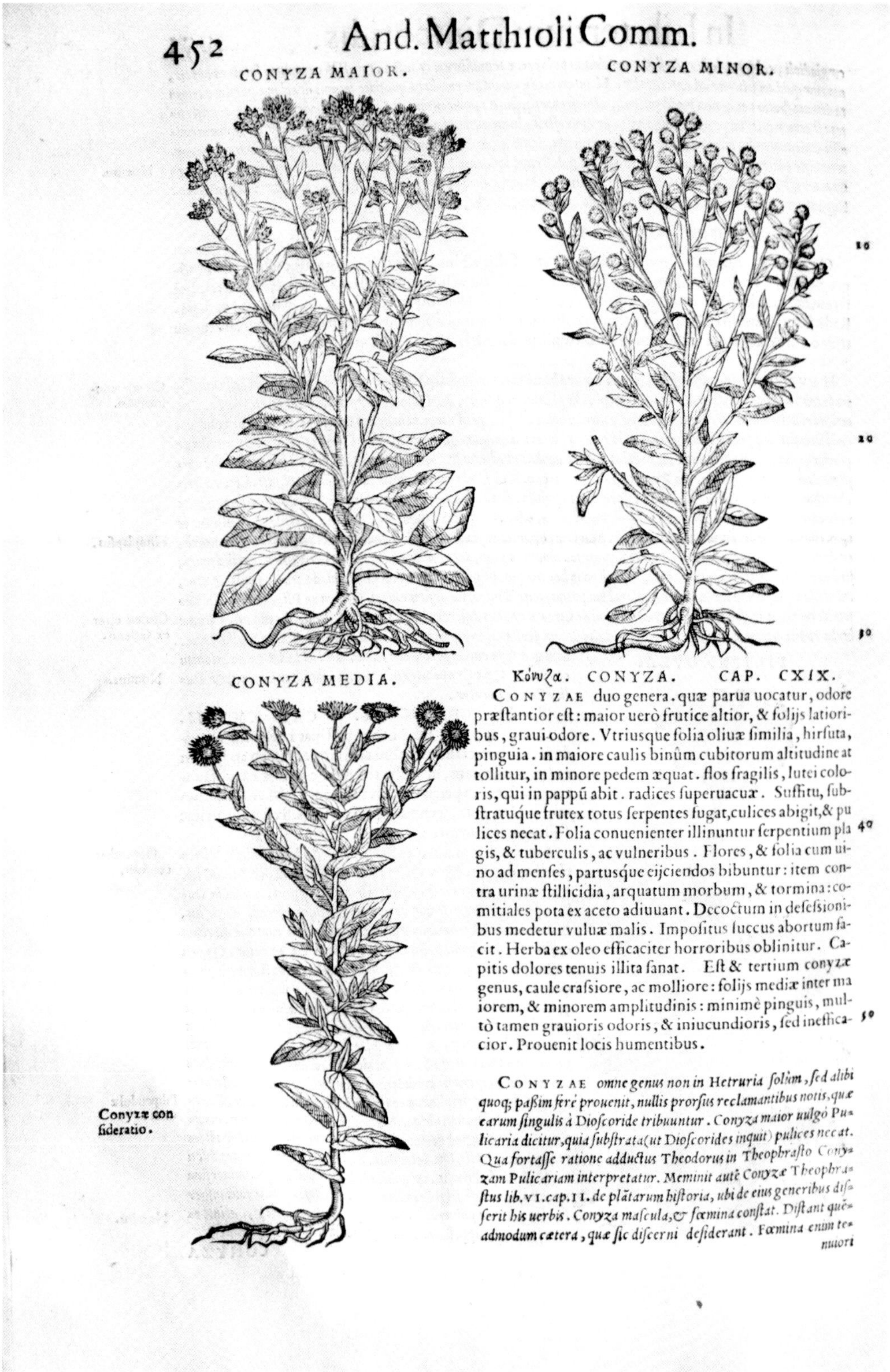

452 And. Matthioli Comm.

CONYZA MAIOR. CONYZA MINOR.

CONYZA MEDIA.

Conyzæ conſideratio.

Κόνυζα. CONYZA. CAP. CXIX.

CONYZAE duo genera. quæ parua uocatur, odore præſtantior eſt: maior uerò frutice altior, & folijs latioribus, graui odore. Vtriusque folia oliuæ ſimilia, hirſuta, pinguia. in maiore caulis binûm cubitorum altitudine attollitur, in minore pedem æquat. flos fragilis, lutei coloris, qui in pappũ abit. radices ſuperuacuæ. Suffitu, ſubſtratuq́ue frutex totus ſerpentes fugat, culices abigit, & pulices necat. Folia conuenienter illinuntur ſerpentium plagis, & tuberculis, ac vulneribus. Flores, & folia cum uino ad menſes, partusq́ue eijciendos bibuntur: item contra urinæ ſtillicidia, arquatum morbum, & tormina: comitiales pota ex aceto adiuuant. Decoctum in defeſsionibus medetur vuluæ malis. Impoſitus ſuccus abortum facit. Herba ex oleo efficaciter horroribus oblinitur. Capitis dolores tenuis illita ſanat. Eſt & tertium conyzæ genus, caule craſsiore, ac molliore: folijs mediæ inter maiorem, & minorem amplitudinis: minimè pinguis, multò tamen grauioris odoris, & iniucundioris, ſed inefficacior. Prouenit locis humentibus.

CONYZAE *omne genus non in Hetruria ſolùm, ſed alibi quoq; paßim ferè prouenit, nullis prorſus reclamantibus notis, quæ earum ſingulis à Dioſcoride tribuuntur. Conyza maior uulgò Pulicaria dicitur, quia ſubſtrata (ut Dioſcorides inquit) pulices necat. Qua fortaſſe ratione adductus Theodorus in Theophraſto Conyzam Pulicariam interpretatur. Meminit autẽ Conyzæ Theophraſtus lib. VI. cap. II. de plãtarum hiſtoria, ubi de eius generibus diſſerit his uerbis. Conyza maſcula, & fœmina conſtat. Diſtant quẽadmodum cætera, quæ ſic diſcerni deſiderant. Fœmina enim tenuiori*

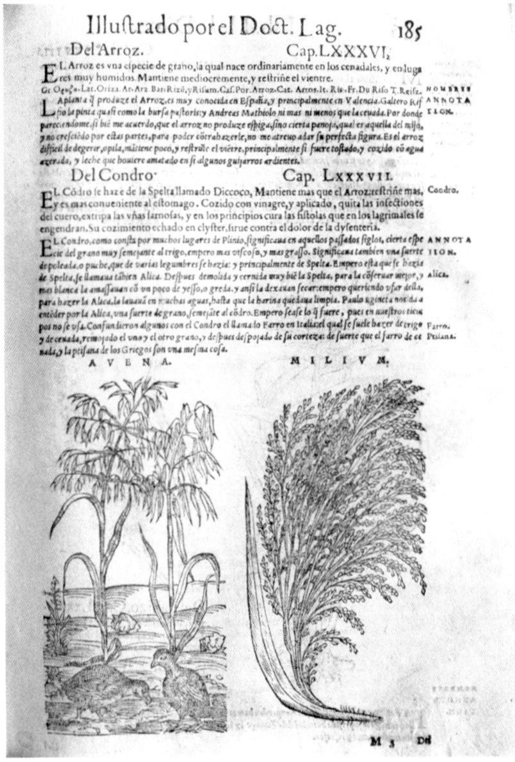

Illuſtrado por el Doct. Lag. 185

Del Arroz. Cap. LXXXVI.

EL Arroz es vna eſpecie de grano, la qual nace ordinariamente en los cenadales, y en lugares muy humidos. Mantiene mediocremente, y reſtriñe el vientre.

Gr. Ὄρυζα. Lat. Oriza. Ar. Arz. Bar. Rizû, y Riſum. Caſ. Por. Arroz. Cat. Arros. It. Riſo. Fr. Du Riſo. T. Reiſz. NOMBRES

LA planta q̃ produze el Arroz, es muy conocida en Eſpaña, y principalmente en Valencia. Galtero Riffio la pinta quaſi como la burſa paſtoris: y Andreas Mathiolo ni mas ni menos que la ceuada. Por donde pareciendome, ſi biẽ me acuerdo, que el arroz no produze eſpiga, ſino cierta panoja, qual es aquella del mijo, y no creſciẽdo por eſtas partes, para poder cõtrahazerle, no me atreuo a dar ſu perfecta figura. Es el arroz difficil de degerir, opila, mãtiene poco, y reſtriñe el viẽtre, principalmente ſi fuere toſtado, y cozido cõ agua azerada, y leche que houiere amatado en ſi algunos guijarros ardientes. ANNOTATION.

Del Condro. Cap. LXXXVII.

EL Cõdro ſe haze de la Spelta llamado Diccoco, Mantiene mas que el Arroz: reſtriñe mas, y es mas conueniente al eſtomago. Cozido con vinagre, y aplicado, quita las infectiones del cuero, extirpa las vñas ſarnoſas, y en los principios cura las fiſtolas que en los lagrimales ſe engendran. Su cozimiento echado en clyſter, ſirue contra el dolor de la dyſenteria. Condro.

EL Condro, como conſta por muchos lugares de Plinio, ſignificaua en aquellos paſſados ſiglos, cierta eſpecie del grano muy ſemejante al trigo, empero mas viſcoſo, y mas graſſo. Significaua tambien vna ſuerte de poleada, o puche, que de varias legumbres ſe hazia: y principalmente de Spelta. Empero eſta que ſe hazia de Spelta, ſe llamaua tãbien Alica. Deſpues demolida y cernida muy biẽ la Spelta, para la cõſeruar mejor, y mas blanca la amaſſauan cõ vn poco de yeſſo, o greda, y anſi la dexauan ſecar: empero queriendo vſar della, para hazer la Alica, la lauauã en muchas aguas, haſta que la harina quedaua limpia. Paulo Egineta nos da a entẽder por la Alica, vna ſuerte de grano, ſemejãte al cõdro. Empero ſeaſe lo q̃ fuere, pues en nueſtros tiempos no ſe vſa. Confundieron algunos con el Condro el llamado Farro en Italia: el qual ſe ſuele hazer de trigo y de ceuada, remojado el vno y el otro grano, y deſpues deſpojado de ſu corteza: de ſuerte que el farro de ceuada, y la ptiſana de los Griegos ſon vna meſma coſa. ANNOTATION. Alica. Farro. Ptiſana.

AVENA. MILIVM.

M 3 Del

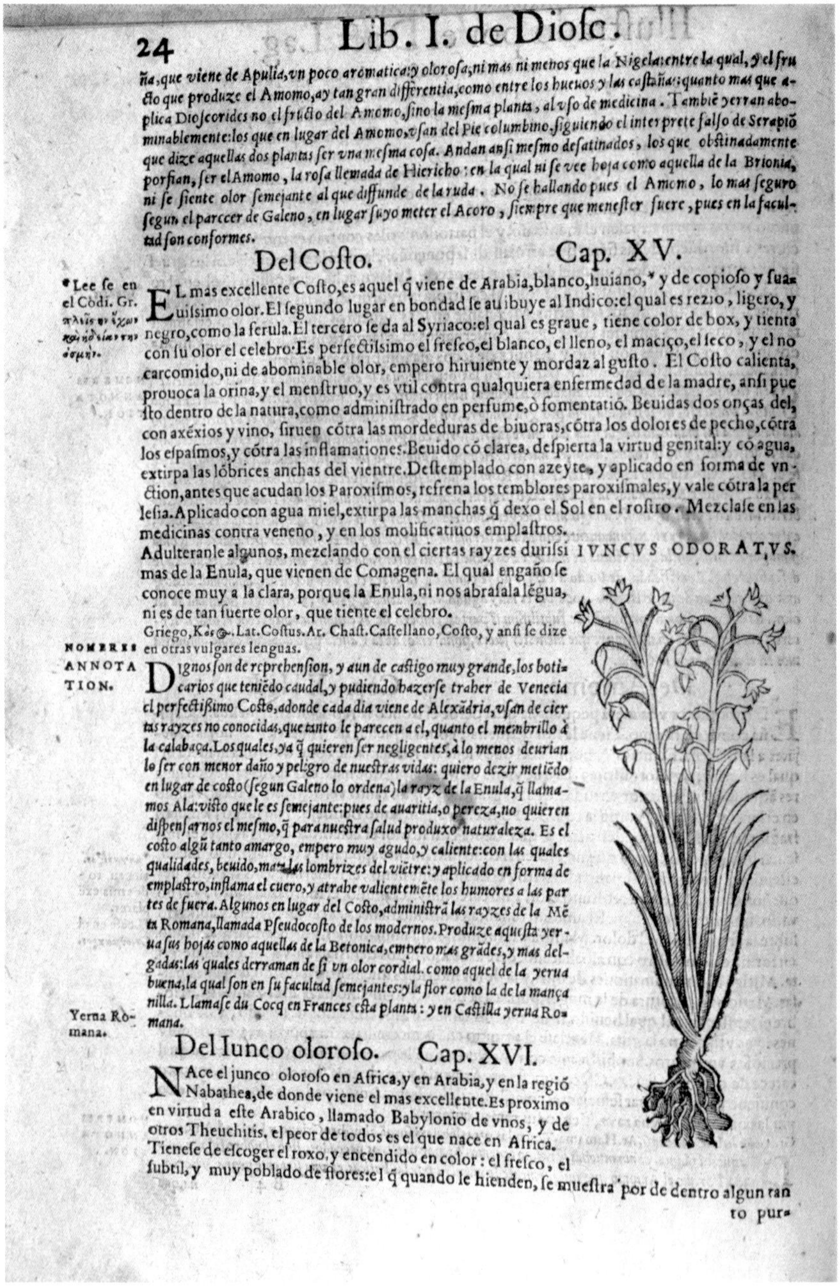

24 Lib. I. de Dioſc.

ña, que viene de Apulia, vn poco aromatica: y oloroſa, ni mas ni menos que la Nigela: entre la qual, y el fructo que produze el Amomo, ay tan gran differentia, como entre los hueuos y las caſtañas: quanto mas que aplica Dioſcorides no el fructo del Amomo, ſino la meſma planta, al vſo de medicina. Tambiẽ yerran abominablemente: los que en lugar del Amomo, vſan del Pie columbino, ſiguiendo el interprete falſo de Serapiõ que dize aquellas dos plantas ſer vna meſma coſa. Andan anſi meſmo deſatinados, los que obſtinadamente porfian, ſer el Amomo, la roſa llamada de Hiericho: en la qual ni ſe vee hoja como aquella de la Brionia, ni ſe ſiente olor ſemejante al que diffunde de la ruda. No ſe hallando pues el Amomo, lo mas ſeguro ſegun el parecer de Galeno, en lugar ſuyo meter el Acoro, ſiempre que meneſter fuere, pues en la facultad ſon conformes.

Del Coſto. Cap. XV.

* Lee ſe en el Cõdi. Gr. πλῆθος ὀσμῆς [illegible]

EL mas excellente Coſto, es aquel q̃ viene de Arabia, blanco, liuiano, * y de copioſo y ſuauiſsimo olor. El ſegundo lugar en bondad ſe atribuye al Indico: el qual es rezio, ligero, y negro, como la ferula. El tercero ſe da al Syriaco: el qual es graue, tiene color de box, y tienta con ſu olor el celebro. Es perfectiſsimo el freſco, el blanco, el lleno, el maciço, el ſeco, y el no carcomido, ni de abominable olor, empero hiruiente y mordaz al guſto. El Coſto calienta, prouoca la orina, y el menſtruo, y es vtil contra qualquiera enfermedad de la madre, anſi pueſto dentro de la natura, como adminiſtrado en perfume, ò fomentatiõ. Beuidas dos onças del, con axénios y vino, ſiruen cõtra las mordeduras de biuoras, cõtra los dolores de pecho, cõtra los eſpaſmos, y cõtra las inflamationes. Beuido cõ clarea, deſpierta la virtud genital: y cõ agua, extirpa las lõbrices anchas del vientre. Deſtemplado con azeyte, y aplicado en forma de vnction, antes que acudan los Paroxiſmos, refrena los temblores paroxiſmales, y vale cõtra la perleſia. Aplicado con agua miel, extirpa las manchas q̃ dexo el Sol en el roſtro. Mezclaſe en las medicinas contra veneno, y en los molificatiuos emplaſtros. Adulteranle algunos, mezclando con el ciertas rayzes duriſsimas de la Enula, que vienen de Comagena. El qual engaño ſe conoce muy a la clara, porque la Enula, ni nos abraſa la lẽgua, ni es de tan fuerte olor, que tiente el celebro.

IVNCVS ODORATVS.

NOMBRES Griego, Κόστος. Lat. Coſtus. Ar. Chaſt. Caſtellano, Coſto, y anſi ſe dize en otras vulgares lenguas.

ANNOTATION. DIgnos ſon de reprehenſion, y aun de caſtigo muy grande, los boticarios que teniẽdo caudal, y pudiendo hazerſe traher de Venecia el perfectiſsimo Coſto, adonde cada dia viene de Alexãdria, vſan de ciertas rayzes no conocidas, que tanto le parecen a el, quanto el membrillo à la calabaça. Los quales, ya q̃ quieren ſer negligentes, à lo menos deurian lo ſer con menor daño y peligro de nueſtras vidas: quiero dezir metiẽdo en lugar de coſto (ſegun Galeno lo ordena) la rayz de la Enula, q̃ llamamos Ala: viſto que le es ſemejante: pues de auaritia, o pereza, no quieren diſpenſarnos el meſmo, q̃ para nueſtra ſalud produxo naturaleza. Es el coſto algũ tanto amargo, empero muy agudo, y caliente: con las quales qualidades, beuido, mata las lombrizes del viẽtre: y aplicado en forma de emplaſtro, inflama el cuero, y atrahe valientemẽte los humores a las partes de fuera. Algunos en lugar del Coſto, adminiſtrã las rayzes de la Mẽta Romana, llamada Pſeudocoſto de los modernos. Produze aqueſta yerua ſus hojas como aquellas de la Betonica, empero mas grãdes, y mas delgadas: las quales derraman de ſi vn olor cordial, como aquel de la yerua buena, la qual ſon en ſu facultad ſemejantes: y la flor como la de la mançanilla. Llamaſe du Cocq en Frances eſta planta: y en Caſtilla yerua Romana. Yerua Romana.

Del Iunco oloroſo. Cap. XVI.

NAce el junco oloroſo en Africa, y en Arabia, y en la regiõ Nabathea, de donde viene el mas excellente. Es proximo en virtud a eſte Arabico, llamado Babylonio de vnos, y de otros Theuchitis. el peor de todos es el que nace en Africa. Tieneſe de eſcoger el roxo, y encendido en color: el freſco, el ſubtil, y muy poblado de flores: el q̃ quando le hienden, ſe mueſtra por de dentro algun tanto pur-

known these plants, Laguna may have had some personal knowledge of them. In other cases, where Laguna diverged from Mattioli on plant identification, he reproduced medieval schematic drawings rather than Mattioli's more realistic ones (fig. 13). This is particularly revealing of Laguna's method of work: he extended Leoniceno's principles from scientific text (for which they were conceived) to scientific illustration. Leoniceno's epistemology was thus fully assimilated and extrapolated to other sectors, going beyond its primary field of application and thus constituting a productive paradigm. At this point, scientific knowledge implied not only an unambiguous designation, but also a specific representation.

These differences between Mattioli's and Laguna's illustrations also help explain why Laguna reproduced Mattioli's representations of plants: just as he recovered philological data from previous works, Laguna borrowed representations as a primary material to be submitted to critical analysis. At the conclusion of this process, previous material was accepted insofar as it corresponded to what was thought to be the original, that is, the rightly identified plant. Again, this shows that Leoniceno's methodology was fully mastered by some of his successors and was a viable tool to be used in new applications.

As for Mattioli's illustrations, the large-scale ones by Giorgio Liberale (c. 1527–1579) introduced in the Czech edition (Prague, 1562; fig. 14), and reprinted in subsequent editions of the work (either in Latin or Italian), are in many cases just slightly revised enlargements of the small ones printed in Latin editions from 1554 onward

12. Andrés de Laguna, *Pedacio Dioscorides Anazarbeo, Acerca de la materia medicinal, y de los venenos mortiferos* (Salamanca: Mathias Gast, 1570), page 185: *Milium*
Biblioteca Nazionale Centrale di Roma

13. Andrés de Laguna, *Pedacio Dioscorides Anazarbeo, Acerca de la materia medicinal, y de los venenos mortiferos* (Salamanca: Mathias Gast, 1570), book 1, chapter 21, page 24: *Iuncus odoratus*
Biblioteca Nazionale Centrale di Roma

(figs. 15 and 16).[100] Whatever their size, Mattioli's representations of plants are not as original or as realistic as scholars have frequently stated. Several of Mattioli's illustrations are reinterpretations of previously used pictorial material. Some correspond to plant representations from Greek manuscripts of *De materia medica* (figs. 17 and 18), while others are comparable to woodcuts from early printed books (figs. 19 and 20). In both cases, however, the drawings were reinterpreted to achieve a more realistic appearance. Here again, the parallel between the treatment of text and image is striking: while Mattioli did recover previous material, he also revised and rearranged it.[101]

Through this revision, Mattioli—and, indirectly, Laguna—made obsolete the ancient system of plant representations (even though both retained some representations produced by this system). The absorption into the text of botanical data based on experience required a transformation of representations. The illustrations had to be not only realistic, as in artistic productions, but also exact, complete, and universally recognizable, like actual botanical specimens. The external world was absorbed into the work (whether in words or in visual representations) and remodeled so as to become an object of knowledge and not merely an exact duplicate of reality.

Conclusions

From antiquity to the dawn of the Renaissance, both botany and pharmacology were characterized by a constant updating of information: the classical heritage—passed from one generation to another and from one culture to another around the shores of the Mediterranean—was regularly revised, updated, and completed on the basis of contemporary needs, resources, and knowledge. There was no rupture between the construction of text and the practice of science. Leoniceno broke this link: not only did he put text before practice, but he also returned the text to the form it had in antiquity, thus canceling all the accretions accumulated over time. In so doing, he caused a temporary stasis of scientific production.

In Germany, however, botanical knowledge proceeded without interruption from

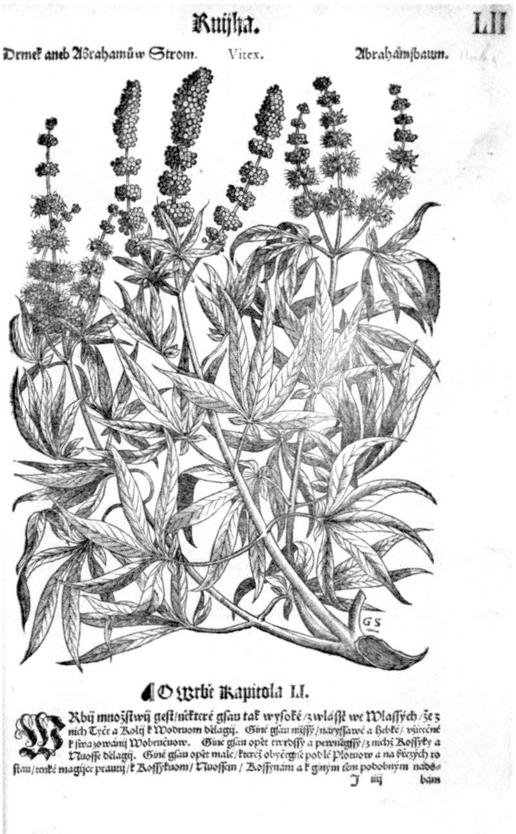

14. Pietro Andrea Mattioli, *Herbarz: ginak Bylinář* (Prague: G. Melantrych, 1562), chapter 51, folio 52r: *Vitex*
Biblioteca Nazionale Centrale di Roma

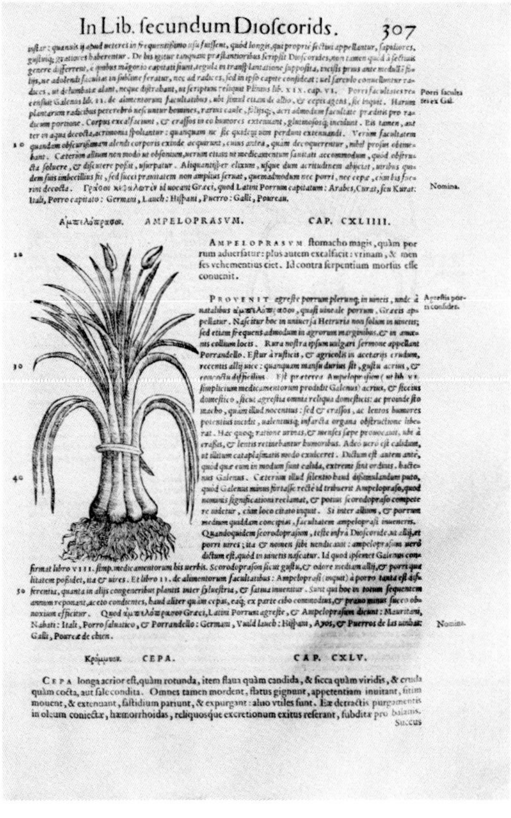

15. Pietro Andrea Mattioli, *Commentarii . . . in libros sex Pedanii Dioscoridis Anazarbei de materia medica* (Venice: Vincentii Valgrisii, 1558), book 2, chapter 144, page 307: *Ampeloprasum*
Biblioteca Nazionale Centrale di Roma

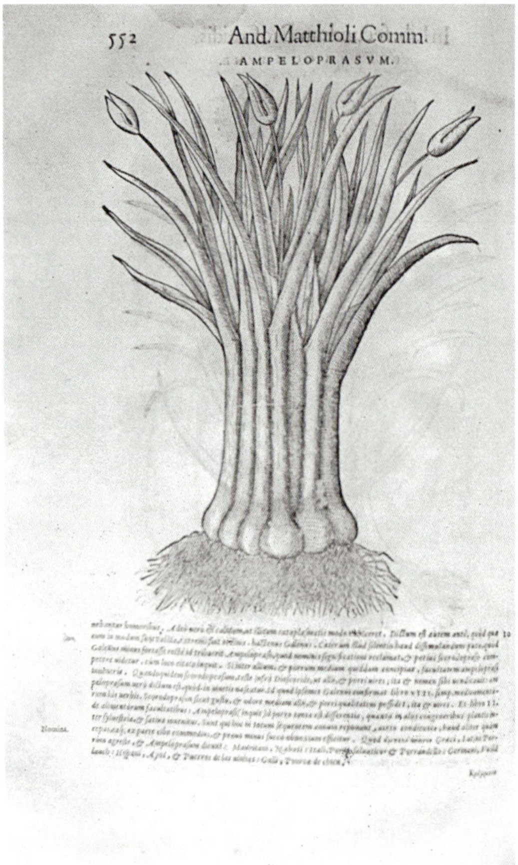

the late medieval period to the Renaissance, even though it evolved more rapidly in thirteen years (from 1529 to 1542) than in previous years. During this period, different scientific models competed. As Brunfels showed, the medieval practice of commentary was still alive, even though it aimed at restoring the knowledge of plants transmitted by classical texts according to Leoniceno's method. Leoniceno thus dealt with these classical works rather than only with their medieval commentaries, even though he recognized that the classical works were known only through their Renaissance Latin translations and commentaries and not in the original Greek. But, at the same time, this model was challenged from another source, the artistic world, in this case by the botanical illustrator Hans Weiditz. The erudite (that is, the Leonicenian) model and artistic renewal did not converge, however, and new artistic contributions can be seen as an interference in the textual pattern. This was not the case with Fuchs, who closely associated textual and iconic analysis, relying in both cases on personal observation—guided, however, by the classical textual heritage.

16. Pietro Andrea Mattioli, *Commentarii in sex libros Pedacii Dioscoridis Anazarbei de materia medica* (Venice: Officina Valgrisiana, 1565), book 2, chapter 144, page 552: *Ampeloprasum*
Biblioteca Nazionale Centrale di Roma

In the centers of scientific activity where Leoniceno's method had more impact, it was Mattioli's and Laguna's achievement to reunite the discursive text and the uses of plants, even though they accomplished the task in different ways. Mainly relying on informants in the field (the Botanical Republic), Mattioli communicated the results of his research to learned physicians. Laguna, instead, traveling throughout Europe (including Germany), was in direct contact with practitioners whose experience he incorporated into his work. And

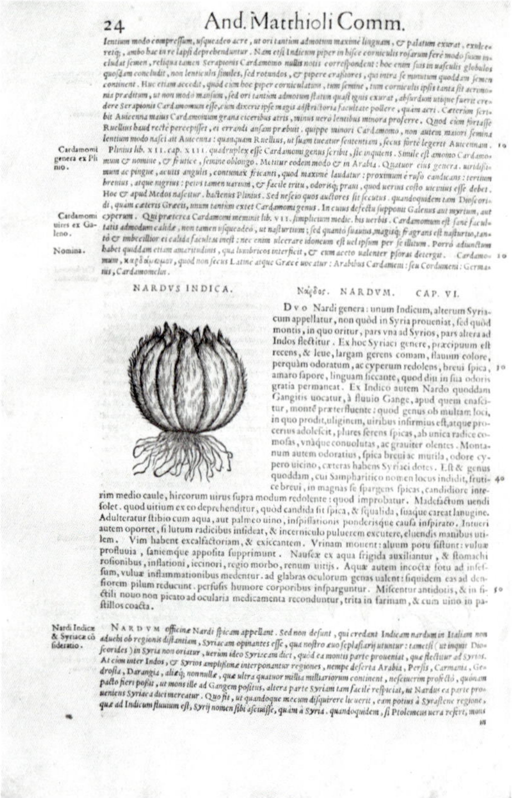
24 And. Matthioli Comm.

NARDVS INDICA. Νάρδος. NARDVM. CAP. VI.

Dvo Nardi genera: unum Indicum, alterum Syriacum appellatur, non quòd in Syria proueniat, sed quòd montis, in quo oritur, pars vna ad Syrios, pars altera ad Indos flectitur. Ex hoc Syriaci genere, præcipuum est recens, & leue, largam gerens comam, flauum colore, perquàm odoratum, ac cyperum redolens, breui spica, amaro sapore, linguam siccante, quod diu in sua odoris gratia permaneat. Ex Indico autem Nardo quoddam Gangitis uocatur, à fluuio Gange, apud quem enascitur, montē præterfluente: quod genus ob multam loci, in quo prodit, uliginem, uiribus infirmius est, atque procerius adolescit, plures ferens spicas, ab unica radice comosas, vnàque conuolutas, ac grauiter olentes. Montanum autem odoratius, spica breui ac mutila, odore cypero uicino, cæteras habens Syriaci dotes. Est & genus quoddam, cui Sampharitico nomen locus indidit, frutice breui, in magnas se spargens spicas, candidiore interim medio caule, hircorum uirus supra modum redolente: quod improbatur. Madefactum uendi solet. quod uitium ex eo deprehenditur, quòd candida sit spica, & squalida, suaque careat lanugine. Adulteratur stibio cum aqua, aut palmeo uino, inspissationis ponderisque causa inspirato. Intueri autem oportet, si lutum radicibus insideat, & incerniculo puluerem excutere, eluendis manibus utilem. Vim habent excalfactoriam, & exiccantem. Vrinam mouent: aluum potu sistunt: uuluæ profluuia, saniemque apposita supprimunt. Nauseæ ex aqua frigida auxiliantur, & stomachi rosionibus, inflationi, iecinori, regio morbo, renum uitijs. Aquæ autem incoctæ fotu ad insessum, uuluæ inflammationibus medentur. ad glabras oculorum genas ualent: siquidem eas ad densiorem pilum reducunt. perfusis humore corporibus insparguntur. Miscentur antidotis, & in fictili nouo non picato ad ocularia medicamenta reconduntur, trita in farinam, & cum uino in pastillos coacta.

17. Pietro Andrea Mattioli, *Commentarii . . . in libros sex Pedanii Dioscoridis Anazarbei de materia medica* (Venice: Vincentii Valgrisii, 1558), book 1, chapter 6, page 24: *Nardus indica*
Biblioteca Nazionale Centrale di Roma

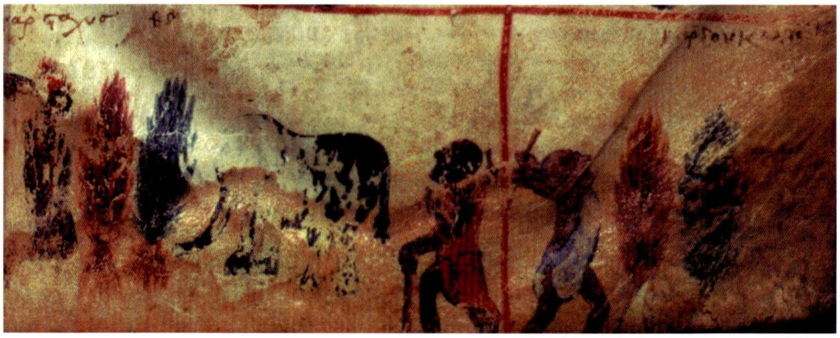

18. MS Athos, Megistis Lavras, Ω 75, Dioscorides, *De materia medica*, mid-eleventh century, folio 6r: *nardos keltikos* (in Greek), folio 6r (detail)
Megistis Lavras Monastery

19. Pietro Andrea Mattioli, *Commentarii . . . in libros sex Pedanii Dioscoridis Anazarbei de materia medica* (Venice: Vincentii Valgrisii, 1558), book 1, chapter 1, page 17: *Iris domestica, Iris sylvestris*

Biblioteca Nazionale Centrale di Roma

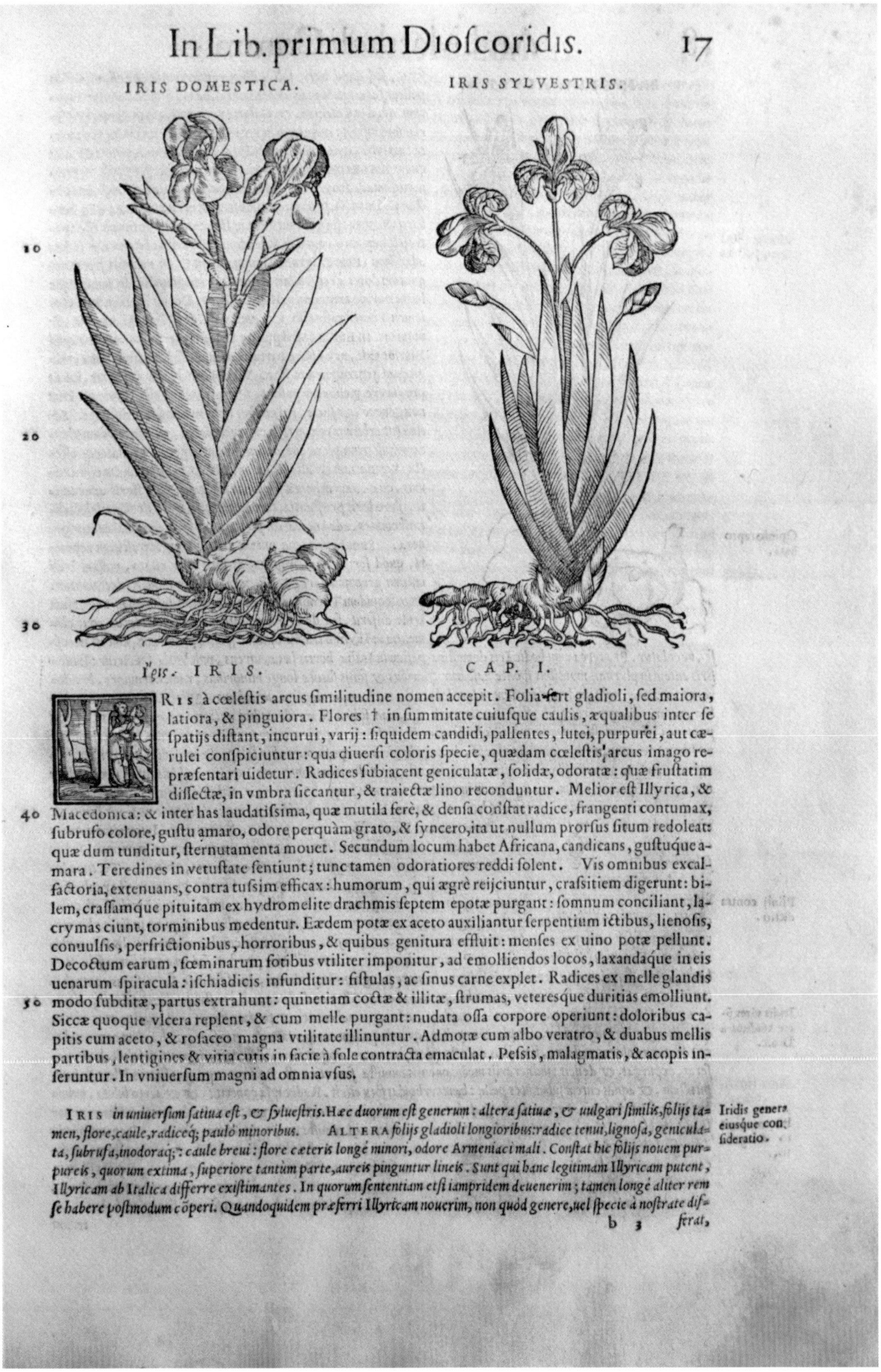

In Lib. primum Dioſcoridis. 17

IRIS DOMESTICA. IRIS SYLVESTRIS.

Ἶρις. IRIS. CAP. I.

IRIS à cœleſtis arcus ſimilitudine nomen accepit. Folia fert gladioli, ſed maiora, latiora, & pinguiora. Flores † in ſummitate cuiuſque caulis, æqualibus inter ſe ſpatijs diſtant, incurui, varij: ſiquidem candidi, pallentes, lutei, purpurei, aut cærulei conſpiciuntur: qua diuerſi coloris ſpecie, quædam cœleſtis arcus imago repræſentari uidetur. Radices ſubiacent geniculatæ, ſolidæ, odoratæ: quæ fruſtatim diſſectæ, in vmbra ſiccantur, & traiectæ lino reconduntur. Melior eſt Illyrica, & Macedonica: & inter has laudatiſsima, quæ mutila ferè, & denſa conſtat radice, frangenti contumax, ſubrufo colore, guſtu amaro, odore perquàm grato, & ſyncero, ita ut nullum prorſus ſitum redoleat: quæ dum tunditur, ſternutamenta mouet. Secundum locum habet Africana, candicans, guſtuque amara. Teredines in vetuſtate ſentiunt; tunc tamen odoratiores reddi ſolent. Vis omnibus excalfactoria, extenuans, contra tuſsim efficax: humorum, qui ægrè reijciuntur, craſsitiem digerunt: bilem, craſſamque pituitam ex hydromelite drachmis ſeptem epotæ purgant: ſomnum conciliant, lacrymas ciunt, torminibus medentur. Eædem potæ ex aceto auxiliantur ſerpentium ictibus, lienoſis, conuulſis, perfrictionibus, horroribus, & quibus genitura effluit: menſes ex uino potæ pellunt. Decoctum earum, fœminarum fotibus vtiliter imponitur, ad emolliendos locos, laxandaque in eis uenarum ſpiracula: iſchiadicis infunditur: fiſtulas, ac ſinus carne explet. Radices ex melle glandis modo ſubditæ, partus extrahunt: quinetiam coctæ & illitæ, ſtrumas, veteresque duritias emolliunt. Siccæ quoque vlcera replent, & cum melle purgant: nudata oſſa corpore operiunt: doloribus capitis cum aceto, & roſaceo magna vtilitate illinuntur. Admotæ cum albo veratro, & duabus mellis partibus, lentigines & vitia cutis in facie à ſole contracta emaculat. Peſsis, malagmatis, & acopis inſeruntur. In vniuerſum magni ad omnia vſus.

IRIS *in uniuerſum ſatiua eſt, & ſylueſtris. Hæc duorum eſt generum: altera ſatiuæ, & uulgari ſimilis, folijs tamen, flore, caule, radiceq; paulò minoribus.* ALTERA *folijs gladioli longioribus: radice tenui, lignoſa, geniculata, ſubruſa, inodoraq;: caule breui: flore cæteris longè minori, odore Armeniaci mali. Conſtat hic folijs nouem purpureis, quorum extima, ſuperiore tantùm parte, aureis pinguntur lineis. Sunt qui hanc legitimam Illyricam putent, Illyricam ab Italica differre exiſtimantes. In quorum ſententiam etſi iampridem deuenerim; tamen longè aliter rem ſe habere poſtmodum cõperi. Quandoquidem præferri Illyricam nouerim, non quòd genere, uel ſpecie à noſtrate differat,* Iridis genera eiusque conſideratio.

b 3

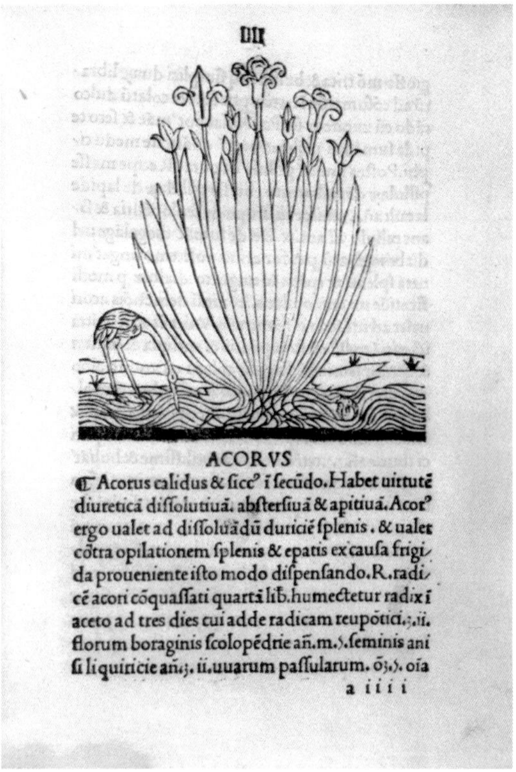

ACORVS

Acorus calidus & sicc' i secūdo. Habet uirtutē diureticā dissolutiuā. absterfiuā & apitiuā. Acor' ergo ualet ad dissoluādū duriciē splenis. & ualet cōtra opilationem splenis & epatis ex causa frigida prouenientē isto modo dispensando. R. radicē acori cōquassati quartā lib. humectetur radix i aceto ad tres dies cui adde radicam reupōtici. ʒ. ii. florum boraginis scolopēdrie aā. m. ʒ. seminis ani si liquiricie aā. ʒ. ii. uuarum passularum. ōʒ. ʒ. oīa

a iiii

20. *Herbarius latinus* (Vicenza: Leonardus de Basilea and Guilielmus de Papia, 1491), folio 3r: *Acorus*
Biblioteca Nazionale Centrale di Roma

he directed this wealth of information to a wider audience than learned physicians.

Thus it is evident that, in its effort to recover the legacy of classical antiquity, humanism did not necessarily provoke an immediate renewal of contemporary science, contrary to the triumphant declarations of the humanists themselves, on the title pages of their works and elsewhere, that they had liberated science from obscurantism.[102]

NOTES

I thank Osvaldo Avallone, director of the Biblioteca Nazionale Centrale di Roma (BNCR), not only for permission to reproduce illustrations from volumes in the library's collections, but also for allowing me to investigate the holdings of the library systematically in close collaboration with the staff, particularly Margherita Breccia Fratadocchi, in charge of the manuscript and rare book room.

1. For a recent synthesis on Theophrastus, see William Forthenbaugh and Johannes van Ophuijsen, "Theophrastos," in *Der neue Pauly: Enzyklopädie der Antike*, vol. 12, part 1, ed. Hubert Cancik and Helmuth Schneider (Stuttgart and Weimar, 2002), 385–393.

2. On botanical knowledge in antiquity, see Geoffrey E. R. Lloyd, *Science, Folklore and Ideology: Studies in the Life Sciences in Ancient Greece* (Cambridge, 1983), 119–135; and, more recently, Luciana Repici, *Uomini capovolti: Le piante nel pensiero dei Greci*, Biblioteca di cultura moderna 1152 (Rome and Bari, 2000). Since botany and pharmacology were associated (see below), see also Alain Touwaide, "Le strategie terapeutiche: i farmaci," in *Storia del pensiero medico occidentale 1, Antichità e Medio-Evo*, ed. Mirko D. Grmek (Rome and Bari, 1993), 353–373 (English translation: "Therapeutic Strategies: Drugs," in *Western Medical Thought from Antiquity to the Middle Ages*, ed. Mirko D. Grmek [Cambridge, Mass., 1998], 259–272, 390–394); and Alain Touwaide, "Healers and Physicians in Ancient and Medieval Mediterranean Cultures," in *Handbook of Medicinal Plants*, ed. Zohara Yaniv and Uriel Bachrach (New York, London, and Oxford, 2005), 155–173.

3. Greek text with English translation: Arthur Hort, *Theophrastus: Enquiry into Plants and Minor Works on Odours and Weather Signs, with an English Translation by Sir Arthur Hort, Bart*, 2 vols. (London and New York, 1916).

4. Greek text with English translation: Benedikt Einarson and George K. K. Link, *Theophrastus: "De causis plantarum,"* 3 vols. (London and Cambridge, Mass., 1976–1990).

5. Georg Wöhrle, *Theophrasts Methode in seinen botanischen Schriften*, Studien zur antiken Philosophie, vol. 13 (Amsterdam, 1985). For medicinal plants, see John Scarborough, "Theophrastus on Herbals and Herbal Remedies," *Journal of the History of Biology* 11 (1978): 353–385.

6. On Pliny, see Klaus Salmann, "Plinius [1] Secundus," in *Der neue Pauly. Enzyklopädie der Antike*, vol. 9, ed. Hubert Cancik and Helmuth Schneider (Stuttgart and Weimar, 2000), 1135–1141.

7. Roger K. French and Frank Greenaway, eds., *Science in the Early Roman Empire: Pliny the Elder, His Sources and His Influence* (London and Sydney, 1986).

8. For a recent synthesis on Dioscorides' biography, including previous literature, see Alain Touwaide, "Pedanios Dioskurides," in Cancik and Schneider 2000, 462–465.

9. In the titles of Greek manuscripts of *De materia medica*, Dioscorides is named Pedanios Dioskoridēs Anazarbeus. In one passage, however, Galen mentions, probably mistakenly, that Dioscorides was from Tarsos.

10. See the edition of the Greek text by Max Wellmann, *Pedanii Dioscuridis Anazarbei de materia medica libri quinque*, 3 vols. (Berlin, 1906–1914; reprint Berlin, 1958). For an English translation, see Lily Y. Beck, *Pedanius Dioscorides of Anazarbus: De materia medica*, Altertumswissenschaftliche Texte und Studien, vol. 38 (Hildesheim, Zürich, and New York, 2005). This translation supersedes the seventeenth-century English version as edited and published in Robert T. Gunther, *The Greek Herbal of Dioscorides: Illustrated by a Byzantine, A.D. 512, Englished by John Goodyer A.D. 1655* (Oxford, 1933) and recently reprinted in updated language by Tess A. Osbaldeston and Robert P. A. Wood, eds., *Dioscorides de materia medica, Being an Herbal with Many Other Medicinal Materials, Written in Greek in the First Century of the Common Era: A New Indexed Version in Modern English* (Johannesburg, 2000).

11. For an analysis of Dioscorides' method, see John M. Riddle, *Dioscorides on Pharmacy and Medicine* (Austin, 1985); Alain Touwaide, "La botanique entre science et culture au Ier siècle de notre ère," *Geschichte der Mathematisch—und Naturwissenschaft in der Antike, 1: Biologie*, ed. Georg Wöhrle (Stuttgart, 1999), 219–252.

12. The inventory is in Marie-Hélène Marganne and Pierre Mertens, "Medici et Medica, 2e édition," *"Specimina" per il Corpus dei Papiri Greci di Medicina*, ed. Isabella Andorlini (Florence, 1997), 9. For a study of the most ancient and longest papyrus (Michigan inv. 3, end of the second century C.E.), "see Rebecca Flemming and Ann-Ellis Hanson, "Dioscorides, *De materia medica* II, 76.2 and 76.7–18," in *Greek Medical Papyri I*, ed. Isabella Andorlini (Florence, 2001), 9–35.

13. For the Greek manuscripts of Dioscorides' text, see *Die Handschriften der antiken Ärzte. Griechische Abteilung*, ed. Hermann Diels (Berlin, 1906), 2:29–35.

14. For an overview of the textual tradition of *De materia medica*, see the introduction to the edition of Dioscorides' Greek text in Wellmann 1906–1914, 2:v–xxiv. The topic needs to be revised, however, since a new inventory of Dioscorides' manuscripts has listed more than 120 items and the history of the text is more complex than Wellmann thought.

15. No inventory of Dioscorides' illustrated manuscripts is currently available. Minta Collins, *Medieval Herbals: The Illustrative Traditions*, The British Library Studies in Medieval Culture (London and Toronto, 2000), is not only incomplete, but is also inaccurate.

16. Among the many titles on plant representations in *De materia medica*, see Aubin-Louis Millin de Grandmaison, "Observations sur les manuscrits de Dioscoride qui sont conservés à la Bibliothèque nationale," *Magasin encyclopédique* 2 (1802): 152–163; Ludwig Choulant, "Über die Handschriften des Dioskorides," *Archiv für die zeichnende Künste* 1 (1855): 56–62; Anton von Premerstein, *De codicis Dioscuridis Aniciae Iulianae, nunc Vindobonensis Med. Gr. 1 historia, forma, scriptura, picturis . . .* (Leiden, 1906); Henri Omont, *Études et notes sur les manuscrits de Dioscoride*, Bibliothèque nationale de France, Paris, manuscript Nouvelles acquisitions françaises 13052; Kurt Weitzmann, "The Greek Sources of Islamic Scientific Illustration," in *Archaeologia Orientalia in memoriam Ernst Herzfeld* (Locust Valley, Calif., 1952), 244–266; Kurt Weitzmann, "Das klassische Erbe in der Kunst Konstantinopels," *Alte und neue Kunst* 3 (1954): 41–59; and Kurt Weitzmann, *Geistige Grundlagen und Wesen der Makedonischen Renaissance*, Arbeitsgemeinschaft für Forschung des Landes Nordrhein-Westfalen 107 (Cologne and Opladen, 1963); these three works are reprinted in Kurt Weitzmann, *Studies in Classical and Byzantine Manuscript Illumination*, ed. Herbert L. Kessler (Chicago and London, 1971), 20–44, 126–150 (in English translation: "The Classical Heritage in the Art of Constantinople"), and 176–223 (in English translation: "The Character and Intellectual Origins of the Macedonian Renaissance"). See also Kurt Weitzmann, *Illustrations in Roll and Codex: A Study of the Origin and Method of Text Illustration* (Princeton, 1947; revised edition 1970); Ann van Buren, "De Materia Medica of Dioscorides," in *Illuminated Greek Manuscripts from American Collections*, ed. Gary Vikan (Princeton, 1973), 66–69; Heide Grape-Albers, *Spätantike Bilder aus der Welt des Arztes: Medizinische Bilderhandschriften der Spätantike und ihre mittelalterliche Überlieferung* (Wiesbaden, 1977); Alain Touwaide, "Un recueil grec de pharmacologie du Xe siècle illustré au XIVe siècle: Le *Vaticanus graecus* 284," *Scriptorium* 39 (1985): 13–56; Emilie Leal, "Un manuscrit illustré du traité de matière médicale de Dioscoride le Paris grec 2180," 2 vols. (degree thesis, Université de Provence, 1997); Alessia Aletta, "Studi e ricerche sul Dioscoride della Pierpont Morgan Library M 652," 2 vols. (Ph.D. diss., Università di Roma, "La Sapienza," 1998). The bibliography on authenticity and origin is too abundant to be fully listed here.

17. On Galen, see Paul Moraux, *Galien de Pergame: Souvenirs d'un médecin*, Collection d'études anciennes (Paris, 1985) and, for a recent synthesis with bibliography, see Vivian Nutton, "Galenos

aus Pergamon," in *Der neue Pauly: Enzyklopädie der Antike*, vol. 4, ed. Hubert Cancik and Helmuth Schneider (Stuttgart and Weimar, 1998), 748–756.

18. Nutton, "Galen ad multos annos," *Dynamis* 15 (1995): 25–40.

19. Karl G. Kühn, ed., *Galeni opera omnia* (Leipzig, 1826), 11:379–892; 12:1–377.

20. Alain Touwaide, "La thérapeutique médicamenteuse de Dioscoride à Galien: Du pharmacocentrisme au médico-centrisme," in *Galen on Pharmacology: Philosophy, History and Medicine. Proceedings of the Fifth International Galen Colloquium, Lille, 16–18 March 1995*, ed. Armelle Debru, Studies in Ancient Medicine, vol. 16 (Leiden, New York, and Cologne, 1997), 255–282.

21. Georg Harig, *Bestimmung der Intensität im medizinischen System Galens: Ein Beitrag zur theoretischen Pharmakologie, Nosologie und Therapie in der galenischen Medizin, Schriften zur Geschichte und Kultur der Antike*, vol. 11 (Berlin, 1974).

22. On Galen's pharmaceutical theory, see also John Scarborough, "Pharmaceutical Theory in Galen's Commentaries on the Hippocratic Epidemics: Some Observations on Roman Views of Greek Drug Lore," *Die hippokratischen Epidemien, Theorie, Praxis, Tradition: Verhandlungen des Ve Colloque international hippocratique, veranstaltet von der Berliner Gesellschaft für Geschichte der Medizin in Verbindung mit dem Institut für Geschichte der Medizin der Freien Universität Berlin, 10.-15. 9. 1984*, ed. Gerard Baader and Rolf Winau, Sudhoffs Archiv, Beiheft 27 (Stuttgart, 1989), 270–282.

23. Karl G. Kühn, ed., *Galeni opera omnia* (Leipzig, 1827), 14:1–209. For a German translation and a study, see Lutz Winkler, "Galens Schrift, De Antidotis: Ein Beitrag zur Geschichte von Antidot und Theriak" (Ph.D. diss., Universität Marburg/Lahn, 1980).

24. On this strategy, see Gilbert Watson, *Theriac and Mithridatism: A Study in Therapeutics* (London, 1966).

25. See Nigel G. Wilson, "The Manuscripts of Theophrastus," *Scriptorium* 16 (1962): 96–99; and Charles B. Schmitt, "Theophrastus in the Middle Ages," *Viator* 2 (1971): 251–270.

26. The bibliography on Byzantine botany and pharmacology is rather limited. See, however, Georg Harig, "Die Galenschrift *De simplicium medicamentorum temperamentis ac facultatibus* und die *Collectiones medicae* des Oreibasios," *NTM Schriftenreihe für Geschichte der Naturwissenschaften, Technik und Medizin* 3 (1967): 3–26; John M. Riddle, "Byzantine Commentaries on Dioscorides," in *Symposium on Byzantine Medicine*, ed. John Scarborough, Dumbarton Oaks Papers 38 (1984): 95–102; and, more recently, the following contributions in *Byzantine Garden Culture*, ed. Antony Littlewood, Henry Maguire, and Joachim Wolschke-Bulmahn (Washington, D.C., 2002): Alice-Mary Talbot, "Byzantine Monastic Horticulture: The Textual Evidence," 37–67; Costas N. Constantinides, "Byzantine Gardens and Horticulture in the Late Byzantine Period, 1204–1453: The Secular Sources," 87–103; Robert Rodgers, "Kêpopoiïa: Garden Making and Garden Culture in the Geoponika," 159–175; John Scarborough, "Herbs of the Field and Herbs of the Garden in Byzantine Medicinal Pharmacy," 177–188.

27. On Arabic medicine and pharmacology, see, for instance, in order of publication, Max Meyerhof, "Esquisse d'histoire de la pharmacologie et botanique chez les Musulmans d'Espagne," *Al-Andalus* 3 (1935): 1–41 (reprinted in Max Meyerhof, *Studies in Medieval Arabic Medicine: Theory and Practice*, ed. Penelope Johnstone [London, 1984]); Manfred Ullmann, *Die Medizin im Islam, Handbuch der Orientalistik*, erste Abteilung: *Der Nahe und der Mittlere Osten*; Ergänzungsband 6, erster Abschnitt (Leiden and Cologne, 1970); Martin Levey, *Early Arabic Pharmacology: An Introduction Based on Ancient and Medieval Sources* (Leiden, 1973); Manfred Ullmann, *Islamic Medicine* (Edinburgh, 1978); Sami K. Hamarneh, *Health Sciences in Early Islam*, ed. Munawar A. Anees (Blanco, Texas, 1984); Hakim M. Said, "Islamic Medicine and the Art of Drug-Making: A Historical Perspective," *Hamdard Medicus* 33 (1990): 43–57; Emilie Savage-Smith, "Medicine," in *Encyclopedia of the History of Arabic Science*, ed. Roshdi Rashed (London and New York, 1996), 3:903–962; Alain Touwaide, "Pharmakologie II, Arabisch-Islamisch Kulturgebiete," in *Der neue Pauly: Enzyklopädie der Antike*, vol. 15, part 2, ed. Manfred Landfester with Hubert Cancik and Helmuth Schneider (Stuttgart and Weimar, 2002), 220–224.

28. Among the abundant literature on the topic, see Dimitri Gutas, *Greek Thought, Arabic Culture: The Graeco-Arabic Translation Movement in Baghdad and Early ʿAbbâsid Society (2nd–4th/8th–10th centuries)* (London and New York, 1998).

29. For a synthesis, see Alain Touwaide, "L'intégration de la pharmacologie grecque dans le monde arabe," in *History of Arabic Medicine*, ed. Luciana R. Angeletti and Alain Touwaide (Rome, 1995), 2:259–289. On Hunayn, see, for instance, Fuat Sezgin, ed., *Hunain ibn Is-haq (d. 873), Texts and Studies*, Manshurat ma had tarikh al-ulum al-arabiyah wa-al-islamliya, al-tibb al-islami, vol. 23 (Frankfurt am Main, 1996).

30. See Hakim M. Said, *Al-Biruni's Book on Pharmacy and Materia Medica* (Karachi, 1973).

31. See Martin Levey, *The Medical Formulary or Aqrâbâdhîn of al-Kindî: Translated with a Study of Its Materia Medica* (Madison, Wis., and London, 1966).

32. On Pre-Salernitan botany and pharmacology, see Augusto Beccaria, *I codici di medicina del periodo presalernitano, secoli IX, X e XI* (Roma, 1956); and Alain Touwaide, "Pharmakologie I: Europa A. Frühmittalter," in Landfester et al. 2002, 215–216.

33. On the assimilation of Arabic pharmacology, see Alain Touwaide, "Pharmakologie I: Europa B. Hochmittelalter," in Landfester et al. 2002, 216–217. On the work carried out in Salerno, see Patricia Skinner, *Health and Medicine in Early Medieval Southern Italy* (Leiden and New York, 1997), and Monica Green, *The Trotula: A Medieval Compendium of Women's Medicine* (Philadelphia, 2001).

34. For a reassessment of the origins of the Salerno medical school, see, among others, Barbara M. Kreutz, *Before the Normans: Southern Italy in the Ninth and Tenth Centuries* (Philadelphia, 1991), 144–147; and Alain Touwaide, "Arabic Medicine in Greek Translation: A Preliminary Report," *Journal of the International Society for the History of Islamic Medicine* 1 (2002): 45–53, esp. 48.

35. The version with d'Abano's commentaries was printed in 1478, at Colle Val d'Elsa.

36. See Michael McVaugh, *Arnaldi de Villanova Opera medica omnia, 2: Aphorismi de gradibus* (Granada and Barcelona, 1975; reprint Barcelona, 1992).

37. The *Synonima* were first printed in 1486. On this work, see Danielle Jacquart, "La coexistence du grec et de l'arabe dans le vocabulaire médical du latin médiéval: L'effort linguistique de Simon de Gênes," in *Transfert de vocabulaire dans les sciences,* ed. M. Groult (Paris, 1988), 177–190. This article is reprinted in Danielle Jacquart, *La science médicale occidentale entre deux renaissances XIIe s.– XVe s.* (Aldershot, 1997).

38. For a similar, though later, case, see Giorgio Vercellin, *Il Canone di Avicenna fra Europa e Oriente nel primo Cinquecento: L'Interpretatio Arabicorum nominum di Andrea Alpago* (Turin, 1991).

39. Nuovo receptario composto dal famosissimo chollegio degli eximii doctori della arte et medicina della inclita cipta di Firenze. The colophon reads *IMPRESSO Nella inclyta Cipta di Firenze per la compagnia del Dragho adi .xxi. di Gennaio .M.CCCCLXXXXVIII.* A facsimile edition was published in *Collana di facsimili a cura della Biblioteca Nazionale Centrale di Firenze,* vol. 2 (Florence, 1968). The year 1498 in the colophon accords with the Florentine calendar. On the Ricettario, see Peter Dilg, "Ricettario Fiorentino," in *Lexikon des Mittelalters,* vol. 7 (Munich, 1995), 808–809.

40. See Alain Touwaide, "Pharmakologie I: Europa, C. Renaissance," in Landfester et al. 2002, 217–220.

41. MS Plut. 74.3, Biblioteca Medicea Laurenziana, Florence.

42. It was probably owned in the fifteenth century by a student of Manuel Chrysoloras (c. 1350–1415) in Florence. On the Medici medical manuscripts borrowed by Poliziano, see E. B. Fryde, *Greek Manuscripts in the Private Library of the Medici, 1469–1510* (Aberystwyth, U.K., 1996), 1:641–644.

43. These copies are MSS Graecus 271 and 272, respectively, Biblioteca Marciana, Venice.

44. Currently MS Graecus 289, Biblioteca Apostolica Vaticana, Vatican City.

45. MS Urbinas graecus 66, Biblioteca Apostolica Vaticana, Vatican City.

46. MS Palatinus graecus 77, Biblioteca Apostolica Vaticana, Vatican City.

47. That copy is now in the Bibliothèque nationale de France, Paris, MS Graecus 2183 (mid-fourteenth century).

48. MS 2659, Biblioteca Universitaria, Universidad de Salamanca.

49. MS Graecus 2224, Bibliothèque nationale de France, Paris.

50. MS Alpha.P.5.17, Biblioteca Regionale e Universitaria, Modena.

51. MS Graecus 2185, Bibliothèque nationale de France, Paris.

52. Codex L 119 sup., Biblioteca Ambrosiana, Milan.

53. On Greek botany and pharmacology during the Renaissance, see Jerry Stannard, "The Graeco-Roman Background of the Renaissance Herbal," *Organon* 4 (1967): 141–146. For Theophrastus' *fortuna,* see Charles B. Schmitt, "Theophrastus," *Catalogus Translationum et Commentariorum: Mediaeval and Renaissance Latin Translations and Commentaries, Annotated Lists and Guides,* vol. 2, ed. Paul Oskar Kristeller (Washington, D.C., 1971), 239–322. For Dioscorides, see Jerry Stannard, "Dioscorides and Renaissance Materia Medica," *Analecta Medico-Historica, I: Materia Medica in the XVIth Century, Proceedings of a Symposium of the International Academy of the History of Medicine Held at the University of Basel, 7th September 1964,* ed. Marcel Florkin (Oxford, 1966), 1–21 (reprinted in Jerry Stannard, *Herbs and Herbalism in the Middle Ages and the Renaissance* [Aldershot, U.K., 1999]). For the print tradition of *De materia medica* during the Renaissance, see John M. Riddle, "Dioscorides," *Catalogus Translationum et Commentariorum: Mediaeval and Renaissance Latin Translations and Commentaries. Annotated Lists and Guides,* vol. 4, ed. F. Edward Kranz and Paul Oskar Kristeller (Washington, D.C., 1980), 1–143. Finally, on Renaissance commentaries on classical botanical works, see Peter Dilg, "Die botanische Kommentarliteratur Italiens um 1500 und ihr Einfluss auf Deutschland," *Der Kommentar in der Renaissance,* ed. August Buck and Otto Herding (Bonn and Bad Godesberg, 1975), 225–252.

54. On Gaza, see Giovanni Salanitro, "Teodoro Gaza traduttore di testi classici," in *Dotti bizantini e libri greci nell'Italia del secolo XV: Atti tai del convengo internazionale, Trento, 22–23 ottobre 1990,* ed. Mariarosa Cortesi and Enrico V. Maltese, Collectanea, vol. 6 (Naples, 1992), 218–225.

55. On Barbaro's botanical research, see Vittore Branca, "Ermolao Barbaro e il suo circolo tra azione civile, fede religiosa, entusiasmo filologico, presperimentalismo scientifico," *Storia della cultura veneta*, vol. 3, part 1, ed. Girolamo Arnaldi and Manlio Pastore Stocchi (Vicenza, 1980), 123–175; reprinted in Vittore Branca, *La sapienza civile: Studi sull'Umanesimo a Venezia*, Biblioteca di lettere italiane, Studi e testi, vol. 50 (Florence, 1998), 59–125.

56. This section of *De expetendis et fugiendis rebus* was republished by Henricus Sybold under the title *De simplicium natura* (Strasbourg, 1537).

57. It was included in volume 4 of the edition of Aristotle's *Opera omnia* published in Venice in 1497 by Aldus Manutius.

58. On Leoniceno's collaboration with Manutius, see Nicolas Mani, "Die griechische *Editio princeps* des Galenos (1525), ihre Entstehung und ihre Wirkung," *Gesnerus* 13 (1956): 29–52 (esp. 38 and notes 34–36); Martin Sicherl, *Handschriftliche Vorlagen der Editio princeps des Aristoteles*, Akademie der Wissenschaften und der Literatur, Mainz, Abhandlungen der Geistes- und Sozialwissenschaftlichen Klasse, Jahrgang 1976, no. 8 (Wiesbaden, 1976), 14, note 33. This study is reprinted in Martin Sicherl, *Griechische Erstausgaben des Aldus Manutius: Druckvorlagen, Stellenwert, kultureller Hintergrund*, Studien zur Geschichte und Kultur des Altertums, Neue Folge, 1 Reihe: Monographien, 10. Band (Paderborn, 1997), 31–113; Philippe Hoffmann, "Un mystérieux collaborateur d'Alde Manuce: l'Anonymus Harvardianus," *Mélanges de l'École française de Rome: Moyen âge, temps modernes* 97 (1985): 45–143, esp. 133–138); Philippe Hoffmann, "Autres données relatives à un mystérieux collaborateur d'Alde Manuce: l'Anonymus Harvardianus," *Mélanges de l'École française de Rome: Moyen âge, temps modernes* 98 (1986): 673–708, esp. 704–708.

59. The treatise was published by Eucharius Cervicornur (i.e., Hirtzhorn) together with the *Anatomice* of Alessandro Benedetti (c. 1450–1512) in 1527, in Cologne.

60. For the treatise on urine analysis, see Giorgio Valla, *De urinae significatione*, published by Henricus Sybold (Strasbourg, 1528?). On epidemiology, see the work published by Simon Bevilacqua: Michael Psellus, *De victu humano* (Venice, 1498). See also two other volumes published by Sybold, *Diocles Carystius, Epistola de secunda valetudine tuenda* (Strasbourg, 1528?) and *De tuenda sanitate per victum . . . Paulus Aeginata De ciborum facultatibus* (Strasbourg, 1529?).

61. Published by Simon Bevilacqua (Venice, 1498).

62. On Valla's scientific activity, see, for example, P. Landucci Ruffo, "Note sulla 'Physiologia' di Giorgio Valla," *Physis* 13 (1971): 13–20, and Vittore Branca, ed., *Giorgio Valla tra scienza e sapienza*, Studi di Gianna Cardenal, Patrizia Landucci Ruffo, Cesare Vasoli; Civiltà Veneziana, saggi, vol. 28 (Florence, 1981).

63. On Leoniceno (in addition to the works cited above), in order of publication, see Edward C. Streeter, "Leoniceno and the School of Ferrara," *Bulletin of the Society of Medical History of Chicago* 1 (1911–1916): 18–22; Lynn Thorndike, *A History of Magic and Experimental Science*, vol. 4 (New York, 1934), 4:594–610; Arturo Castiglioni, "The School of Ferrara and the Controversy of Pliny," in *Science, Medicine, and History: Essays on the Evolution of Scientific Thought and Medical Practice Written in Honour of C. Singer*, ed. E. Ashword Underwood, 2 vols. (Oxford, 1953), 1:269–279; M. Santoro, "La polemica pliniana fra il Leoniceno e il Collenuccio," *Filologia romanza* 3 (1956): 162–205; Fridolf Kudlien, "Zwei medizinisch-philologische Polemiken am Ende des 15. Jahrhunderts," *Gesnerus* 22 (1965): 85–92 (see especially 88–92); William F. Edwards, "Nicolò Leoniceno and the Origins of Humanist Discussion of Method," in *Philosophy and Humanism: Renaissance Essays in Honor of Paul Oskar Kristeller*, ed. Edward P. Mahoney (Leiden, 1976), 283–305; Daniela Mugnai Carrara, "Profilo di Nicolò Leoniceno," *Interpres* 2 (1979): 169–212; Jerome J. Bylebyl, "The School of Padua: Humanistic Medicine in the Sixteenth Century," in *Health, Medicine and Mortality in the Sixteenth Century*, ed. Charles Webster (Cambridge, 1979), 335–370 (see especially 339–342); Daniela Mugnai Carrara, "Fra causalità astrologica e causalità naturale: Gli interventi di Nicolò Leoniceno e della sua scuola sul morbo gallico," *Physis* 21 (1979): 37–54; and, by the same author, "Una polemica umanistico-scolastica circa l'interpretazione delle tre dottrine ordinate di Galeno," *Annali dell'Istituto e Museo di storia della scienza di Firenze* 8 (1983): 31–57; Roger K. French, "Pliny and Renaissance Medicine," in *Science in the Early Roman Empire: Pliny the Elder, His Sources and Influence*, ed. Roger K. French and Frank Greenaway (London and Sydney, 1986), 252–281; Daniela Mugnai Carrara, "La polemica 'de cane rabido' di Nicolò Leoniceno, Nicolò Zocca e Scipione Carteromaco: Un episodio di filologia medico-umanistica," *Interpres* 9 (1989): 196–236; and "Nicolò Leoniceno e Giovanni Mainardi: Aspetti epistemologici dell'Umanesimo medico," *Alla Corte degli Estensi: Filosofia, arte e cultura a Ferrara nei secoli XV e XVI: Atti del Convegno internazionale di studi, Ferrara, 5–7 marzo 1992*, ed. M. Bertozzi (Ferrara, 1994), 19–40; Giovanna Ferrari, *L'esperienza del passato: Alessandro Benedetti filologo e medico umanista*, Biblioteca di Nuncius, vol. 22 (Florence, 1996), especially 256–296; Vivian Nutton, "The Rise of Medical Humanism: Ferrara, 1464–1555," in *Renaissance Studies* 11 (1997): 2–19; and "Hellenism Postponed: Some Aspects of Renaissance Medicine, 1490–1530," *Sudhoffs Archiv* 81 (1997): 158–170; Alain Touwaide, *"Loquantur ipsi ut velint . . . modo quis serpens sit tirus . . . . non ignorent:* Leoniceno's Contribution to Renaissance Epistemological Approach to Scientific Lexicology," in *Medical Latin from the Late Middle Ages to the Eighteenth Century*, ed. Wouter Bracke and Herwig Deumens (Brussels, 2000), 151–173.

64. See Daniela Mugnai Carrara, *La biblioteca di Nicolò Leoniceno, tra Aristotele e Galeno: Cultura e libri di un medico umanista*, Accademia Toscana di Scienze e Lettere, "La Colombaria," Studi 118 (Florence, 1991).

65. For a reprint of the Latin text (not without errors) and Italian translation, see Loris Premuda, ed. and trans., *Nicolò Leoniceno, De Plinii in medicina erroribus* (Milan, 1958). On this work and the polemic it provoked, in addition to the bibliography cited above, see Charles G. Nauert, "Humanists, Scientists and Pliny: Changing Approaches to a Classical Author," *American Historical Review* 84 (1979): 72–85; Giovanna Ferrari, "Gli errori di Plinio: fonti classiche e medicina nel conflitto tra Alessandro Benedetti e Nicolò Leoniceno," in *Sapere e/è potere: Discipline, dispute e professioni nell'università medioevale e moderna*, ed. A. Cristiani (Bologna, 1990), 2:173–204.

66. On Pliny's *fortuna* during the Renaissance, see Charles G. Nauert, "Caius Plinius Secundus," in Kristeller 1980, 297–422.

67. For new approaches to humanism, see Ronald G. Witt, *"In the Footsteps of the Ancients": The Origins of Humanism from Lovato to Bruni* (Leiden, Boston, and Cologne, 2000); and Robert Black, *Humanism and Education in Medieval and Renaissance Italy: Tradition and Innovation in Latin Schools from the Twelfth to the Fifteenth Century* (Cambridge, 2001).

68. On polemics against Arabs, see Peter Dilg, "The Antiarabism in the Medicine of Humanism," *La diffusione delle scienze islamiche nel medioevo europeo* (Rome, 1987), 269–289. Arabic medicine (particularly Avicenna's *Qanun*) continued to be taught for a long time, however. See Nancy Siraisi, "Renaissance Commentaries on Avicenna's Canon, Book 1, Part 1, and the Teaching of Medical Theoria in the Italian Universities," *History of Universities* 4 (1984): 47–97; Nancy Siraisi, *Avicenna in Renaissance Italy: The Canon and Medical Teaching in Italian Universities after 1500* (Princeton, 1987). On the polemics against medicine and pharmacology, see Peter Dilg, "Materia medica und therapeutische Praxis um 1500: zum Einfluss der arabischen Heilkunde auf den europäischen Arzneischatz," in *Kommunikation zwischen Orient und Okzident: Alltag und Sachkultur* (Vienna, 1994), 353–377.

69. James Hankins, "Renaissance Crusaders: Humanist Crusade Literature in the Age of Mehmet II," *Dumbarton Oaks Papers* 49 (1995): 111–207.

70. See, for instance, the letter to Sultan Mehmet II (1432–1481) from Pope Pius II (1405–1464), critical edition and English translation by Albert Bacca, *Epistola ad Mahomatem II, Epistle to Mohammed II, Aeneas Silvius Piccolomini* (New York, 1990); and, more recently, critical edition with German translation and commentary, Reinhold F. Glei and Markus Köhler, *Pius II, Papa Epistola ad Mahumetem: Einleitung, kritische Edition, Übersetzung*, Bochumer Altertumswissenschaftliches Colloquium 50 (Trier, 2001). On Pope Pius II, Enea Silvio Piccolomini, see Charles-Edouard Naville, *Enea Silvio Piccolomini: L'uomo, l'umanista, il pontefice, 1405–1464* (Locarno, 1984). The crusades and religious polemic did not, however, preclude contacts between the Ottoman world and the West. For Ottoman imitation of Western arts, see, for example, Lisa Jardine and Jerry Brotton, *Global Interests: Renaissance Art between East and West* (Ithaca, N.Y., 2000), or, for Byzantine intellectuals in dialogue with the Ottoman occupiers, see Pedro Bádenas de la Peña, "Corrientes conciliadoras de intelectuales griegos en la corte del Gran Turco," *Erytheia* 20 (1999): 197–208.

71. *Nicolaus Leonicenus Hieronymo Menochio* (1529 edition), 215.

72. On the polemical nature of Leoniceno's editorial activity, see Lorenzo Minio-Paluello, "Attività filosofica-editoriale aristotelica dell'Umanesimo," in *Umanesimo europeo e umanesimo veneziano*, Civiltà europea e civiltà veneziana: Aspetti e problemi, vol. 2 (Florence, 1963), 245–262, esp. 251–253. This study is reprinted in Lorenzo Minio-Paluello, *Opuscula: The Latin Aristotle* (Amsterdam, 1972), 483–500, esp. 489–491.

73. On this edition, see Annaclara Cataldi-Palau, *Gian Francesco d'Asola e la tipografia aldina: La vita, le edizioni, la biblioteca dell'Asolano* (Genoa, 1998), 612, no. 38.

74. On this work, see most recently Peter Godman, *From Poliziano to Machiavelli* (Princeton, 1998), and Anthony Grafton, *Bring Out Your Dead: The Past as Revelation* (Cambridge, Mass., and London, 2001), 1–15.

75. *Epistolae medicinales in quibus multa recentorium errata et antiquorum decreta reserantur . . .*, published by Bernardinus de Odonino (Ferrara, 1529). On Manardo, see the contributions in *Atti del Convegno internazionale per la celebrazione del V centenario della nascita di Giovanni Manardo, 1462–1536, Ferrara, 8–9 dicembre 1962* (Ferrara, 1963), particularly Loris Premuda, "Un discepolo di Leoniceno tra filologia ed empirismo: G. Manardo e il 'libero esame' dei classici della medicina in funzione di più spregiudicati orientamenti metodologici," 43–56.

76. *Examen omnium simplicium medicamentorum, quorun in officinis usus est . . .*, published by Joannes and Franciscus Frelones (Lyon, 1537, with numerous reissues). On Brasavola, see Vivian Nutton, "Medicine, Diplomacy and Finance: The Prefaces to a Hippocratic Commentary of 1541," in *New Perspectives on Renaissance Thought: Essays in the History of Science, Education and Philosophy in Memory of Charles B. Schmitt*, ed. John Henry and Sarah Huyyon (London and Naples, 1990), 230–243.

77. Giovanni Pozzi, "Appunti sul *Corollarium* del Barbaro," in *Tra latino e volgare: Per Carlo Dionisotti*, Medioevo e Umanesimo, vols. 17–18 (Padua, 1974), 2:619–640.

78. See, for example, the *Herbarius Latinus* published by Simon Bevilacqua (Venice, 1499).

79. Venice was an important center of drug trade and hence of pharmaceutical activity. See Richard Palmer, "Pharmacy in the Republic of Venice in the Sixteenth Century," in *The Medical Renaissance of the Sixteenth Century*, ed. Andrew Wear, Roger K. French, and Iain M. Lonie (Cambridge, 1985), 100–117; Richard Palmer, "Medical Botany in Northern Italy in the Renaissance," *Journal of the Royal Society of Medicine* 78 (1985): 149–157; *Di sana pianta: Erbari e taccuini di sanità: Le radici storiche della nuova farmacologia*, ed. R. Bussi [exh. cat., Abbazia Benedettina di Praglia, 1988] (Modena, 1988).

80. On the gardens in Pisa, see Francesco Garbari, Lucia Tongiorgi Tomasi, and Alessandro Tosi, *Giardino dei semplici: L'Orto botanico di Pisa dal XVI al XX secolo* (Pisa, 1991). On the gardens in Padua, see *L'Orto botanico di Padova 1545–1995*, ed. A. Minelli (Venice, 1995).

81. See Michael Reinecke, *Galen und Vesal: Ein Vergleich der anatomish-physiologischen Schriften*, Episteme kai therapeia: Studien zur Medizin als Wissenschaft und Heilkunst, vol. 1 (Muenster, 1995); Roger French, *Dissection and Vivisection in the European Renaissance* (Aldershot, U.K., 1999). On Renaissance anatomy, see also Andrea Carlino, *La fabbrica del corpo: Libri e dissezione nel Rinascimento* (Turin, 1994); English translation: *Books of the Body: Anatomical Ritual and Renaissance Learning* (Chicago, 1999).

82. On Renaissance botanical illustration, see Sachiko Kusuwaka, "Illustrating Nature," in *Books and the Sciences in History*, ed. Marina Frasca-Spada and Nick Jardine (Cambridge, 2000), 90–113.

83. For a synthesis on the rise of Greek studies in Germany during the Renaissance, see *Graecogermanica: Griechischstudien deutscher Humanisten*, ed. Dieter Harlfinger (Weinheim and New York, 1989).

84. See the case of Ulrich von Hutten (1488–1523), for example, highlighted in Lewis Jillings, "The Eagle and the Frog: Hutten's Polemic against Venice," *Renaissance Studies* 2 (1988): 14–26.

85. Max H. Fish, *Nicolaus Pol Doctor 1494* (New York, 1947), 28.

86. This reflects Dioscorides' statement on the necessity to know plants at all stages of their development. See Wellmann 1906–1914, praefatio, § 7 (1:4, lines 7–15).

87. On Otto Brunfels, see Jerry Stannard, "Brunfels, Otto," *Dictionary of Scientific Biography* (New York, 1970), 2:535–538; on *Herbarum vivae eicones*, see the essay by Pamela Smith in this volume.

88. On Weiditz, see Walter Rutz, *Pflanzenaquarelle des Hans Weiditz aus dem Jahre 1529: Die Originale zu den Holzschnitten im Brunfels'schen Kräuterbuch* (Bern, 1936).

89. For a general approach to the Reformation, see Erika Rummel, *The Confessionalization of Humanism in Reformation Germany*, Oxford Studies in Historical Theology (Oxford, 2000); and, for a study of Reformation genesis, Anthony Levi, *Renaissance and Reformation: The Intellectual Genesis* (New Haven and London, 2002). For the specific case of astronomy, see Peter Barker, "The Role of Religion in the Lutheran Response to Copernicus," in *Rethinking the Scientific Revolution*, ed. Margaret J. Osler (Cambridge, 2000), 59–88.

90. For Fuchs' biography, see most recently Frederick G. Meyer, Emily W. E. Trueblood, and John L. Heller, *The Great Herbal of Leonhart Fuchs, De historia stirpium commentarii insignes, 1542: Notable Commentaries on the History of Plants*, 2 vols. (Stanford, 1999); Brigitte Baumann, Helmut Baumann, and Susanne Baumann-Schleihauf, eds., *Die Kräuterbuchhandschrift des Leonhart Fuchs* (Stuttgart, 2001); and Gerd Brinkhus and Claudine Pachnicke, eds., *Leonhart Fuchs (1501–1566): Mediziner und Botaniker: Ausstellung im Stadtmuseum Tübingen, 21. Juni bis 16. September 2001* (Tübingen, 2001).

91. See Luigi Samoggia, "Manardo e la Scuola umanistica-filologica tedesca con particolare riguardo a Leonardo Fuchs," in *Atti del Convegno internazionale per la celebrazione del V centenario della nascita di Giovanni Manardo, 1462–1536. Ferrara, 8–9 dicembre 1962* (Ferrara, 1963), 241–251; and, by the same author, *Le ripercussioni in Germania dell'indirizzo filologico-medico Leoniceniano della scuola ferrarese per opera di Leonardo Fuchs*, Quaderni di Storia della Scienza e della Medicina, vol. 4 (Ferrara, 1964). See also *Errata recentiorum medicorum. LX numero, adjectis eorundem confutationibus, in studiosorum gratiam, iam primum aedita* (1530). Italian translation: A. Morricone and V. Pedicino, *Leonhart Fuchs, Sessanta errori dei medici contemporanei: Con l'aggiunta delle confutazioni per uso degli studiosi* (Rome, 1963).

92. See Georg Harig, "Leonhart Fuchs und die theoretische Pharmakologie der Antike," *NTM-Schriftenreihe für Geschichte der Naturwissenschaften, Technik und Medizin* 3 (1966): 74–104.

93. See Sachiko Kusukawa, "Leonhart Fuchs on the Importance of Pictures," *Journal of the History of Ideas* 58 (1997): 403–427.

94. On Mattioli, see, among others, Jerry Stannard, "P. A. Mattioli and Some Renaissance Editions of Dioscorides," in *Books and Libraries at the University of Kansas* 4 (1966): 1–5; and "P. A. Mattioli: Sixteenth-Century Commentator on Dioscorides," *Bibliographical Contributions* 1 (1969): 59–81 (both studies reprinted in Jerry Stannard, *Herbs and Herbalism in the Middle Ages and the Renaissance* [Aldershot, U.K., 1999]); Bruno Zanobio, "Mattioli," *Dictionary of Scientific Biography* 9 (1974), 178–180; Tiziana Pesenti, "Il 'Dioscoride' di Pier Andrea Mattioli e l'editoria botanica," in *Trattati di prospettiva, architettura militare, idraulica e altre discipline* (Vicenza, 1985), 61–103; Franco Pedrotti, "Pietro Andrea Mattioli e il Trentino," in *I giardini dei semplici e gli orti botanici della Toscana*, ed. Sara Ferri and Francesca Vannozzi (Perugia, 1993), 193–200;

Franco Ottaviani, "Quattro generazioni di medici trentini (1539–1658)," in *I Madruzzo e l'Europa, 1539–1569: I principi vescovi di Trento tra Papato e Impero*, ed. Laura Dal Prà (Milan, 1993), 673–677; Sara Ferri, ed., *Pietro Andrea Mattioli, Siena 1501–Trento 1578, la vita, le opere* (Perugia, 1997).

95. *Libri cinque della historia, & materia medicinale tradotti in lingua volgare italiana . . . con amplissimi discorsi, et comenti, et dottissime annotationi, et censure del medesimo interprete. Da cui potra ciascuno facilmente acquistare la vera cognitione de semplici, non solamente scritti da Dioscoride, ma da altri antichi, & moderni scrittori, & massimamente da Galeno . . .* (Venice, 1544).

96. The reprints were published from 1549 onward. Later editions have been republished in facsimile. See, for example, the facsimile of the 1563 edition under the title *I discorsi di M. Pietro And. Matthioli senese . . . nei sei libri di Pedacio Dioscoride Anazarbeo, Della materia medicinale . . .* (Todi, 1989). The main translation of Mattioli's work is the Latin one, published from 1554 onward under the title *Commentarii, in libros sex Pedacii Dioscorides Anazarbei, De medica materia . . .* (Venice, 1554).

97. For the concept of the Botanical Republic, see Paula Findlen, "The Formation of a Scientific Community: Natural History in Sixteenth-Century Italy," in *Natural Particulars: Nature and the Disciplines in Renaissance Europe*, ed. Anthony Grafton and Nancy Siraisi, Dibner Institute Series in the History of Science and Technology (Cambridge, Mass., and London, 1999), 370–400. For another view on the intellectual context, see Richard Palmer, "Medical Botany in Northern Italy in the Renaissance," *Journal of the Royal Society of Medicine* 78 (1985): 149–157.

98. *Pedacio Dioscorides Anazarbeo acerca de la materia medicinal y de los venenos mortiferos . . .* (1st ed., Anvers, 1555). The work has been reissued in many reprints and new editions. The translation and commentary of the so-called sixth and seventh books of Dioscorides (inauthentic) have been recently republished by Antonio Gamoneda, *Andrés Laguna, Libro de los venenos* (Madrid, 1995). Regarding Laguna, the monumental study by Cesar Dubler remains fundamental: Cesar E. Dubler, *La "Materia médica" de Dioscórides, transmisión medieval y renacentista*, 6 vols. (Barcelona, 1953–1959). See especially volume 4 for Laguna's biography. Published more recently (but with many errors) are Miguel Ángel González Manjarrés, *Andrés Laguna y el humanismo médico*, Estudios de historia de la ciencia y de la técnica, vol. 15 (Valladolid, 2000), and Miguel Ángel González Manjarrés, *Entre la imitación y el plagio: Fuentes e influencias en el Dioscórides de Andrés Laguna* (Segovia, 2000).

99. For an attempt to identify this manuscript, see Antonio Guzmán Guerra, *El Dioscorides de Laguna y el manuscrito de Paez de Castro* (Madrid, 1978). The question remains open, however.

100. On the illustration of Mattioli's treatise by Liberale, see William Patrick Watson, Sandra Raphael, and Ian Bain, *The Mattioli Woodblocks* (London and Amsterdam, 1989).

101. Matteoli's representation of *Nardus indica* (fig. 17) and that in the Athos manuscript of *De materia medica* (fig. 18) correspond in depicting a small, dense, bushy plant (vertical in the Greek manuscript and more rounded in Matteoli's work), but neither is accurate. Despite his claim that he had personally observed every plant described in his work, it seems unlikely that Matteoli had actually seen an example of *Nardus indica*, especially since Renaissance botanists spent considerable time attempting to identify it. At least two derivatives of the Athos manuscript can be traced to northern Italy in the late fifteenth and early sixteenth centuries. One of these may have provided Matteoli, or the artist who worked for him, with an image to be "reinterpreted" in a way that would appear original.

102. For a different view, see Vivian Nutton, "Greek Science in the Sixteenth-Century Renaissance," in *Renaissance and Revolution: Humanists, Scholars, Craftsmen and Natural Philosophers in Early Modern Europe*, ed. J. V. Field and Frank A. J. L. James (Cambridge, 1993), 15–28.

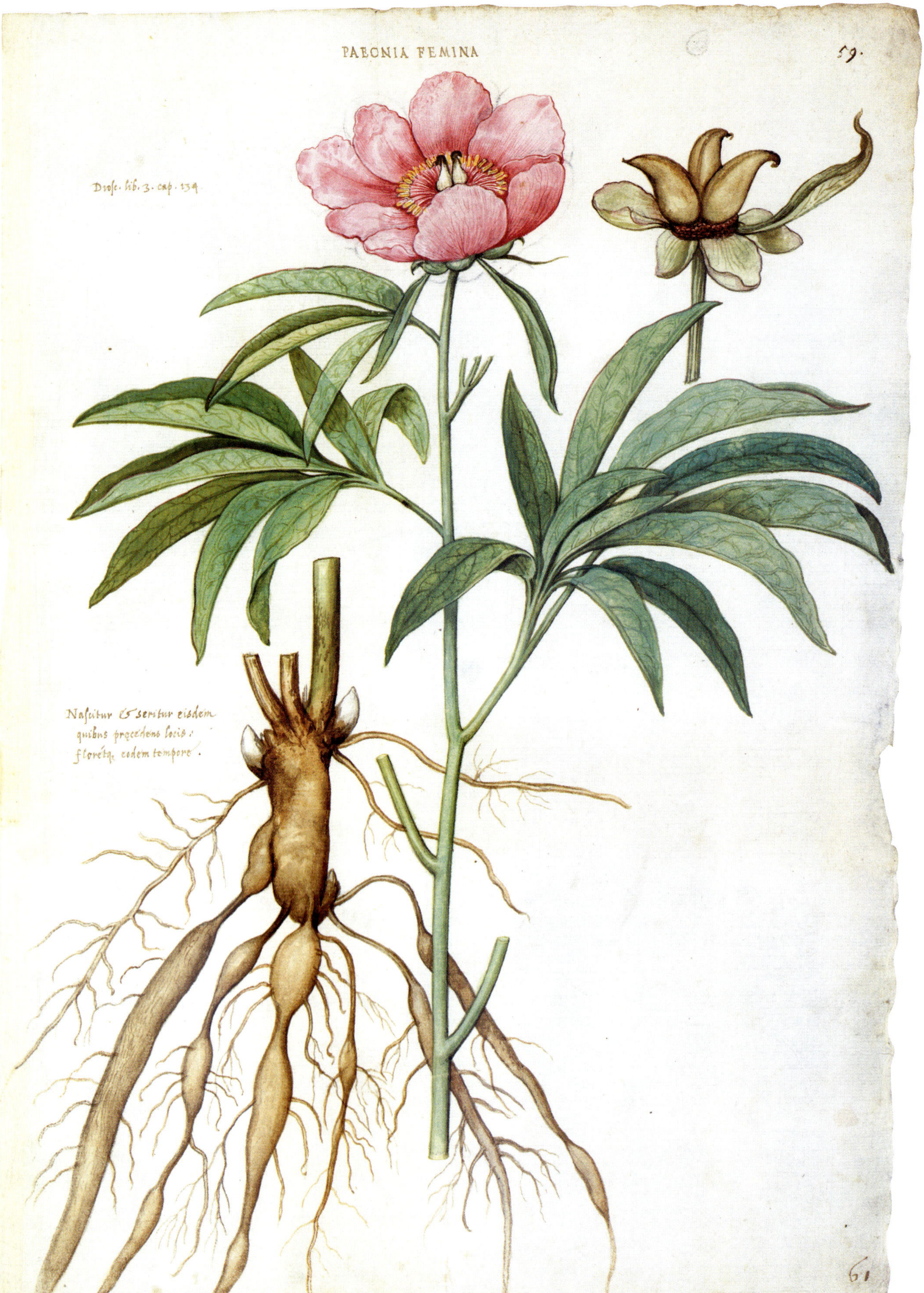
PAEONIA FEMINA
59.
Drosc. lib. 3. cap. 134
Nascitur & seritur eisdem
quibus praecedens locis:
floretq. eodem tempore.
61

CLAUDIA SWAN
Northwestern University

The Uses of Botanical Treatises in the Netherlands, c. 1600

The period that extended from the introduction of the movable-type press in the last decades of the fifteenth century to the middle of the seventeenth century was, in northern and southern Europe alike, an era of illustrated natural history: some of its most familiar products are encyclopedic herbals laden with pictures, generally in the form of woodcuts and, later, engravings. Throughout the early modern period, botanical treatises played a key role in the scholarly experience and the study of the natural world. Enterprising publishers across Europe issued herbals in which recently recorded specimens hailing from the New World were catalogued and illustrated alongside indigenous European plants that had not already been charted by such classical luminaries as the Greek naturalists Theophrastus (third century B.C.E.) and Dioscorides (first century B.C.E.), Galen (second century C.E.), and the Roman author Pliny the Elder (d. 79 C.E.). Numerous plant varieties not described by these classical authors were "discovered" by European medical professionals and natural historians and were factored into a growing body of botanical knowledge that was disseminated, emended, and perpetuated in print.[1] Alongside published works, unpublished treatises—consisting of images alone or a combination of text and images—also attest to the ways and means by which early modern natural historians charted and made sense of the plant world. Natural history treatises were repositories of information gathered by reading and observing, and through correspondence and discussion with fellow scholars of the natural world. The improvements these writers made over their classical precedents resulted from an ongoing process of revision, a process that was as much a factor of the production of these treatises as it was of their reception and use.

Much has been written about the many catalogues of the natural world published and disseminated in rapid succession throughout Europe at that time. The numerous early modern botanical, zoological, and ichthyological volumes are significant not only for the numbers in which they were produced and their ambitious scope, but also for the nature of the descriptive means they employed. In general, these texts consist of verbal descriptions accompanied by morphological images—catalogues of a sort unthinkable in the absence of print technology. Vagaries of appearance were recorded, (medicinal) effects were noted where known, and the discovery of new members of the various classes of nature—animal, vegetable, and mineral—was declared. Most important, emphasis was consistently placed on the description of the specimens under review. The case of botany is exemplary. The realization that varieties of indigenous European flora had not been recorded by ancient authorities—Theophrastus, Galen,

Anonymous, *Paeonia femina* (*Paeonia officinalis*; peony), c. 1590, from Libri Picturati, A.25, folio 59, watercolor
Jagiellon University Library, Kraków

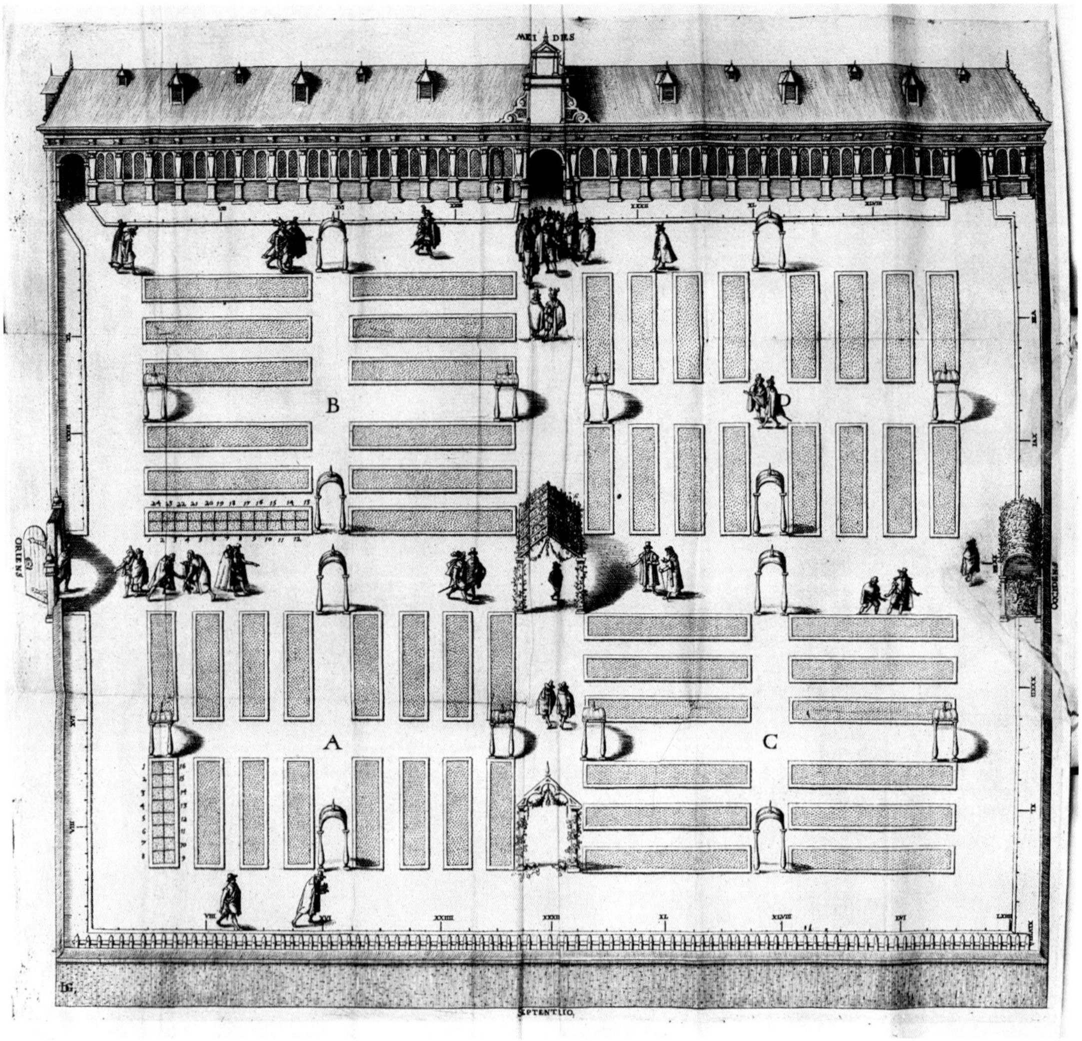

1. Jacques de Gheyn II, *Plan of the Leiden University Garden*, 1601, engraving
National Herbarium of the Netherlands, Leiden

and Dioscorides first among them—provided an incentive to produce the first generation of modern herbals, and print technology offered the means to disseminate, emend, and test new descriptions of specimens on an international scale. While specimens from outside Europe were initially registered in terms of their pharmacological uses, observation was ultimately emancipated from practical exigencies and acquired disciplinary status through the study of similarities and differences between appearances and internal structures.[2] In the initial flush of descriptive botany, the visual record played a critical role. The dissemination by means of print technology of what historian William M. Ivins, Jr., memorably termed "exactly repeatable pictorial statements" enabled documentation of the varieties of the natural world unknown to the ancients and provided stable points of reference for cataloguers and later taxonomists.[3]

A variety of botanical treatises—printed, painted, folio size, field-guide format, pedagogic, and personal—were produced in the Netherlands in the later sixteenth and early seventeenth centuries. This essay surveys a number of Dutch botanical treatises, offering a sort of group portrait of some of the types available, paying special attention to the different ways in which they were used and the approaches to the natural world from which they grew and which, in turn, they fostered. Because most of the authors and purveyors of treatises discussed here were affiliated with the botanical garden of Leiden University—the nexus of botanical study in the northern Netherlands c. 1600—and because the garden fostered modes of observation and study reflected in the treatises, we begin within its walls. Over the course of about four decades on either side of the turn of the seventeenth century, different kinds of botanical knowledge and different forms of interest in the plant world were cultivated at Leiden. The range of approaches to the plant world and its investigation can, in turn, be traced to different botanical treatises produced there.

The Botanical Garden at Leiden

The primary arena of botanical study in the northern Netherlands at the turn of the seventeenth century was a plot of land in the heart of the city of Leiden that measured 30 by 40 meters (roughly 98 by 131 feet)—the Leiden University garden (figs. 1, 2).[4] Established in 1574 as the first university founded in what would become the Dutch Republic, Leiden University attracted a tremendous constellation of scholars; it also housed the principal botanical garden north of Paris. Founded in 1593 to accommodate instruction in materia medica, the Leiden garden grew rapidly in scope and renown. By 1599, a long gallery, called in seventeenth-century documents an ambulacrum, had been constructed at the western edge of the garden to house the more delicate foreign plants and to exhibit a range of related objects, from maps and prints to the kinds of natural specimens that were part and parcel of scientific collections at the time: dried blowfish, coral, crocodiles, ethnographic items, and a collection of minerals (fig. 2; note the items depicted in the lower register). The Leiden garden, much like the coeval university anatomical theater, served as a locus for teaching medicine (one of the principal disciplines taught at the young university), allowing for the kind of eyewitness, or autoptic, experience that became a hallmark of early modern science and its instruction.[5] Classes were taught on site in both the garden and the theater and in

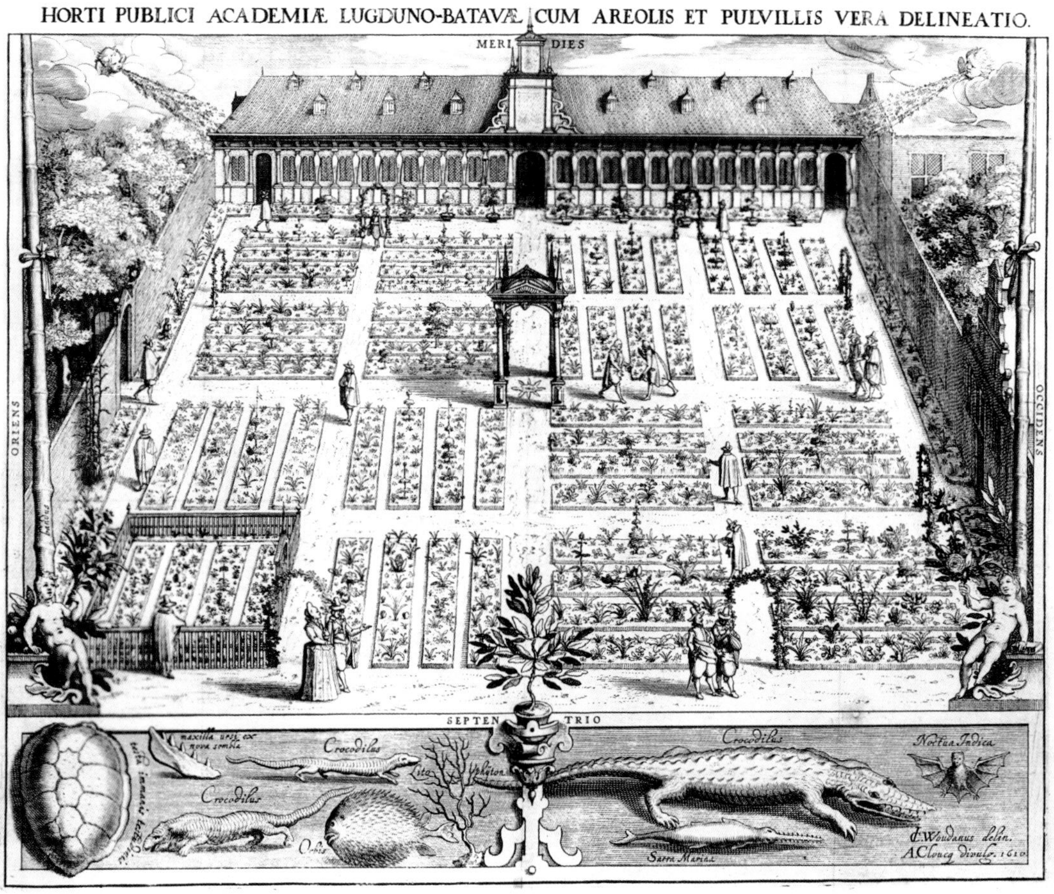

2. Willem Swanenburgh after Jan Corneliszn Woudanus (Jan Corneliszn van't Woudt), *Leiden Garden,* 1610, engraving
Rijksmuseum, Amsterdam

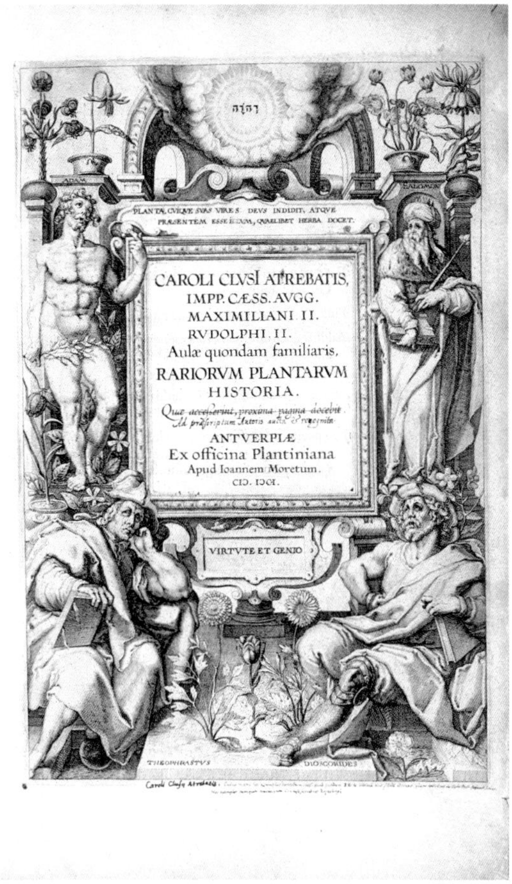
CAROLI CLVSI ATREBATIS,
IMPP. CÆSS. AVGG.
MAXIMILIANI II.
RVDOLPHI II.
Aulæ quondam familiaris.
RARIORVM PLANTARVM
HISTORIA.
ANTVERPIÆ
Ex officina Plantiniana
Apud Ioannem Moretum.
CIↃ. IↃCI.

3. Title page from Carolus Clusius, *Rariorum plantarum historia* (Antwerp, 1601), engraving
Amsterdam University Library (UvA), Special Collections

the presence of the subjects of instruction, whether plants or anatomical specimens. By virtue of its contents, its keepers, and the circulation of information it inspired and supported, the Leiden University garden became the fulcrum for the production of botanical knowledge at the turn of the seventeenth century in the Netherlands.

The *Rariorum plantarum historia*

One of the best-known Dutch natural history treatises dating from the turn of the seventeenth century is the *Rariorum plantarum historia* (Leiden, 1601) by the Flemish botanist Carolus Clusius (Charles de l'Escluse; 1526–1609) (figs. 3, 4). This weighty tome was published nearly a decade into Clusius' tenure as director of the Leiden University garden. Having distinguished himself by publishing a number of works on regional flora (of Austria and Spain), as well as by translating early texts on New World materia medica into Latin and by serving as imperial gardener in Vienna, Clusius was a fine catch for the Dutch university, whose garden he directed from 1593 until his death in 1609. The *Rariorum* is exemplary of the generously illustrated, widely distributed editions issued throughout the later sixteenth century by the Antwerp publisher Christophe (Christoffel) Plantin (1520–1589). Along with Rembertus Dodonaeus (Rembert Dodoens; 1517–1585) and Matthias Lobelius (Mathieu de l'Obel, 1538–1616), both of whom also wrote monumental herbals in the later sixteenth century that Plantin printed with copious woodcut illustrations of individual plants, Clusius is frequently hailed as one of the fathers of Netherlandish botany.[6]

Like his colleagues Dodonaeus and Lobelius, Clusius assembled his *Rariorum* much as he had composed the flora before it and much as he assembled his later *Exoticorum libri decem* (Leiden, 1605)—by compiling individual descriptions, both verbal and visual, of specimens culled from the natural world (figs. 5, 6). The descriptions were based on information about specimens provided by a network of colleagues, friends, and fellow amateur naturalists, who are cited throughout the text. Clusius' use of images in his publications is both

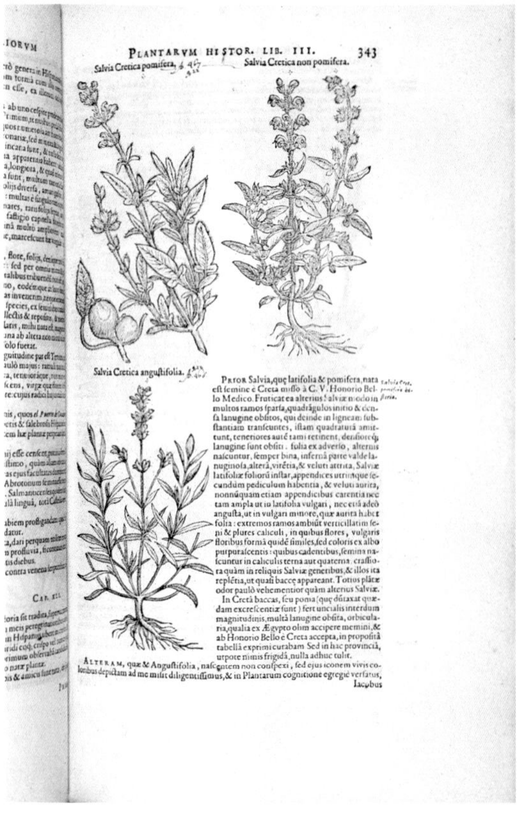
PLANTARVM HISTOR. LIB. III. 343

4. Anonymous, *Salvia cretica pomifera*, *Salvia cretica non pomifera*, and *Salvia cretica angustifolia* (varieties of aromatic sage), from Carolus Clusius, *Rariorum plantarum historia* (Antwerp, 1601), volume 3, page 343, woodcuts
University Library, Leiden, 755 A3

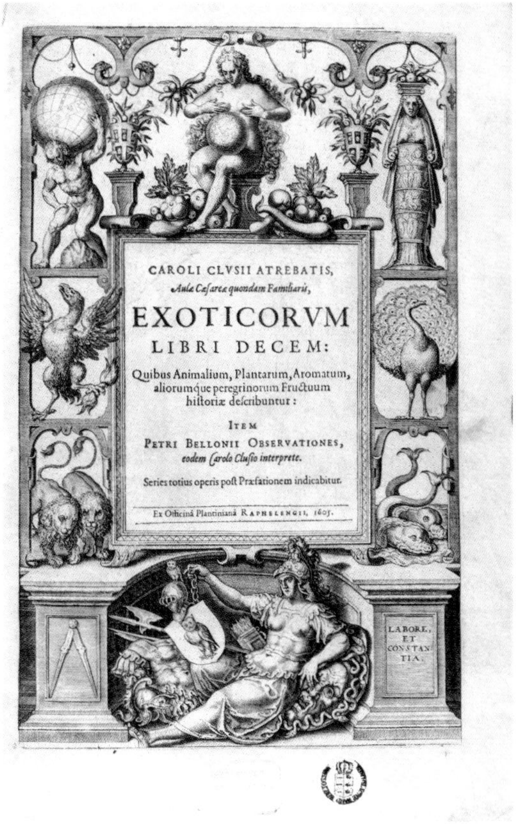
CAROLI CLVSII ATREBATIS,
Aulæ Cæsareæ quondam Familiaris,
EXOTICORVM
LIBRI DECEM:
Quibus Animalium, Plantarum, Aromatum, aliorumque peregrinorum Fructuum historiæ describuntur:
ITEM
PETRI BELLONII OBSERVATIONES,
eodem Carolo Clusio interprete.
Series totius operis post Præfationem indicabitur.
Ex Officinâ Plantinianâ Raphelengii, 1605.

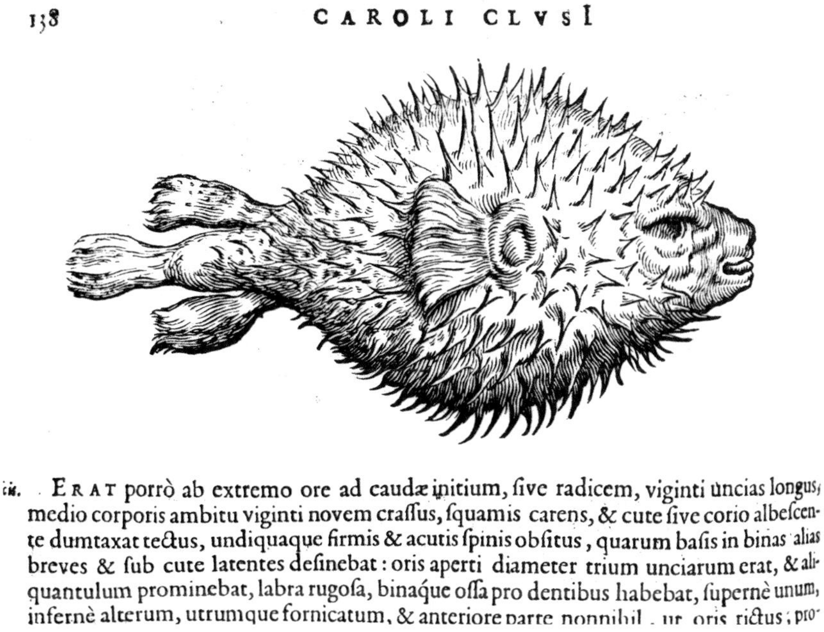
138 CAROLI CLVSI

ERAT porrò ab extremo ore ad caudæ initium, sive radicem, viginti uncias longus, medio corporis ambitu viginti novem crassus, squamis carens, & cute sive corio albescente dumtaxat tectus, undiquaque firmis & acutis spinis obsitus, quarum basis in binas alias breves & sub cute latentes desinebat: oris aperti diameter trium unciarum erat, & aliquantulum prominebat, labra rugosa, binaque ossa pro dentibus habebat, supernè unum, infernè alterum, utrumque fornicatum, & anteriore parte nonnihil . ur oris rictus; pro-

5. Title page from Carolus Clusius, *Exoticorum libri decem* (Leiden, 1605), engraving
Amsterdam University Library (UvA), Special Collections

6. Anonymous, *Spiny Pufferfish*, from Carolus Clusius, *Exoticorum libri decem* (Leiden, 1605), page 138, woodcut
National Herbarium of the Netherlands, Leiden

noteworthy and consonant with contemporary science and its products: that most of his entries on individual plants or other specimens are illustrated with a woodcut reflects a conviction, amplified in his text, that images convey information crucial to description and to some form of rudimentary classification.[7]

In his *Rariorum,* Clusius supplemented his descriptions of plants and flowers with considerations as to their relative scale, the time of the year in which they bloom, and their provenance; but the characteristics most crucial to their classification are those that can be observed in the immediate presence of the specimen—those that can be recorded pictorially.[8] In his groundbreaking chapter on tulips, Clusius describes a variety of dwarf tulip within the category of the "intermediates"; it blossoms between the "early" and the "late" varieties. This class is described generally as follows:

> The dwarf [intermediate] tulip is not more than a foot high, usually even less, and in its leaves and flower it strongly resembles the early tulip. All its segments are pointed, but the outer ones are much longer, externally dull red but at the outermost margins greenish; the inner segments are of a brilliant, fiery red throughout. The claws are yellow and radiating, but marked with a jet-black patch in such a way that the latter appears encircled by a mere golden aureole and bears some likeness to an eye; the filaments and their anthers are blackish. It should be noted that its bulbous root is woolly; the outer membrane enveloping and covering the substance of the bulb is so tightly filled with an abundance of dense, white, soft stuffing that it must form a very soft resting place for the bulb.[9]

From the opening sentence of his description, Clusius encourages the reader to imagine the plant as described. The dwarf tulip resembles the early tulip in its overall appearance; and in the more specific rendering of the appearance of this flower, the reader is led from part to part by gradations and shifts of color. At the center of the plant and of the description (while looking into the tulip from above, observing the appearance of the golden aureole), the reader

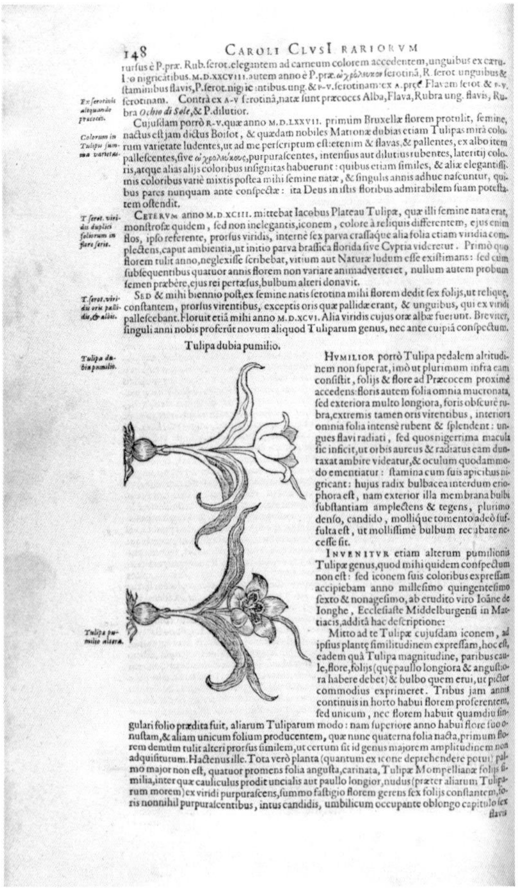

7. Anonymous, *Tulipa dubia pumilio* (dwarf tulip), from Carolus Clusius, *Rariorum plantarum historia* (Antwerp, 1601), page 148, woodcut
Amsterdam University Library (UvA), Special Collections

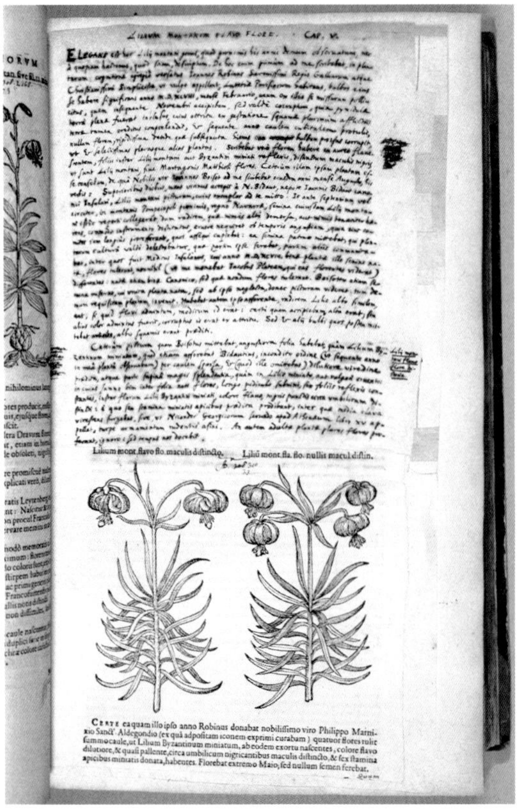

8. Page with handwritten insert, from Carolus Clusius' personal copy of *Rariorum plantarum historia* (Antwerp, 1601), pages 134–135, pen and ink and woodcut
University Library, Leiden, 755 A3

encounters, in an almost specular manner, "some likeness to an eye." Clusius helpfully provides an image to corroborate the verbal evidence he summons (fig. 7), which offers at least partial confirmation of the substance of his description.

Within the context of late sixteenth-century natural history, the combination of text and image here is typical, and Clusius' descriptive method exemplifies the industry standard. Verbal and visual descriptions are coupled throughout natural history writing of the time, in which external, observable characteristics are adduced in attempts to describe and classify the specimens recorded. Throughout Clusius' *Rariorum* and *Exoticorum,* as in most contemporary natural history publications, images served to complement verbal description; where images were the only available evidence, they also served as a basis for description, as well as for classification.[10] The close affiliation between word and image in such treatises attests to the authors' commitment to a morphological account of the natural world. Clusius and his published works embody the effort, undertaken throughout Europe at the time, to understand the natural world by describing it.

A copy of the *Rariorum* preserved in the Leiden University Library contains Clusius' emendations to the published text. Clusius continued to accumulate information about the plants represented in his catalogue after its publication in 1601: the entries grew and changed and, by means of cutting and pasting, were relocated as well (figs. 8, 9).[11] These pages from Clusius' own redacted copy of the *Rariorum* further exemplify the sort of early modern natural history treatise represented by the *Rariorum* and the *Exoticorum* in their published forms. The catalogues of the natural world produced by the fathers of Netherlandish botany and published by Plantin and his heirs were accumulations—more or less elaborately illustrated collections—of information about the natural world. In the texts of both the *Rariorum* and the *Exoticorum,* Clusius made frequent reference to individuals across Europe who had sent him specimens, as well as to his contemporaries in whose collections, homes, or shops he had observed the specimens he described. Clusius' treatises, to

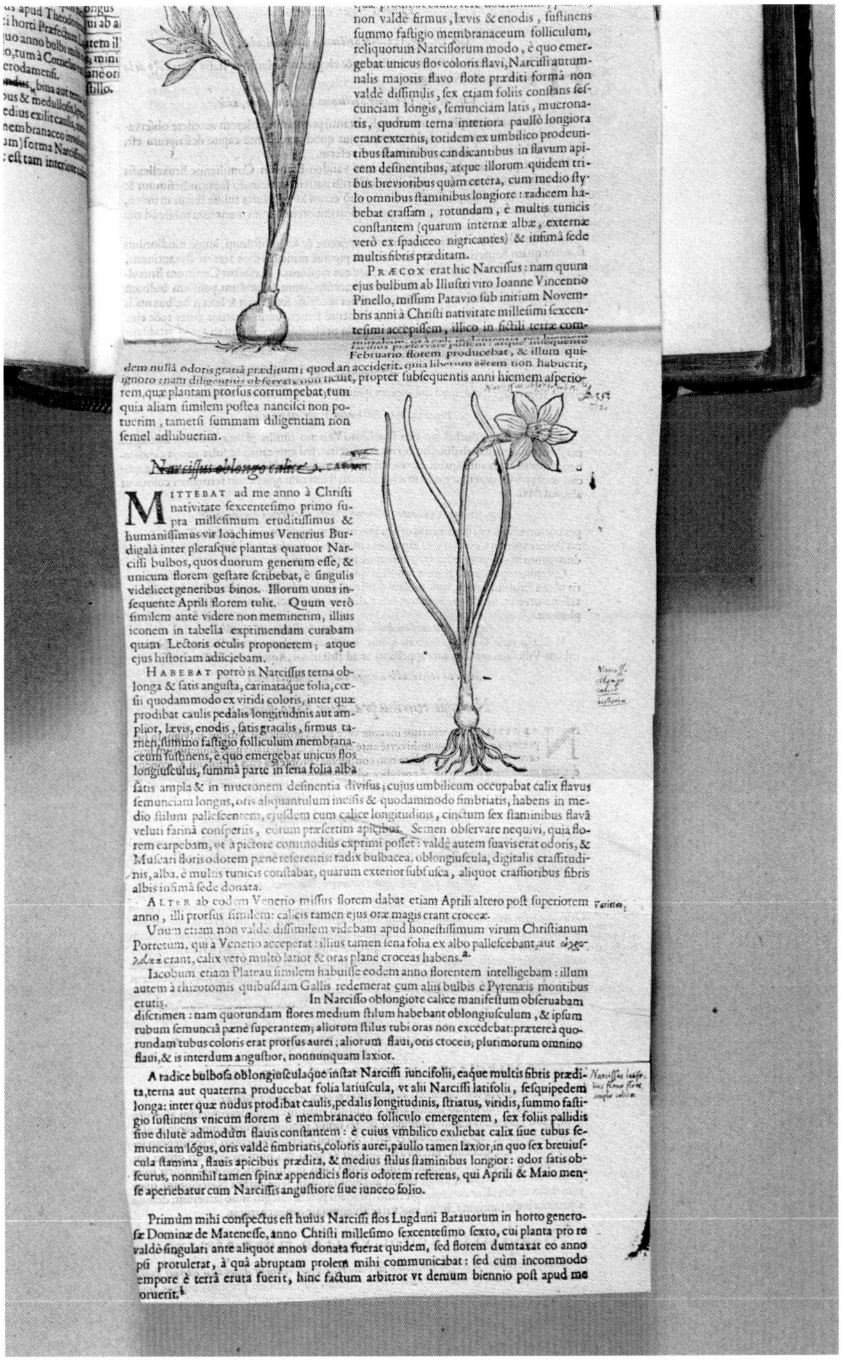

9. Page with printed insert, from Carolus Clusius' personal copy of *Rariorum plantarum historia* (Antwerp, 1601), pages 158–159, woodcut
University Library, Leiden, 755 A3

an even greater extent than those of Lobelius or Dodonaeus, were almost entirely devoid of information about the (medicinal) uses of the individual specimens represented; consider, again, the example of the dwarf tulip. Clusius typically offered descriptions of the morphology and the source or origin of the items he catalogued. Not trained as a medical doctor and largely uninterested in the practice of medicine, Clusius structured his botanical treatises as collections of specimens of the natural world, whose distinguishing characteristics he described and compared. His emendations consist of additions to the record—in the form of units of text and image. By cutting and pasting, he could reassemble entries without substantial changes to their overall structure, which served more or less as a container. The blocks of text and image could be shuffled and new ones inserted, and the overall catalogue would retain its catchall form.

Raphelengius' Codex

Another volume in the Leiden University Library offers a vivid example of a different but related sort of botanical treatise—a more interactive work that served as a resource in the comprehension of the natural world and in the construction of further models of its organization. This codex consists of roughly two hundred pages of woodcuts excerpted from the publications of the great Netherlandish trio of authors on the plant world: Dodonaeus, Lobelius, and Clusius. Here, the illustrations have been cut and pasted in an original order. In some cases the woodcuts are replaced by pen-and-ink and watercolor drawings or accompanied by drawings affixed to pages along with pasted-in woodcuts (figs. 10, 11). Compiled by Franciscus Raphelengius II (Frans van Ravelingen the Younger, d. 1628; grandson of Christophe Plantin), heir to the publisher's works and himself an amateur botanist, this volume is a minor masterwork of cutting and pasting. According to the introductory text in Raphelengius' hand, the volume grew out of his attempt to collate the nomenclature and illustrations used by his predecessors.[12] This codex is a fascinating document for many reasons, not the least of which is that it exemplifies the appropriation of existing botanical treatises by authors and publishers in the early modern era. The slow and steady process of classification carried out here with knife, glue, and pen constitutes an extreme example of scholarly or scientific use of such treatises, in which the process of interactive reading becomes a practice of cutting and pasting. Aside from representing the initial stages of a treatise that

Raphelengius may have intended to publish, the codex was also an exercise in studying the plant world. As he states in the introduction, he "certainly . . . gained a great deal of knowledge about the vegetable kingdom through this exercise."[13]

In some instances, Raphelengius supplemented the woodcuts he garnered from the publications of the three fathers of Netherlandish botany with drawings that must also have been used as the basis (the designs) for woodcuts that were published by Clusius in his *Rariorum* (compare figs. 4 and 11). The genealogy of these images is too complex to unravel fully here, but it is clear that the botanical enterprises represented by Clusius and Raphelengius, respectively, built on a common stock of materials; botany was a cumulative undertaking. Throughout the sixteenth century, publishers reused existing woodblocks in subsequent publications on the plant world; the effort to produce new images for each edition would have been all too time-consuming and expensive. In the particular case of these Leiden books, we may surmise that Raphelengius came into the possession of watercolors that Clusius may have commissioned himself, but in any case these were the designs for the woodcuts that illustrated his treatises.[14] We must bear in mind that a direct association between the two men can be traced via Plantin, Clusius' publisher and Raphelengius' grandfather. There are also instances, in Raphelengius' paste-book, of drawings that appear as woodcuts in Clusius' own redacted copy of the *Rariorum* but were not previously published; they do not appear in other copies of the 1601 volume (see figs. 10 and 12). The correspondences between drawings and woodcuts are too striking to ignore; moreover, for each case in which a relationship can be traced, the woodcut is a mirror image of the drawing, making it all the more likely that the drawings served as models for the prints. Why did Raphelengius include drawings as well as woodcuts in his codex? Conversely, why did he fail to include the woodcuts to which Clusius clearly had access when he revised the *Rariorum*? The answers to both questions may ultimately remain elusive. Still, Raphelengius' use of images is symptomatic of one mode of participation in natural history. By reshuffling existing sources and supplementing them with new ones, Raphelengius aimed to clear a path out of the confusion that had accrued to botanical works of the time. In a handwritten preface to future owners of his work, he declared that he set out to rectify errors and distractions that had resulted from the fact that, in the books his grandfather had published, "Lobelius and Dodonaeus had made use of the same illustrations more than once and often given them the same botanical name, but also often a different and several times a completely opposite name."[15] Where the

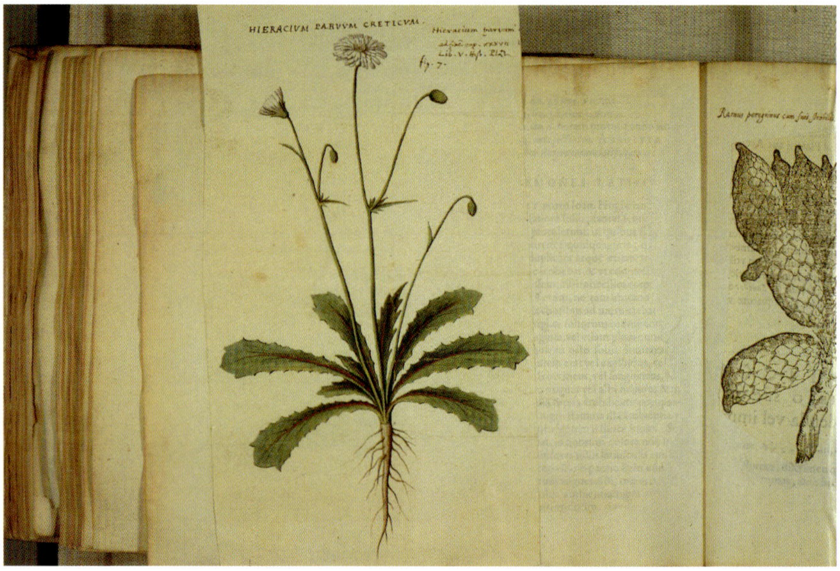

10. Anonymous, *Hieracium parvum creticum* (hawkweed), inserted in Franciscus Raphelengius' codex, c. 1618, watercolor
University Library, Leiden, MS BPL 948, fol. 97v

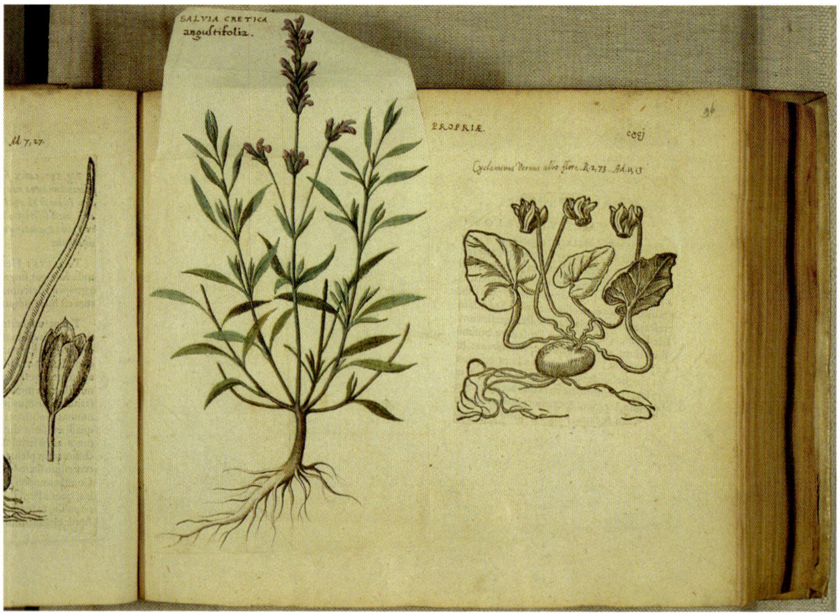

11. Anonymous, *Salvia cretica angustifolia* (aromatic sage), inserted in Franciscus Raphelengius' codex, c. 1618, watercolor
University Library, Leiden, MS BPL 948, fol. 46r

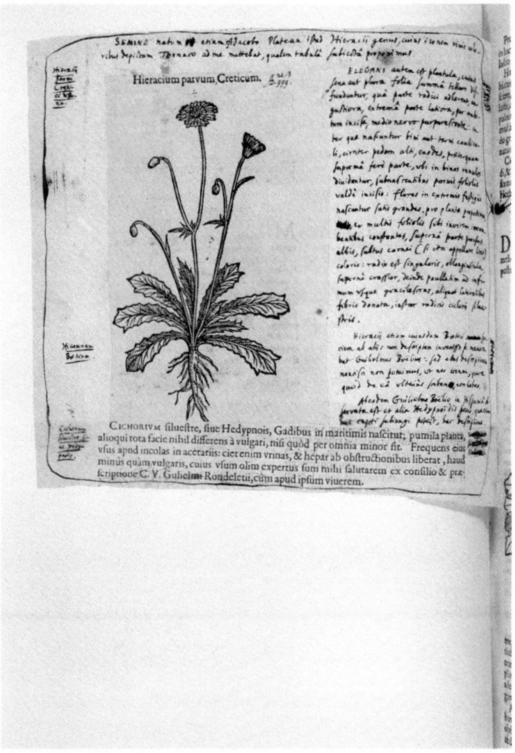

12. Anonymous, *Hieracium parvum creticum* (hawkweed), inserted in Carolus Clusius' personal copy of *Rariorum plantarum historia* (Antwerp, 1601), pages cxlii–cxliii, woodcut
University Library, Leiden, 755 A3

dependence on images had once allowed confusion to build (where single woodcuts were reused to represent multiple varieties of plants, for example), in Raphelengius' hands, that same dependence on images served to clarify such confusion.

The Contributions of German Natural History Writers

The examples of botanical treatises cited thus far all participate in what had, by the turn of the seventeenth century, become a relatively conventional mode of writing natural history. Committed to images for the morphological information they could convey, this empirical endeavor is exemplified by illustrated treatises. As early as the third decade of the sixteenth century, German authors spearheaded the production of illustrated texts aimed to offer as much morphological information as could be conveyed in brief verbal descriptions and woodcut illustrations. Otto Brunfels (1488–1534) was the first of several authors to publish extensive accounts of local flora that included systematically descriptive images.[16] In its organization and text, Brunfels' *Herbarum vivae eicones ad naturae imitationem* (Strasbourg, 1532–1536) differed little from its classical sources (especially Pliny and Dioscorides), but the images—the very subject of the book's title—heralded an entirely new form of engagement with nature. In 1542 Brunfels' compatriot Leonhard Fuchs (1501–1556) published *De historia stirpium,* in which roughly 550 plants were recorded and illustrated (fig. 13). Fuchs outlined his descriptive project in a page-long qualifying subtitle, where he explained that his (verbal) descriptions of the habitats, nature, and medicinal properties of plants were accompanied by the most artful and expressive illustrations, made *ad naturam* (from life).[17] Like Brunfels, who referred to the images of plants in his book as "portrayed with great diligence and artifice," Fuchs advertised the artistic quality of the woodcuts he published. The degree of artistry was closely monitored, however. Fuchs specified that "shading and other less crucial things with which painters sometimes strive for artistic glory" have been discouraged in the interest of making "the

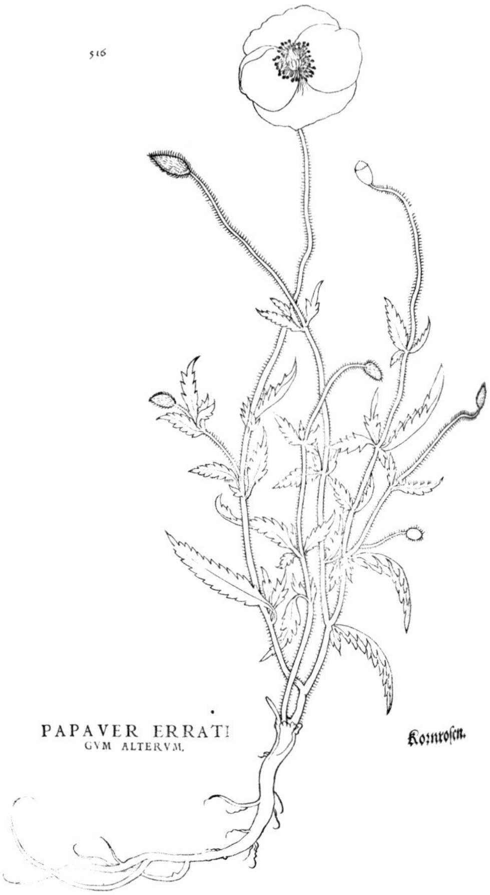

13. Anonymous, *Papaver errati* (poppy), from Leonhard Fuchs, *De historia stirpium commentarii insignes* (Basel, 1542), page 516, woodcut
Newberry Library, Chicago

pictures correspond [more] to the truth."[18] Apparently Fuchs was concerned that signs of individual authorship not occlude faithful transcriptions of the natural world and, indeed, most natural history illustrations of this era were unsigned.[19]

Although complaints about the ability of static images to capture the variable forms of nature continued to be lodged, a characteristically pictorial natural history asserted itself in the sixteenth and seventeenth centuries.[20] These images served a variety of uses, among them description, identification, instruction, substitution for the real item (in collections, for example, where images supplanted unavailable specimens), and elicitation of wonder. Both Clusius' *Rariorum* (like his later *Exoticorum*) and Raphelengius' pastiche codex represent a form of engagement with the natural world that is central to early modern efforts to catalogue it.

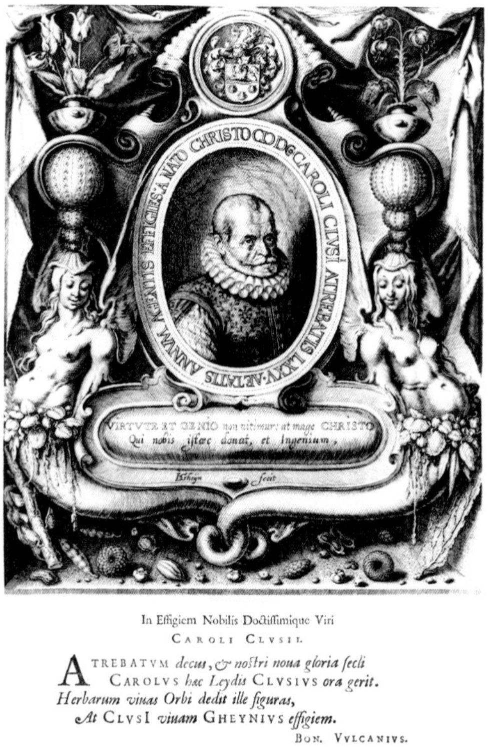

14. Jacques de Gheyn II, *Portrait of Carolus Clusius*, 1601, engraving
Amsterdam University Library (UvA), Special Collections

Distinctions among the Dutch Treatises

While similarities among Dutch treatises around the turn of the seventeenth century abound, distinctions among them can also be drawn. In this author's view, they can best be drawn along practical lines. That is to say, as the remainder of this essay aims to demonstrate, the uses of treatises—in the library as opposed to the field and garden—are captured in their form. Whereas the encyclopedic format of Clusius' (and Plantin's other botanical) publications made them eminently suitable for consultation at a table or behind a desk and encouraged collation with other published sources, and whereas the interactivity encouraged by the Raphelengius volume was solitary, intellectual, and devoted to the task of classification, other coeval treatises attest to a different approach to the forms of nature described.

Clusius, although the first director of the Leiden University garden, was not the university's first choice (fig. 14). The position was first extended to Bernardus Paludanus (Berent ten Broecke, 1550–1633), the prominent physician of the northern port town of Enkhuizen, who was also an avid collector of naturalia. Nevertheless, Paludanus—whose links to overseas trade and whose interest in natural history were renowned throughout Europe—declined the post. Clusius had been informally approached by friends in Leiden who wished him to be nominated, but he put them off by stating firmly in private correspondence that nothing could convince him to take on public responsibility to teach or even to manage a public garden.[21] Meanwhile, a second candidate was discussed by the trustees of the university—Dirck Outgaertszn Cluyt (Theodorus Clutius, 1546–1598), a pharmacist in Delft (fig. 15). Like Paludanus, Cluyt possessed a fine collection of dried specimens, which he would have been required to bring with him to Leiden had he become director of the garden. However, Cluyt's lack of an academic degree prevented the trustees from offering him the position. After lengthy negotiations, Clusius was hired in the fall of 1592, and he moved to Leiden from Frankfurt in the fall of 1593.[22]

The official offer of employment made to Clusius specified that he would be exempt from teaching any formal classes but would be required to give daily lessons in the garden in the summer, "when the plants are vital and fresh," during the course of which he would name the plants and relate their histories and properties. In the winter, he was expected to give lessons twice weekly on "spices, stones, soils, metals, and other things used by doctors."[23] It was suggested that if Clusius found the latter task too imposing, he might share this responsibility with the professors who taught the subject publicly at the time—Pieter Pauw (1564–1617) or Gerard de Bont (1536–1599). In his response to these stipulations, Clusius insisted that he would under no conditions in his sixty-sixth year commence teaching, for he had never taught in his life. He offered to accompany students on botanical walks, since he acknowledged that it was important to study plants in their native habitats, but insisted that he be excused from the responsibility of teaching in the winters since he knew nothing, he wrote, of metals.[24]

Clusius certainly added luster to the Leiden *hortus;* he was in no way, however, the representative of the revived study of materia medica that the trustees and mayors had sought as director of the garden. Throughout his tenure at Leiden, Clusius' efforts were focused on accumulating plants and other artifacts for the collections of the university. Teaching was never his province. Moreover, Clusius' early years in Leiden were particularly trying on account of explicit tensions between him and his colleague Pieter Pauw, professor of anatomy and botany (fig. 16). It may have been, as documents suggest, that Pauw was an arrogant, jealous man and Clusius an unbending crank. But the tension between them was also broader and more intellectual. Clusius was not trained in medicine and adamantly refused to play an active role in teaching it. Determining or teaching the medicinal properties of plants was never foremost among his interests. The title of his *Rariorum plantarum historia,* published at Leiden in 1601, is indicative of his botanical interests, which favored the aesthetic or, at least, morphological qualities of plants over their medicinal properties or uses.[25]

Others who studied the plant world at Leiden—Pieter Pauw and the pharmacist Dirck Cluyt among them—did embody reformed medical study. For Pauw, the Leiden *hortus* served as an arena for teaching—as a *hortus medicus* or *hortus academicus.* The difference between Clusius' study of the plant world and Pauw's engagement in it is borne out in documents, in images, and in treatises. As their roles evolved, it became painfully obvious that Clusius would not carry out any of the day-to-day operations of the garden, and in 1594 the second candidate for the directorship—Dirck Cluyt—was hired as *hortolanus,* "to promote and maintain the university garden" and "to serve and further medical studies."[26] Cluyt, like Pauw, was vitally engaged in the cultivation of and instruction in the materia medica,

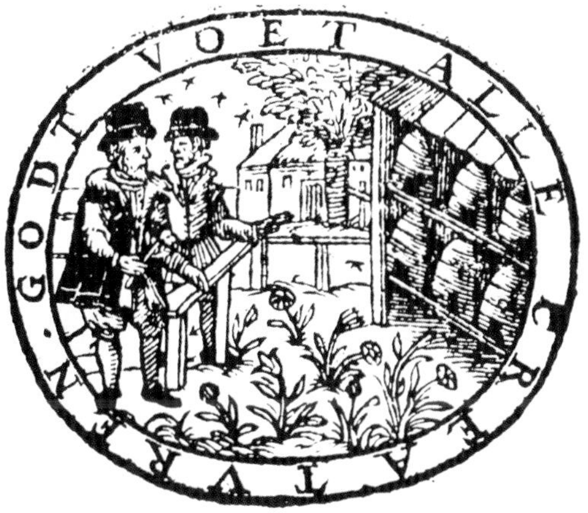

15. Anonymous, vignette showing Dirck Outgaertszn Cluyt and Carolus Clusius in the Leiden Garden, from Dirck Outgaertszn Cluyt, *Van de byen, haer wonderlyke oorspronk, natuer, eygenschap, krachtighe, onghehoorde en seltsame werken* (Leiden, 1597), frontispiece, woodcut
National Herbarium of the Netherlands, Leiden

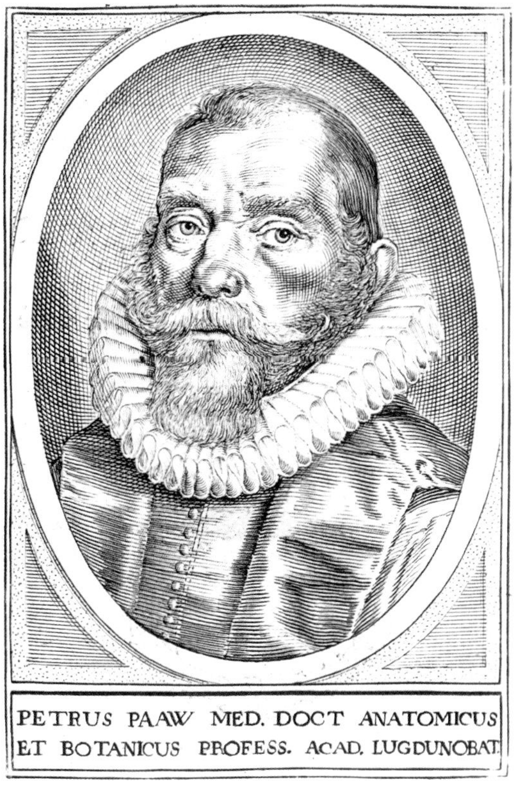

16. Anonymous, *Pieter Pauw,* from *Succenturiatus anatomicus* (Leiden, 1616), engraving
Amsterdam University Library (UvA), Special Collections

while Clusius acted as figurehead for this new arena of medical instruction. Clusius' orientation differed effectively from that of the agents who mediated the garden's use. In turn, these differences are reflected in the botanical treatises produced by and for these various agents.

A number of volumes of watercolors were used in the course of medical instruction at Leiden in the last decade of the sixteenth century. Through such use, students learned to identify plants for their medicinal purposes. Among these volumes were botanical watercolors owned by Cluyt, who was prefect of the university garden from 1594 until his death four years later.[27] A document submitted by the medical students at Leiden to the trustees (*curatoren*) of the university in 1598 cites "six painted books of all sorts of herbs and flowers [painted] from the life, which serve us instead of the garden in winter." In the wake of Cluyt's sudden and premature death in 1598, the students rallied to ensure that the volumes would remain accessible to them. The students' formal request detailed the potential loss of valuable academic resources resulting from Cluyt's death. Referring to Cluyt as "our most sedulous and intelligent herbalist . . . a very clever, experienced, and diligent simplicist necessary to this university and the likes of which are known to no other," they requested that the university name Cluyt's eldest son, Outgaert, who was only twenty-one at the time, as his father's successor. By doing so, the university would be able to preserve the collections and the volumes of images as well as "numerous chests with at least 4,000 *simplicia* [simples], which [Outgaert's] father had carefully collected throughout his entire life up to the age of fifty-two."[28] These items, the students stated, were in the home of the deceased, and his widow intended to pass them on to Outgaert in order to help secure her son's professional success. Along with this formidable collection of dried specimens, the volumes of images—the winter garden—were deemed indispensable to the course of study. The students' request to the trustees indicates that Cluyt's stock of watercolors was used to teach the identification of plants during the winter months, when the renowned botanical garden was dormant. In this respect, the watercolors shifted the terms of the standard medical curriculum, conventionally divided between *lectura* (reading) and *ostensio* (demonstration), so that texts were studied in one season to prepare students for practical exercises in the presence of actual specimens in another (i.e., anatomical texts were studied in warmer months and cadavers in the winter; botanical texts were the subject of winter classes, and live plant specimens were studied in the garden during warmer months).[29] By dint of substitution, Cluyt's watercolors of plants and flowers provided information about specimens whose availability was otherwise limited by the cycle of the seasons and the whims of weather. The images effectively bridged the distance between the study of pharmacological texts and the study of nature, between *lectura* and *ostensio*. The text of the students' petition has substantial bearing on a broader discussion of the practice of medical study at the close of the sixteenth century, as well as on the uses of naturalistic representation at this point in its history. It also has particular significance for the history of the role of images within scientific curricula during this period.

Elsewhere, I have argued that the watercolors Cluyt is known to have owned may be identical with a body of works currently housed in the Jagiellon University Library, in Kraków (essay frontispiece, and figs. 17 and 18).[30] These watercolors, known as the Libri Picturati watercolors by virtue of their longtime classification under that rubric, are extraordinary documents of early modern investment in naturalistic pictorial records of the natural world. The drawings in the Libri Picturati represent a wide array of specimens of the plant world, ranging from ferns, mosses, and lichens to fruit and nuts (in both fresh and dried form), and from spices (such as those sixteenth-century European imports that inspired wonder and pharmaceutical speculation) to ornamental plants, many of them bulbous, that were soon to become all the rage among collectors and gardeners. Varieties of orchids, lilies, and narcissi are depicted, as are vegetables and fungi, grasses and trees.[31]

The plants are attentively rendered in color, and their forms are organized in such a way that their various surfaces—the backs

17. Anonymous, *Paeonia femina* (*Paeonia officinalis*; peony), c. 1590, Libri Picturati, A.25, folio 59, watercolor
Jagiellon University Library, Kraków

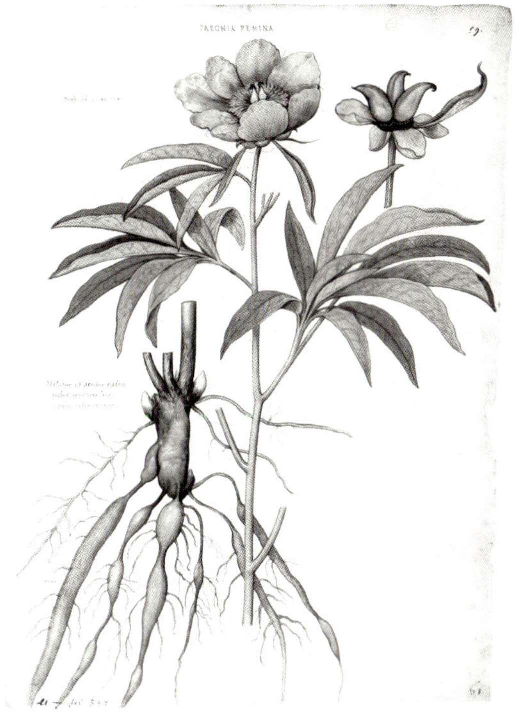

depicted are cut so that the entire plant can be shown and drawn to scale. In many cases, then, the drawings mimic the compositions of dried herbals, where stems are bent or cut in order to fit a given specimen onto a page; they are eminently legible insofar as there is little loss of figures into rendered depth and portions of plants are not obscured by others. In all, more than six hundred specimens are painstakingly labeled and inscribed.

With some exceptions, these drawings were never published. (A very few are identical to woodcuts that appeared in Plantin publications.) Internal evidence suggests that if they were not intended solely for instructional purposes, they nonetheless suited that purpose ideally. A significant number of the Libri Picturati watercolors bear inscriptions, and many of them are in at least two hands (though they may well have been written by a single individual). These inscriptions can be separated into four sorts, each of which dates to a different period. What concern us here are the first sort—in what has been called a "professional" hand—which record the names of the plants; and the second sort—in the same hand—which record citations of published

and fronts of leaves and the variegated forms of stems and petals—are fully visible.[32] Nearly without exception, the root structures are described as carefully as any of the other features of the plants, and are shown in full. The roots are firm and tuberous in some instances, and in others wispy, fibrous root elements dangle or spread, according to their actual growth patterns. In some cases, as on A.22.71 and A.22.72 (*Iris silvestris* and *Iris maritima*), the tuber is represented as if it had been cut, which serves to show the interior structure and color of the root. In a number of cases, plants are depicted within the same image at various stages of maturity. For example, two sorts of cherries, *Cerasa acida* and *Cerasa dulcia*, are each represented by a substantial portion of a branch, cut to show its inner structure, which bears fruit (A.20.80v). Separated by a few millimeters from the fruit-bearing branches are twigs in blossom. The means by which a plant is represented in both its flowering and its fruit-bearing stages is employed consistently; a barely perceptible break in a branch separates it from the other portion of the plant, shown at a different stage of maturation. One additional feature of these drawings that merits comment is that many of the stems of the plants thus

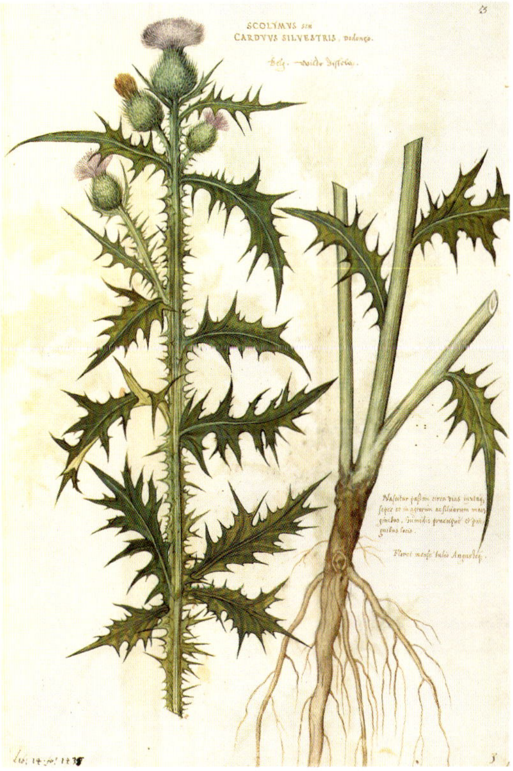

18. Anonymous, *Carduus sylvestris* (*Carduus crispus*; creeping thistle), c. 1590, Libri Picturati, A.27, folio 41, watercolor
Jagiellon University Library, Kraków

authors relevant to the plants depicted. The names of the plants depicted are recorded in majuscules at the upper center of the sheets in Greek or Latin or both. Directly below, the abbreviations for a variety of European languages (Italian, French, German, and Flemish being the most common) are followed by the plant names in those languages. In a number of cases there are also inscriptions—generally located in the lower portion of the sheets—that describe the natural habitats of the plants depicted and indicate when they flower or seed. These inscriptions are, again, in the same hand as the names and the citations. Frequently, these lengthier inscriptions specify the relative scarcity or abundance in the North of the plants depicted, contain references to "our gardens," and indicate that the plants depicted were the subject of study by medical students. Oddly enough, the fact that these inscriptions include references to "our gardens" or that any of them includes a reference to medical students has been passed over in silence by scholars.[33]

The authors cited by book and section number in the "professional" hand of the first sort of inscriptions are classical—Dioscorides, Galen, Theophrastus, Pliny. In the later sixteenth century, these classical authors were studied most conscientiously by those engaged in the reform of pharmaceutical practice. Moreover, the lengthier inscriptions that accompany many of the sheets (which specify where the plants depicted grow and mature) seem to indicate that these images were structured to accommodate a specific form of study. The information (both visual and verbal) they convey conforms to the kind that authors such as Clusius worked to compile. However, there is no extensive cross referencing to contemporary authors, nor are broad taxonomic issues addressed in the inscriptions. The images and their inscriptions simply serve to identify the specimens described and, in so doing, would have made them ideally suited to direct consultation. In my judgment, the form of study that these sheets accommodated was the same form practiced by members of the medical community, whose interests ranged from the strictly pharmaceutical to the incipient botanical. The consistent references to "the gardens" or "our gardens" in the inscriptions suggest that the core group of watercolors under discussion represented a local project. They are at once the record of the contents of these gardens (or garden, if the collective plural can refer to various plots of a single garden) and the record of a broad selection of natural specimens. They are structured in such a way that medical students, for example, whose study of pharmacology necessarily involved the study of the identification of plants and of the stages of their maturation, could indeed make use of these images as a virtual garden.

The *Hortus publicus academiae Lugduno-Batavae*

The single horticultural or botanical publication authored by Pieter Pauw, *Hortus publicus academiae Lugduno-Batavae* (1601), is also symptomatic of a close relationship between a particular kind of treatise and the use of the garden at Leiden.[34] It is, perhaps, a stretch to refer to the *Hortus publicus* as a treatise, for the only text it contains is a preface to its users. Pauw's book otherwise consists of pages upon pages of diagrams made up of blank rectangular boxes (fig. 19). In his preface, Pauw explains that students of plants were to take the small book with them to the Leiden garden and to fill in the rectangles with the names of the plants growing in the corresponding plots of the garden. Like the catalogue of the Padua garden after which it was modeled, the *Hortus publicus* is an open catalogue of the contents of the garden: the series of printed, numbered quadrants corresponds to the quadrants of the garden as it was planted in 1601. The engraved plan of the Leiden garden by Jacques de Gheyn II (1565–1629) (see fig. 1) was commissioned by Pauw in 1600 and, folded in quarters, was inserted into copies of the *Hortus publicus*.[35] According to Pauw's preface, the book was to be used by students, among others, who could fill in the names of the plants growing in the garden as they were demonstrated during the course of instruction. The print of the garden by de Gheyn functions as a highly legible plan of the garden as it appeared in 1601. This image would have served as a key for the series of printed rectangles within the body of the book that were to be filled in by its readers and users. Active phytographs translated the users' experience of the plots (which de Gheyn represented in his bird's-eye view) into the spatially coordinated charts, or tables, of the plants contained in the *Hortus publicus*.[36]

The *Botanologicon*

By way of conclusion, and to further evoke what was involved in early modern study of the plant world and how botanical treatises were used, it is important to examine the *Botanologicon,* by the German botanist and physician Euricius Cordus (Heinrich Ritze; 1484–1535), published in Cologne in 1534. Cordus offers a brief but telling description of a fundamental and widely practiced approach to observing the plant world. The mode of observation he evinces informs, in ways I have attempted to illustrate in the foregoing, the production and use of botanical treatises and botanical knowledge at Leiden around 1600. Cordus' *Botanologicon* is a lovely and largely ignored source of incisive commentary on how medical botany was practiced and what was at stake in its practice. Much of the text is a description of a botanical expedition that the individual study of plants and the particular form of attention "good botany" should cultivate; throughout the text, Cordus is viciously critical both of arrogant medical doctors and of unlearned medical practitioners. In addition, the *Botanologicon* contains a number of references—both explicit and implicit—to the role of images in the practice of botany.[37] Composed as a colloquy between the author and four fellow medical students, it opens at Cordus' home, and, after some general discussion, the team of friends sets out to "botanize." Cordus encourages them outdoors, noting that he

> . . . will follow [his] usual practice, just as if none of you were here, and take along a book or two. I take great pleasure in going into the countryside, and in comparing all sorts of herbs and plants that grow in various locales and about which I have read at home, with the images stored in my memory and observing them; and sometimes I am able to ask their

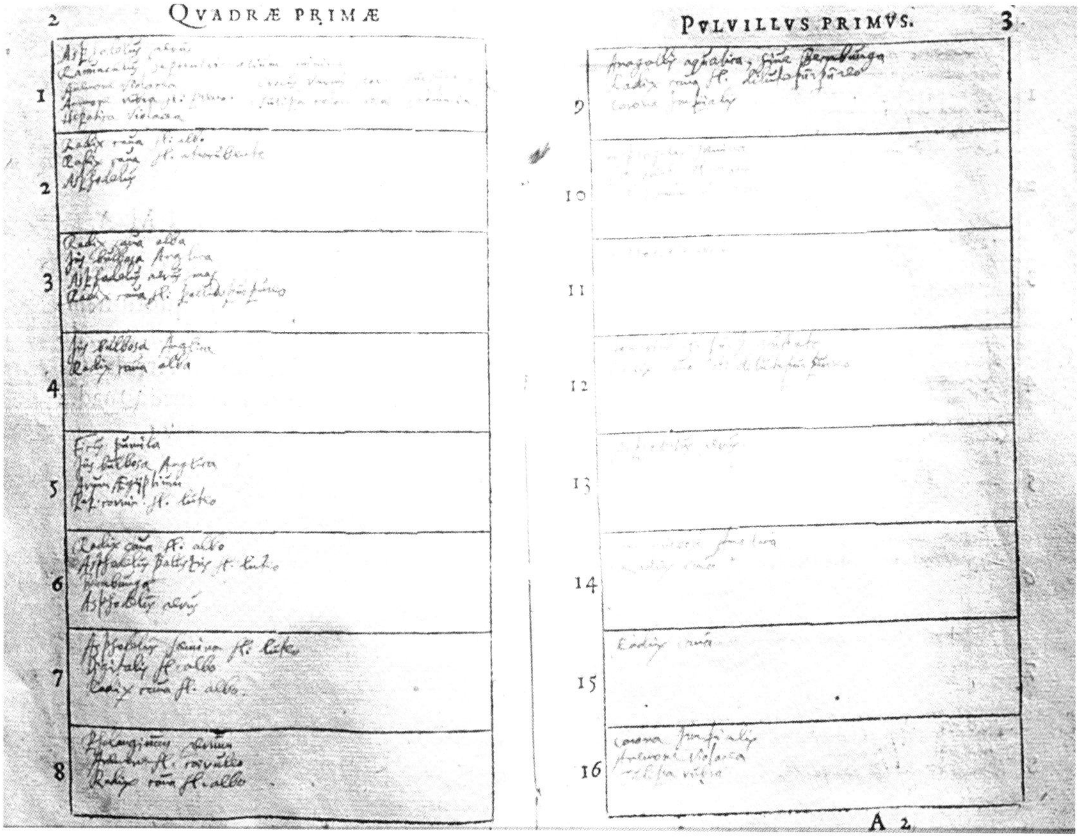

19. Anonymous, diagram from Pieter Pauw, *Hortus publicus academiae Lugduno-Batavae* (Leiden, 1601), pages 52–53, engraving
National Herbarium of the Netherlands, Leiden University Branch

properties or their names from the old wives I meet along the way. On this basis—after comparing all of them with their descriptions—I am the better able to judge them clearly and come to as accurate a conclusion as possible about them. . . . If it please you, we will first enter this my little garden by the house. . . .[38]

Cordus goes out into the fields with book in hand (note the privacy or absorption characteristic of this practice, implicit in his saying that he, "will follow [his] usual practice, just as if none of you were here." By way of comparing his reading of such texts and the pictures of their contents, which he carries in his memory, with the actual plants he encounters, and by using these treatises as a point of departure for gathering more information en route ("their properties or their names from the old wives I meet along the way"). Cordus effectively summarizes the use of such treatises in the accumulation of knowledge of the plant world, by means of virtual and actual cutting and pasting of information. "Botanizing," or "herborizing," was essential for sixteenth-century naturalists. It came to be practiced in botanical gardens and in the presence of herbaria (collections of dried plants) as well. At its heart lay the autoptic experience of nature and the process of learning it by collating one's experience with one's prior knowledge—and, indeed, with images—of the plants at hand. Cordus offers one account of the technique: texts were juxtaposed with fresh herbs, images of plants, and information gleaned from those, such as herbalists and shepherds, who plied their trade in the woods and fields. By assimilating and processing this information, the assiduous botanist worked to identify the specimens he or she might encounter. The ends of such identification were first and foremost medical, which is to say pharmaceutical. One needed, in Cordus' view as in the views of many of his contemporaries, to know the makings of medicine in order to practice it. In the pursuit of knowledge of the natural world and the ability to distinguish its elements and their properties, images such as those featured in so many publications of the time would have played a crucial role. Indeed, Cordus and his friends take along a "Dioscorides minor" and two volumes of Brunfels—which must be the first two volumes of Otto Brunfels' *Herbarum vivae eicones*. Moreover, Cordus describes a mode of observation whose primary arena is the garden: he and his fellow students walk out into the garden before heading out into the fields to go botanizing.

The Leiden garden was the nexus of botanical study and representation in the Netherlands because it fostered observation of a sort that is closely aligned with the production of books and other volumes on the plant world, and because the authors and purveyors of those treatises were affiliated with it. What I hope to have demonstrated in this essay is that these records of the observation of the plant world bear the traces of practices and concerns—distinct, but not incompatible, interests. The uses of botanical treatises—whether printed volumes by Clusius or volumes of watercolors owned and deployed by Cluyt—are clearly various. And where use depends on composition (in ways that I have only begun to map out), where function and form seem to follow from one another, it is especially important to attend to the shifting priorities and uses represented by the individuals discussed here.

The longer history of pictorial natural history in print began in the late fifteenth century in Germany and came to fruition in the publications of Brunfels and Fuchs. We have seen that the fathers of German botany set an example amply expanded by their later Netherlandish counterparts. Around the turn of the seventeenth century in Holland, a number of key aspects of the cumulative enterprise that would come to be called botany were in active play. The genealogy of plant histories such as those authored by Dodonaeus, Lobelius, and Clusius—the so-called fathers of Netherlandish botany—cannot be written without acknowledging the central role of Christophe Plantin. As we have seen, however, his work bore on unpublished treatises (such as those produced by his grandson Raphelengius) as well. The practice of botany appears to have fostered the production of similar sorts of books, whether published or unpublished. To a large extent, it is the role of images in the context of descriptive accounts of the plant world that remains consistent in spite of the changing context, from such published treatises as Clusius' 1601 and 1605 volumes, the *Rariorum* and the *Exoticorum*, to the paste-books assembled later by Raphelengius. Woodcuts and drawings alike served the purposes of contemporary efforts at taxonomy in standing in for the subjects represented. The mode of observation recorded in and supported by Pieter Pauw's *Hortus* and described earlier by Cordus in his *Botanologicon* involved a relay between autoptic experience and accumulated knowledge (conveyed in the form of texts) for the ends of identification. For all of the stability of these patterns, which informed the production and the use of natural history treatises around the turn of the seventeenth century, there are also important distinctions—not necessarily of kind, but certainly of degree. The uses of botanical treatises in the Netherlands around 1600 pertain to broader conventions but at the same time reflect local and divergent concerns: insofar as these concerns shaped the nascent field of botany, the treatises invoked here speak volumes—not just about plants, but about their respective uses as well.

NOTES

1. For general information on early modern printed botanical treatises, see Agnes Arber, *Herbals, Their Origin and Evolution: A Chapter in the History of Botany, 1470–1670*, ed. William T. Stearn (1912; 3rd rev. ed., Cambridge, 1986); Wilfred Blunt and William T. Stearn, *The Art of Botanical Illustration, An Illustrated History* (1950; reprint New York, 1994); and David Landau and Peter Parshall, *The Renaissance Print: 1470–1550* (New Haven, 1994), especially "Printed Herbals and Descriptive Botany," 245–259. On Netherlandish publications, see *Botany in the Low Countries (End of the 15th Century–ca. 1650)* [exh. cat., Plantin-Moretus Museum] (Antwerp, 1993).

2. On "the epistemological precedence experienced by botany," see Michel Foucault, *The Order of Things: An Archaeology of the Human Sciences* (New York, 1970), 137; and F. David Hoeniger, "How Plants and Animals Were Studied in the Mid-Sixteenth Century," in *Science and the Arts in the Renaissance*, ed. John W. Shirley and F. David Hoeniger (Cranbury, N.J., 1985), 130–148, especially 137. On the shift from pharmacology to botany, see Arber 1986; the excellent work by Armand Louis, *Geschiedenis van de Plantkunde: Eerste Periode, De Plantkennis in dienst van Farmacie en Landbouw, Van Oudheid tot Renaissance* (Ghent, 1977; Antwerp, 1993); Alan G. Morton, "Renascence of Botany in Europe: From Herbal to Flora (1483 to 1623)," chap. 5 in *History of Botanical Science* (London, 1981); Karen M. Reeds, *Botany in Medieval and Renaissance Universities* (New York, 1991); and Jerry Stannard, "The Herbal as a Medical Document," *Bulletin of the History of Medicine* 43 (1969): 212–220.

3. William M. Ivins, Jr., *Prints and Visual Communication* (1953; reprint Cambridge, 1969), 3. Without prints, Ivins writes, "we should have very few of our modern sciences, technologies, archaeologies, or ethnologies—for all of these are dependent, first or last, upon information conveyed by exactly repeatable visual or pictorial statements." See also Louis Marin, "Mimésis et description: Ou de la curiosité à la méthode de l'âge de Montaigne à celui de Descartes," in *Documentary Culture: Florence and Rome from Grand Duke Ferdinand I to Pope Alexander VII*, ed. Elizabeth Cropper, Giovanna Perini, and Francesco Solinas (Bologna, 1992), 23–47. Even though the images employed in the majority of these efforts changed markedly in character over time, becoming relatively elaborate when engraved rather than simply reproduced as woodcuts, they were all purported to have been done *ad vivum*. This claim had already been made in the text of the early *Gart der Gesundheit* (Mainz, 1484); these woodcuts are vastly more schematic than those published by Brunfels and his followers, all of whom invoke the same phrase. See Landau and Parshall 1994, 245–246; and Claudia Swan, "*Ad vivum, naer het leven:* From the Life; Considerations on a Mode of Representation," *Word & Image* 11 (1995): 353–372.

4. On the Leiden University garden, see Hesse Veendorp and Lourens Gerhard Marinus Baas-Becking, *Hortus Academicus Lugduno Batavus 1587–1937: The Development of the Gardens of Leyden University* (Haarlem, 1938); W. K. H. Karstens and Herman Kleibrink, *De Leidse Hortus: Een Botanische Erfenis* (Zwolle, 1982); *The Authentic Garden: A Symposium on Gardens*, ed. Leslie Tjon Sie Fat and Erik de Jong (Leiden, 1991), especially Erik de Jong, "Nature and Art: The Leiden Hortus as 'Musaeum,'" 37–60. See also Claudia Swan, "Medical Culture at Leiden University ca. 1600: A Social History in Prints," *Nederlands kunsthistorisch jaarboek* 52 (2002): 216–239.

5. See Reeds 1991; Nancy Siraisi, *Avicenna in Renaissance Italy: The Canon and Medical Teachings in Italian Universities after 1500* (Princeton, 1987); Paula Findlen, *Possessing Nature: Museums, Collecting, and Scientific Culture in Early Modern Italy* (Berkeley, 1994), esp. 241–287; and Brian Ogilvie, *The Science of Describing: Natural History in Renaissance Europe* (Chicago, 2006).

6. The standard reference on Clusius remains Friedrich Wilhelm Tobias Hunger, *Charles de l'Escluse (Carolus Clusius): Nederlandsch kruidkundige 1526–1609*, 2 vols. (The Hague, 1927 and 1943). See also Gyula Istvanffi, *Études et commentaires sur le code de L'Escluse, augmentés de quelques notices biographiques* (Budapest, 1900); Henriette Bosman-Jelgersma, "Clusius en Clutius," *Farmaceutisch tijdschrift voor België* 58 (1981): 41–45; Florence Hopper,

"Clusius' World: The Meeting of Science and Art," in Tjon Sie Fat and de Jong 1991, 13–36.

7. See Claudia Swan, "From Blowfish to Flower Still Life Painting: Classification and Its Images ca. 1600," in *Merchants and Marvels: Commerce, Art and the Representation of Nature in Early Modern Europe*, ed. Pamela Smith and Paula Findlen (New York, 2002), 109–136; and Brian Ogilvie, "The Many Books of Nature: Renaissance Naturalists and Information Overload," *Journal of the History of Ideas* 64 (January 2003): 29–40.

8. More generally, on the representation of variable qualities of plants, see also David Freedberg, "The Failure of Colour," in *Sight and Insight: Essays on Art and Culture in Honour of E. H. Gombrich at 85*, ed. John Onians (London, 1994), 245–262; and Sachiko Kusukawa, "Leonhart Fuchs on the Importance of Pictures," *Journal of the History of Ideas* 58 (1997): 403–427.

9. Carolus Clusius, *Rariorum plantarum historia* (Leiden, 1601), book 2, 147, as translated by W. Van Dijk, *A Treatise on Tulips by Carolus Clusius of Arras* (Haarlem, 1951), 50.

10. See Swan 2002 and Ogilvie 2006.

11. Leiden University Library, MS 755 A3. Clusius may well have intended to publish a revised edition; *Curae posteriores* (Antwerp, 1611), published two years after his death, included previously unpublished material relevant both to his *Rariorum* and to his *Exoticorum libri decem: quibus animalium, plantarum, aromatum, aliorumque peregrinorum fructuum historiae describuntur* (Leiden, 1605).

12. Leiden University Library, MS BPL 948. To the best of my knowledge the volume has been cited in only one publication to date, namely Antwerp 1993, cat. no. A1, 86–87.

13. As cited in Antwerp 1993, 86.

14. The pamphlet catalogue of Clusius' library, auctioned in Leiden after his death in 1609, contains a notice that, aside from the books, "maps [and] drawings in vivid colors of plants, flowers, fungi, fruits, animals, etc." would be sold: "Monitio ad Lectorem. Peracta Librorum auctione, in ijsdem aedibus [Pauli Stochij] habebitur auctio Plantarum rariorum hortuli Caroli Clusij; qua etiam vendentur Cartae Geographicae, designationes plantarum, florum, fungorum, fructuum, bestiarum &c. vivis coloribus: olea item & liquores partim naturales, partim artificiales: Supellex praeterea numismatum, & aliquot manufactorum peregrinorum, & imprimis fructuum, radicum, seminumque Exoticorum: mineralium insuper complurium: terrarum sigillatarum: & permulta alia similis curiositatis." *Catalogus librorum bibliothecae clarissimi viri Caroli Clusii . . .* (Leiden, 1609), folio 4r.

15. Leiden University Library BPL 948, verso of title page (original in Latin).

16. Otto Brunfels, *Herbarum vivae eicones ad naturae imitationem, summa cum diligentia et artificio effigiatae, una cum effectibus earundem, in gratiam veteris illius, & jamjam renascentis herbariae medicinae. Quibus adjecta ad calcem, Appendix isagogica de usu & administratione simplicium* (Strasbourg, 1532–1536), 3 vols. (sometimes bound as one volume). The inclusion of roots, the description of surface texture, and the effort to show leaves and flowers from a variety of angles distinguish Brunfels' work from all previous publications of the sort. Emphasis on the specific characteristics of individual specimens and the inflections of shadows were criticized, and avoided in later publications, demonstrating that the literal demands of working from life were tempered by the necessity of communicating visual information. See Landau and Parshall 1994, 254–255.

17. *De historia stirpium commentarii insignes, maximis impensis et vigiliis elaborati . . .* (Basel, 1542). For Fuchs and his publications, see Frederick G. Meyer, Emily Emmart Trueblood, and John L. Heller, eds., *The Great Herbal of Leonhart Fuchs: De historia stirpium commentarii insignes, 1542 (Notable Commentaries on the History of Plants)*, 2 vols. (Stanford, 1999); Brigitte Baumann, Helmut Baumann, and S. Baumann-Schleihauf, *Die Kräuterbuchhandschrift des Leonhart Fuchs* (Stuttgart, 2001); and Thomas A. Sprague and Ernest Nelmes, *The Herbal of Leonhart Fuchs* (London, 1931).

18. Fuchs 1542, folio 7v.; see also Arber 1986, 206.

19. James Ackerman suggests that Fuchs "allowed or encouraged" the three illustrators to include their self-portraits "in compensation" for restraining "the urge . . . to express themselves at the cost of accuracy"; see Ackerman's "Early Renaissance 'Naturalism' and Scientific Illustration," in his *Distance Points: Essays in Theory and Renaissance Art and Architecture* (Cambridge, 1991), 185–207.

20. David Topper, "Towards an Epistemology of Scientific Illustration," in *Picturing Knowledge: Historical and Philosophical Problems Concerning the Use of Art in Science*, ed. Brian Baigrie (Toronto, 1996), 215–249.

21. Clusius wrote: "Nulla ratione adduci possum ut publicum docendi munus, vel adeo publici horti curam suscipiam." As cited in Philip C. Molhuysen, *Bronnen tot de geschiedenis der Leidsche universiteit*, 7 vols. (The Hague, 1913–1924), 1:193*; see also 70. On Paludanus, see esp. Friedrich Wilhelm Tobias Hunger, "Bernardus Paludanus (Berent ten Broecke), 1550–1633: Zijn verzamelingen en zijn werk," in *Itinerario: Voyage ofte schipvaert van Jan Huygen van Linschoten naer oost ofte Portugaels Indien, 1579–1592*, ed. Hendrik Kern, 8 vols. (The Hague, 1910–1957), 3 (1934): 249–268; and Florike Egmond, "Een mislukte benoeming: Paludanus en de Leidse universiteit," in *Souffrir pour parvenir: De wereld van Jan Huygen van Linschoten*, ed. Roelof van Gelder et al. (Haarlem, 1998), 51–64.

22. On Clusius' appointment, see principally Hunger 1927, 1:166.

23. Molhuysen 1913–1924, 1:205*.

24. Interestingly, Clusius argued the importance of this kind of botanical study not on medicinal grounds but because, he said, one must know where plants are to be placed in a garden when transplanted. Molhuysen 1913–1924, 1:231*.

25. See Hopper 1991.

26. Molhuysen 1913–1924, 1:294*. The document begins: "Ten opsichte van des voors. D. Clusii zwackheyt, ongesteltenisse ende onvermogentheyt als om andere redenen ende insichten den selven 't behulp van een goet naerstich ende getrou persoon heeft moeten werden bijgevoegt. . . ." See also Bosman-Jelgersma 1981, 41–45.

27. On Dirck Cluyt, see Henriette A. Bosman-Jelgersma, "Dirck Outgaertszn Cluyt," *Farmaceutisch Tijdschrift voor België* 53 (1976): 525–548; and, by the same author, "Dirck Cluyt: De eerste Leidse hortulanus," *Jaarboekje voor geschiedenis en oudheidkunde van Leiden en Omstreken* 83 (1991): 75–88. The watercolors, now in the Jagiellon University Library in Kraków (Libri Picturati A.16–A.30), are the subject of Claudia Swan, "Lectura-Imago-Ostensio: The Role of the Libri Picturati A.16–A.30 in Botanical Instruction at the Leiden University," in *Natura-Cultura: L'interpretazione del mondo fisico nei testi e nelle immagini,* ed. Giuseppe Olmi, Lucia Tongiorgi Tomasi, and Attilio Zanca (Florence, 2000), 189–214; Claudia Swan, *The Clutius Botanical Watercolors: Plants and Flowers of the Renaissance* (New York, 1998); and Luis Ramón-Laca, "Charles de l'Écluse and the Libri Picturati A.16–30," *Archives of Natural History* 28 (2001): 195–243.

28. The relevant portion of the document reads: "Booven seeckerheyt van deze menichvuldighe kennis hebben wy ten huise van sijn [Outgaert's] vader gevonden veel kaskens met wel ruym 4000 simplicia, die sijn vader [Dirck Outgaertszn Cluyt] door sijn gants leeven tot 52 jaer toe neerstich vergadert heeft, sonder eenighe kosten daer aen te sparen, de welcke hy soo wel verstaet als sijn vader selfs deede, voorts oock ses geschilderde boucken naer t leven van alderhande cruiden en bloemen, dewelcke ons in den winter in plaetse van den thuin wesen mochte, twelck de moeder ende broeders al voor Augero bewaren willen, verhoopende hem daermede elders tot eeren te brengen." Molhuysen 1913–1924, 1: bijlage nr. 325, 380*.

29. On the seasonal division of the curriculum, see Swan 2002.

30. See Swan 1998 and Swan 2000.

31. The present order of the folios is not original; many of the sheets are also numbered in brown ink in the upper and lower right-hand corners.

32. The rendering of these sheets is more or less consistent and, pending further research, may be attributed to two or, at the most, three hands. None of the sheets of botanical specimens is signed.

33. See Swan 1998 and Swan 2000.

34. The extended title of the small book is *Hortus publicus academiae Lugduno-Batavae eius ichnographia, descriptio, usus. Addito quas habet stirpium numero, & nominibus Operâ Petri Pawi Horti Praesidi.* A copy Pauw presented to the university, in which he himself recorded all of the plants present in the garden in February 1601, is in the collection of Leiden University (Hortus Botanicus). In 1603, a revised edition of the catalogue was printed in Leiden by Ioannes Patus (Ex. Officinâ Ioannis Patii, Academ. Lugduno-Bat. Typographi); the page size is smaller and there are minor changes to the text. The catalogue contains a total of 176 pages. See Swan 2002.

35. See Swan 2002. That de Gheyn's engraving of the garden was intended for inclusion in the book when it was first published is clear from a marginal note in the 1601 edition that reads, "ad ea quæ sequuntur, inspicienda erit Horti ichnographia, inserta pagina" (folio 7v); this marginal note does not occur in the 1603 edition.

36. When Jan Janszn Orlers, in his *Beschrijvinge der stadt Leyden* (Leiden, 1614), describes the university garden, de Gheyn's print serves as the point of reference for its dimensions and layout, even though the text is illustrated with a reduced copy of an engraving by Willem Swanenburgh after Jan Corneliszn Woudanus (fig. 2). In effect, de Gheyn's plan became the plan by which the garden was known. For the reduced version, see Orlers 1614, 143–144.

37. Peter Dilg, *Das Botanologicon des Euricius Cordus: Ein Beitrag zur botanischen Literatur des Humanismus* (Marburg, 1969). *Botanologicon* is distinct in structure and voice from Brunfels, Fuchs, and others, but was produced in their immediate context and addresses their projects both by name and in subtler ways as well. See Claudia Swan, "The Uses of Realism in Early-Modern Illustrated Botany," in *Visualizing Medieval Medicine and Natural History, 1220–1550,* ed. Jean A. Givens, Karen Meier Reeds, and Alain Touwaide (Aldershot, U.K., 2006).

38. *Euricii Cordi Botanologicon* (Cologne, 1534), 26–27; adapted from Dilg 1969, 147.

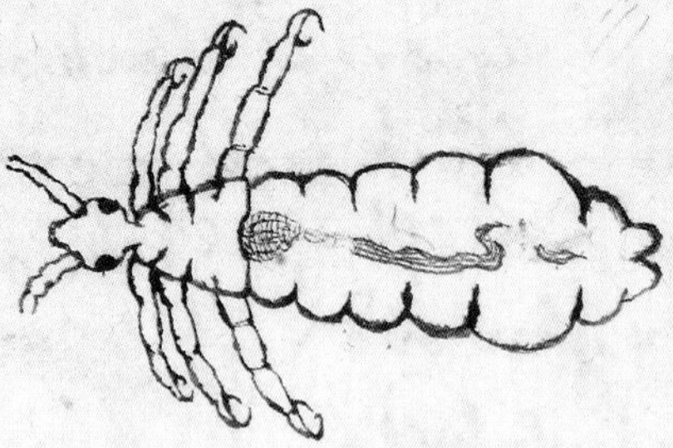

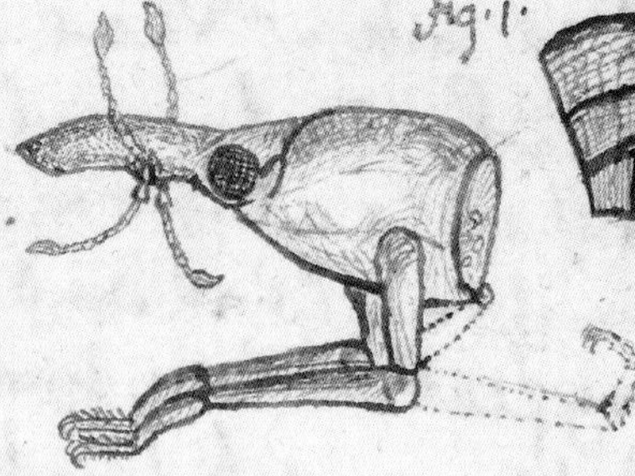

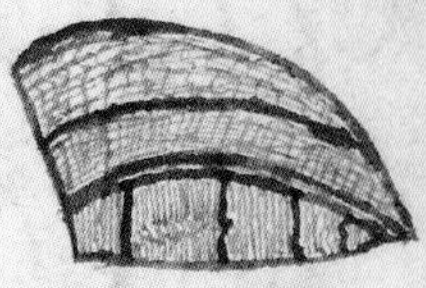

C.

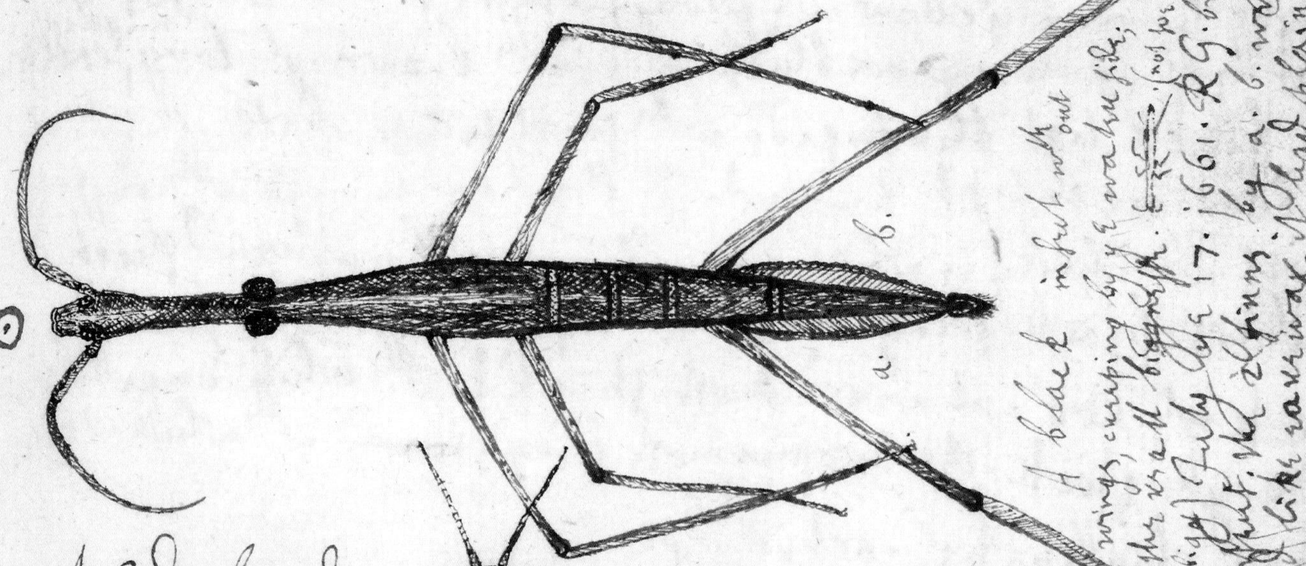

A black insect with out wings, creeping by ye water side; the exact bigness (not perfect) July ye 17. 1661. R.G. being present, the eyes by a 6. were like wax. I had plainly some eyes, or at least ye similitude.

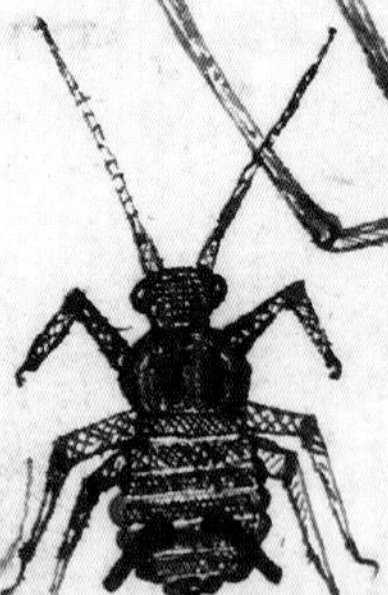

A sad coloured green flye ye wings being pulled off, the exact bigness was . Aug. 24. 1660. E.P. being present.

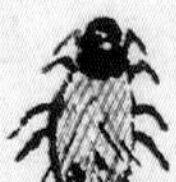

R: bigness: A mite. July 28. 1660. D.C. being present.

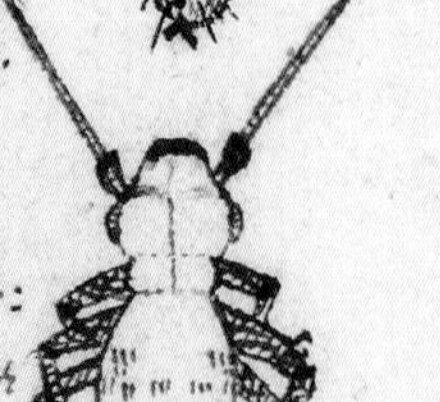

Another sort of mite creeping on rotten wood, this picture is imperfect in the legges, it was of a pale tawny col. E.T. ocul. testis. S.S.
I suppose ye forme of ye legges all being double joynted, and forked at ye end like S.S. they be not in their naturall position.

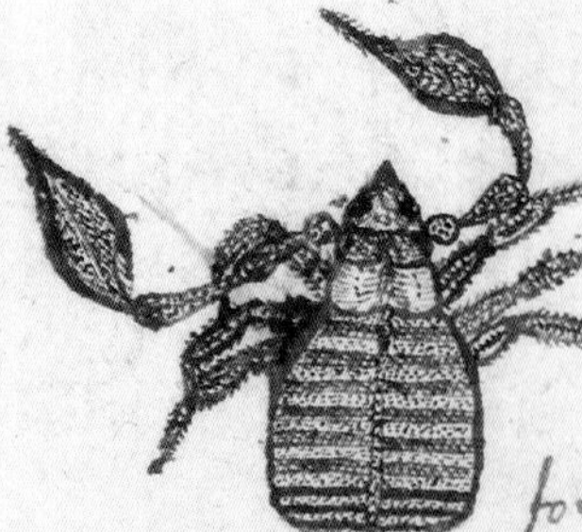

A Kind of Tick found creeping upon paper, it was drawn dead, the right side is perfect only the 2d joynt of the four leggs should have been shortest ye 3d ye longest, it was of browne colour, the legges and nose full of hairs, all the little spots shining like gold. it was drawn (exact) April. 11. 1661. R.H.

JANICE NERI
Boise State University

Between Observation and Image: Representations of Insects in Robert Hooke's Micrographia

Robert Hooke and others, drawings and descriptive annotations of various insects, from John Covel, *Natural History and Commonplace Notebook* (London, c. 1660–1713), ink
The British Library, MS Add. 57495, folio 113v

The English polymath Robert Hooke (1635–1703) can be credited with a number of mathematical and mechanical inventions, including the equation describing elasticity known as Hooke's Law, and with originating the term "cell" in biology. But it is Hooke's *Micrographia*, published in London in 1665, that is considered a landmark in the history of scientific illustration.[1] His spectacular illustrations of plants, insects, astronomical bodies, and mechanical objects have long been praised for their artistic merit, scrupulous accuracy, and careful attention to detail. Containing thirty-eight copperplate engravings, *Micrographia* stands as a testament to Hooke's talents as an observer and illustrator. Little attention has been paid, however, to the visual techniques and traditions Hooke used to translate what he saw through the lens of the microscope into two-dimensional images.

In examining the sequence of events leading up to the publication of *Micrographia*, as well as the circumstances of the book's production, this essay will reveal how Hooke labored to present himself as an authoritative and trustworthy observer. This persona, like Hooke's images and specimens, was a carefully crafted construct intended to support the authority and authenticity of Hooke's observations and, in turn, to create a sense of trust among readers that his images were accurate reflections of those observations.

In addition to considering the published images in *Micrographia*, this essay discusses several recently discovered early sketches of microscopical observations of insects by Hooke, sketches that shed important light on his early work with the microscope and his working practices as a scientist and a scholar. In this essay, I assert that the illustrations in *Micrographia* were central to Hooke's efforts to present the microscope as an instrument that provided access to a hidden world of marvels and wonders. As the gatekeeper to this hidden world, Hooke both controlled access to and shaped perceptions of these wonders through the skilled crafting and manipulation of images. The serenely arranged objects and elegantly posed creatures that are featured in Hooke's images can also be understood within the broader context of the Royal Society's efforts to promote the New Philosophy as a source of rational, useful knowledge that supported, rather than undermined, stability and peace in Restoration society. Hooke's carefully constructed images and authorial persona worked together to produce the impression that *Micrographia* offered readers straightforward, transparent descriptions of natural phenomena. The persona of the distanced, unbiased observer was designed to support Hooke's own status as a trustworthy member of the Royal Society while supporting the society's desire to avoid associations with contentious political and religious issues.

As the son of the curate of a parish church on the Isle of Wight, Hooke came from a respectable but modest family. His early years were marked by tragedy; his father died when he was thirteen years old. Hooke used the small inheritance he received to finance his move to London around 1648, where he served an apprenticeship in the workshop of the Dutch portrait painter Peter Lely (1618–1680) and soon after attended Westminster School. Because his inheritance would not have been enough to cover all of the expenses he incurred during this time, it is believed that Hooke may have received financial assistance from a family friend to help fund his education. When he later attended Oxford, he also received financial assistance.[2] Thus, although his intelligence and talents were widely recognized from an early age, his humble background would not have gone unnoticed by his peers and teachers. The issue of Hooke's social standing would be an important factor in his later interactions with members of the Royal Society. Furthermore, he may have suffered a facial disfigurement from a childhood bout with smallpox, another factor that may have contributed to his social status as an outsider.

The Origins of *Micrographia*

The origins of *Micrographia* are rooted in a series of microscopical drawings of insects made by Christopher Wren (1632–1723) in the late 1650s or early 1660s, which Wren presented as a gift to King Charles II sometime in early 1661. Contemporary accounts by visitors to the king's cabinet indicate that Wren's drawings included illustrations of a flea, a louse, and the wing of a fly.[3] The drawings caused a troublesome situation for the Royal Society shortly after Wren made his gift to the king. Charles II was so delighted by the drawings that he requested more of them from Wren. He made his request not directly to Wren, however, but through the Royal Society, of which Wren was a founding member. Henry Powle then wrote to Wren at Oxford informing him of the king's request, indicating that Wren should fulfill the commission without delay: "[H]e doth expect an Account of this from you shortly."[4] On 17 May 1661, Paul Neile and Robert Moray wrote to Wren on behalf of the Royal Society with the same request, directing Wren "to delineate by the Help of the Microscope the Figures of all Insects, and small living creatures you can light upon, as you have those you presented to his Majesty."[5]

The king's request had placed the Royal Society in an awkward position. Rather than presenting the original set of drawings through the Royal Society, Wren had made his gift directly to the king. The members had not been given the opportunity to approve or discuss the gift before it was made, nor had they enjoyed the privilege of viewing the drawings before they became the king's property, and access to them was restricted to a select few. Now, however, the Royal Society was responsible for carrying out the king's request that Wren provide him with more drawings. Although the society's members would have been eager to carry out the king's wishes and thereby gain his favor, their ability to induce Wren to complete the project was limited. Although he was a well-respected founding member of the society, Wren's activity and attendance at meetings had become somewhat sporadic after he took up his appointment as the Savilian Professor of Astronomy at Oxford in May of 1661. Apparently Wren did not act on the royal assignment, for on 3 July the Royal Society asked Henry Powle to speak with him about his progress on "the pictures of small insects by the microscope."[6]

There is no record of Wren's response to Powle's inquiry, but by August of 1661 Wren had indicated to the Royal Society that he would not be able to make the drawings for the king. Wren's withdrawal from the project is recorded in a letter about the matter sent to him by Robert Moray. Rather than leave the king's request unfulfilled, Moray and the other members of the society had decided to reassign the task to Robert Hooke. On 13 August, Moray wrote to Wren: "[I]n Compliance with your Desire to be eased of the further Task of drawing the Figures of small Insects by the Help of the Microscope, we have moved his Majesty to lay his demands on another, *Vander Diver*; and we have also persuaded Mr. Hook, to undertake the same thing."[7] Thus Moray's text indicates that two people had been asked to work on the project; unfortunately the identity of the person referred to as "Vander Diver" remains unknown.

The politics of the situation were grave. The Royal Society naturally stood to gain a great deal of favor through Wren, whose work greatly pleased the king, but they also risked disfavor if Wren did not comply with the king's wishes. Furthermore, they could not be assured that they would reap the benefits of Wren's talents if Wren did not choose to associate his gift with the Royal Society; however, they were almost guaranteed to lose favor with the king if they failed to carry out his request. Robert Moray, who acted as the Royal Society's unofficial liaison to the court, seems to have successfully negotiated this potentially damaging situation by convincing the king to pass the task on to Hooke. There is no further mention of the project in the Royal Society's minutes, however, nor is there any record that the society presented the king with drawings by Hooke.[8] No further reference to microscopical drawings is made in the Royal Society's minutes until March of 1663, almost two years after the episode involving Wren's drawings.

Hooke's Role in the Royal Society

At the time he was asked to produce microscopical drawings of insects for the king, Robert Hooke had a markedly different standing in the Royal Society from the more central role he would later occupy as its curator of experiments. Hooke was not yet a member of the society in 1661 and began attending meetings only after he was appointed its curator in November 1662. He was certainly known to the membership before 1661, as he had been employed by Robert Boyle in Oxford since 1657 and had been acquainted with many of the society's founding members since the start of his studies at Oxford in 1655.[9] Hooke's name first appears in the minutes of the society on 10 April 1661, when members decided to discuss his tract on capillary action at their next meeting.[10] Although Hooke had worked closely with Boyle in Oxford on the design and construction of the air pump, it is not likely that he was present in London for Boyle's demonstrations of the device at the Royal Society meetings during the early 1660s. Indeed, Steven Shapin and Simon Schaffer have argued that the problem of the repeated failures of the air-pump demonstrations at these meetings was solved only when Hooke was appointed curator; the frustrations of the failed demonstrations seem to have played a part in the society's decision to appoint a curator of experiments.[11] As Shapin and J. A. Bennett have shown in separate studies, Hooke's status in the Royal Society remained deeply ambiguous throughout his tenure as curator and later as secretary. In later years, Hooke would move on a daily basis between the different social worlds of gentlemen virtuosi and those skilled in trades, and although his ability to conduct experiments and operate instruments was rarely questioned, he was never fully accepted as a social or a professional equal by the other members of the Royal Society.[12]

Hooke's status in the early 1660s was more marginal than it was during his later years with the society, but it was also less ambiguous. As noted earlier, Hooke's family was not wealthy, and he relied on assistance for his attendance at Westminster School and at Oxford. Thus, the fact that Hooke was not a person of independent financial means would have been well known to the members of the Royal Society in the early 1660s. As a paid assistant to Robert Boyle and later a paid employee of the society, he would also have occupied a social position distinct from that of a gentleman. A gentleman, in the context of the Royal Society, could be trusted to offer his unbiased opinion precisely because he was a person of independent financial means. Members of the Royal Society were, for the most part, men from the upper levels of society—gentlemen and nobles who were, as Shapin has shown, "free" men and thus considered trustworthy because of their social standing. "Unfree" men, such as technicians and servants who expected remuneration for their efforts and ideas, were subject to the control of others and therefore considered less trustworthy.[13] When Hooke became an official member of the Royal Society, while retaining his paid position, he held an uncertain status among the society's members. His work on *Micrographia* in the early 1660s would coincide with the transition he was making from the relatively stable social position of paid assistant to the more ambiguous and problematic role of paid employee and fellow of the Royal Society. This dual role and its associated problems would come to characterize many of Hooke's later interactions with the society and its members. The procedures that Hooke and the Royal Society devised during this period to bring the *Micrographia* into print also reflected Hooke's increasingly ambiguous status within the organization.

Hooke had been curator for six months when the members of the society decided, in March of 1663, that he should "prosecute his microscopical observations, in order to publish them."[14] From this point on, Hooke presented microscopical observations at meetings on an almost weekly basis, showing his first drawing—of "common moss"—on 8 April.[15] Hooke continued to present microscopical drawings at meetings throughout 1663, showing his last drawing on 12 December.[16] The drawings he presented between March and December of 1663 would become the basis for most of the illustrations in *Micrographia.* The society's minutes do not record who actually proposed that a publication on the microscope should be pursued, but it is clear that the members intended to monitor closely the book's progress and, to some extent, its content.

For the most part, Hooke's drawings pleased the members, but on occasion they did withhold their approval, as in the case of the drawing of a spider with six eyes. On 29 April Hooke presented two microscopical drawings, "one of a mine of diamonds usual in flints; the other a spider appearing to have six eyes: but the latter was not yet perfectly drawn." The spider with six eyes was not satisfactory because the image Hooke presented was not judged to be "perfectly drawn," and this deficiency does not seem to have been rectified, since the insect never did appear in print.[17]

Hooke was generally allowed to choose his own subject matter for observations, but on occasion he would be ordered to investigate items of interest to other members. For this reason, much of the time he spent making microscopical investigations during the month of June was devoted to the search for spiders living in sage leaves. On 3 June, for example, he was ordered to "observe by a microscope, whether there be any cavities in sage leaves for little spiders to lodge in," and at the next meeting, on 10 June, he reported that he had seen none. Despite this negative report, the interest in spiders in sage leaves persisted. On 17 June, Hooke was ordered to continue looking for such spiders and, on 1 July, the order was issued to him again. No illustration of spiders in sage leaves appears in *Micrographia,* and thus it is probable that Hooke never found the creatures, despite his repeated searches.[18]

It is interesting to note that the meeting at which Hooke was ordered to begin the unfruitful series of sage-leaf investigations was also the meeting at which he was elected a fellow of the society. Thus, at the moment at which Hooke had achieved the recognition and respect of the members of the society in being elected to their ranks, he also found himself under obligation to carry out duties that highlighted his subservient role within the group. It is not difficult to imagine Hooke bridling at the suggestion that he may have failed to see the spiders in the sage leaves and resenting the time spent making the additional observations when he had many other pressing duties to fulfill as curator. On the other hand, he may have felt that there was good reason to continue the investigations. Although this example points to the collaborative nature of the activities of the early Royal Society, it is important to note that these collaborations were not always carried out among equals. Unlike other members of the Royal Society, Hooke could be ordered to take a course of action that he might not otherwise have chosen to pursue; conversely, a request that might not have been appropriate to make of a regular member of the society—one that entailed an implicit questioning of that person's judgment and abilities—was appropriate to make of Hooke because he was an employee.

Hooke and Wren

Many scholars have assumed that Robert Hooke and Christopher Wren worked together in making microscopical drawings for *Micrographia* and that the book represents a collaboration between the two men. Hooke and Wren met in Oxford during the

1650s and were associated with the natural philosophers, mathematicians, and other scholars in the circle of John Wilkins, Seth Ward, and Robert Boyle. Wren had been making microscopical drawings since the early 1650s, and Hooke is believed to have begun his work with microscopes some time during the late 1650s.[19] Hooke's and Wren's collaborations in the years after 1666 are well documented, as the two worked together closely on a number of architectural projects after the Fire of London.[20] Knowledge of their working relationship during the period of the preparation and publication of *Micrographia,* however, is not as firmly established.

Wren's association with *Micrographia* stems primarily from two contemporary sources. The first of these is an account of a conversation that occurred in September of 1655 between Wren and the Polish-born gentleman scientist Samuel Hartlib (c. 1600–1662), in which Wren told Hartlib of a book he was preparing "with pictures of 'Observ[ationes] Microscop[icae].'"[21] The drawings that Wren gave the king in 1661 may have been related to the publication Wren had already been planning in 1655, but no further references to Wren and a book on microscopy occur until 1663. In August of 1663 Henry Power informed readers of the preface to his *Experimental Philosophy* that they could "expect shortly from Doctor Wren, and Master Hooke . . . Cuts and Pictures drawn at large, and to the very life of these and other Microscopical Representations."[22] Power (c. 1623–1668) was a corresponding member of the Royal Society who resided in Halifax, and his *Experimental Philosophy* included a section on microscopical observations that contained several small illustrations. In June of 1663 Power attended a meeting of the Royal Society in London and presented his own microscopical observations. At the same meeting, Hooke, Wren, and John Wilkins were "appointed to join together for more observations of the like nature."[23] At the time of Power's visit to the Royal Society, Hooke had already made seven presentations of microscopical drawings to the society and was thus well along with his preparations for *Micrographia.* It is possible that Power mistakenly assumed that Wren was working on the book with Hooke, since Wren had been in attendance when Power presented his microscopical observations and was a member of the group assigned to make similar investigations on that day. However, there is no further mention in the minutes of such a collaboration among Hooke, Wren, and Wilkins, or of any presentations of drawings by Wren. Hooke continued to present his microscopical drawings at meetings throughout the year, and is the only person noted as engaging in such activities during the period preceding *Micrographia*'s publication.

Wren has also been linked to *Micrographia* on the basis of the drawings he presented to Charles II in 1661. Because those drawings are known to have included illustrations of a flea and a louse—two of the more memorable creatures appearing in *Micrographia*—and because there is no record that Hooke presented drawings of these particular insects to the Royal Society, it has been argued that Wren provided the illustrations for these and several other insects in *Micrographia.*[24] But although Wren's early microscopical drawings and observations clearly influenced the Royal Society members' decision to direct Hooke to work on a book of microscopical observations, it is unlikely that the drawings Wren made in 1661 were used as the basis for any of the engravings in the book. For one reason, Hooke would not have had access to Wren's drawings in the king's cabinet, nor is there any record that a representative of the Royal Society approached either Wren or Charles II with such a request. It is more likely that Hooke produced his own versions of these insects for *Micrographia,* perhaps acting upon the advice of Wren or other members of the society, but did not present these drawings at any meetings. Hooke did not present any microscopical drawings after December 1663, but he continued to work on *Micrographia* in the following months. He seems to have been nearly finished with the book in June of 1664, at which time it was decided that Lord Brouncker (William Brouncker, 1620–1684; first president of the Royal Society) would review Hooke's manuscript and pass it on to other members of the society for their review before approving it for publication.[25]

Hooke himself referred to Wren's early drawings in the preface to *Micrographia,* writing that "Dr. Wren . . . was the first that attempted any thing of this nature: whose original draughts do now make one of the Ornaments of that great Collection of Rarities in the Kings Closet."[26] It is significant that Hooke does not specifically state in this passage that Wren contributed material to *Micrographia,* although he did take pains to recognize Wren's earlier work in the field. Thus there is no reason to believe that Hooke would not have acknowledged Wren's contributions to the book if any had, in fact, been made. Indeed, it is likely that both Hooke and the Royal Society would have liked very much for Wren to work on the publication, since Wren had already distinguished himself in the field and had been closely associated with the microscope since the 1650s. However, in spite of these circumstances, it is unlikely that Wren was involved with the book in anything more than an advisory capacity. Although Wren had been planning to publish an illustrated book of microscopical observations in 1655, he seems to have ceased working actively in this area after declining the king's request for additional microscopical drawings in 1661. So although *Micrographia* was shaped by the various collaborative relationships that existed between Hooke and the members of the Royal Society, it remains largely the product of Hooke's labor, and as such reflects many of Hooke's particular experiences, ideas, and beliefs.

Crafting Natural Appearances: Creating Calm and Order in the Microworld

In his preface to *Micrographia,* Robert Hooke wrote that all that was required for studying nature was "a sincere hand, and a faithful eye, to examine and to record the things themselves as they appear."[27] In this well-known passage, Hooke implies that with his words and pictures he acted as the microscope's amanuensis, simply recording whatever appeared through the lens of the instrument. Rather than taking Hooke at his word, however, let us instead consider the techniques he used to craft his images, specimens, and observations to create "natural" appearances. In so doing, we will see that between the moment of looking through the lens and the act of creating the image more was involved than a seamless transfer of information from the eye to the hand of the observer, as Hooke would have us believe. In composing the words and images of *Micrographia,* Hooke not only rendered visible a world that could be seen only by means of the microscope, but also filled that world with a tantalizing collection of strange and unusual creatures and objects. From a chaotic and confusing array of disparate observations, Hooke constructed for his readers a quiet, ordered, and comprehensible world of hidden wonders.

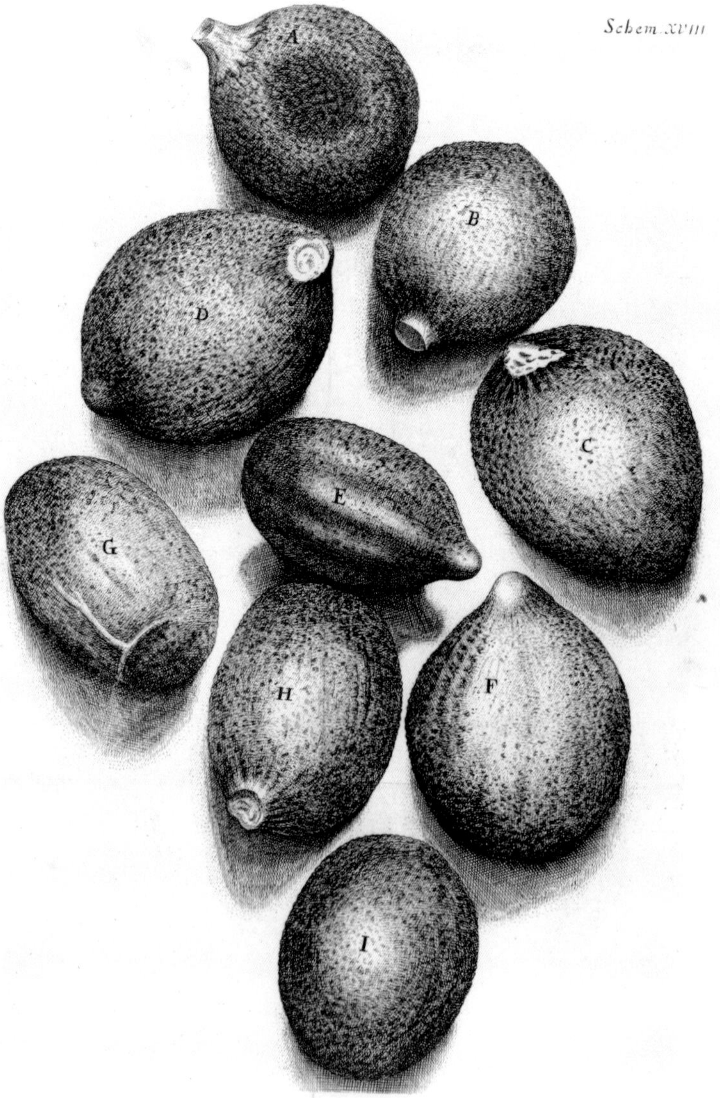

1. Robert Hooke, *Thyme Seeds,* from *Micrographia* (London, 1665), schema 18, engraving
Research Library, The Getty Research Institute, Los Angeles

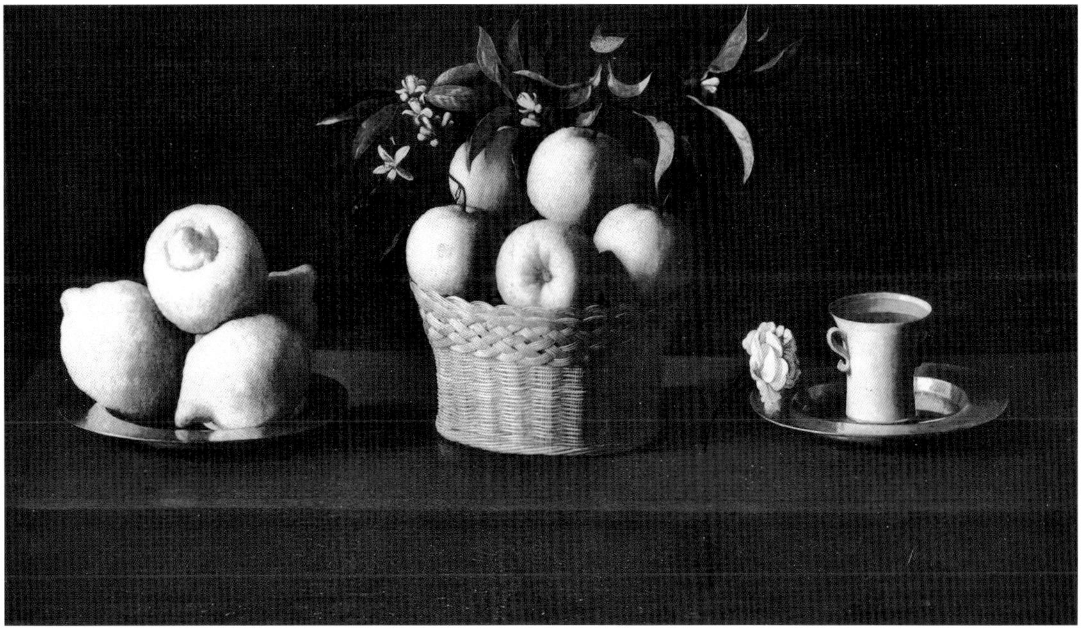

2. Francisco de Zurbarán, *Lemons, Oranges, and a Rose,* 1633, oil on canvas
The Norton Simon Museum, Pasadena; photograph The Norton Simon Foundation

In order to present microscopical observations in a familiar format for seventeenth-century audiences, Hooke drew upon a number of sophisticated image-making traditions such as still-life painting, manuscript illumination, and natural history illustration as he prepared the images for *Micrographia.* Although Hooke never worked professionally as an artist, he had been apprenticed for a short time to the royal portrait painter Peter Lely, and he maintained an interest in art throughout his life.[28] In schema 18 of *Micrographia,* Hooke presents a magnified view of thyme seeds (fig. 1); in the corresponding text he describes the seeds as "pretty fruits" that resemble oranges and lemons, writing that "the Grain affords a very pretty object for the *Microscope,* namely, a Dish of Lemmons plac'd in a very little room."[29] Hooke's statement recalls the banquet and "laid-table" compositions typical of seventeenth-century still-life painting, such as Francisco de Zurbarán's *Lemons, Oranges, and a Rose* of 1633 (fig. 2) or Willem Kalf's *Chinese Bowl and Nautilus Beaker* of 1660 (fig. 3). The vivid colors and unusual textures and shapes, as well as the rarity, of citrus fruits made them a favorite subject for seventeenth-century still-life painters. Like the still-life paintings by Zurbarán and Kalf, Hooke's engraving of thyme seeds arrayed on a tablelike surface has the crisp delineation and well-illuminated composition that facilitate quiet contemplation of the objects thus portrayed. Still-life painting offered Hooke a pictorial strategy for translating fleeting impressions of surfaces and fragments into images of physical objects in a three-dimensional space. Like oranges and lemons in a still life, Hooke's thyme seeds engaged his viewers' eyes in a delightful exploration of the unusual texture and shapes of rare objects.[30]

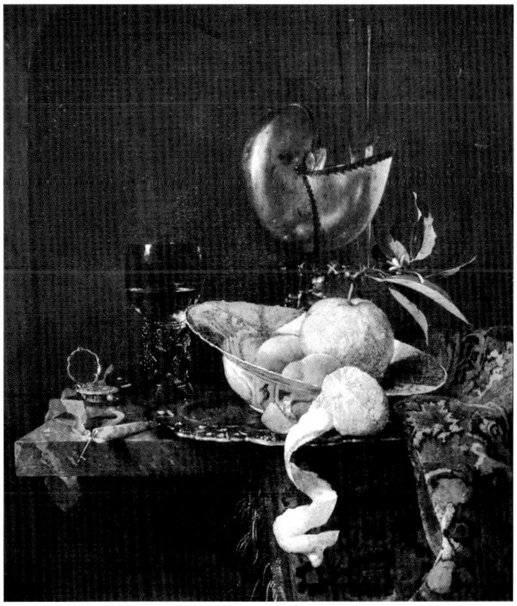

3. Willem Kalf, *Still Life with Chinese Bowl and Nautilus Beaker,* 1660, oil on canvas
© Museo Thyssen-Bornemisza, Madrid

Throughout *Micrographia,* Hooke described the objects he viewed through the microscope as "pretty" and "beautiful"; these aesthetic qualities were important criteria for Hooke in choosing which objects to view with the microscope, and more important, which objects to illustrate in *Micrographia.* Hooke's choices were informed by principles similar to those governing illustrated natural histories and cabinets of curiosities, in particular, the focus on rare, unusual, and beautiful forms shared by many early modern European collectors and artists, such as the Italian Renaissance artist Bartolomeo Bimbi (see the essay by Hans Hubert in this volume). In his description of purslane seeds (fig. 4), for example, Hooke noted that the seeds had "very notable shapes, appearing through the *Microscope* shap'd somewhat like a *nautilus* or *Porcelane* shell. . . . The order, variety, and curiosity in the shape of this little seed, makes it a very pleasant object for the *Microscope.*"[31] Hooke also noted that, when viewed through the microscope, the seeds of the Venus looking glass, or corn violet (fig. 5), seemed "to be the Cabinet of Nature, wherein are laid up its Jewels . . . nor indeed is there in any part of the Vegetable so curious carvings, and beautiful adornments, as about the seed."[32] Couched in the language of collecting and natural history, Hooke's text suggests that the world revealed by the microscope was made up of chambers filled with beautiful and unusual objects, much like a cabinet of curiosities. The illustrations in *Micrographia* also recall the conventions of natural history and botanical illustration in early modern Europe. Many of the images present objects against white backgrounds containing few contextual elements, aside from an occasional shadow, as in the illustration of an ant in schema 32 (fig. 6).

Although *Micrographia* was not intended as a treatise on insects, such creatures occupied much of Hooke's attention, appearing in fifteen of the thirty-eight plates, or "schemas," of the book. Hooke's use of the term "schema" to identify his plates indicates that he approached his images in a diagrammatic manner and implies the study or visual dissection of the objects portrayed. Insects were closely associated with the microscope from the time of its invention in the early seventeenth century and appear in the earliest published microscopical image from 1630, an engraving of a bee and bee parts by Francesco Stelluti (1577–1652), a member of the Accademia dei Lincei. Insects had drawn the attention of European artists and naturalists since the sixteenth century, and by the 1660s they had become frequent subjects of study for microscopists.[33] Prior to the Royal Society's decision to have Hooke publish a book of microscopical investigations, several members had met on a regular basis to discuss the topic of the generation of insects.[34] The intricate structures of insects, as well as their size, made them ideal subjects for viewing through the microscope, but these small, delicate creatures also posed a great challenge to Hooke. Insects were easily damaged and killed in the process of preparing them for viewing, and for this reason were difficult to represent in "natural" poses, as was the case with the delicate body of the ant pictured by Hooke in schema 32 (fig. 6). Hooke wrote that the ant was "a creature, more troublesome to be drawn, then any of the rest, for I could not, for a good while, think of a way to make it suffer its body to ly quiet in a natural posture."[35] Hooke eventually discovered that he could incapacitate the ant by submerging it in spirits, a technique he favored because "if I killed it, its body was so little, that I did often spoile the shape of it, before I could thoroughly view it: for this is the nature of these minute Bodies, that as soon, almost as ever their life is destroyed, their parts immediately shrivel, and lose their beauty."[36] Using this method to preserve the ant's lifelike qualities was perhaps too effective, for after about an hour, "upon a sudden, as if it had been awaken out of a drunken sleep, it suddenly reviv'd and ran away."[37]

Natural history and botanical illustration offered Hooke another pictorial strategy for translating microscopical observations into representations of physical objects. Whereas still-life painting provided Hooke with a means of presenting multiple objects in a three-dimensional space, natural history and botanical illustration allowed him to focus attention on a single object or creature. Hooke's illustration of the ant, at a scale that is many times actual size, highlights the insect's spiky limbs and brittle structures by silhouetting them against a stark white background, effectively transforming the seemingly familiar ant into an exotic and mysterious alien creature.

Two contemporary works of natural history make appearances in Hooke's text: *Historia naturalis Brasiliae,* by Willem Piso (1611–1678) and *A True and Exact History of the Island of Barbados,* by Richard Ligon (1585–1662).[38] In his written account of the ant pictured in schema 32, Hooke refers readers who wish to learn more about the "seemingly rational actions" of ants to Ligon's book.[39] His reference to Ligon's book—rather than to a book dealing specifically with insects, such as Thomas Moffet's *Theatrum insectorum* or Ulisse Aldrovandi's *De animalibus insectis*[40]—implies that Hooke conceived of his insects as exotic and strange creatures akin to those described by travelers to the New World. With its focus on a single specimen and its lively, inquisitive stance, Hooke's ant can also be understood as a distant relative of Albrecht Dürer's *Stag Beetle* of 1505 (fig. 7). But while Hooke's image reflects a similar concern with conveying the movements and character of a living specimen, it is also based on the combined observation of living, dead, and dismembered ants.

For Hooke, the images in *Micrographia* functioned as organizational tools for arranging and ordering the fragmented and confusing observations he amassed while preparing and viewing his specimens. To assist him in this task, he adopted conventions of specimen illustration from the traditions of natural history and botanical illustration as well as those of late medieval manuscript illumination. In the examples of natural history illustration discussed above, individual specimens are presented against plain backgrounds, a technique that encourages both artist and viewer to understand the plant or animal depicted as an isolated physical object. Similarly, in the illustrated borders of illuminated manuscripts, such as the *Book of Hours of Catherine of Cleves* (c. 1440), the surface of the page serves as a repository for mussel shells and a small

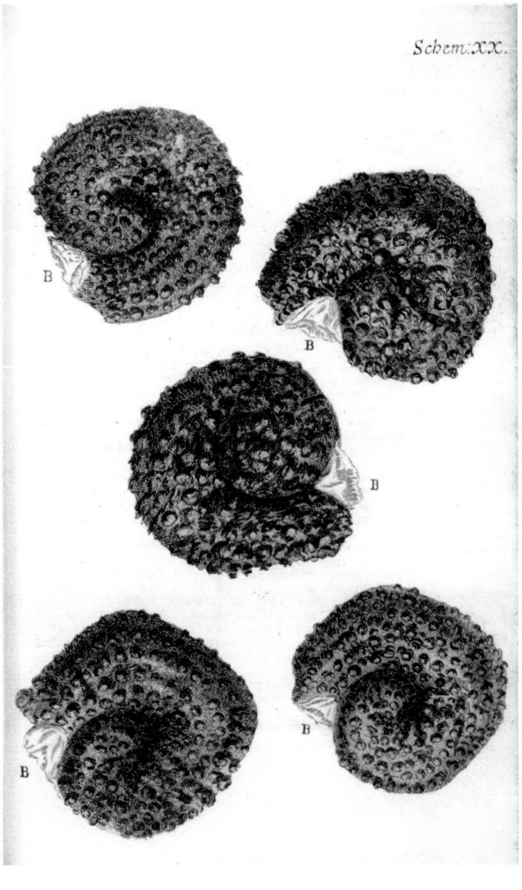

4. Robert Hooke, *Purslane Seeds,* from *Micrographia* (London, 1665), schema 20, engraving
The Huntington Library, San Marino, California

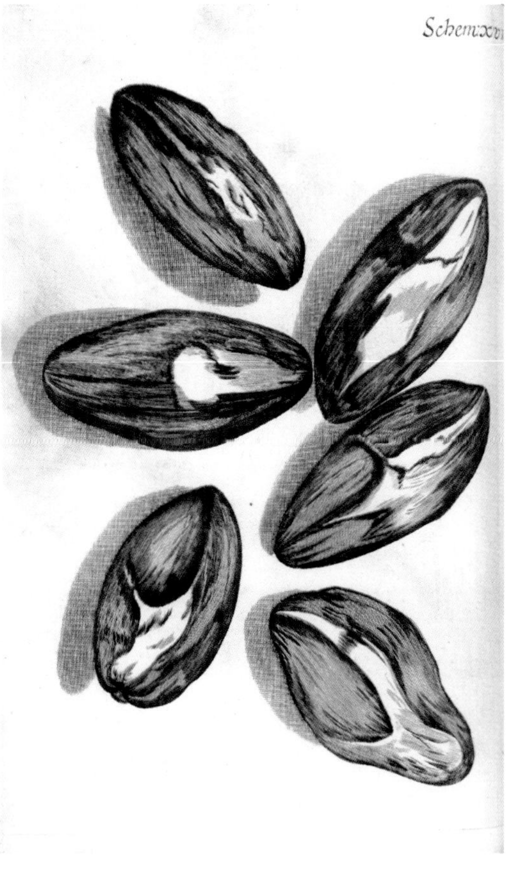

5. Robert Hooke, *Corn Violet Seeds,* from *Micrographia* (London, 1665), schema 17, engraving
The Huntington Library, San Marino, California

crab (fig. 8), which are presented in much the same manner as precious, jewel-like objects in a curiosity cabinet. Each of these pictorial traditions offered Hooke varying means of using images to organize information and to transform obscured forms and uncertain knowledge into gleaming objects and solid bodies.

As the example of Hooke's ant reveals, making observations and producing illustrations for *Micrographia* involved far more than simply recording, as he put it, "the things themselves as they appear." Insects served as "pretty objects" for the microscope only if they could be made visible to the instrument; "natural appearances" required careful crafting of both specimen and image. I have argued thus far that Hooke used a variety of pictorial strategies to show that the microscope provided access to a "hidden world" of strange, unusual, and beautiful objects and creatures. Presenting the microworld *as* a world was one of Hooke's central aims in constructing the images in *Micrographia.* The illustrations were crucial to his overall effort to portray the microworld as a place of order, quiet, and calm rather than a confusing array of incomprehensible fragments. Consequently, the images minimize the disjointed experience of viewing objects through the microscope and betray none of the messiness, tedium, and uncertainty experienced by early users of microscopes.

Hooke built up his images from numerous observations made from multiple vantage points, under varying lighting conditions, and with lenses of differing powers. Similarly, his specimens required a great deal of manipulation and preparation in order to make them visible through the microscope. Nowhere in *Micrographia,* however, do we find visual traces of mutilation or violence, although insects were of course dismembered, pinned, and killed in the course of preparing and dissecting specimens. Instead Hooke almost always presents his insects as whole, living creatures. In schema 24, for example, viewers are confronted with the intense stare of the eye of a fly (fig. 9). Although only a small part of the fly's body is visible, one does not sense that this is merely a fragment of a body. The fly seems to be peering out from the darkness in a lively and compelling manner, suggesting that its body lies hidden in the surrounding shadows. As with the other illustrations in the *Micrographia,* an atmosphere of calm and quiet prevails as the insect serenely presents itself to the viewer for study and contemplation. The image successfully masks the violence of the removal of the fly's head for microscopical observation. In another illustration of a fly (fig. 10), Hooke presents the insect alongside an enlarged view of one of its wings. The fly is shown as

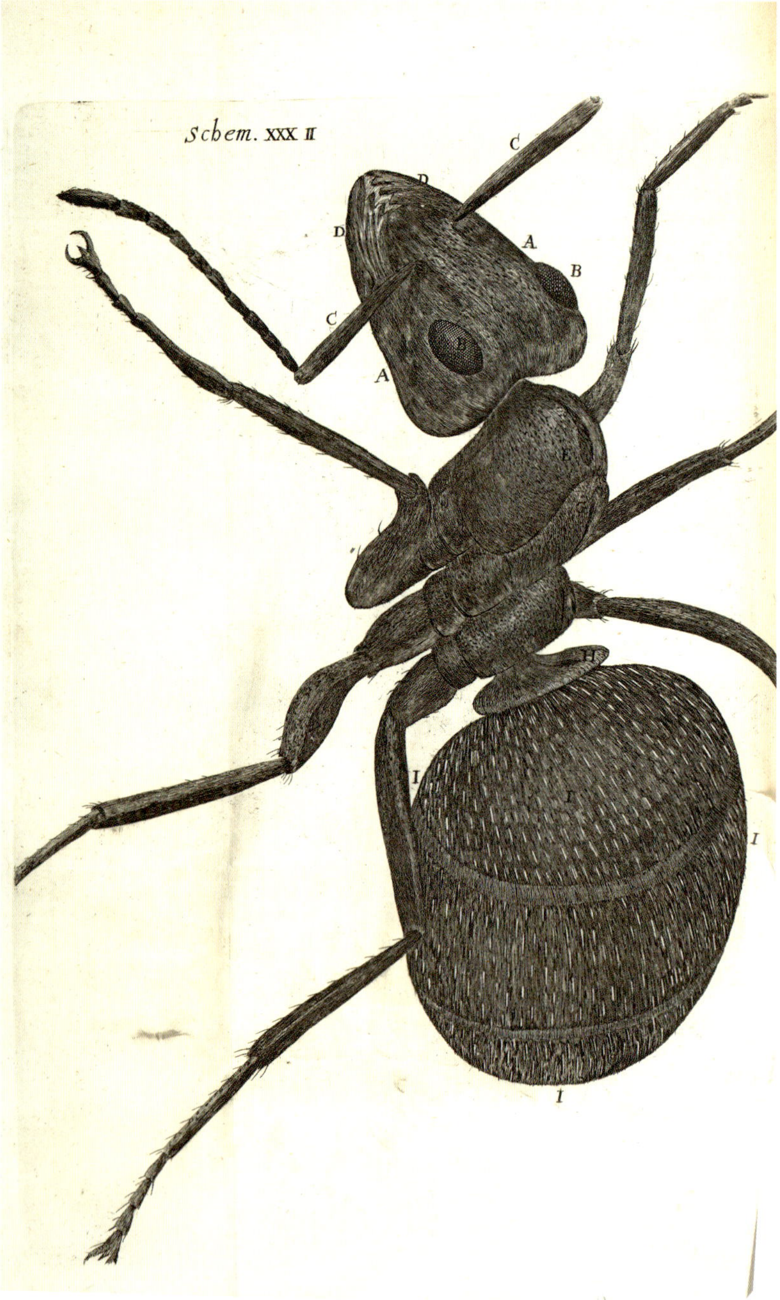

6. Robert Hooke, *Ant,* from *Micrographia* (London, 1665), schema 32, engraving
The Huntington Library, San Marino, California

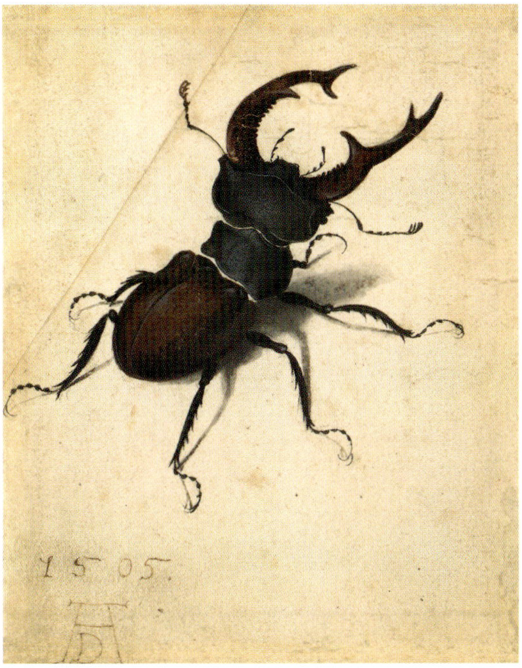

7. Albrecht Dürer, *Stag Beetle,* 1505, gouache and watercolor
The J. Paul Getty Museum, Los Angeles

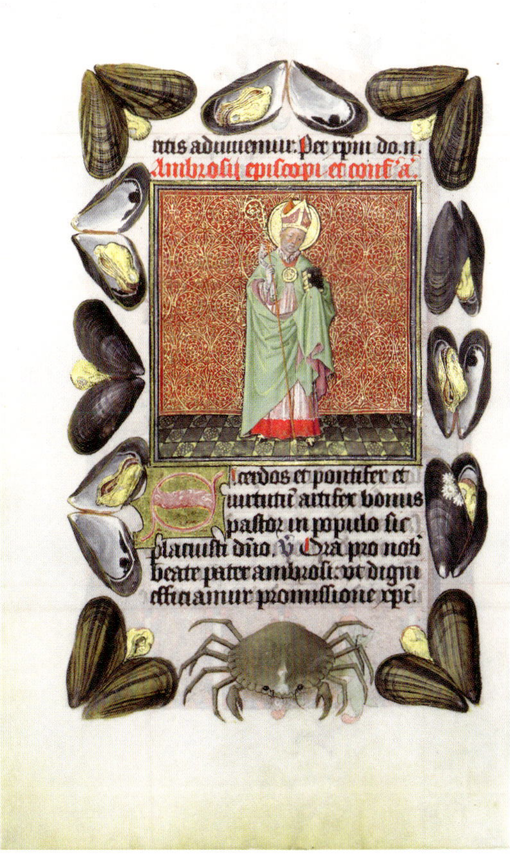

8. Master of Catherine of Cleves, illuminated manuscript with border of mussel shells and crab, from *Book of Hours of Catherine of Cleves,* c. 1440, tempera on vellum
The Pierpont Morgan Library, New York, purchased on the Belle da Costa Greene Fund with the assistance of the Fellows, MS M. 917, folio 244

an intact, living specimen, gently cradling its own severed wing in a peaceful embrace. Hooke's written account describes his dissection of the fly and his observations of its internal organs, but these observations are not illustrated in the image. "Nor was the inside of this creature less beautiful then its outside, for cutting off part of the belly, and then viewing it . . . I found, much beyond my expectation, that there were abundance of branchings of Milk-white vessels, no less curious then the branchings of veins and arteries in bigger terrestrial Animals. . . ."[41] Again, Hooke has crafted the image to reveal his specimen as a beautiful and noble creature, carefully concealing the destruction of its body that was necessary to observe it.

Hooke's desire to show the microworld as a place of calm and order paralleled the aim of the members of the Royal Society to be perceived as purveyors of useful knowledge gained through reasoned inquiry. The Royal Society was eager to show that it contributed to the orderly workings of society and that its activities would not undermine the smooth functioning of the newly restored monarchy. John Harwood has argued that *Micrographia* was "designed to persuade Restoration society that the New Philosophy addressed important intellectual and social problems."[42] The project emerged from the social and political concerns of the Royal Society at a time when the group was becoming increasingly aware of the printing medium as a way to establish and protect its public image.[43] In *Micrographia,* Harwood argues, Hooke presented an idealized picture of himself and the New Philosophy and emphasized "the existence of an interpretive community committed to serious questions and likely to achieve significant, highly visible results."[44] Michael Aaron Dennis has also argued that political concerns shaped the form and content of *Micrographia,* particularly the problems of public indifference and ecclesiastical criticism that the early Royal Society faced. "Responding to the charge that the experimental philosophy was part of the atheism, materialism, and enthusiasm unleashed during the Civil War and Interregnum, [*Micrographia*] demonstrated that the new optical instruments generated 'safe' knowledge clearly distinguished from the claims of atheists and enthusiasts, the elusive outcasts of

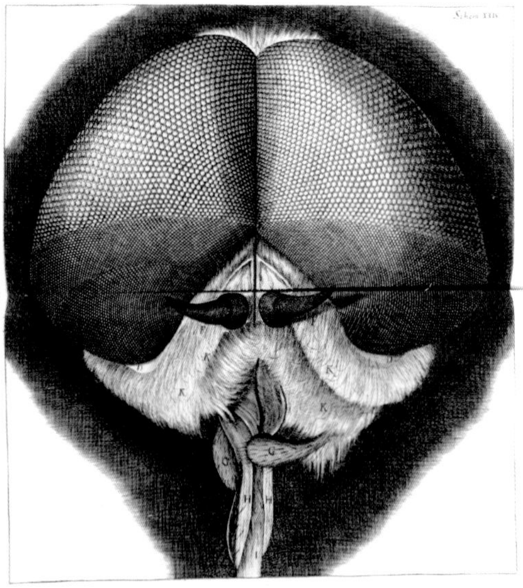

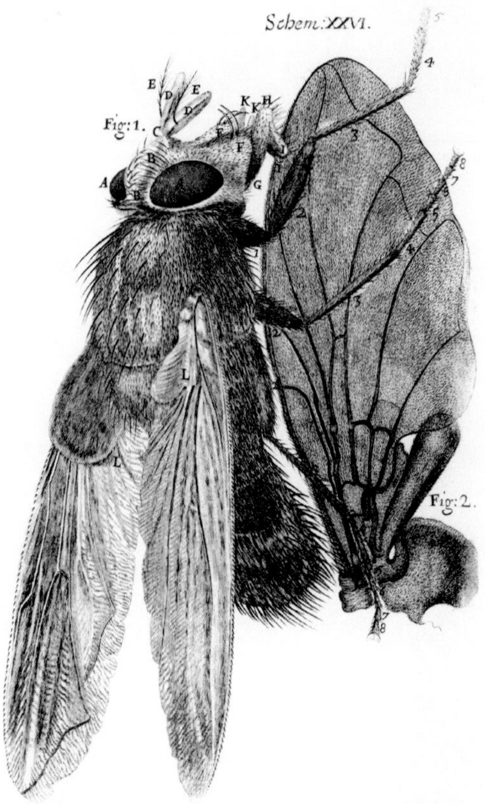

9. Robert Hooke, *Eye of a Fly*, from *Micrographia* (London, 1665), schema 24, engraving
The Huntington Library, San Marino, California

10. Robert Hooke, *Blue Fly*, from *Micrographia* (London, 1665), schema 26, engraving
The Huntington Library, San Marino, California

Restoration society."[45] Dennis argues that interpretation was a central concern of both the Reformation and the New Philosophy. While Protestant reformers were involved in a conflict over the meaning and significance of the Bible, the New Philosophy was involved in a contest over the meaning and theological status of the Book of Nature. Religious enthusiasts and atheists alike were attacked by the restored episcopacy and monarchy because they were believed to bear most of the responsibility for the chaos of the Civil War and Interregnum. "Cast with such suspect company, the experimental philosophers vigorously attempted to differentiate themselves from these 'subversive' readings of the Book of Nature while demonstrating the ability of their own interpretive practice—representation—to undermine the claims of both atheists and enthusiasts. A 'reformation in Philosophy' promised a philosophy appropriate to the needs of church and state, one that would aid the 'Mighty King' in maintaining 'an Empire over the best Invisible Things of this World, the Minds of Men.'"[46]

Dennis sees the interpretive practice of representation that Hooke advocated in *Micrographia* as the visual expression of disciplined seeing. "Observers practicing disciplined seeing were transparent conduits through which an audience *re-viewed* the observer's private experience. With disciplined seeing, Hooke denied that there was a translation of private experience into public knowledge. . . . In *Micrographia* the reader sees through Hooke to 'the things themselves,' forever *re-presented* on the printed page."[47] In denying that communicating microscopical observations required an act of translation on the part of the author, Hooke sought to avoid the troublesome political and religious issues raised by critics of the New Philosophy. As we have seen, however, he used numerous pictorial strategies to translate his observations into comprehensible visual images for his audience. He carefully crafted the deceptively straightforward illustrations in *Micrographia* by using sophisticated image-making techniques designed to convince viewers that the objects and creatures found in the microworld were similar to those found in the macroworld.

Making Images and Making Observations: Hooke's Early Drawings of Insects

Let us now turn to Hooke's early microscopical observations, to investigate some of the processes by which he created his written and visual accounts and how his published accounts worked to conceal the very circumstances of their production. Hooke's early activity with the microscope and his preparatory work for *Micrographia* are known primarily through the minutes of the Royal Society meetings at which he presented drawings and observations. Although a listing for "Dr Rob. Hookes *Micrographia* with Fig. And the Original Drawing" was included in the sale catalogue of Hooke's library published after his death, the drawings are not known to have survived, and it has been suggested that most of them were destroyed in the Fire of London in 1666.[48] Only one preparatory drawing for *Micrographia* is known to exist, a drawing of figures found in frozen urine and water, which is published as figures 1 and 6 of schema 8.[49] Outside of the detailed accounts given in the text of *Micrographia*, little has been known about Hooke's working methods during the period he was making observations and creating the images for the book. Recently, however, I have found evidence for attributing to Robert Hooke a series of sketches that shed important light on his early working practices as well as on the development of his visual and verbal observations in *Micrographia*.

The sketches appear in a notebook, now in the collection of the British Library, that once belonged to Dr. John Covel (1638–1722), an English naturalist who served as the chaplain to the English ambassador to Constantinople from 1670 to 1677.[50] The notebook contains numerous drawings, by Covel, of the plants and insects he observed while living in Turkey.[51] Many years after his return to England, Covel passed his notebook on to the noted collector and botanist James Petiver (1663/1664–1718). Covel seems to have given the notebook to Petiver in order to obtain his help in identifying the Turkish plants and insects. Petiver's written comments, which appear throughout the notebook, usually signed "Petiver" or "Pet.," can be dated to some time after 1707 but were most likely written after his trip to the Netherlands in 1711.[52] The notebook is a fascinating example of the role of visual images in facilitating the exchange of information about the natural world in early modern Europe. In his annotations, Petiver often addresses Covel directly, in a conversational manner, and offers praise and encouragement for Covel's work. Commenting on a plant Covel drew on 10 April 1675, Petiver wrote, "You have been very happy in the accurate delineation of the different Leaves of this peculiar and to me unknown plant. PET."[53] Petiver's comments show that he gained his knowledge of Middle Eastern plants and insects from reading published works by travelers and botanists who had visited the area, as well as through the study of specimens and drawings he obtained by way of his broad network of correspondents. He writes of receiving drawings of plants that grow in Tuscany from the botanist Bruno Tozzi (1656–1743) and of receiving specimens of insects from the Levant.[54] The written entries and annotations show that both Covel and Petiver share Hooke's affinity for unusual and beautiful specimens. Covel includes the following written description with a drawing of a beetle: "The English in Turkey call this a Buffalo bugge: it is of a shining Azure or blewish green like the feathers of a drakes neck, bravely checker'd or fretted on the back. . . . [I]t is set in gold, and silver and worn as a great amulet by the Turkes. The first that ever I found (june 4.72 at belgrade) as I was peeping upon it pist or squirted on my left eye. . . . This I found at Pera. May 20th. It lived till this day the 30th. I perct its 2d & 3d trunk with a pin to hold him fast, and yet he lived." Petiver concurs with Covel's assessment of the beetle's appearance, writing, "I have observed of this wingless genus 3 or 4 sorts in England one of a copperish green [pocked] like this but not so beautiful. Pet."[55]

The earliest entries in the notebook date from 1660 and 1661, predating Covel's trip to Turkey by ten years. These entries consist of sketches of insects and inscriptions in Hooke's hand and include several unmistakable links to *Micrographia*. It is not presently known how or why the notebook passed from Hooke to Covel; the bindings appear to be original and it does not appear that any pages have been added or removed. Covel and Hooke were students at Cambridge and Oxford, respectively, during the 1650s, but there is no record of their knowing or working with one another, and Covel was never a member of the Royal Society. Although the early provenance of the notebook is as yet unclear, the written and visual records it contains are nevertheless important documents of Hooke's working practices. Hooke's entries in the notebook reveal how closely the making of images and the practice of observation were linked in his early work with the microscope. They also show how he worked and reworked his written and visual observations over the course of several years.

Most of Hooke's contributions to the notebook appear on folio 113v (fig. 11), a page that contains sketches of magnified views of insects accompanied by written entries.[56] Several of the sketches are dated, and from these dates it appears that the entries on folio 113v were made over a one-year period, between 28 July 1660 and 17 July 1661. The insect pictured in the lower right corner of the page, labeled "A Kind of Teek," is accompanied by an inscription that is signed and dated "April 11. 1661. R.H." The entries on this page are written in a different hand from Covel's and Petiver's entries and are consistent with known examples of Hooke's handwriting.[57] In the lower left corner of the page, the legs of the insect referred to as "another sort of mite creeping on rotten wood" are labeled with a small letter *s* in two places. These letters are used to direct readers to a particular aspect of the drawing referred to in the written inscription, which reads, "I supplyed the forme of the leggs all being double joynted, and forked at the end like S.S. they not being in theire naturall position." This technique of labeling similar body parts with the same key letter is used quite frequently by Hooke in *Micrographia,* as seen, for example, in figure 1 of schema 27, an illustration "Of the Water-Insect or Gnat" (fig. 12). Figure 2 of schema 27 illustrates another pictorial technique used on folio 113v, that of the dotted or "prickt" line. Hooke had used a dotted line to represent the leg of the dissected insect pictured at the upper right of folio 113v (see fig. 11), in order to show the position of the legs after it had died.[58] Conversely, in figure 2 of schema 27 of *Micrographia,* this same dotted line is used to indicate the movement of the living insect's swimming appendage as it propels itself through water, rather than the position of its body when dead.[59]

The sketches on folio 113v afford a glimpse into the development of Hooke's visual strategies for representing the microworld that would eventually be fully elaborated in *Micrographia.* One of the fundamental visual problems Hooke needed to work out was that of scale. Because there were no stationary markers or recognizable landmarks in the microworld, early

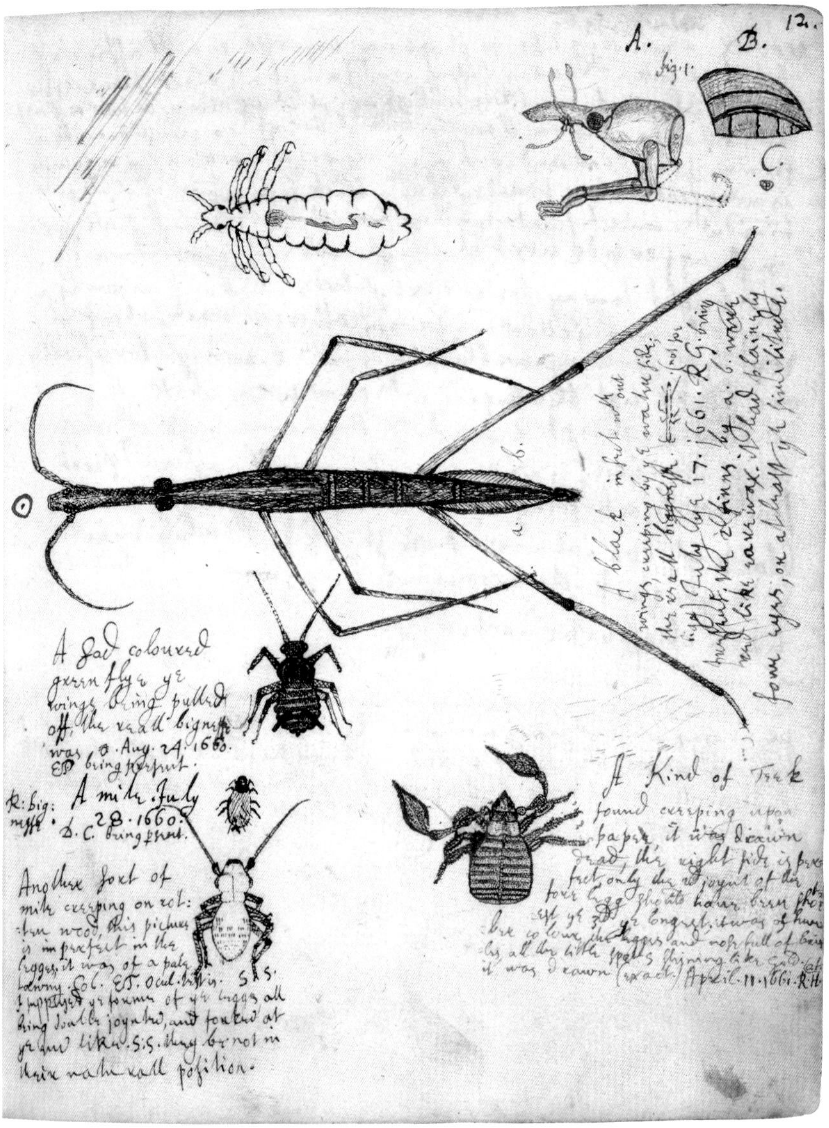

11. Robert Hooke and others, drawings and descriptive annotations of various insects, from John Covel, *Natural History and Commonplace Notebook* (London, c. 1660–1713), ink
The British Library, MS Add. 57495, folio 113v

users of microscopes were limited in their ability to communicate the relative sizes of objects and the distances between them when viewed through the microscope. In the sketches in the Covel notebook, this problem was solved by including a small drawing or mark to indicate the insect's actual size, or "real bignesse." For example, to the left of the drawing labeled "A mite. July 28. 1660" is a small dot marked "R. bignesse." Similarly, within the descriptive text for the insect pictured in the center of the page, following the phrase "the reall bignesse was," is a small drawing of the insect at actual scale. This schematic strategy seems to have been lost on James Petiver, since his annotation for the drawing indicates that he may not have been aware that he was viewing a magnified image. Petiver's comment, which appears opposite the sketch on folio 114, is directed to Covel, who he assumed was the artist. In this passage, Petiver refers Covel to his own work on English beetles that appears in *The Monthly Miscellany: or, Memoirs for the Curious* (London, 1707–1709). "A Tipula or water spider, this has been observed at London but very rare. PET. vid. Scarab. Anglic. in the Memoirs for the Curious. id." Petiver obviously did not read the inscription accompanying the drawing and thus missed the basic information that the image represents a magnified view of a very small insect, rather than a life-size or slightly larger-than-life-size image of a larger insect.

In not realizing that he was looking at a magnified view, Petiver also missed what was most remarkable about the image: it revealed forms usually hidden from the eye, which could be seen only by examination through a microscope or with a magnifying lens. In *Micrographia,* Hooke left no doubt that what his readers were seeing in the illustrations were magnifications. Not only did the title of the book clearly state the subject matter and the text include detailed descriptions of the microscope, but the illustrations themselves almost always announced their status as magnified images. Hooke often enclosed the objects he presented within a round frame, thus offering viewers an evocation of the experience of looking through the lens of a microscope. For many of the insect illustrations, Hooke created large-format images that sometimes exceeded the size of the printed page and had to be folded into the book, as he did for his images of the flea and the louse (fig. 13). As noted earlier, Hooke almost always presented whole insects rather than dissected or disembodied parts of insects. Thus, in addition to furthering his overall effort to present the microworld as a place of calm and order, this strategy allowed viewers to comprehend at once the subject matter of the illustrations. By deftly weaving together elements of the familiar and the unfamiliar, Hooke signaled to his audience that they were seeing greatly enlarged views of very small creatures. A very early stage in the development of this technique for representing microscopical observations is evident in the sketches on folio 113v of the Covel notebook (see fig. 11). The dates accompanying the sketches show that the insects were depicted on an increasingly larger scale with each observation. The earliest dated image is also the smallest, that of the mite dated 28 July 1660. The next dated image on the page is that of "a sad coloured green flye, ye wings being pulled off," which appears directly above the mite of 28 July.[60] The latest dated image is the sticklike insect that spans almost the entire width of the page.

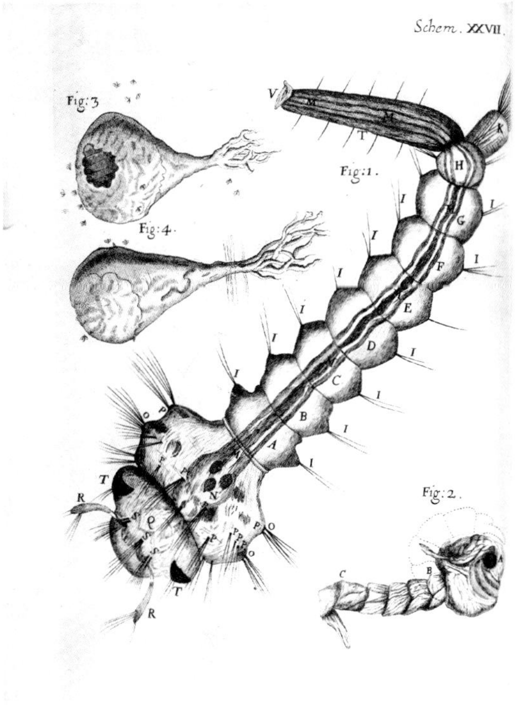

12. Robert Hooke, *Water Insect,* from *Micrographia* (London, 1665), schema 27, engraving
The Huntington Library, San Marino, California

This insect possesses some of the qualities of the monumental images of insects found in *Micrographia,* but judging from Petiver's annotation, it was not quite large enough to signal its status as a magnified image.

The most direct link between the sketches in the Covel notebook and the plates in *Micrographia* is the insect pictured in the lower right corner of folio 113v (fig. 11), which is accompanied by an inscription dated 11 April 1661 and signed with the initials "R. H." The inscription reads as follows:

> A Kind of Teek found creeping upon paper, it was drawn dead, the right side is perfect, only the 2d joynt of the four legg should have been shorter the 3d the longest. it was of humber coloure the legges and nose full of bristles, all the little spotts shining like Gold. [Cat] it was drawn (exact) April 11. 1661. R. H.

The insect in this sketch corresponds to the "crab-like insect" that appears as figure 2 of schema 33 of *Micrographia* (fig. 14). In Hooke's longer and more detailed account of this insect in *Micrographia,* he notes that he had observed the insect on only one occasion. From this statement, then, it is clear that Hooke reworked his visual and written accounts of the insect from memory and possibly by consulting the sketches in the Covel notebook. In observation 51 of *Micrographia,* Hooke gives the following account of the crablike insect: "Reading one day in September I chanced to observe a very small creature creep over the Book I was reading, very slowly; having a Microscope by me, I observ'd it to be a creature of very unusual form. . . ." He continues: "This creature, though I could never meet with more then one of them, and so could not make so many examinations of it as otherwise I would, I did notwithstanding, by reason of the great curiosity that appear'd to me in its shape, delineate it."[61] In the written account published in *Micrographia,* the surface that the insect creeps across has been transformed from paper into the page of a book. Hooke uses the word "reading" twice in the published version, portraying himself as a scholar sitting quietly in his study, immersed in the world of learning but at the same time alert to the workings of nature. With his microscope at his side, Hooke presents himself as ready to conduct observations carefully and thoughtfully whenever an interesting phenomenon might present itself.

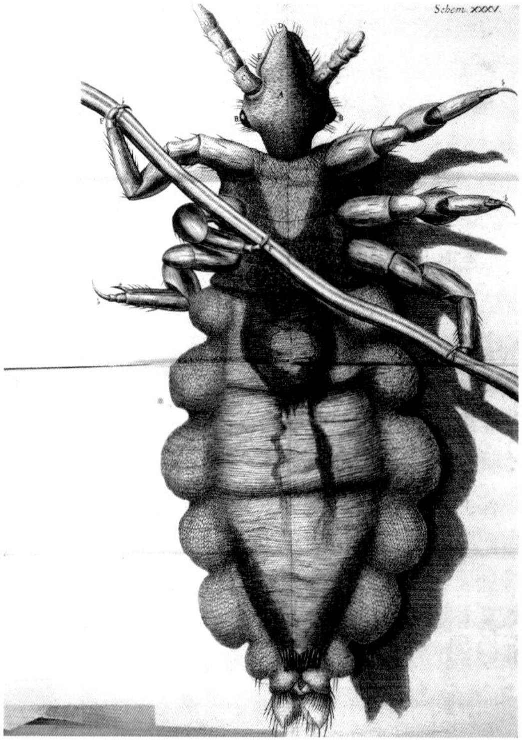

13. Robert Hooke, *Louse,* from *Micrographia* (London, 1665), schema 35, engraving
The Huntington Library, San Marino, California

The fully realized portrait of the crablike insect that appears in the engraving has also been substantially reworked from the version that appears in the notebook sketch, which was presumably made at some point closer in time to Hooke's actual observation of the insect. In the sketch, the insect is shown to possess only six legs, but in the published version it has eight legs. As noted above, Hooke did not make any further observations of the insect and thus may have included the additional legs on the basis of his knowledge of the anatomy of other species, such as spiders or crabs. As is typical of other such illustrations in *Micrographia,* the crablike insect is shown on a large scale, and every surface of its body is described in minute detail. Hooke's early sketch does not contain the same level of surface detail, so it is likely that he again filled in the "missing" information from his knowledge of other insects. Another source he may have consulted in composing the published version of the image was Willem Piso's illustrated book on the natural history of Brazil, *Historia naturalis Brasiliae,*

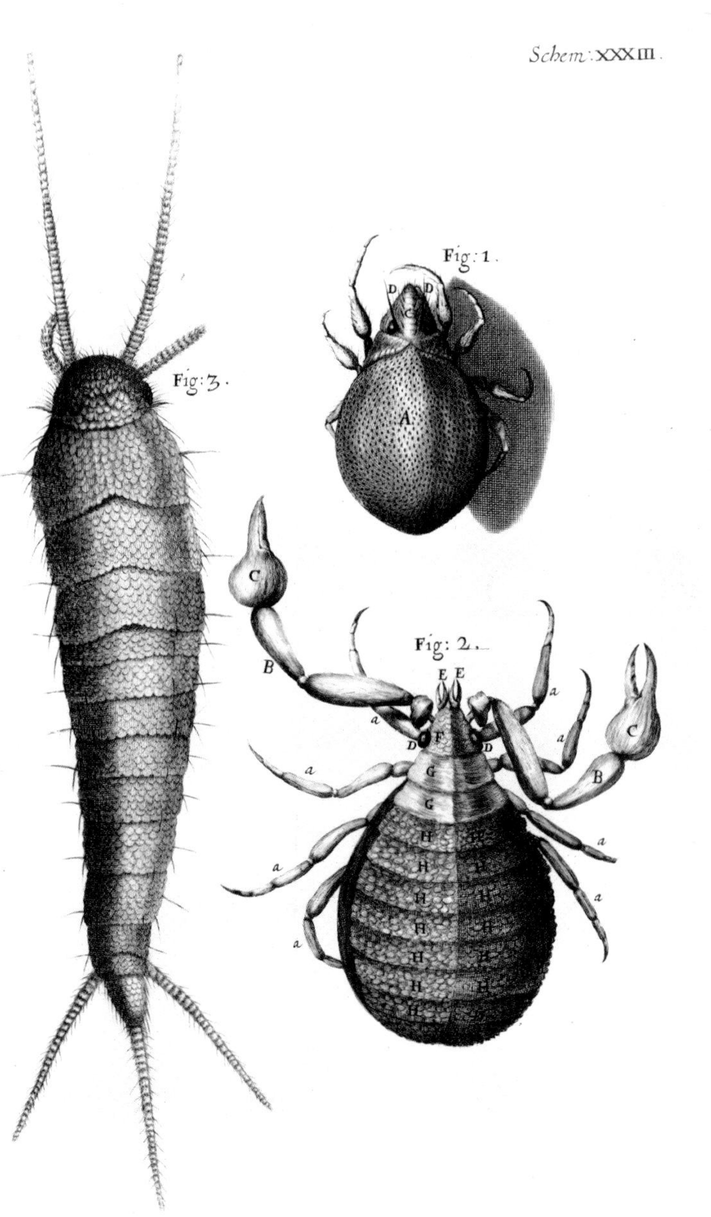

14. Robert Hooke, *Crablike Insect*, from *Micrographia* (London, 1665), schema 33, engraving
Research Library, The Getty Research Institute, Los Angeles

which Hooke referred to and quoted from in observation 43, "Of the Water-Insect or Gnat." Hooke may have used the illustrations of crabs that appear in Piso's book as a model for constructing his image of the crablike insect, for the insect that is represented in schema 33 is represented using the same convention of depicting the "claws" as slightly open (fig. 15).[62]

Constructing the Persona of a Scientific Observer

The differences between the early sketch and notes and the published version in *Micrographia* show that, just as Hooke labored to create specimens with "natural appearances," so too did he labor over the crafting of his visual and verbal accounts. In presenting himself as a solitary observer, however, Hooke not only substantially reworked his early observations, but also eliminated certain pieces of information from them. The written inscriptions accompanying the sketches in the Covel notebook indicate that Hooke did not conduct his observations alone. Several colleagues, or co-observers, are noted as having witnessed the observations with Hooke. A person referred to only as "DC" was present for the earliest observation, that of the mite dated 28 July 1660, while two other people—referred to as "ET" and "RG" respectively—are listed as being present for observations made on 24 August 1660 and 17 July 1661. The undated sketch of a "mite creeping on rotten wood" also lists "ET" as a witness. In the verbal observations in *Micrographia,* Hooke makes no mention of conducting microscopical investigations with co-observers. Steven Shapin has noted the appeal of the persona of the solitary observer for Hooke and other early modern intellectuals. "The presentation of the philosopher's persona as hermit was [not only] a way of understanding . . . who the philosopher was and what might be expected of him, but also a way of warranting claims to knowledge. A man so abstracted from the world was a man free of the hold of its idols and in immediate contact with reality, divine or mundane."[63] As Shapin also points out, the image of the scholarly hermit did not necessarily reflect Hooke's actual experience or social status. However, to gain the respect and trust of his fellow members of the Royal Society, Hooke may have found it prudent to present himself as a solitary and authoritative observer who did not need to rely on others to confirm or support his observations. As noted earlier, Hooke occupied an unusual social position within the Royal Society, and while in some ways he embodied the ideal of the mechanical philosopher in his ability to conduct experiments and construct and operate instruments, he was never fully accepted as a social or professional equal by the society's other members. A possible candidate for the "RG" listed as a witness to the observation of the sticklike insect in the Covel notebook is Ralph Greatorex (1625–1712), a London instrument maker who worked with Hooke on building the air pump for Robert Boyle during the late 1650s.[64] Greatorex and other instrument makers had many dealings with members of the Royal Society but were never elected to their ranks.[65] Hooke may not have wanted to be identified with the world of instrument makers as he endeavored to prove his trustworthiness and intellectual capability as a newly elected member of the society.

Another reason why Hooke may have minimized the presence of his co-observers in his written accounts was his desire to avoid difficult questions about whose observations were being presented and how the viewpoints of multiple observers were or could be reconciled. Potentially disruptive issues such as these may have undermined Hooke's credibility and that of the microscope itself, and had to be avoided if Hooke was to be perceived as the microscope's amanuensis and if his images were to be accepted as exact transcriptions of his observations. Shapin has shown that "virtual witnessing" was both central to Robert Boyle's experimental program during the late 1650s and early 1660s and an essential component in the process by which matters of fact were generated within the community of natural philosophers. Virtual witnessing was made possible through the dissemination of experimental reports that were rich in circumstantial detail, allowing readers to witness an experiment without directly experiencing it or replicating it themselves. Although Shapin describes virtual witnessing as a "literary technology"—generated primarily through written reports—visual images also played a role in the process. Shapin argues that visual representations functioned as mimetic devices in Boyle's texts, facilitating virtual witnessing in their density of circumstantial detail, and because they "imitated reality and gave the viewer a vivid impression of the experimental scene."[66] For Hooke, visual images also created a vivid impression for viewers through, among other things, their density of circumstantial detail. But unlike the illustrations in Boyle's texts, Hooke's images of his microscopical observations were not meant to function as mimetic devices for

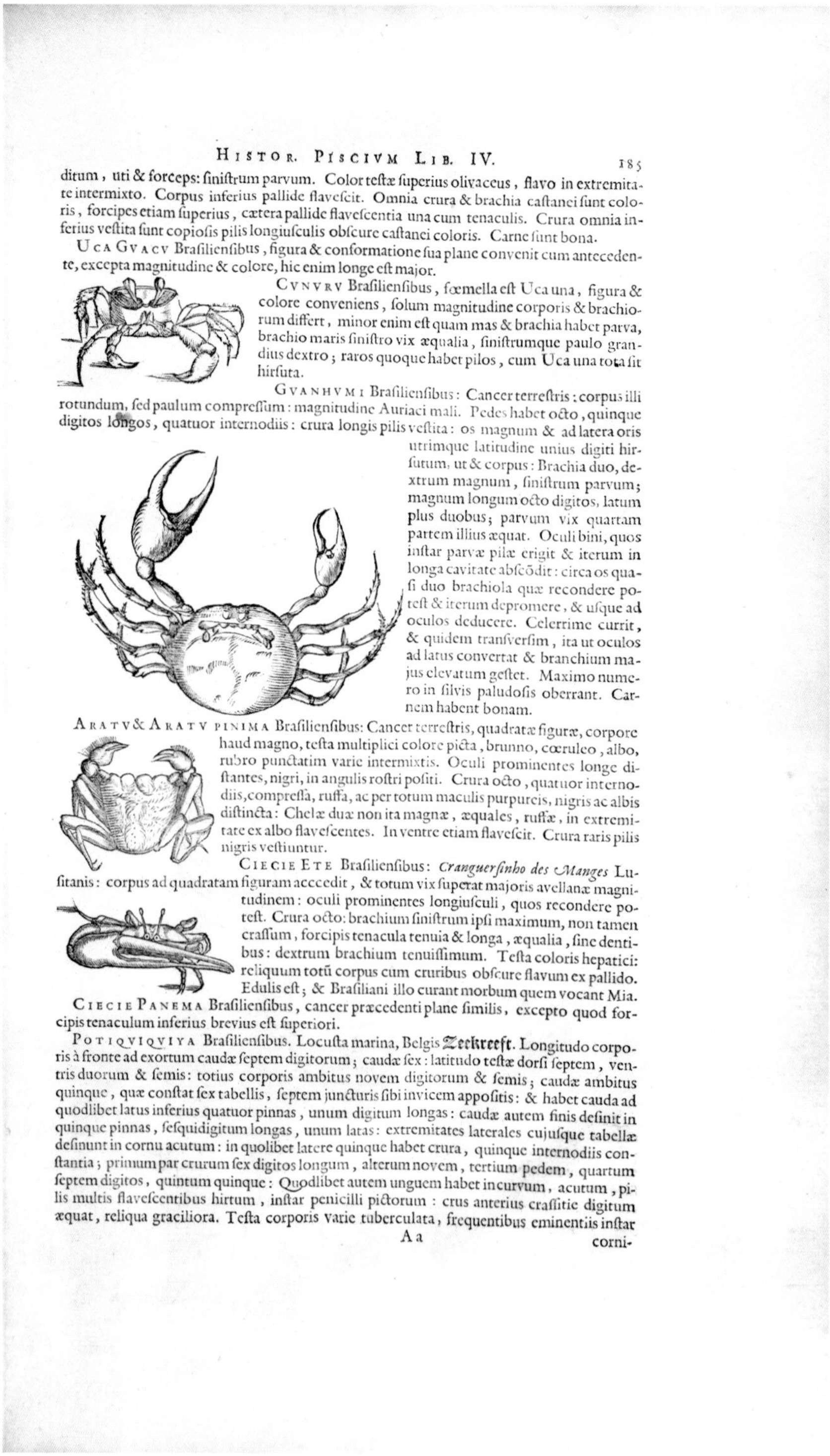
HISTOR. PISCIVM LIB. IV. 185

ditum, uti & forceps: ſiniſtrum parvum. Color teſtæ ſuperius olivaceus, flavo in extremitate intermixto. Corpus inferius pallide flaveſcit. Omnia crura & brachia caſtanei ſunt coloris, forcipes etiam ſuperius, cætera pallide flaveſcentia una cum tenaculis. Crura omnia inferius veſtita ſunt copioſis pilis longiuſculis obſcure caſtanei coloris. Carne ſunt bona.

UCA GUACU Braſilienſibus, figura & conformatione ſua plane convenit cum antecedente, excepta magnitudine & colore, hic enim longe eſt major.

CUNURU Braſilienſibus, fœmella eſt Uca una, figura & colore conveniens, ſolum magnitudine corporis & brachiorum differt, minor enim eſt quam mas & brachia habet parva, brachio maris ſiniſtro vix æqualia, ſiniſtrumque paulo grandius dextro; raros quoque habet pilos, cum Uca una tota ſit hirſuta.

GUANHUMI Braſilienſibus: Cancer terreſtris: corpus illi rotundum, ſed paulum compreſſum: magnitudine Auriaci mali. Pedes habet octo, quinque digitos longos, quatuor internodiis: crura longis pilis veſtita: os magnum & ad latera oris utrimque latitudine unius digiti hirſutum, ut & corpus: Brachia duo, dextrum magnum, ſiniſtrum parvum; magnum longum octo digitos, latum plus duobus; parvum vix quartam partem illius æquat. Oculi bini, quos inſtar parvæ pilæ erigit & iterum in longa cavitate abſcōdit: circa os quaſi duo brachiola quæ recondere poteſt & iterum depromere, & uſque ad oculos deducere. Celerrime currit, & quidem tranſverſim, ita ut oculos ad latus convertat & branchium majus elevatum geſtet. Maximo numero in ſilvis paludoſis oberrant. Carnem habent bonam.

ARATU & ARATU PINIMA Braſilienſibus: Cancer terreſtris, quadratæ figuræ, corpore haud magno, teſta multiplici colore picta, brunno, cœruleo, albo, rubro punctatim varie intermixtis. Oculi prominentes longe diſtantes, nigri, in angulis roſtri poſiti. Crura octo, quatuor internodiis, compreſſa, ruffa, ac per totum maculis purpureis, nigris ac albis diſtincta: Chelæ duæ non ita magnæ, æquales, ruffæ, in extremitate ex albo flaveſcentes. In ventre etiam flaveſcit. Crura raris pilis nigris veſtiuntur.

CIECIE ETE Braſilienſibus: *Cranguerſinho des Manges* Luſitanis: corpus ad quadratam figuram accedit, & totum vix ſuperat majoris avellanæ magnitudinem: oculi prominentes longiuſculi, quos recondere poteſt. Crura octo: brachium ſiniſtrum ipſi maximum, non tamen craſſum, forcipis tenacula tenuia & longa, æqualia, ſine dentibus: dextrum brachium tenuiſſimum. Teſta coloris hepatici: reliquum totū corpus cum cruribus obſcure flavum ex pallido. Edulis eſt; & Braſiliani illo curant morbum quem vocant Mia.

CIECIE PANEMA Braſilienſibus, cancer præcedenti plane ſimilis, excepto quod forcipis tenaculum inferius brevius eſt ſuperiori.

POTIQUIQUIYA Braſilienſibus. Locuſta marina, Belgis Zeekreeft. Longitudo corporis à fronte ad exortum caudæ ſeptem digitorum; caudæ ſex: latitudo teſtæ dorſi ſeptem, ventris duorum & ſemis: totius corporis ambitus novem digitorum & ſemis; caudæ ambitus quinque, quæ conſtat ſex tabellis, ſeptem juncturis ſibi invicem appoſitis: & habet cauda ad quodlibet latus inferius quatuor pinnas, unum digitum longas: caudæ autem finis deſinit in quinque pinnas, ſeſquidigitum longas, unum latas: extremitates laterales cujuſque tabellæ deſinunt in cornu acutum: in quolibet latere quinque habet crura, quinque internodiis conſtantia; primum par crurum ſex digitos longum, alterum novem, tertium pedem, quartum ſeptem digitos, quintum quinque: Quodlibet autem unguem habet incurvum, acutum, pilis multis flaveſcentibus hirtum, inſtar penicilli pictorum: crus anterius craſſitie digitum æquat, reliqua graciliora. Teſta corporis varie tuberculata, frequentibus eminentiis inſtar

Aa corni-

15. Willem Piso, images of various crustaceans, from *Historia naturalis Brasiliae* (Leiden and Amsterdam, 1648), 185, woodcut
William Andrews Clark Memorial Library, University of California, Los Angeles

his audiences, creating a mental image of a demonstration of a device such as the air pump. Instead, the drawings Hooke presented at meetings of the Royal Society and later published in *Micrographia* functioned as demonstrations in themselves.

Whereas the air pump could be brought into the public rooms of the Royal Society and experiments with it performed for all to see, demonstrations of the microscope could not be "performed" in this manner. Because of the nature of the instrument, microscopical observations were conducted by individuals in private settings. The public nature of demonstrations at the Royal Society meant that Hooke and others generally presented "completed" phenomena—in other words, experiments or instruments they were sure would work. The potential for embarrassment was great if an instrument or demonstration were to fail. Experiments in which the outcome was uncertain thus usually took place in private.[67] The physical setting of the Royal Society's meetings would also have precluded individuals' making their own observations with a microscope. The president sat at a central table, and members would sit in rows surrounding him. Instruments were set up in advance on the central table or at a table to the side of the room. It is unlikely that members would have wished to make their own observations during meetings, even if it were physically feasible to do so, because receiving instruction on the use of the microscope might have been viewed as an activity inappropriate to their elevated social status. Gentlemen did not engage in manual labor or crafts, precisely the activities that Hooke was hired to handle. In the context of the microscope, therefore, images were more effective than verbal or written descriptions in facilitating the process of virtual witnessing that was so central to generating facts in Restoration science.

For Hooke's images to be trusted as exact descriptions of his observations, there could be no doubt that he had made the observations himself and that the images represented precisely what he himself had seen. If Hooke's co-observers in making the early sketches were indeed fellow technicians and instrument makers, their very presence in Hooke's "official" published accounts might have cast doubt upon Hooke's authority and his capacity to act as a trustworthy conduit of microscopical observations. For, as Steven Shapin has also pointed out, although technicians and operators enjoyed the type of direct and unmediated access to experimental phenomena that was highly valued by natural philosophers, "directness of experience did not in itself confer epistemic value on technicians' understandings." Whether such experience was recognized by employers depended on the "moral texture of social relations."[68] Given Hooke's ambiguous social and professional standing within the Royal Society, it was important for him to eliminate elements from his visual and written accounts that might suggest instability in his authority. The entries in the Covel notebook show that, in his earlier accounts, Hooke was more likely to acknowledge the uncertainty of knowledge gained through microscopical observation. In describing the insect pictured in the center of folio 113v, he writes that "it had plainely some eyes, or at least the similitude." Such uncertainty is rarely present in the written accounts published in *Micrographia.* Instead, Hooke shifts uncertain knowledge from himself to less experienced or less skilled practitioners. In the preface to *Micrographia,* he notes that the microscope can produce confusing images and that in viewing objects through the microscope "there is much more difficulty to discover the true shape, then of those visible to the naked eye. . . ."[69] Hooke goes on to describe how careful he is in making observations, unlike other users of microscopes such as Henry Power. "The Eyes of a Fly in one kind of light appear almost like a Lattice, drilld through with an abundance of small holes; which probably may be the Reason, why the Ingenious Dr. Power seems to suppose them such."[70] Hooke found the eye of a fly to instead be "shap'd into a multitude of small Hemispheres."[71] In *Micrographia,* Hooke presents himself as an authority on the microscope, fully in command of the instrument and in possession of all the skills necessary for obtaining accurate observations.

The sketches in the Covel notebook record some of the earliest microscopical observations Hooke made, working with several

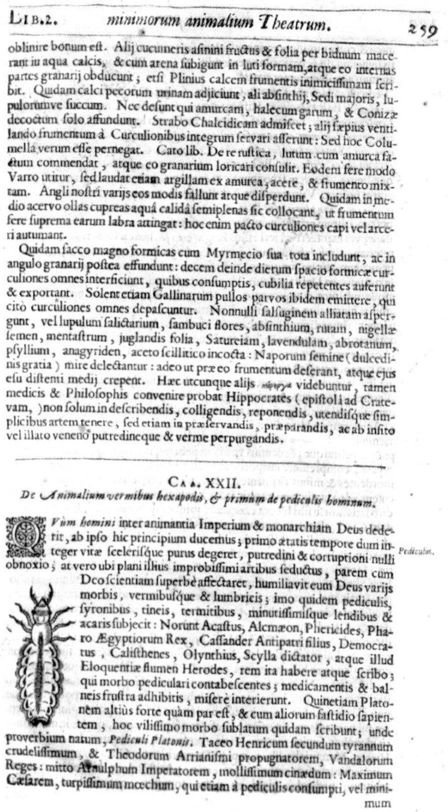

LIB. 2. *minimorum animalium Theatrum.* 259

oblinire bonum est. Alij cucumeris asinini fructus & folia per biduum macerant in aqua calcis, & cum arena subigunt in luti formam, atque eo internas partes granarij obducunt; etsi Plinius calcem frumentis inimicissimam scribit. Quidam calci pecorum urinam adjiciunt, ali absinthij, Sedi majoris, lupulorumve succum. Nec desunt qui amurcam, halecum garum, & Conizæ decoctum solo affundunt. Strabo Chalcidicam admiscet; alij sæpius ventilando frumentum à Curculionibus integrum servari asserunt: Sed hoc Columella verum esse pernegat. Cato lib. De re rustica, lutum cum amurca factum commendat, atque eo granarium loricari consulit. Eodem fere modo Varro utitur, sed laudat etiam argillam ex amurca, acere, & frumento mixtam. Angli nostri varijs eos modis fallunt atque disperdunt. Quidam in medio acervo ollas cupreas aquâ calidâ semiplenas sic collocant, ut frumentum fere suprema earum labra attingat: hoc enim pacto curculiones capi vel arceri autumant.

Quidam sacco magno formicas cum Myrmecio sua tota includunt, ac in angulo granarij postea effundunt: decem deinde dierum spacio formicæ curculiones omnes interficiunt, quibus consumptis, cubilia repetentes auferunt & exportant. Solent etiam Gallinarum pullos parvos ibidem emittere, qui citò curculiones omnes depascuntur. Nonnulli salsuginem alliatam aspergunt, vel lupulum salictarium, sambuci flores, absinthium, rutam, nigellæ semen, mentastrum, juglandis folia, Satureiam, lavendulam, abrotanum, psyllium, anagyriden, aceto scillitico incocta: Naporum semine (dulcedinis gratia) mire delectantur: adeo ut præ eo frumentum deserant, atque ejus esu distenti medij crepent. Hæc utcunque alijs πάρεργα videbuntur, tamen medicis & Philosophis convenire probat Hippocrates (epistol i ad Cratevam,) non solum in describendis, colligendis, reponendis, utendisque simplicibus artem tenere, sed etiam in præservandis, præparandis, ac ab insito vel illato veneno putredineque & verme perpurgandis.

CAP. XXII.

De Animalium vermibus hexapodis, & primum de pediculis hominum.

Quum homini inter animantia Imperium & monarchiam Deus dederit, ab ipso hic principium ducemus; primo ætatis tempore dum integer vitæ scelerisque purus degeret, putredini & corruptioni nulli obnoxio; at vero ubi plani illius improbissimi artibus seductus, parem cum Deo scientiam superbè affectaret, humiliavit eum Deus varijs morbis, vermibusque & lumbricis; imo quidem pediculis, syronibus, tineis, termitibus, minutissimisque lendibus & acaris subjecit: Norunt Acastus, Alcmæon, Phericides, Pharo Ægyptiorum Rex, Cassander Antipatri filius, Democratus, Calisthenes, Olynthius, Scylla dictator, atque illud Eloquentiæ flumen Herodes, rem ita habere atque scribo; qui morbo pediculari contabescentes; medicamentis & balneis frustra adhibitis, miserè interierunt. Quinetiam Platonem altiùs forte quàm par est, & cum aliorum fastidio sapientem, hoc vilissimo morbo sublatum quidam scribunt; unde proverbium natum, *Pediculi Platonis.* Taceo Henricum secundum tyrannum crudelissimum, & Theodorum Arrianismi propugnatorem, Vandalorum Reges: mitto Arnulphum Imperatorem, mollissimum cinædum: Maximum Cæsarem, turpissimum mœchum, qui etiam à pediculis consumpti, vel minimum

Pediculus.

16. Thomas Moffet, *Louse,* from *Insectorum sive minimorum animalium theatrum [Theatrum Insectorum]* (London, 1634), 259, woodcut
The Huntington Library, San Marino, California

co-observers. Though in some cases his sketches are but a hasty set of pen scratches, they are nevertheless crucial to our understanding of the images in *Micrographia.* Hooke's illustrations for *Micrographia* are intricately constructed compositions that were the products of several years of work and many reworkings of the material; they are not simply direct transcriptions of observations, but complex translations of those observations. The sketches, on the other hand, record an important early stage in the process of making microscopical observations and demonstrate that this process was inseparable from making the final images.

These twin processes can also be seen at work in Hooke's illustration of the louse, one of the most striking and best-known images from *Micrographia* (see fig. 13). Folio 113v of the Covel notebook (see fig. 11) contains a sketch that is perhaps one of Hooke's first efforts at depicting the louse and suggests that he may have used an earlier image of a louse as a model. In its size and format, Hooke's sketch resembles an illustration that appears in Thomas Moffet's *Theatrum Insectorum* of 1634 (fig. 16). Where the Moffet louse hints at the presence of internal organs with a jagged line along its abdomen, Hooke's sketch renders these forms in greater detail. In addition, folio 112v of the Covel notebook (fig. 17) includes a separate sketch of the internal organs of an insect that could be the louse. If this is the case, the sketch of the louse on folio 113v shows that Hooke combined his observations of the dissected internal organs with the Moffet model to produce an image that begins to approximate the finished version in schema 35 in *Micrographia.*

In the engraving, Hooke uses the transparent membrane of the insect's body as a window to its internal structures, uniting in a single image information gained through dissection with observations of the creature's external appearance. The louse's anthropomorphized forelegs curl around the hair of its human host with biceps, forearm, and pincers, and rather than aggressively confronting the viewer, this stoic sentry quietly offers us a fascinating glimpse of the workings of nature. The image thus supports one of Hooke's central aims in the *Micrographia:* to present the microscope as a means of accessing a world of fascinating objects and creatures rather than merely offering a chaotic jumble of fragments and dead insects.

While Hooke had important social, political, and professional reasons for constructing the microworld as a place of calm and order, he may also have had personal ones for creating such sympathetic images of insects. While he was working on *Micrographia,* he was also called upon to perform respiratory experiments on live dogs for the Royal Society. Hooke was so distressed by the animals' suffering that he refused to repeat the experiments when requested to do so by members of the society, though he was eventually compelled to carry them out.[72] It is in this context that Hooke's representation of insects in *Micrographia* as whole—as living creatures occupying a peaceful world—can also be understood. Both the words and images of *Micrographia* were constructed to hide the confusion and messiness involved in making microscopical

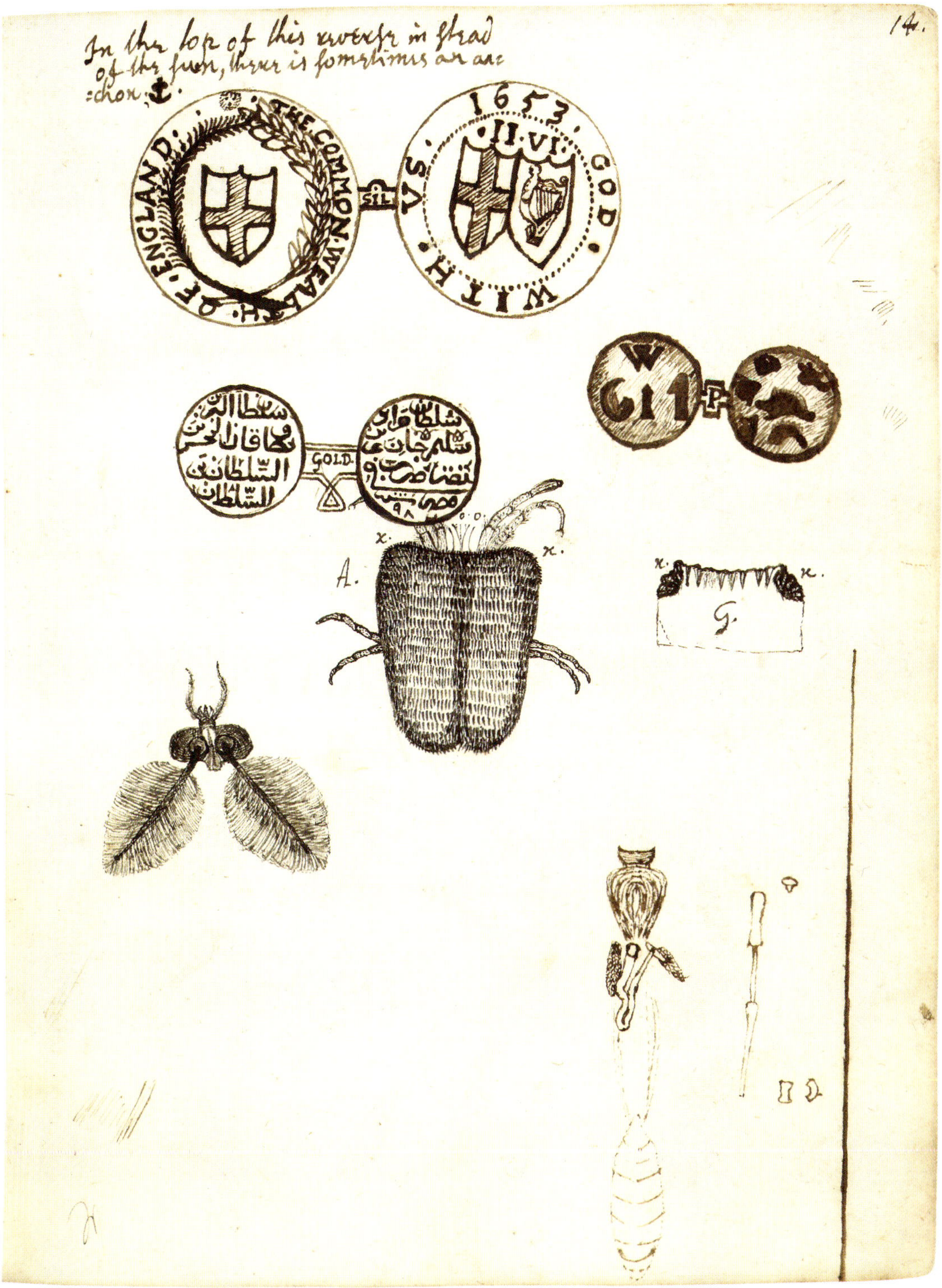

17. Robert Hooke and others, folio with drawings of various coins and insect parts, including a sketch of the internal organs of an insect, from John Covel, *Natural History and Commonplace Notebook* (London, c. 1660–1713), ink
The British Library, MS Add. 57495, folio 112v

observations, and in that respect the book served as an idealized model of the proper way to investigate the natural world. In his account of the water insect, Hooke argues that insects are the ideal subjects and microscopes the ideal instruments; together, they allow observers to study nature

> . . . according to her usual course and way, undisturbed, whereas, when we endeavor to pry into her secrets by breaking open the doors upon her, and dissecting and mangling creatures whil'st there is life yet within them, we find her indeed at work, but put into such disorder by the violence offer'd, as it may be easily imagin'd, how differing a thing we should find if we could, as we can with a *Microscope,* in these smaller creatures, quietly peep in at the windows, without frightening her out of her usual byas.[73]

NOTES

1. *Micrographia: Or Some Physiological Descriptions of Minute Bodies Made by Magnifying Glasses, with Observations and Inquiries Thereupon* (London, 1665), hereafter cited as Hooke 1665. On Robert Hooke's life and work, see J. A. Bennett et al., eds., *London's Leonardo: The Life and Work of Robert Hooke* (Oxford, 2003); Stephen Inwood, *The Man Who Knew Too Much: The Strange and Inventive Life of Robert Hooke, 1635–1703* (London, 2002); Lisa Jardine, *The Curious Life of Robert Hooke: The Man Who Measured London* (New York, 2004).

2. Hooke's inheritance was between fifty and one hundred pounds, according to Inwood 2002, 8. These details of Hooke's biography are taken from Michael Cooper, "Hooke's Career," in Bennett et al. 2003, 1–8; Inwood 2002, 6–17; and Jardine 2004, 21–56.

3. Balthasar de Monconys (1611–1665), a French diplomat and art connoisseur, related that, on his visit to the king's cabinet, he saw these drawings, along with Wren's lunar globe (see note 5): "où je vis la Lune, que M. Rene à faite de relief en carton, suivant le dessein d'Hevelius, et des desseins à la plume d'un Poû, d'une Pûce de la teste, et d'une Aille d'une Mouche, fait par le Microscope" (where I saw the Moon that Monsieur Wren had made in cardboard relief, following the design of [the astronomer, Johannes] Hevelius, and some pen-and-ink drawings of a louse, a flea, and the wing of a fly, made by the use of a Microscope). Balthasar de Monconys, *Journal des voyages de Monsieur de Monconys . . . publié par le sieur de Liergues son fils,* 3 vols. (Lyon, 1665–1666), 2:82. Christiaan Huygens (1629–1695), a Dutch mathematician and physicist, also reported seeing Wren's insect drawings in the king's cabinet; see J. A. Bennett, *The Mathematical Science of Christopher Wren* (Cambridge, 1982), 73. The location of the insect drawings Wren presented to Charles II is not presently known. Harwood reports searching for them without success in the collections of the Royal Library, Windsor; the Natural History Museum (London); and the British Library; see John T. Harwood, "Rhetoric and Graphics in *Micrographia,*" in *Robert Hooke: New Studies,* ed. Michael Hunter and Simon Schaffer (Woodbridge, Suffolk, U.K., and Wolfeboro, N.H., 1989), 122 n. 15.

4. Henry Powle (1630–1692) was a founding member of the Royal Society. Powle's letter to Wren is published in Christopher Wren, *Parentalia; or, Memoirs of the Family of the Wrens; Viz., of Mathew, Bishop of Ely, Christopher, Dean of Windsor, Etc. But Chiefly of Sir Christopher Wren . . . in Which Is Contained, Besides His Works, a Great Number of Original Papers and Records . . . Compiled by His Son Christopher* (1750; reprint, Farnborough, Hants, U.K., 1965), 210. Hereafter cited as Wren 1750.

5. Wren 1750, 210. Along with this request for drawings, the king also asked that Wren construct a lunar globe for him. Wren's activities involving the lunar globe and the problems it raised for the Royal Society

are discussed in Adrian Tinniswood, *His Invention So Fertile: A Life of Christopher Wren* (London, 2001), 77–79, and in Bennett 1982, 40–41. Wren would eventually complete the lunar globe, again presenting the gift directly to the king, in August of 1661.

6. Thomas Birch, *The History of the Royal Society of London for Improving of Natural Knowledge from Its First Rise*, 4 vols. (1756–1757; reprint, New York, 1968), 1:33.

7. Wren 1750, 211.

8. In July of 1663 the Royal Society was preparing for a visit from Charles II, and Hooke was directed to prepare a "handsome book" of his microscopic drawings for the king's entertainment. Hooke was in the midst of his work on *Micrographia* at this time, and the drawings he was asked to present were probably those he had been showing at meetings since March of that year. There is no record that the visit from Charles II actually occurred. See Birch 1756–1757, 1:272.

9. Hooke did not receive a salary from the Royal Society until 1664, when he was appointed professor of "the history of the trades" at Gresham College. Birch 1756–1757, 1:453.

10. Birch 1756–1757, 1:21. The publication referred to at this meeting is Robert Hooke, *An Attempt for the Explication of the Phaenomena, Observable in an Experiment Published by the Honourable Robert Boyle* (London, 1661).

11. Shapin and Schaffer point out that Hooke was appointed curator in November of 1662 and "from this moment, significantly, the pump in London worked better. Within six months it had been completely rebuilt along the lines of Boyle's machine in Oxford, a machine whose existence was partly due to Hooke himself. Thus it was not until Hooke was given responsibility for this pump . . . that the London pump, now reconstructed, could work reliably." See Steven Shapin and Simon Schaffer, *Leviathan and the Air-Pump: Hobbes, Boyle, and the Experimental Life* (Princeton, 1985), 235. On Hooke's work with Robert Boyle, see esp. 36–38, 231, and 250.

12. For further discussion of Hooke's status within the Royal Society, see J. A. Bennett, "Robert Hooke as Mechanic and Natural Philosopher," *Notes and Records of the Royal Society of London* 35 (1980): 33–48, esp. 34; Steven Shapin, "Who Was Robert Hooke?" in Hunter and Schaffer 1989, 253–285, esp. 253–254, 256, 263, and 283–285.

13. Steven Shapin, "The House of Experiment in Seventeenth-Century England," *Isis* 79 (1988): 396.

14. Birch 1756–1757, 1:213.

15. The image of moss would later be published as schema 13 in *Micrographia*. In Harwood's chronology of Hooke's observations, the moss is listed as Hooke's fourth observation. However, the first three observations listed in Harwood's chronology did not involve presentations of microscopic drawings. Hooke later made drawings of the first three items listed by Harwood and published illustrations of them in *Micrographia*, but the initial presentations of these items involved verbal rather than visual reports. See Harwood 1989, 124–125.

16. Harwood's chronology lists one further presentation, of petrified wood, on 24 August 1664. This presentation did not involve a drawing; see also note 15.

17. Birch 1756–1757, 1:231. Harwood's chronology states that the six-eyed spider was illustrated in schema 31 of *Micrographia*, but the spider pictured in that plate has only two eyes.

18. Birch 1756–1757, 1:250, 255, 262, and 270.

19. On Wren's early work with the microscope, see Bennett 1982, 73. Historian A. Rupert Hall dates Hooke's early work with the microscope to this period as well and speculates that Hooke and Wren worked together on microscopy in Oxford during the 1650s; see A. Rupert Hall, *Hooke's "Micrographia," 1665–1965* (London, 1966), 6–7.

20. On later collaborations between Wren and Hooke, see Margaret 'Espinasse, "Hooke as Surveyor and Architect," in *Robert Hooke* (London, 1956), 83–105; Bennett 1982, esp. 19, 41–43, 50–56, 61–70, 73–74, 87–88; Tinniswood 2001, 98–99, 150–151; and Lisa Jardine, "Monuments and Microscopes: Scientific Thinking on a Grand Scale in the Early Royal Society," in *Notes and Records of the Royal Society of London* 55 (2001): 289–308.

21. Recorded in Hartlib's *Ephemerides* for 1655, as quoted in H. W. Turnbull, "Samuel Hartlib's Influence on the Early History of the Royal Society," in *Notes and Records of the Royal Society of London* 10 (1953): 114. Bennett mistakenly gives the year for this conversation as 1665, the year in which *Micrographia* was published (and three years after Hartlib's death). See Bennett 1982, 73.

22. Henry Power, *Experimental Philosophy* (London, 1664). See also Harwood 1989, 122–129.

23. Birch 1756–1757, 1:266. John Wilkins (1614–1672) was one of the founders of the Royal Society and warden of Wadham College, Oxford. While Hooke was studying at Oxford he attended scientific meetings organized by Wilkins and later praised him in the preface to *Micrographia*.

24. Maxwell E. Power, "Sir Christopher Wren and the *Micrographia*," *Transactions of the Connecticut Academy of Arts and Sciences* 36 (1945): 37–44.

25. Birch 1756–1757, 1:442.

26. Hooke 1665, preface, n.p.

27. Hooke 1665, preface, n.p.

28. Hooke's apprenticeship to Lely is reported by Richard Waller in *The Posthumous Works of Robert Hooke* (1705; reprint, New York and London, 1969), iii. Hooke's diary entries in later years record purchases of paintings, prints, and drawing materials, as well as his acquaintance with a number of artists, including Lely, his former painting-master; see Leona Rostenberg, *The Library of Robert Hooke: The Scientific Book Trade of Restoration England* (Santa Monica, Calif., 1989), 8–9.

29. Hooke 1665, 153.

30. Svetlana Alpers also notes the relationship between Hooke's illustration of thyme seeds and Dutch still-life paintings of lemons and other objects. Alpers argues that both types of images exhibit a "fragmenting approach," one that aims to draw the viewer's attention to the individual features of an object. Although the isolation of individual features is characteristic of Hooke's images, his overall pictorial aim was to build unified compositions and self-contained worlds from these fragments, rather than to create a series of fragmented, isolated views. See Svetlana Alpers, *The Art of Describing: Dutch Art in the Seventeenth Century* (Chicago, 1983), 85.

31. Hooke 1665, Observation 31, 156.

32. Hooke 1665, 152.

33. Francesco Redi published his illustrations of insects based on microscopic investigations shortly after Hooke's *Micrographia* was published. See Francesco Redi, *Esperienze intorno alla generazione degli insetti* (Florence, 1668). The original drawings for Redi's publication have recently been identified in the collections of the Biblioteca Nazionale Centrale in Florence; see Lucia Tongiorgi Tomasi and Paolo Tongiorgi, "Il naturalista e il cappellano: Osservazione della natura e immagini 'dal naturale,'" in Francesco Redi, *Natura e immagine: Il manoscritto di Franceso Redi sugli insetti delle galle*, ed. Walter Bernardi et al. (Pisa, 1997). See also Lucia Tongiorgi Tomasi, "L'infinitamente piccolo: immagini al microscopio di Redi e al tempo di Redi," in *Franceso Redi: Un protagonista della scienza moderna*, ed. Walter Bernardi and Lucia Guerrini (Florence, 1999).

34. Harwood 1989, 128.

35. Hooke 1665, 203.

36. Hooke 1665, 203.

37. Hooke 1665, 204.

38. Willem Piso et al., *Historia naturalis Brasiliae* (Leiden and Amsterdam, 1648); Richard Ligon, *A True and Exact History of the Island of Barbados* (London, 1657). Harwood has shown that Hooke copied the illustration of an aquatic beehive that appears as figure 3 of schema 27 in *Micrographia* from another work by Piso, *Indiae utriusque re naturali et medici* (Amsterdam, 1658), 113; see Harwood 1989, 142, n. 52.

39. Hooke 1665, 204.

40. Moffet (1553–1604) was a physician and member of Parliament. The full title of his book, published posthumously in 1634, is *Insectorum sive minimorum animalium theatrum*. Aldrovandi (1522–1605) was an Italian naturalist who helped to establish the botanical gardens in Bologna, one of the first in Europe. The full title of his treatise on insects is *De animalibus insectis libri septem, cum singulorum iconibus ad vivum expressis* (Bologna, 1602).

41. Hooke 1665, 184.

42. Harwood 1989, 121.

43. Harwood 1989, 131.

44. Harwood 1989, 142.

45. Michael Aaron Dennis, "Graphic Understanding: Instruments and Interpretation in Robert Hooke's 'Micrographia,'" *Science in Context* 3 (1989): 312.

46. Dennis 1989, 313.

47. Dennis 1989, 324.

48. The sale catalogue listing is quoted from Geoffrey Keynes, *A Bibliography of Dr. Robert Hooke* (Oxford, 1960), 23. The sale catalogue is reprinted in Rostenberg 1989, 143–221.

49. Royal Society, Classified Papers 20.7. The drawing is reproduced in Harwood 1989, 126, plate 9a. Hooke discusses figure 1 on page 88 of *Micrographia* and discusses figure 6 on page 93.

50. John Covel, Natural History Notebook and Commonplace Book: 1660–1713, British Library, Add. 57495. Hereafter referred to as BL Add. 57495.

51. On Covel, see *Extracts from the Diaries of Dr. John Covel, 1670–1679*, vol. 2 of *Early Voyages and Travels in the Levant*, ed. J. Theodore Bent (New York, 1964); and Jean-Pierre Grélois, ed., *Dr. John Covel, Voyages en Turquie, 1675–1677* (Paris, 1998).

52. On folio 114 of the notebook, Petiver refers to his own publication, *The Monthly Miscellany: Or, Memoirs for the Curious* (London, 1707–1709), and on folio 2 he refers to having seen a specimen, while in Holland, of an insect drawn by Covel. Petiver's only known visit to Holland occurred in 1711. For more on Petiver's life and collecting activities, see Raymond Phineas Stearns, "James Petiver, Promoter of Natural Science, c. 1663–1718," *Proceedings of the American Antiquarian Society* 62 (1952): 243–365.

53. BL Add. 57495, f. 58v.

54. BL Add. 57495, ff. 32v and 2.

55. BL Add. 57495, f. 96.

56. Identifications of the insects and other animals pictured in this folio, along with full transcriptions of the annotations, can be found in Janice L. Neri, "Some Early Drawings by Robert Hooke," *Archives of Natural History* 32 (2005): 41–47.

57. Hooke's lowercase *e* is particularly distinctive.

58. A lengthy written description of this drawing on a separate page of the notebook describes the steps taken to immobilize, dissect, and observe the insect. The treatment of the legs is described as follows: "I layd his leggs in the posture you see there drawn, but befour long they were crumpled as you may see prickt." BL Add. 57495, f. 118v.

59. For Hooke's written description of this illustration, see Hooke 1665, 187.

60. Here, Hooke seems to have used the word "sad" to mean "dark" or "deep," referring to the color green.

61. Hooke 1665, 208.

62. Piso et al. 1648, 185.

63. Shapin 1989, 259.

64. On the work of Greatorex and Hooke on the air pump, see Shapin and Schaffer 1985, 379. On Greatorex, see A. V. Simcock, "An Equinoctal Ring Dial by Ralph Greatorex," in *Making Instruments Count: Essays on Historical and Scientific Instruments Presented to Gerard L'Estrange Turner,* ed. R. G. W. Anderson, J. A. Bennett, and W. F. Ryan (Aldershot, Hampshire, U.K., and Brookfield, Vt., 1993), 201–215.

65. Michael Hunter, *The Royal Society and Its Fellows, 1660–1700: The Morphology of an Early Scientific Institution* (London, 1994), 11.

66. Steven Shapin, "Pump and Circumstance: Robert Boyle's Literary Technology," *Social Studies of Science* 14 (1984): 492.

67. In Hooke's case, experiments took place in his rooms. For a discussion of the domestic setting of laboratories and of the physical layout of the Royal Society's meeting rooms, see Shapin 1988, 391.

68. Steven Shapin, *A Social History of Truth: Civility and Science in Seventeenth-Century England, Science and Its Conceptual Foundations* (Chicago, 1994), 381.

69. Hooke 1665, preface, n.p.

70. Hooke 1665, preface, n.p.

71. Hooke 1665, 175.

72. Hooke described the experiment on a live dog and, in a letter to Robert Boyle dated 10 November 1664, confessed his unwillingness to perform the experiment again. See "Letter from Hooke to Boyle, 10 Nov. 1664," in R. T. Gunther, *Early Science in Oxford,* vol. 6:216–218, 217, as cited in Andreas-Holger Maehle and Ulrich Tröhler, "Animal Experimentation from Antiquity to the End of the Eighteenth Century: Attitudes and Arguments," in *Vivisection in Historical Perspective,* ed. Nicholas Rupke (London and New York, 1987), 43 n. 76. Nevertheless, Hooke did perform the experiment again for the Royal Society, with Richard Lower, on 10 October 1667. For a discussion of the 1664 experiment and Hooke's reaction, see Maehle and Tröhler 1987, 23. For a brief discussion of the experiment done with Lower, see Andreas-Holger Maehle, "Literary Responses to Animal Experimentation in Seventeenth- and Eighteenth-Century Britain," *Medical History* 34 (1990): 27–51, 31.

73. Hooke 1665, 186.

masarlos se echaran quatro arrobas / q con los segundos
q han de entrar masados se acresciente vna arroba de agua
fresca sobre las dos q quedaron del primero cozimiento & se torne
al fuego & cueza hasta que se espese & se torne como arrope / o
miel & estando asi se ha de quitar del fuego & dexarlo
asentar & despues cuelanlo por vn cedaço de cerdas no muy
ralo porque o vso seguiente q se limpio el licor o balsamo
artificial & ponen aparte lo limpio en sus botes / o re
domas / e vntan las llagas / o desgarraduras & avnque fal
te carne en la herida / la restaña la sangre & cura las lla
gas maravillosamente. & algunos dizen ahora q es mejor
q el balsamo del guacanax & esta muy esperimentado. la ho

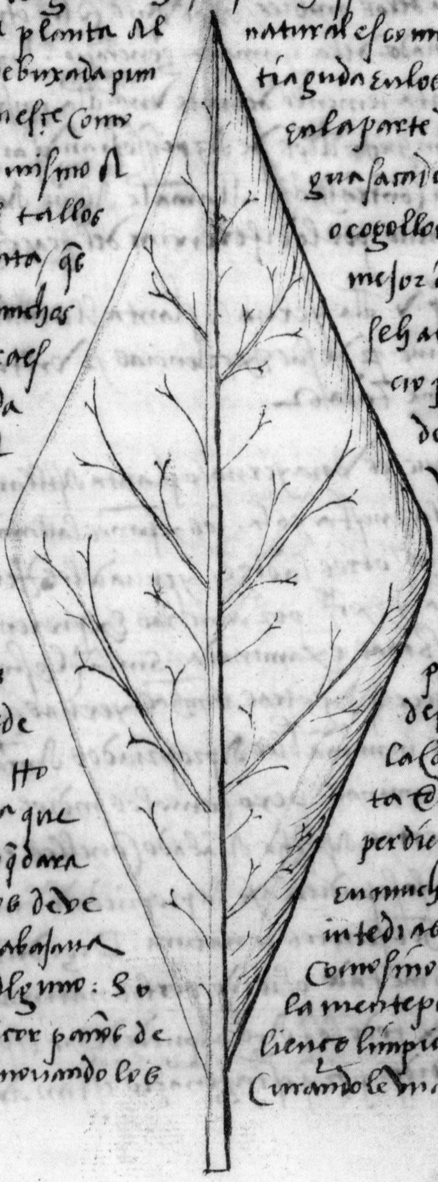

ja verdadera desta planta al natural es como esta q
aqui esta debuxada puntiaguda en los estremos
asi donde se fenesce como en la parte del pezon.
hazen asi mismo a guas sacado por alqui
tara de los tallos o cogollos della cima
desta planta / q es mejor q agua ar
diente & muchas se hallan bien
con ella. Acaesc en poco tiempo ha
q vna rueda de vna sierra
ta tomo a vn negro
la pierna por la pa
torcilla al luen
go & no de traves / por
q no le ha pro hueso a
guno mas desgarro le mu
cha parte de la carne / in
churada & tto ta de tal mano
q se pensaua que perdiera la pierna
o la vida / o q dara en mucha manqdad
& en menos de de inte dias estuuo
bueno & tan sano como si no ouiera te
nido mal alguno: Solo la mente poniendo le
con este licor paños de lienços limpios vntados
en el & renouando los curando le vna o dos vezes

a 5

Esta por hazer segun aqui esta

JESÚS CARRILLO CASTILLO
Universidad Autónoma de Madrid

The Eyes of the New Pliny: The Use of Images in Gonzalo Fernández de Oviedo's Historia general y natural de las Indias

Goaconax Leaf, from Gonzalo Fernández de Oviedo, *Historia general y natural de las Indias*, book 11, chapter 4, pen and ink
The Huntington Library, San Marino, California, HM 177, volume 2, folio 74r

The main concern of my analysis of early descriptions of American nature has been the search for genealogies of modern science that provide an alternative to a linear history. The goal has been to find a connection between scientific explorations in the New World and the simultaneous formation of discourses of empire in Europe.[1] Rough and unsophisticated as they may appear to readers today, the early Spanish texts concerning America—by such authors as Gonzalo Fernández de Oviedo y Valdés (1478–1557), Pedro Cieza de León (c. 1518–1560), Francisco López de Gómara (c. 1511–c. 1566), Bartolomé de Las Casas (1484–1566), and Bernal Díaz del Castillo (c. 1492–1581)—were neither spontaneous responses to the new nor mere continuations of former historiographic traditions. Besides being conveyors of differences, these treatises were also different as texts in themselves. As Francesco Patrizi noted in his 1560 treatise on the writing of history, the texts about "i viaggi de' Portughesi & de' Castigliani all'Indie & al Nuovo Mondo" (the voyages of the Portuguese and the Castilians to the [West] Indies and the New World) were essentially hybrid, at once descriptive and narrative, synchronic and diachronic.[2] This hybrid and all-embracing genre, which was to be known in the late sixteenth century as "historia natural y moral," was never to develop into more specialized and objective modes of addressing natural issues because the facts (the processes of discovery and conquest) and the people and things (the objects of these actions) were all part of one and the same discourse, and a highly ideological one at that.

The genre could simply be dismissed as a failed attempt, or another monstrous outcome, alien to the normal birth of modern natural history from the mating of late medieval herbals with the humanist rediscovery of ancient knowledge of nature. These writings, however—which emerged from the core of Western tradition and claimed to derive their authority from its highest institutions (the Spanish Empire and the Roman Catholic Church)—were the first to disseminate, in printed form, comprehensive accounts of American natural phenomena. They were also the first to include images based on empirical observation. It is important to remember that Oviedo's early summary of the findings later documented in the *Historia general y natural de las Indias*, for example, was published in Toledo—the center of Charles V's empire—in 1526.[3] But these remarks are not so much meant to claim a Spanish priority in the matter from the typically chauvinistic point of view of a provincial historiography. Instead, they are intended to point out the interest of these writings—involved as they were in a process of imperialist conquest and domination—in the formation of a world view in which modern natural history would finally find its place.

The aim of my ongoing research on Oviedo's *Historia general y natural de las Indias* (1535–1549) has been to trace the path he took in producing these written descriptions and images of the American natural world and to resolve once and for all the answer to the question of why he embarked on this project in the first place. First, it must be said that it took some time for America to become an object of representation, even for the Spaniards who had been living there for several decades. In 1528, while working as a mining supervisor in Nicaragua, and for fifteen years after his first arrival in America, Oviedo had been writing a treatise on heraldry.[4] Earlier, during a stay in Panama, he had written a chivalric novel, *Claribalte,* probably the first novel produced in the New World (fig. 1).[5]

Oviedo's enormous effort to make America a subject of historical writing was due at least in part to his need to gain recognition within the highly competitive field of official history, in which he had striven for years to find a relevant place. As outlined in the introduction to the 1526 summary, his main intention was to add a chapter to a larger historical project on the many kingdoms of Charles V (fig. 2).

If we read the terms of his official appointment as chronicler of the newly discovered territories in 1532, very little is said about the need to describe the natural world. We must wait more than forty years—until the reign of Philip II, when Nicolas de Ovando was president of the Council of the Indies, for an official order on these matters to be delivered and carried out. Besides, Oviedo obtained no financial aid from either the council or the crown for the printing of the first part of his work in 1535,[6] and the final attempt to print the remaining two parts had to be canceled when their author and only sponsor died in 1557. Oviedo's claim of merely performing an official duty when he was so carefully describing the plants and animals of America was thus to a great extent theatrical, but it also provided his ambitious editorial project the prestige of imperial authority. One could say that Oviedo, a lifelong courtier with experience as a notary, recognized the power of the state

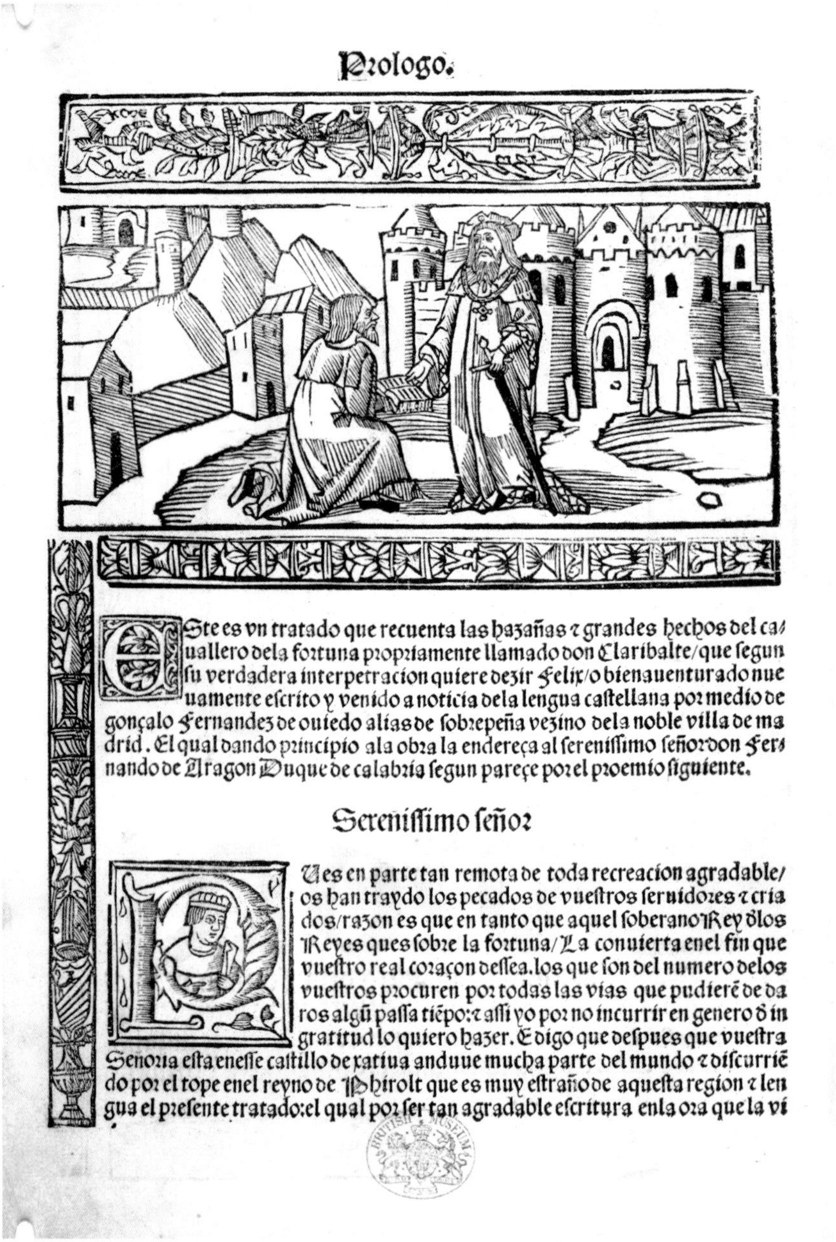
Prologo.

Este es vn tratado que recuenta las hazañas ⁊ grandes hechos del cauallero dela fortuna propriamente llamado don Claribalte/ que segun su verdadera interpetracion quiere dezir Felix/ o bienauenturado nueuamente escrito y venido a noticia dela lengua castellana por medio de gonçalo Fernandez de ouiedo alias de sobrepeña vezino dela noble villa de madrid. El qual dando principio ala obra la endereça al serenissimo señor don Fernando de Aragon Duque de calabria segun pareçe por el proemio siguiente.

Serenissimo señor

Pues en parte tan remota de toda recreacion agradable/ os han traydo los pecados de vuestros seruidores ⁊ criados/ razon es que en tanto que aquel soberano Rey dlos Reyes ques sobre la fortuna/ La conuierta enel fin que vuestro real coraçon dessea. los que son del numero delos vuestros procuren por todas las vias que pudieré de daros algũ passa tiẽpo: ⁊ assi yo por no incurrir en genero d ingratitud lo quiero hazer. E digo que despues que vuestra Señoria esta enesse castillo de xatiua anduue mucha parte del mundo ⁊ discurriẽdo por el tope enel reyno de Phirolt que es muy estraño de aquesta region ⁊ lengua el presente tratado: el qual por ser tan agradable escritura enla ora que la vi

1. Opening page of the prologue, from Gonzalo Fernández de Oviedo, *Libro del muy esforzado cavallero de la fortuna, propiamente llamado Claribalte* (Valencia, 1519), woodcut
The British Library

to produce truth and to stimulate beliefs years before the state actually began using that power in a systematic way. In addition, it should be noted that Oviedo personally signed the letter to the president of the council that appeared at the beginning of each copy of the 1535 edition, in order to validate the "official" character of the book's contents (fig. 3).

Who, then, was Oviedo's real audience? The publication of both *Claribalte* and the *Historia general y natural de las Indias* in expensive folio editions indicates that his

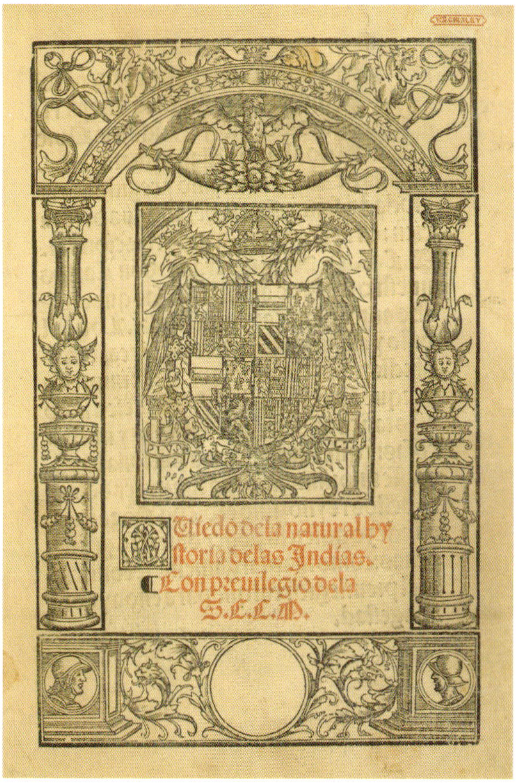

2. Title page from Gonzalo Fernández de Oviedo, *De la natural hystoria de las Indias* (Toledo, 1526), woodcut
Library of Congress

intended readers included wealthy gentlemen and members of the high clergy. Clive Griffin's studies on the printing industry in Spain during the first half of the sixteenth century, however, show that Oviedo's project was probably too ambitious for the Spanish market.[7] The publication of the fifty books and nearly three thousand manuscript folia containing more than one hundred original images on American themes would have been far too expensive even for the Council of the Indies, even if the members had ever considered financing the project. The money obtained from sales could by no means have surmounted the costs. Only the direct intervention of the king could have made publication possible, but Oviedo's repeated attempts at obtaining royal support—his *Libro de la Cámara del Príncipe don Juan* of 1548[8] and his *Quinquágenas* of 1555[9] were personally dedicated to Prince Philip—were fruitless.

It is thus obvious that for Oviedo, the *Historia general y natural de las Indias* was, above all, a question of personal prestige. It is also apparent that his quest for recognition was more important than any financial benefits he might have realized from selling the book—and those would have been nothing compared to the profits of his many businesses in Santo Domingo. Despite the highly nationalistic tone of his writings, the fame and recognition for which he strove so earnestly were of a universal character. For the same reason that he claimed to be the spokesperson of the Spanish Empire, Oviedo was also addressing an international audience. But this audience was not exactly the homogeneous audience of Latin-reading scholars from Scotland to Poland, whose theoretical approach Oviedo was said to despise. Rather, it was the expanding heterogeneous public, reading in the vernacular.

In 1526 Oviedo handed Andrea Navagero, ambassador of the Republic of Venice, a manuscript of the summary of the *Historia general y natural de las Indias,* to be translated into Italian even before it was published in Toledo. This version was published twice in Venice, in 1534 and 1556,[10] and was translated into English by Richard Eden in 1555, becoming the most widely known of Oviedo's texts. That same year, a French translation of the first part of the *Historia,*

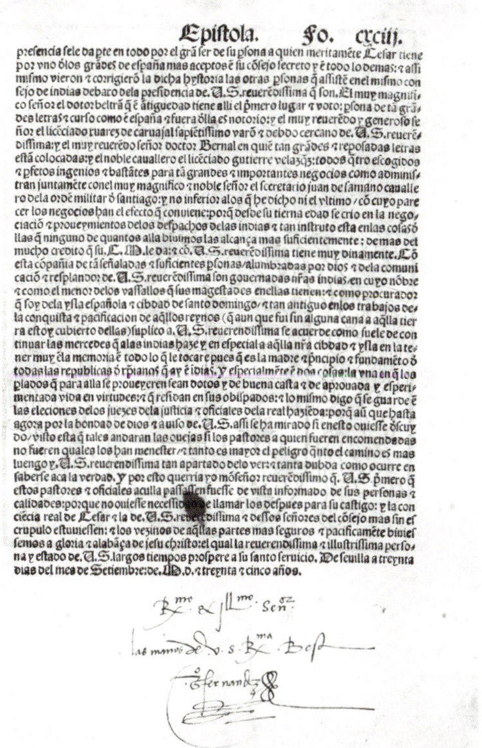
Epistola. Fo. cxciij.
presencia sele da pte en todo por el grã ser de su psona a quien meritamẽte Cesar tiene
por vno dlos grãdes de españa mas aceptos ẽ su cõsejo secreto y ẽ todo lo demas: ⁊ assi
mismo vieron ⁊ corrigierõ la dicha hystoria las otras psonas q̃ assistẽ enel mismo con
sejo de indias debaxo dela presidencia de. V.S. reuerẽdissima q̃ son. El muy magnifi-
co señor el dotor beltrã q̃ ẽ ãtiguedad tiene alli el pmero lugar ⁊ voto: psona de tã grã-
des letras ⁊ curso como ẽ españa ⁊ fuera dlla es notorio: y el muy reuerẽdo y generoso se
ñor el licẽciado xuarez de caruajal sapiẽtissimo varõ ⁊ debdo cercano de. V.S. reuerẽ-
dissima: y el muy reuerẽdo señor doctor Bernal en quiẽ tan grãdes ⁊ reposadas letras
estã colocadas: y el noble cauallero el licẽciado gutierre velazq̃z: todos q̃tro escogidos
⁊ pfetos ingenios ⁊ bastãtes para tã grandes ⁊ importantes negocios como adminis-
tran juntamẽte conel muy magnifico ⁊ noble señor el secretario juan de samano caualle
ro dela ordẽ militar d santiago: y no inferior alos q̃ he dicho ni el vltimo / cõ cuyo pare
cer los negocios han el efecto q̃ conuiene: porq̃ desde su tierna edad se crio en la nego-
ciaciõ ⁊ proueymientos delos despachos delas indias ⁊ tan instruto esta enlas cosas d
llas q̃ ninguno de quantos alla biuimos las alcança mas suficientemente: demas del
mucho credito q̃ su. C.M. le da: ⁊ cõ. V.S. reuerẽdissima tiene muy dinamente. Cõ
esta cõpañia de tã señaladas ⁊ suficientes psonas / alumbradas por dios ⁊ dela comuni
caciõ ⁊ resplandor de. V.S. reuerẽdissima son gouernadas nr̃as indias / en cuyo nõbre
⁊ como el menor delos vassallos q̃ sus magestades enellas tienen: ⁊ como procurador
q̃ soy dela ysla española ⁊ cibdad de santo domingo / ⁊ tan antiguo enlos trabajos de-
la conquista ⁊ pacificacion de aq̃llos reynos (q̃ aun que fui sin alguna cana a aq̃lla tier
ra estoy cubierto dellas) suplico a. V.S. reuerendissima se acuerde como suele de con
tinuar las mercedes q̃ alas indias haze y en especial a aq̃lla nr̃a cibdad ⁊ ysla en la te-
ner muy ẽla memoria ẽ todo lo q̃ le tocare pues q̃ es la madre ⁊ pncipio ⁊ fundamẽto d
todas las republicas d xpianos q̃ ay ẽ idias. Y especialmẽte ẽ dos cosas: la vna en q̃ los
plados q̃ para alla se proueyeren sean dotos y de buena casta ⁊ de aprouada y esperi-
mentada vida en virtudes: ⁊ q̃ residan en sus obispados: ⁊ lo mismo digo q̃ se guarde ẽ
las eleciones delos juezes dela justicia ⁊ oficiales dela real haziẽda: porq̃ aũ que hasta
agora por la bondad de dios ⁊ auiso de. V.S. assi se ha mirado si enesto ouiesse dscuy
do / visto esta q̃ tales andaran las ouejas si los pastores a quien fueren encomendadas
no fueren quales los han menester / ⁊ tanto es mayor el peligro q̃nto el camino es mas
luengo y. V.S. reuerendissima tan apartado delo ver: ⁊ tanta dubda como ocurre en
saberse aca la verdad. Y por esto querria yo mõseñor reuerẽdissimo q̃. V.S. pmero q̃
estos pastores ⁊ oficiales aculla passassen fuesse de vista informado de sus personas ⁊
calidades: porque no ouiesse necessidad de llamar los despues para su castigo: y la con
ciẽcia real de Cesar ⁊ la de. V.S. reuerẽdissima ⁊ dessos señores del cõsejo mas sin es
crupulo estuuiessen: ⁊ los vezinos de aq̃llas partes mas seguros ⁊ pacificamẽte biuies
semos a gloria ⁊ alabãça de jesu christo: el qual la reuerendissima ⁊ illustrissima perso-
na y estado de. V.S. largos tiempos prospere a su santo seruicio. De seuilla a treynta
dias del mes de Setiembre: de. M.D. ⁊ treynta ⁊ cinco años.

3. Letter to the president of the Council of the Indies, signed by the author, from Gonzalo Fernández de Oviedo, *Historia general y natural de las Indias, con privilegio imperial* (Seville, 1535), folio 193r, woodcut
Rare Books Division, New York Public Library, Astor, Lenox and Tilden Foundations

bearing traces of Oviedo's close supervision, was issued by the printing house of Michel Vasconsan.[11] As Giovanni Battista Ramusio observed in his introduction to the third volume of his famous *Navigazioni e viaggi,* numerous readers were eagerly awaiting the remaining two issues of Oviedo's work, the first part of which Ramusio was presenting in his collection of travels. That Oviedo himself was behind the international dissemination of his work is confirmed by the fact that, as early as 1540, he had obtained papal permission to publish the book in all Christian territories.[12]

As stated earlier, Oviedo's ideal audience was not necessarily Spanish and not necessarily a scholarly one. Instead, he assumed that his readers accepted Spain as the legitimate heir to the Roman Empire and that they, as he did, read voraciously in the vernacular about matters of history, travel, and nature. This audience, which he identified as being an "orthodox Christian readership," was addressed not as an abstraction, but as a mass of individuals—the consensus of whom he frequently invoked from the pages of the *Historia*—eager to learn about both the contingencies of everyday life and the novelties coming from distant countries.

Oviedo's constant call to readership was essential to his representation of nature. For in the dominant rhetorical approach to knowledge that he systematically applied to his writings, there was no significant difference between the customary interpretation of things and the nature of things themselves. According to the combination of Aristotelian and Augustinian themes that Oviedo used to support his position, humankind was naturally biased toward nature, which was "naturally" available to knowledge and experience by the human intellect. There was thought to be a spontaneous understanding—*entendimiento*—of things typically identified with vision which, ideally, did not serve any specific goal or purpose. For Oviedo, this altruistic approach to nature was not only morally good; it also made humans more God-like.[13]

For Oviedo, then, the main source of knowledge was believed to be the spontaneous understanding of things derived from direct experience. But experience could lead to truth only after being subjected to a process of representation, through its translation into the words and ideas that made it comprehensible, both to oneself and to others. The representation of experience was particularly necessary in the New World, where chaos and disorder seemed to overwhelm the capacity to understand. As Oviedo wrote in the *Historia general y natural,* all Spanish conquistadors should follow the example of Captain Rodrigo Ranjel, a member of Hernando de Soto's expedition to Florida, who systematically recorded all his daily sensations and emotions in order to make sense of an otherwise alienating stream of life experiences. Oviedo compared the benign effects of such a practice to those of Christian confession, as well as to the satisfaction derived from communicating one's experiences back home.[14]

By no means did this indicate a disregard for differences from one part of the world to another. As a matter of fact, among the Spanish writers of the sixteenth century, Oviedo was one of the main champions of American difference. Nature, according to his scheme, was essentially the same across the globe but it was, to no less a degree, radically different in its local accidents. For Oviedo, "natural" had two clear meanings: On the one hand, it meant the normal order of things in the world according to the Aristotelian definition. On the other hand, it meant the particular and unrepeatable arrangement of those things in a specific location, as recognized by its "natural" inhabitants.

The appeal to novelty, a notion that Oviedo constantly cited in order to attract readers, was legitimate only insofar as it led to recognizing the contingent diversity of things within the essential continuity of the natural world. But by the same token, the impulse to recognize, typical of a conqueror's mentality, had to be measured against the local differences of things, of which only sustained experience in situ could provide a proper account. American nature was thus simultaneously open and restricted to Oviedo's readers, and it was the negotiation between these two contradictory qualities that formed the basis of his rhetoric of representation.

The ideal language with which to approach natural matters was, according to Oviedo, the plain and direct language of gentlemanly conversation, rather than the affected jargon of scholars. But the colloquial language to which Oviedo was referring was not just the transparent and natural conveyor of a kind of knowledge derived from experience; it was, more precisely, the language—the Castilian language—of navigators, soldiers, captains, and Oviedo himself, in which such truth was being revealed. This is important to note since, behind the defense of an apparently open natural language, there was an undercurrent of linguistic exclusion. His defense of the vernacular, to which he devoted a whole chapter in his *Historia general y natural de las Indias,* was not only directed against the academic milieu that might question his proficiency in natural matters. It also had a clearly nationalist aspect, participating in a debate that had begun during the reign of King Ferdinand, when Italian historians such as Lucio Marineo Siculo (c. 1444–c. 1533) were competing with Spanish authors for appointments as royal chroniclers. According to the local applicants, the history of the Spanish kingdoms could never be trusted if written by an Italian, because, as everyone knew, the Italians naturally took a republican perspective rather than an imperial point of view.[15]

Oviedo's exclusionism applied not only to academics and foreign writers but also to any possible competitor who had not actually experienced the things described in situ, as Oviedo had done. Such was the case with his predecessor, Pietro Martire d'Anghiera, known as Peter Martyr (1457–1526), whose writings Oviedo discredited time and again for having a necessarily distorted perception of American nature because d'Anghiera merely compiled secondhand accounts written by various explorers from the comfort of his position at the royal court. Finally, not even the individual sailor or soldier who had actually seen or discovered a particular plant or animal could be completely trusted, because, according to Oviedo, most of them were influenced by their own special interests and ambitions and lacked the objectivity of the civil servant who had been officially assigned to such a duty, as he had been.

His strategy of systematic exclusions ultimately indicated that he, Fernández de Oviedo, royal chronicler of the Indies, was the only legitimate bearer of the supreme task of representing nature. It is not surprising, then—when one analyzes the linguistic instruments that Oviedo used to carry out his project—that one finds an eclectic fabric of different discursive methods underlying the transparent surface of natural language that he claimed to use.

The emphatic empirical quality of Oviedo's discourse is one of these complex constructions. The way in which he described natural species, and the very reason *why* he described them, is dependent upon the fact that Oviedo had seen them or, even more so, the fact that he was actually looking at them as he described their qualities for his readers' eyes. *Autopsia,* the act of seeing with one's own eyes, had been the favorite vehicle for discourses on distant countries since at least the time of the Greek historian Herodotus (484–432 B.C.E.). The central subject matter of the autoptic narrative is experience, since the object of that experience is beyond the reader's visual field as it is being described. The autoptic method is a perfect instrument by which a writer can evoke the presence—*energeia*—of an object and engender belief in the existence of something absent, but such writing does not comment on anything about the object unless it has affected the narrator in one way or another. The autoptic approach thus provides a perfect control on the representational process, but it is a poor conveyor of information. It is a rhetorical trope at work in different methods of description, having no descriptive function in itself.

Oviedo was a master of different genres of representation involving autoptic elements, genres that qualified the object as much as they qualified the writer. The Petrarchan subjective discourse on absence and the ekphrasis of precious objects was among these techniques. As discussed earlier, Oviedo was among the Spanish avant-garde in adopting Italian literary trends, which he had the opportunity to learn about during his time in the Italian courts of the Sforza, Gonzaga, and Borgia as well as in Aragón in the late fifteenth century. A close friend of the Petrarchan poet Serafino dell'Aquila and an amateur sonnet writer himself, Oviedo was an expert in evoking an object by recalling its fragments and the intense subjective sensations invoked by its presence. Indeed, it was the absence of the natural species in Spain and its incommensurability in Spanish terms that triggered the process of representation.[16] The fetishization of the object that was the basis of Petrarchan poetry was also an essential element of artistic ekphrasis, a genre that Oviedo applied systematically to one of his later works, a manuscript collection of more than two hundred biographies titled *Batallas y quinquágenas* (1535–1552). At the end of each biography, following the convention of providing an aesthetic experience, Oviedo added a detailed and vivid description of a medal or a picture illustrating the *impresa* of the person being described, produced by one of the best silversmiths or painters of the period, including the Spanish painter Pedro Berruguete and the Italian Andrea Mantegna. The description of an extraordinary bird, brought to Oviedo from the American mainland by the expeditionary captain Fray Andrés de Urdaneta, provides a good illustration of the use of such methods applied to the description of nature in the *Historia general y natural de las Indias:*

> The main feathers of both body and tail have a beautiful and nice brownish tawny color. The tail may have four feathers, each four hand-spans long. In their thickest part (near their root) they are the size of a big pinhead and, from there to the end, they become thinner and thinner to the point where they look like two threads, while the brown becomes darker and blackish. They are as rough as a saw when touched with your fingers and have neither hair nor other feathers, only very small ones near the roots. The rest of it is rough and thin, as I said. . . . Its head is as big as that of a thrush, and its feathers are yellow with a shade of orange. The collar is golden green, extremely colorful, covered with very dense, short, and very black feathers that look little longer than velvet, from which the beak starts, as big, straight, and pointed as that of a magpie. I can neither praise nor even describe accurately the wings of this bird. . . . I must confess that no painter will be able to paint it following only what I said; but as I read it while looking at the bird and giving thanks to God, I think I have said something. From my point of view this is the most remarkable bird, for its feathers and gentleness, that I have ever seen, and the one that I have admired the most.[17]

The use of autoptic elements in these descriptions goes beyond the realm of forensic rhetoric—the discourse on truth—that has usually been considered the origin of scientific language, and instead enters the domain of epideptics—the discourse on value—that provides the object of experience with enough cultural substance to be worthy of representation in the first place. This was particularly necessary when dealing with a subject—like America—that was devoid of fixed qualities and stable meanings. But the use of these techniques had a further function. The use of autoptic discourse not only fetishized the object of representation, but also remarked on its outstanding difference, a quality that challenged the existing terms of description and provoked a representational catastrophe. This catastrophe placed the subjective experience of the narrator—Oviedo—at the center of the discourse, at the same time that it defined the nature of the object in terms of its differences. This emphatically subjectivist system of representation was instrumental in the development of early colonial discourses since, on the one hand, it inaugurated a new and distinct place of utterance—the New World—from which Oviedo derived his authority as a direct witness. On the other hand, it simultaneously generated a new and different object—the New World—that Oviedo was seeing and describing. This expansive logic of territorialization was based on, but also bound to, the individual agents of conquest and colonization.

Oviedo's appeal to the visual—indeed, his very use of images—has to be understood as a trope to refer to the exteriority of these new objects with regard to language and convention. This is particularly patent in his famous description of the pineapple:

> Neither the painting of my pen nor that of my words can specifically represent the *blason* [i.e., appearance] of this fruit properly, or provide full satisfaction. This example cannot be described in its particulars without the use of the brush and the drawing, and even colors may be required in order both to render, at least in part, and to transmit what I do and say, so that the sight of readers can participate in this truth. Nevertheless, I shall put it here as best as I can, as badly drawn as explained.

It will be enough for those who have seen this fruit, and they may add the rest.[18]

The use of images in this example, as in most descriptions of flora and fauna that appear in the first (1535) edition of the *Historia general y natural,* has to do with the precise function of invoking the drama of communicating differences. It is indexical with regard to the experience of the chronicler without being fully iconic in regard to the species being described. Whereas Oviedo's text is extremely vivid and evocative, the images he uses are very often schematic and abstract figures whose existence and meaning depend strictly on the verbal description. The text justifies the inclusion of the image in each case and determines its specific function. In Oviedo's treatise, the woodcut illustration appears on the printed page at the precise line where the reference to that image is made. In this regard, Oviedo differs from other natural historians of the sixteenth century, such as the ichthyologist Guillaume Rondelet (1507–1566), the ornithologist Pierre Belon (c. 1517–1564), or the encyclopedist and naturalist Conrad Gesner (1516–1565). Such precursors established a distinct separation between the encyclopedic information provided by the text and the naturalism of the illustrations, which, according to Gesner, were meant to convey the illusion of the subject's actual presence.[19]

A good example of the subjection of the figurative to the verbal is the illustration of the leaf of the *perebecenuc* (fig. 4). The highly abstract shape and shading in this picture achieve precise meaning only through their connection with the text: "That shade that you see on the edges of these leaves I have drawn should be interpreted as being purple."[20] The visual appearance of these images was limited to conveying more specific meanings—an object's size, shape, texture, or tonality—that were contained in the text. The schematic character of the woodcuts avoided any superfluous elements that might distract the reader from noticing their more precise characteristics. The introduction of the image was usually justified as a way to make the eyes of his readers "participate" *(participar),* in the double sense of having a part in and partaking of a reality that existed beyond their reach. In other cases, the illustrations were said to reproduce the *forma* or *figura*—the shape—of the phenomenon in order to make more intelligible—*dar a entender*—that which would otherwise remain obscure. Indeed, most botanical illustrations of the 1535 edition reproduce a leaf or, more precisely, the shape of a leaf, instead of the whole plant. Such images would thus remind the readers that full vision and, therefore, full understanding pertained exclusively to the chronicler, the one who had actually seen the object. Thus it was through their introduction

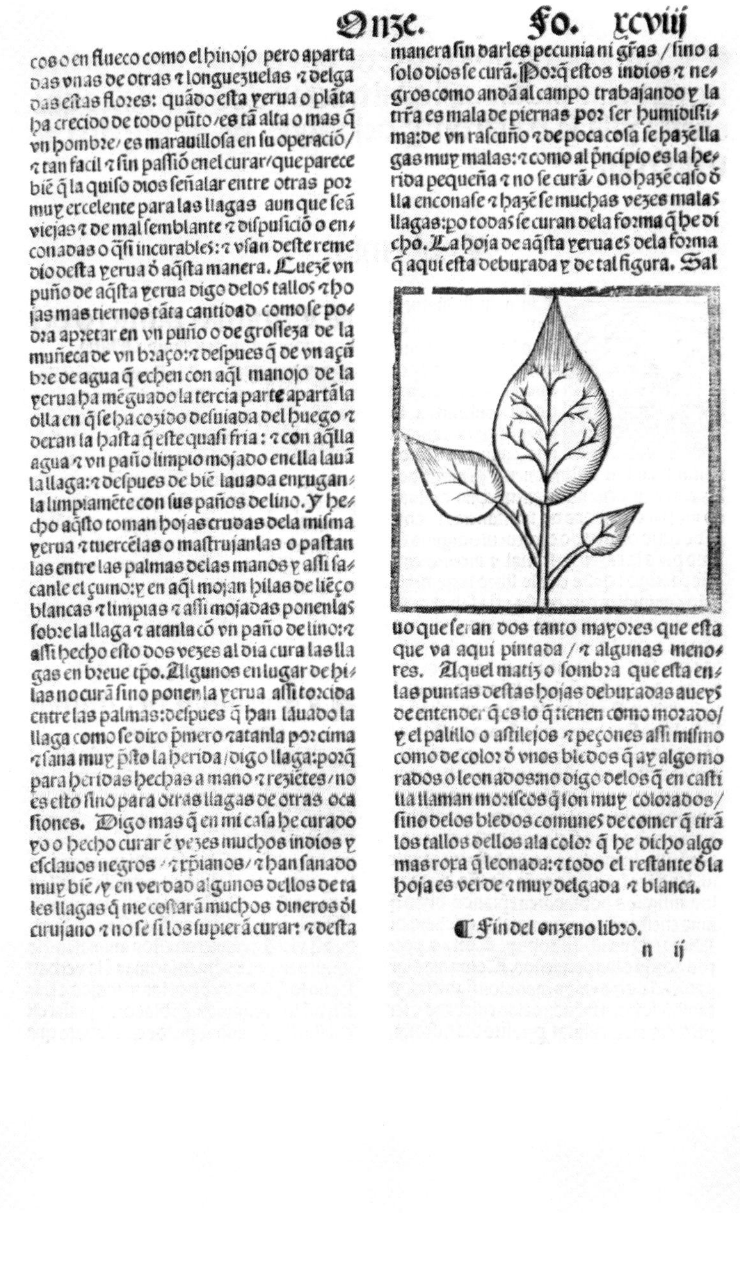
Onze. Fo. xcviij

cos o en flueco como el hinojo pero apartadas vnas de otras ⁊ longuezuelas ⁊ delgadas estas flores: quãdo esta yerua o plãta ha crecido de todo pũto / es tã alta o mas q̃ vn hombre / es marauillosa en su operaciõ / ⁊ tan facil ⁊ sin passiõ enel curar / que parece biẽ q̃ la quiso dios señalar entre otras por muy excelente para las llagas aun que seã viejas ⁊ de mal semblante ⁊ disposiciõ o enconadas o q̃si incurables: ⁊ vsan deste remedio desta yerua dẽ aq̃sta manera. Cuezẽ vn puño de aq̃sta yerua digo delos tallos ⁊ hojas mas tiernos tãta cantidad como se podra apretar en vn puño o de grosseza de la muñeca de vn braço: ⁊ despues q̃ de vn açũbre de agua q̃ echen con aq̃l manojo de la yerua ha mẽguado la tercia parte apartãla olla en q̃ se ha cozido desuiada del huego ⁊ dexan la hasta q̃ este quasi fria: ⁊ con aq̃lla agua ⁊ vn paño limpio mojado enella lauã la llaga: ⁊ despues de biẽ lauada enxuganla limpiamẽte con sus paños de lino. Y hecho aq̃sto toman hojas crudas dela misma yerua ⁊ tuercẽlas o mastrujanlas o pastan las entre las palmas delas manos y assi sacanle el çumo: y en aq̃l mojan hilas de liẽço blancas ⁊ limpias ⁊ assi mojadas ponenlas sobre la llaga ⁊ atanla cõ vn paño de lino: ⁊ assi hecho esto dos vezes al dia cura las llagas en breue tp̃o. Algunos en lugar de hilas no curã sino ponen la yerua assi torcida entre las palmas: despues q̃ han lãuado la llaga como se dixo p̃mero ⁊ atanla por cima ⁊ sana muy p̃sto la herida / digo llaga: porq̃ para heridas hechas a mano ⁊ reziẽtes / no es esto sino para otras llagas de otras ocasiones. Digo mas q̃ en mi casa he curado yo o hecho curar ẽ vezes muchos indios y esclauos negros / ⁊ xp̃ianos / ⁊ han sanado muy biẽ / y en verdad algunos dellos de tales llagas q̃ me costarã muchos dineros dl cirujano ⁊ no se si los supierã curar: ⁊ desta manera sin darles pecunia ni gr̃as / sino a solo dios se curã. Porq̃ estos indios ⁊ negros como andã al campo trabajando y la tir̃a es mala de piernas por ser humidissima: de vn rascuño ⁊ de poca cosa se hazẽ llagas muy malas: ⁊ como al p̃ncipio es la herida pequeña ⁊ no se curã / o no hazẽ caso dlla enconase ⁊ hazẽ se muchas vezes malas llagas: po todas se curan dela forma q̃ he dicho. La hoja de aq̃sta yerua es dela forma q̃ aqui esta debuxada y de tal figura. Saluo que seran dos tanto mayores que esta que va aqui pintada / ⁊ algunas menores. Aquel matiz o sombra que esta enlas puntas destas hojas debuxadas aueys de entender q̃ es lo q̃ tienen como morado / y el palillo o astilejos ⁊ peçones assi mismo como de color dẽ vnos bledos q̃ ay algo morados o leonados: no digo delos q̃ en castilla llaman moriscos q̃ son muy colorados / sino delos bledos comunes de comer q̃ tirã los tallos dellos ala color q̃ he dicho algo mas roxa q̃ leonada: ⁊ todo el restante dla hoja es verde ⁊ muy delgada ⁊ blanca.

Fin del onzeno libro.

n ij

4. *Perebecenuc Leaf,* from Gonzalo Fernández de Oviedo, *Historia general y natural de las Indias,* book 11, chapter 5 (Seville, 1535), folio 98r, woodcut
Rare Books Division, New York Public Library, Astor, Lenox and Tilden Foundations

to the precarious character of representation that readers could grasp the different nature of things in the New World. The two schematic images that Oviedo included, which illustrate the leaves of two varieties of yucca, show this concept very clearly.[21] The thread that linked the eyes of the chronicler with the understanding of his readers was essentially incompatible with the multiple gazes that an autonomous, lifelike reproduction of the object might have provoked.

But images were not only the markers of difference. As noted earlier, Oviedo defended the "naturalness" of American nature as much as he defended its local differences, his text being filled with descriptions of both native inhabitants and Spanish colonists using and eating those species in the course of everyday life. The figure of the iguana at the end of the chapter dedicated to it in book 13 (on fish) is a perfect example of this "normalizing" role of images (fig. 5). In this chapter, Oviedo deployed his best skills to provide a rich and vivid account of this American reptile. The length of his account is a consequence of both the ubiquitous presence of the iguana in the life of the Spanish colonists and its visual ambiguity with regard to traditional zoological classifications.[22] The iguana being both marine and terrestrial, Oviedo decided to transfer its description in a later revision of his treatise to book 12, which deals with animals. Oviedo's iguana was not clearly meat or fish, and it was definitely not the Plinian crocodile that Peter Martyr d'Anghiera had assumed it was in his *Decadas de Orbe Novo.* Oviedo solved the apparent paradox of the iguana by stressing its role in the realm of everyday life. It was in connection to this collective experience that Oviedo included an image of the iguana at the end of the chapter.

The schematic quality of this image indicates neither ignorance nor a lack of sensibility with regard to the iconic value of images. When moving into a different literary domain, Oviedo often seized the opportunity to show his taste for sophisticated and delicate images. Examples can be found in the many drawings of artistocratic *imprese* (heraldic crests) that illustrate an earlier project, an ambitious treatise on heraldry, which he later included in the *Batallas y quinquágenas* (figs. 6, 7, 8).[23] Simplification of these images made them easier to control, both with regard to the correct interpretation of his account and through the later editions of his book. The case of Ramusio's Italian

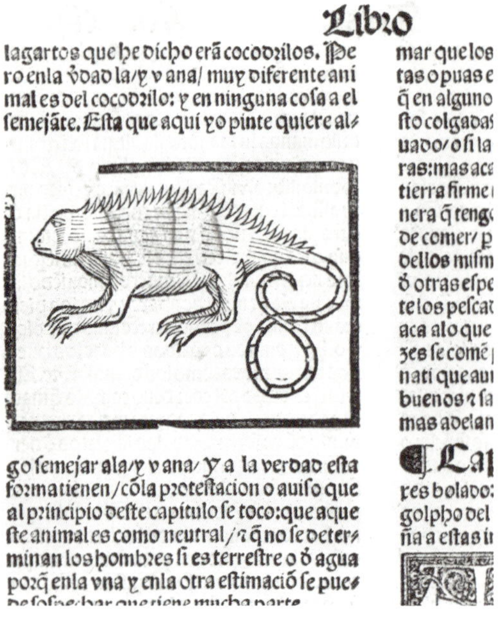
Libro

lagartos que he dicho erã cocodrilos. Pe
ro enla vdad la/y v ana/ muy diferente ani
mal es del cocodrilo: y en ninguna cosa a el
semejãte. Esta que aquí yo pinte quiere al-
go semejar ala/y v ana/ y a la verdad esta
forma tienen/cõla protestacion o auiso que
al principio deste capitulo se toco: que aque
ste animal es como neutral/ ⁊ q̃ no se deter-
minan los hombres si es terrestre o d agua
porq̃ enla vna y enla otra estimació se pue-

5. *Iguana,* from Gonzalo Fernández de Oviedo, *Historia general y natural de las Indias,* book 13, chapter 3 (Seville, 1535), folio 103v, woodcut
Rare Books Division, New York Public Library, Astor, Lenox and Tilden Foundations

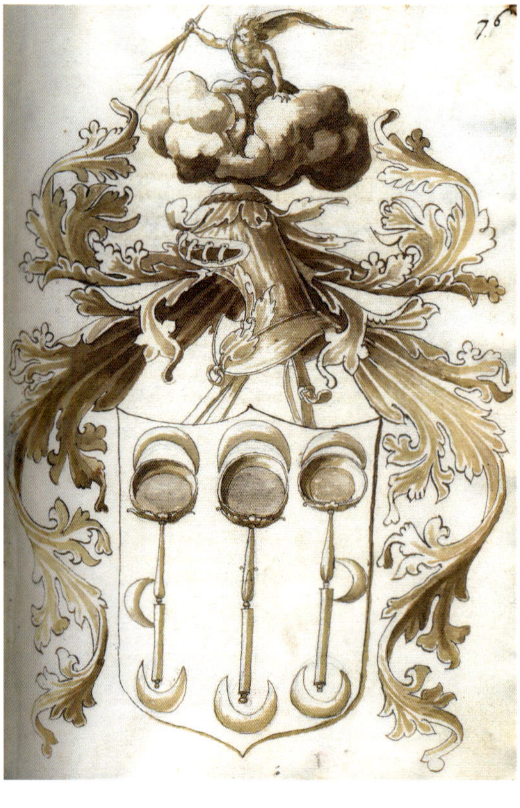

6. *Heraldic Crest of Antonio de Padilla,* from Gonzalo Fernández de Oviedo, *Batallas y quinquágenas* (1535–1552), batalla 2, quinquágena 1, dialogue 39, pen and ink
Biblioteca Universitaria de Salamanca, MS 359, folio 76r

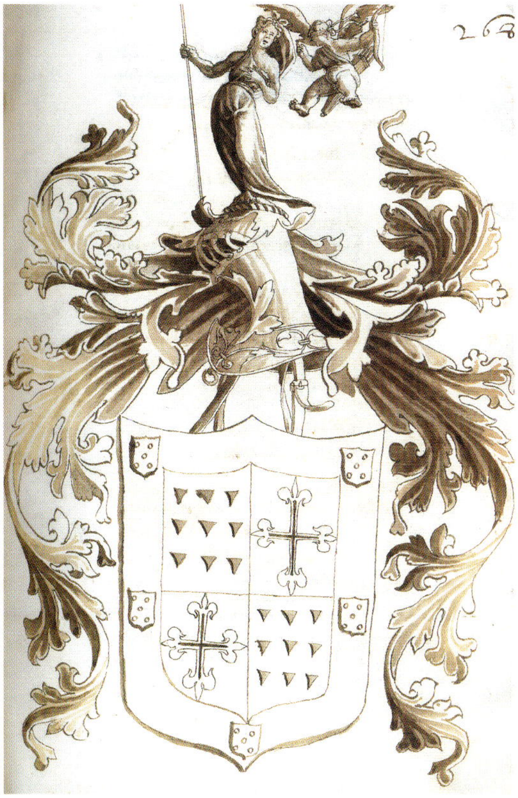

7. *Heraldic Crest of Enrique de Acuña,* from Gonzalo Fernández de Oviedo, *Batallas y quinquágenas* (1535–1552), batalla 2, quinquágena 1, dialogue 40, pen and ink
Biblioteca Universitaria de Salamanca, MS 359, folio 268r

edition of Oviedo's work is very clear in this respect. Whereas Oviedo's pictures were conceived as a schematic visual support of a vivid verbal description (figs. 9, 10), Ramusio's version conceived the American themes as autonomous counterparts of the written account (figs. 11, 12).

The illustrations in the Italian edition represented an American landscape where indigenous peoples—not the Spaniards appearing in Oviedo's text—interacted with the objects or species described. In the pictures of both the ceiba tree and the *plátano* leaf, the figures of the indigenous peoples serve the double function of being participants in a scene of American life and of providing a visual reference to show the relative sizes of the phenomena described. In both cases, the precise information provided by the text was thus rendered irrelevant. Ignoring the subordination of pictures to the narration of Oviedo's own experience, Ramusio produced images of America in which the voice of the chronicler was totally absent. The best example of this disjunction is Ramusio's representation of a Native American making a fire as having a beard—in keeping with the traditional iconography of the savage and the barbarian (fig. 13). But in so doing, he openly contradicted Oviedo's explicit statement that indigenous men of the Indies were rarely bearded.

Oviedo's use of autoptic discourse changed during the lengthy process of producing his treatise. As mentioned above, immediately after the publication of the first part of the *Historia general y natural de las Indias* in 1535, Oviedo started preparing a new and more ambitious text. The project not only included the remaining parts of the work (2 and 3), but also offered an expanded version of part 1, already published. This—as Ramusio lamented in a letter to the Italian physician Girolamo Fracastoro—could not be published despite Oviedo's desperate efforts to obtain funding and official support.[24] After three years of unsuccessful attempts at finding the necessary funding, Oviedo decided to go back to Santo Domingo in 1549, leaving four volumes of manuscripts ready for printing in Seville. It took more than three hundred years to publish them, and, in the meantime, important parts of the manuscript were lost. Fortunately,

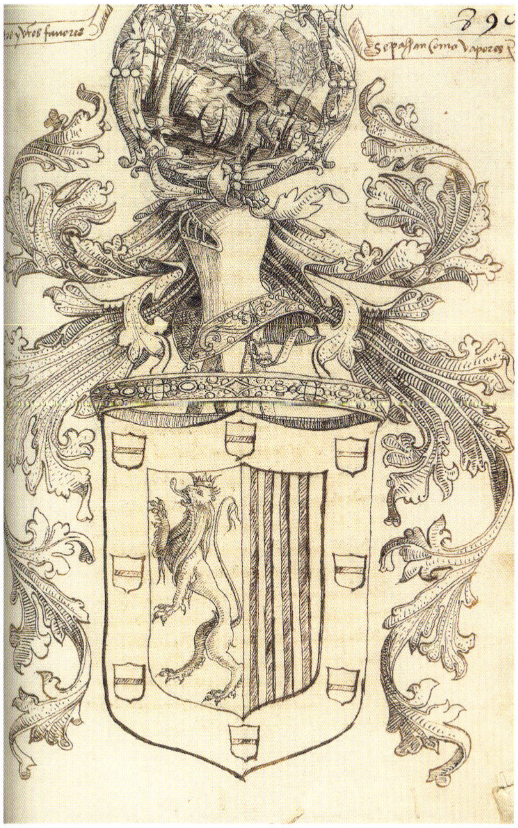

8. *Heraldic Crest of Rodrigo Ponce de León,* from Gonzalo Fernández de Oviedo, *Batallas y quinquágenas* (1535–1552), batalla 2, quinquágena 1, dialogue 10, pen and ink
Biblioteca Universitaria de Salamanca, MS 359, folio 890r

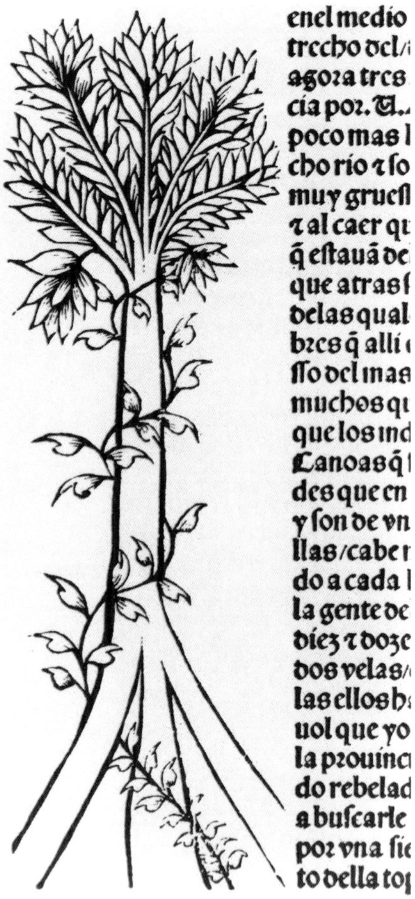

enel medio
trecho del/
agora tres
cia por. T
poco mas l
cho rio ⁊ so
muy gruess
⁊ al caer q
q̃ estauā de
que atras
delas qual
bres q̃ alli
sso del mas
muchos q
que los ind
Canoas q̃
des que en
y son de vn
llas/cabe
do a cada l
la gente de
diez ⁊ doze
dos velas/
las ellos b
uol que yo
la prouinci
do rebelad
a buscarle
por vna si
to della to

9. *Ceiba Tree,* from Gonzalo Fernández de Oviedo, *De la natural hystoria de las Indias* (Toledo, 1526), folio 40r, woodcut
Library of Congress

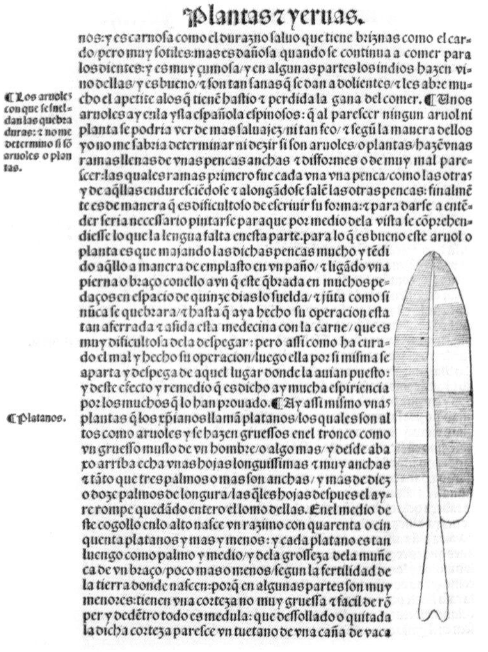

Plantas ⁊ yeruas.

nos: y es carnosa como el durazno saluo que tiene briznas como el cardo/pero muy sotiles: mas es dañosa quando se continua a comer para los dientes: y es muy çumosa/y en algunas partes los indios hazen vino dellas/y es bueno/⁊ son tan sanas q̃ se dan a dolientes/⁊ les abre mucho el apetito alos q̃ tienē hastio ⁊ perdida la gana del comer. ¶ Unos aruoles ay enla ysla española espinosos: q̃ al parescer ningun aruol ni planta se podria ver de mas saluajez/ni tan feo/⁊ segū la manera dellos yo no me sabria determinar ni dezir si son aruoles/o plantas/hazē vnas ramas llenas de vnas pencas anchas ⁊ disformes o de muy mal parescer: las quales ramas primero fue cada vna vna penca/como las otras y de aq̃llas endurresciēdose ⁊ alongādose salē las otras pencas: finalmēte es de manera q̃ es dificultoso de escriuir su forma: ⁊ para darse a entēder seria necessario pintarse para que por medio dela vista se cōprehendiesse lo que la lengua falta enesta parte. para lo q̃ es bueno este aruol o planta es que majando las dichas pencas mucho y tēdido aq̃llo a manera de emplasto en vn paño/⁊ ligādo vna pierna o braço con ello avn q̃ este q̃brada en muchos pedaços en espacio de quinze dias lo suelda/⁊ jūta como si nūca se quebrara/⁊ hasta q̃ aya hecho su operacion esta tan aferrada ⁊ asida esta medecina con la carne/que es muy dificultosa dela despegar: pero assi como ha curado el mal y hecho su operacion/luego ella por si misma se aparta y despega de aquel lugar donde la auian puesto: y deste efecto y remedio q̃ es dicho ay mucha espiriencia por los muchos q̃ lo han prouado. ¶ Ay assi mismo vnas plantas q̃ los xp̃ianos llamā platanos/los quales son altos como aruoles y se hazen gruessos enel tronco como vn gruesso muslo de vn hombre/o algo mas/y desde abaxo arriba echa vnas hojas longuissimas ⁊ muy anchas ⁊ tāto que tres palmos o mas son anchas/y mas de diez o doze palmos de longura/las q̃les hojas despues el ayre rompe quedādo entero el lomo dellas. Enel medio deste cogollo enlo alto nasce vn razimo con quarenta o cinquenta platanos y mas y menos: y cada platano es tan luengo como palmo y medio/y dela grosseza dela muñeca de vn braço/poco mas o menos/segun la fertilidad de la tierra donde nascen: porq̃ en algunas partes son muy menores: tienen vna corteza no muy gruessa ⁊ facil de rōper y de dētro todo es medula: que dessollado o quitada la dicha corteza paresce vn tuetano de vna caña de vaca

¶ Los aruoles con que se sueldan las quebraduras: ⁊ no me determino si sō aruoles o plantas.

¶ Platanos.

10. *Leaf of the Plátano Tree,* from Gonzalo Fernández de Oviedo, *De la natural hystoria de las Indias* (Toledo, 1526), folio 92v, woodcut
Library of Congress

the curious inquisitor of Seville, Andrés Gascó, made a copy of most of the original manuscripts a little after Oviedo's death, so that, when the historian Amador de los Rios (1818–1878) began to publish his edition of Oviedo's work in 1851, he had a quite complete view of the texts. Scholars today are not so fortunate, since a considerable part of Oviedo's manuscripts were sold under obscure circumstances in the 1920s. The Huntington Library in San Marino, California, acquired some of them, while others were lost, perhaps forever. For these missing portions of the manuscript, we must rely on the copies made by the Spanish historian Juan Bautista Muñoz (1745–1799) at the end of the eighteenth century.[25]

During the fifteen years that it took Oviedo to prepare his new edition for printing, many elements of both structure and content were added, eliminated, or transformed. These changes did not occur as the result of a new program, but were the outcome of a continuous process of correction, traces of which can be observed in the original manuscripts.

Although it did not contain the four hundred images announced by Ramusio, the new version of the *Historia general y natural de las Indias* did include many more pictures, and more elaborate ones, than either the summary of 1526 or the 1535 edition of part 1. Most of these illustrations were displayed in the revised edition of part 1 because the decision was made that all the books on natural history should be published together, instead of separately in three geographical sections, as originally conceived. This compilation would have made the enlarged version of part 1 of the *Historia general y natural de las Indias* the largest and most richly illustrated natural history book ever printed. To accomplish this task, Oviedo tried to solve technical problems by giving precise instructions to the wood engravers in nearly every case. A good example is the image of the *cigaraca* (*Monotropa uniflora*, commonly called ghost plant or Indian pipe), shown in book 11, beside which Oviedo noted that "the five little stems should be twice as thin as those in the drawing" (fig. 14).

The new role of images in the enlarged version of the *Historia general y natural de*

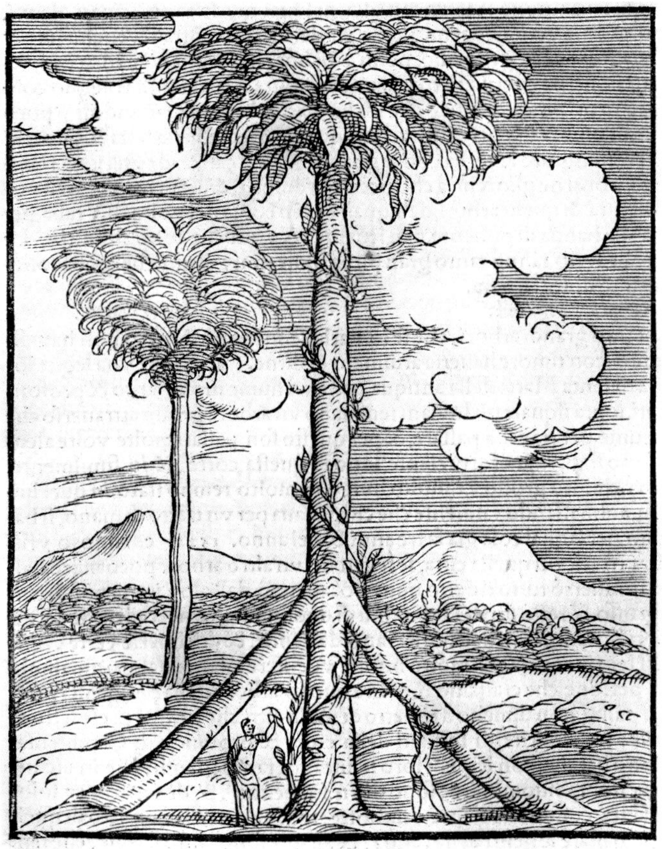

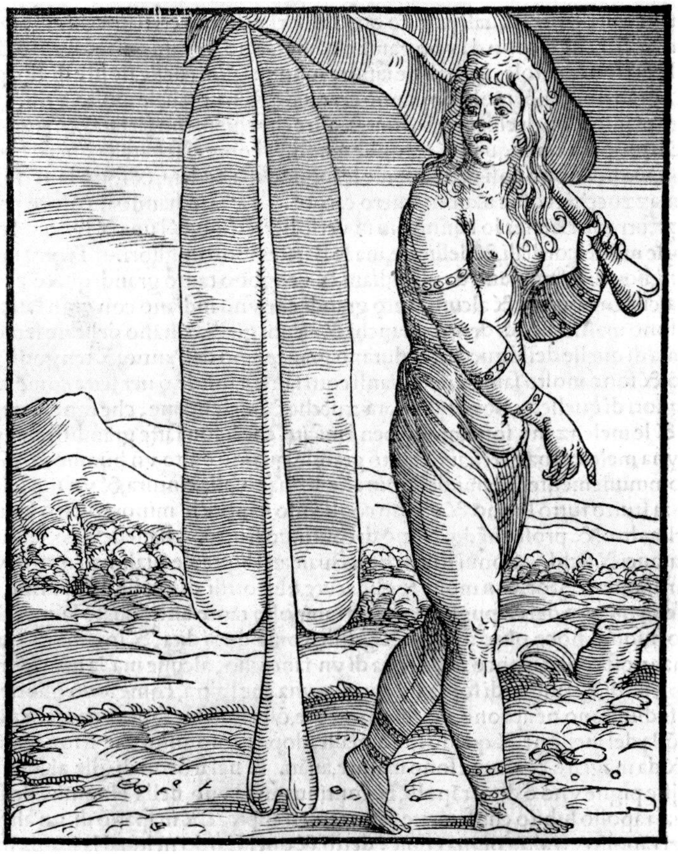

11. *Ceiba Tree,* from Gonzalo Fernández de Oviedo, *Sommario delle Indie Occidentali,* in Giovanni Battista Ramusio, *Navigazioni e viaggi,* volume 3 (Venice, 1556), woodcut
University of Illinois at Urbana-Champaign Library

12. *Leaf of the Plátano Tree,* from Gonzalo Fernández de Oviedo, *Sommario delle Indie Occidentali,* in Giovanni Battista Ramusio, *Navigazioni e viaggi,* volume 3 (Venice, 1556), woodcut
University of Illinois at Urbana-Champaign Library

las Indias should be understood as part of a global transformation of Oviedo's descriptive techniques. This shift is particularly evident where Oviedo dealt with difficult cases, such as the existence of mermen, asserted by the companions of Captain Alonso de Santa Cruz in Mar del Plata, or the description of the volcano Masaya in Nicaragua that perhaps evokes the marvels and monsters described in less authoritative texts. In both instances, Oviedo put subjectivity aside in order to adopt the distant and impassive style of an official report. Here, for example, is his description of the Masaya crater:

> I say that at the bottom and the deepest end of this wheel, as I saw it, there was a fire as liquid as water, or whatever matter it was. Its color was livelier than burning ashes and it could be said that this matter seemed more fiery than any fire could be. It covered all the ground and the lower part, boiling not everywhere at the same time, but moving from place to place, the boiling bursting incessantly again and again. In those areas in which the boiling had stopped the soil was covered on top by a kind of cracked fabric which showed through its fissures that it was liquid fire, like water underneath. Every now and then, all that matter rose up impetuously throwing up many drops which would then fall again in the same matter or fire. In my estimation they rose more than one stadium.[26]

The adoption of official language necessitated a displacement of the narrator from the leading role that he had played in the description. In the account of the expedition to the volcano, his experience was merely instrumental, as a vicarious agent of the empire. This withdrawal of the subjective voice of the chronicler equally affected the use of images. Although the precise passage in book 42, dealing with Oviedo's expedition to the Masaya volcano, is absent from the original manuscript, the copy ordered by the inquisitor Andrés Gascó around 1565 does contain this text, including two illustrations authenticated by the initials "A. G." as the patron of the transcription (figs. 15, 16). These initials would be those either of Andrés himself or of his nephew Antonio.

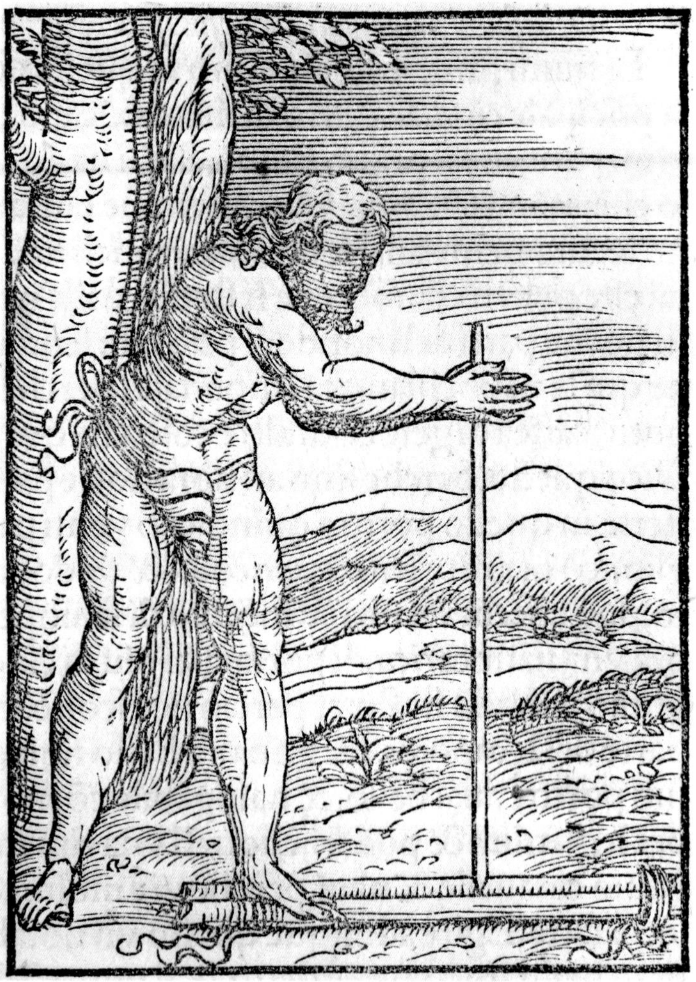

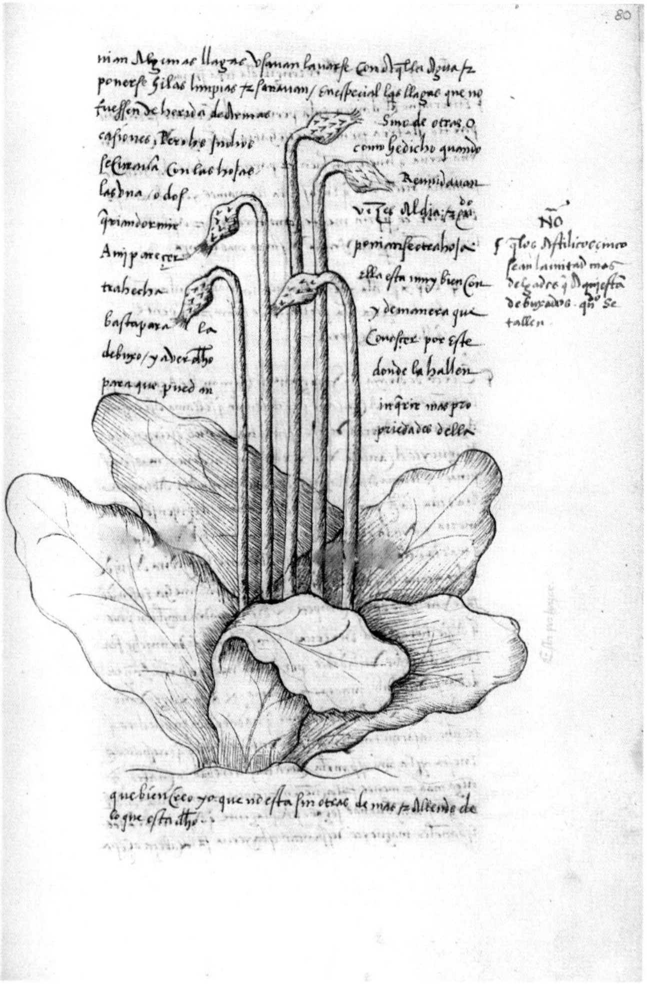

The illustration relevant to this discussion is a diagram of the volcano consisting of three circles, one within the other (fig. 16). The inner one, schematically shaded to suggest depth, bears the word *pozo* (well or hole), indicating that it refers to the crater itself. The second is called *la plaza de abaxo,* the lower plateau. The outermost circle bears the legend "Este es el circulo alto de la abertura del monte para abaxo" (This is the upper circle of the mountain where the crater goes down). Four small labels with the cardinal points orient the picture. The importance of this diagram lies in the fact that it belongs to a procedure for acquiring information that entailed, as Oviedo described, the chronicler's trip to the site and the careful observation of the phenomenon: "Estuve más de dos horas, e aun cuasi hasta las diez del día de Sancta Ana gloriosa, mirando lo que he dicho" (I spent more than two hours, nearly until ten, observing what I have just described, the day of glorious Saint Anne). Finally, a schematic drawing of the crater allows him to convey this information to his readers: "debujando la forma deste monte con papel, como aquí lo he puesto" (drawing the shape of this mountain on a piece of paper, in the same way I have done here).

In this case the illustration neither indicates an everyday experience nor gives visual proof that would support the verisimilitude of a personal account. This illustration is a diagram. Whereas the use of images as evidence of something else prevents them from becoming conveyors of information by themselves, diagrams are bearers of meaning par excellence. Unlike the other illustrations provided for the 1535 edition of the *Historia general y natural de las Indias,* these illustrations are evidence of personal

13. *Native American Making a Fire,* from Gonzalo Fernández de Oviedo, *Sommario delle Indie Occidentali,* in Giovanni Battista Ramusio, *Navigazioni e Viaggi,* volume 3 (Venice, 1556), woodcut
University of Illinois at Urbana-Champaign Library

14. *Cigaraca,* from Gonzalo Fernández de Oviedo, *Historia general y natural de las Indias,* book 11, chapter 9, pen and ink
The Huntington Library, San Marino, California, HM 177, volume 2, folio 80r

experience—more than two hours of observation, as Oviedo informs the readers—which is offered in support of the value of the drawing and not the other way around. This drawing is neither a mere suggestion of an experience, as was the case of the image of the iguana, nor a mere reflection of the vision of the chronicler, as in most of the schematic figures illustrating the 1535 text. The volcano drawing is now the final cause of and the justification for the whole process of experience.

This representational shift had radical consequences for the use of botanical images. The most obvious difference with regard to the 1535 edition is that the former hierarchy visually subordinating the image to the text has been overturned. In the illustration of the *perorica*, the "lily of the mainland," the plant occupies the center of the page while the text is organized around its silhouette (fig. 17).

This placement was not new, of course, in book illustration, but Oviedo's choice of this method indicates a clear shift in the respective values of word and image as conveyors of information. The implications of this shift are more evident when analyzing the new ways in which texts refer to images in the manuscripts. The most effective illustration contained in the 1535 edition is probably that of the leaf of the *goaconax* (*Liquidambar styraciflua*, or sweetgum) that covers most of one of the two columns of the printed page (fig. 18). Oviedo created here the vivid effect of an absent artifact by identifying the moment of writing with that of autopsy: "having the same natural one before me." Using the size of the illustration on the page to indicate the size of the actual specimen reinforced the sense of immediacy: "I believe this image is very well made and it has the same measure." As in Pliny the Elder's account of the origins of painting, the effect of realism was produced by the evocation of an original act of viewing—that is, Oviedo looking at the plant. This, then, would be followed by a second and subordinate act: the reader looking at a picture of a *goaconax* leaf that is the same size as the real leaf. Despite the highly communicative power of the image and the overall effect of vividness achieved, the representation of the *goaconax* leaf corresponds to the rhetoric of translation and mediation that still depends on the chronicler's narration of the autoptic experience.

When, a few years later, Oviedo revised the chapter on the *goaconax*, he drastically reduced his previous emphasis on both autopsy and the size identity between image

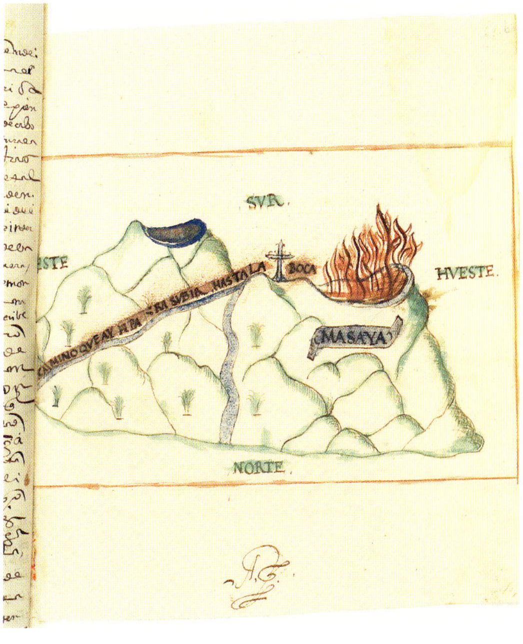

15. Schematic illustration of Masaya volcano, from Gonzalo Fernández de Oviedo, *Historia general y natural de las Indias*, book 42, chapter 6, pen and ink
Biblioteca del Palacio Real de Madrid, MS II-3042, folio 63r

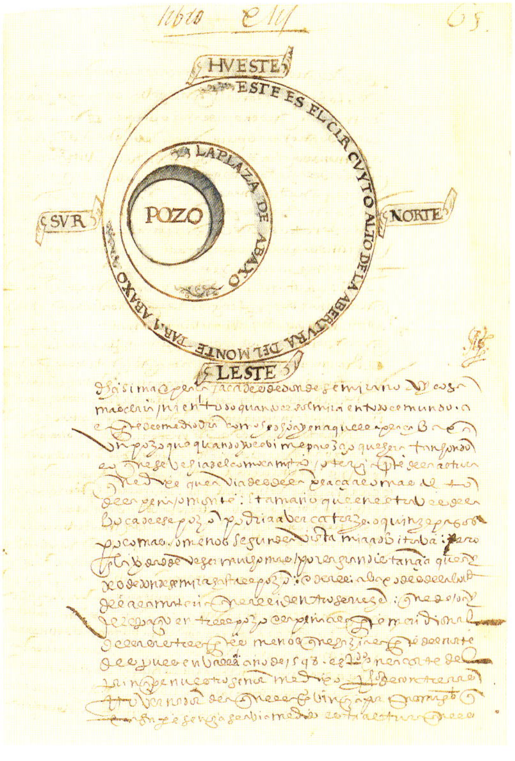

16. Diagram of crater, Masaya volcano, from Gonzalo Fernández de Oviedo, *Historia general y natural de las Indias*, book 42, chapter 6, pen and ink
Biblioteca del Palacio Real de Madrid, MS II-3042, folio 65r

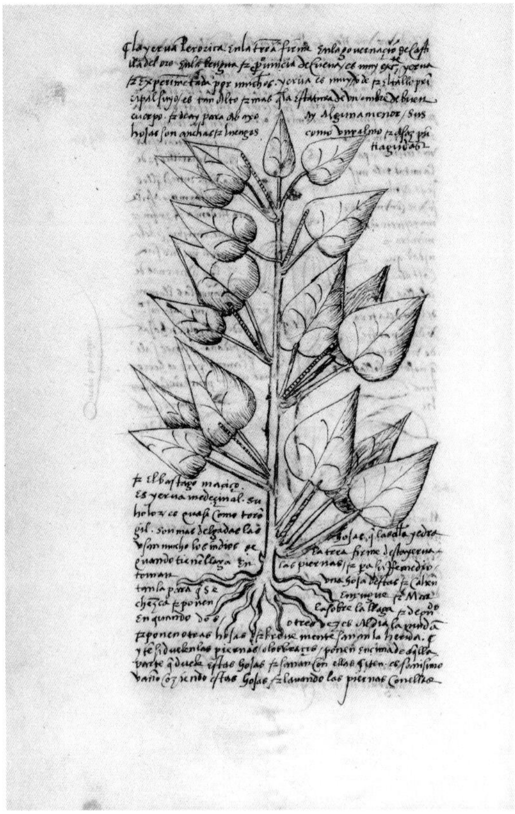

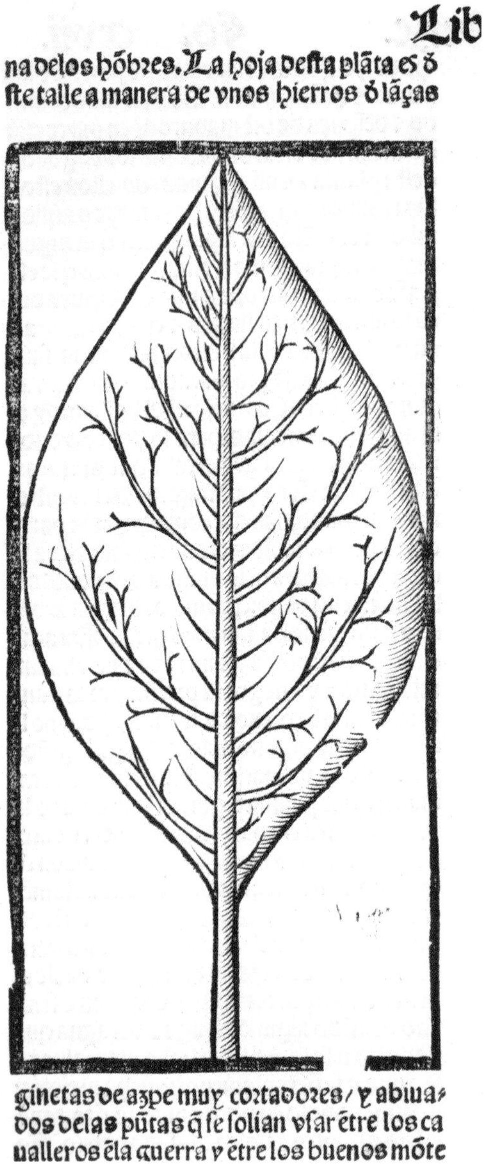

17. *Perorica,* from Gonzalo Fernández de Oviedo, *Historia general y natural de las Indias,* book 11, chapter 8, pen and ink
The Huntington Library, San Marino, California, HM 177, volume 2, folio 79r

18. *Goaconax Leaf,* from Gonzalo Fernández de Oviedo, *Historia general y natural de las Indias* (Seville, 1535), book 11, chapter 4, folio 97v, woodcut
Rare Books Division, New York Public Library, Astor, Lenox and Tilden Foundations

and specimen (fig. 19). Instead, he presented the illustration to readers as an accurate portrait of the natural phenomenon: "the true and natural leaf of this plant is like the picture that has been drawn here." The illustration does not allude indirectly and metonymically to the real, it is *like* the real. The function of conveying likeness, performed now by the picture, entails an apparent autonomy with regard to the text. This works as a mere indicator of the resemblance conveyed by the image, a fact that does not depend on the text, but on the image itself. The chronicler has the detached role of recording and certifying this fact as a notary would do.

The picture of the *perorica* (see fig. 17) included in one of the new chapters of book 11 appears without any introduction in the text. Its presence does not need any verbal justification since its lifelikeness was considered enough to validate its function. In this case the image represents not a fragment or a leaf, but a complete specimen, roots included, as if it were an herbal illustration. The text refers to the image only a posteriori and, instead of justifying its presence, merely comments on it. Whereas the 1535 illustration of the *perebecenuc* leaf was provided as didactic support to the text, the image of the *perorica* was provided to explain the different parts of a figurative organism that stood independently as the true portrait of the American specimen.

The sharp separation of word and image, as seen in Oviedo's treatises, involved the simultaneous elimination of subjective narrative as a plausible mediator between the two worlds. This shift in Oviedo's economy of representation can be related to his increasing skepticism about the ability of

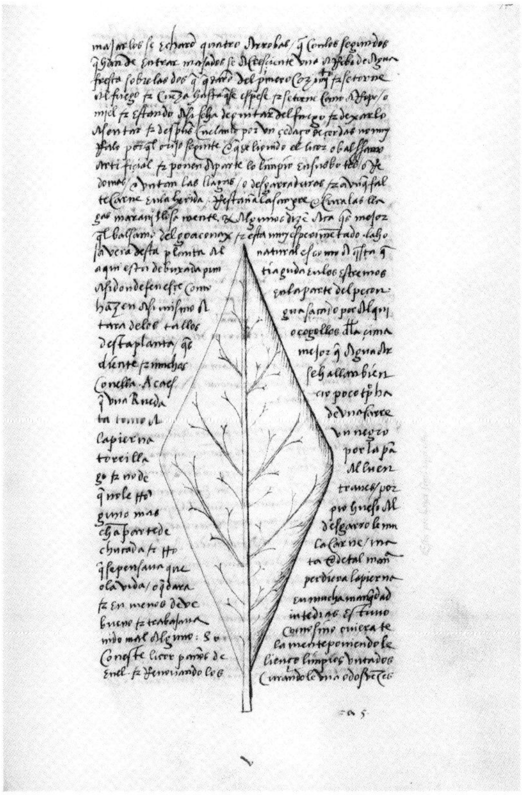

19. *Goaconax Leaf,* from Gonzalo Fernández de Oviedo, *Historia general y natural de las Indias,* book 11, chapter 4, pen and ink
The Huntington Library, San Marino, California, HM 177, volume 2, folio 74r

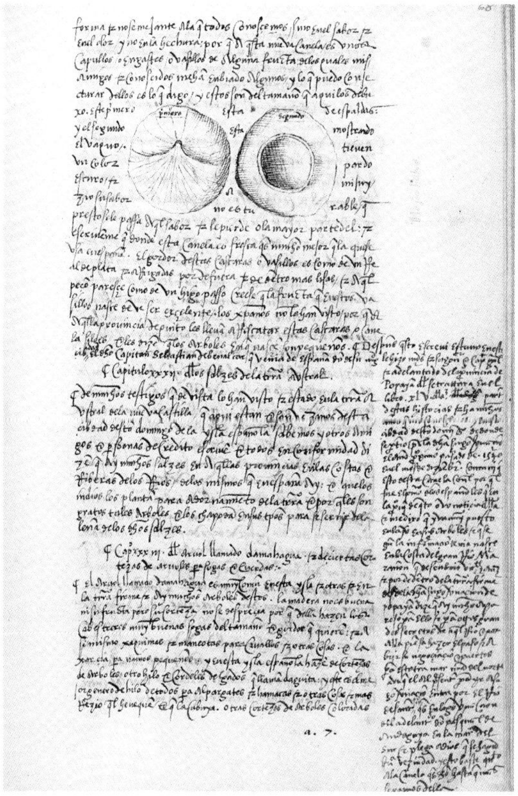

20. *Quito Cinnamon Nut,* from Gonzalo Fernández de Oviedo, *Historia general y natural de las Indias,* book 11, chapter 31, pen and ink
The Huntington Library, San Marino, California, HM 177, volume 2, folio 68r

individuals to make sense of the world from their own subjective perspectives. Oviedo perceived this as at once a defeat in a moral sense and a victory on behalf of the new imperial system. Although he felt nostalgia for the days in which a man was the master of his own fate, he eagerly observed that the new methods of accumulating and processing data were far more effective for an expanding world than the old method, which depended on the intellectual capacities of isolated individuals. The practical mind of Oviedo, an officer of the empire, conceived the representation of American nature as independent from its complete assimilation and mental comprehension. The use of an image as evidence of personal or collective experience was overshadowed by its newly discovered power to record the very appearance of little fragments of a wide and heterogeneous universe. Differences in newly discovered objects no longer posed a danger to the comprehension of the real meaning of things or served as mere invitations to curiosity. The representation of differences was, rather, the proof and the sign of a positive intervention over the newly unveiled surface of the world. The empire, the only agency capable of this conquest, could be embodied by single individuals through a formal process of desubjectivization, in which the agents of empire became instruments of observation and taxation.

Oviedo's representational tableau was no longer that of mediation and translation. The act of possession that he performed on the page did not involve the impossible reduction of the new phenomenon to the realm of the familiar, but instead subjected it to a much more effective sort of control through observation, dissection, and measurement. In his description of so-called Quito cinnamon (*quinela de Quito; Wintera granatensis,* a variety of winter's bark), Oviedo illustrated a specimen brought to him by some anonymous soldiers. Not yet satisfied with the conventional portrait of the species, Oviedo showed the nut of the new cinnamon from both the inside and the outside (fig. 20). By showing the fruit cut in two and depicting the exterior of one half and the interior of the other, he made it completely visible for his readers.

As in contemporary herbals of the fifteenth century, accurate images of natural specimens served as essential instruments of identification; but, unlike their European parallels, Oviedo's images did not correspond to a stable list of medicinal properties. The visibility of American phenomena did not depend upon the conventional process of cognition. In the case of the cinnamon nut, the illustration portrayed something that, as Oviedo reminded his readers, still required further inquiry. The image, as in the case of the *cigaraca* (see fig. 14), anticipated the actual knowledge of the new species and became the best promise of its complete possession in the future: "I believe it is a good portrait and it should be sufficient to recognize it. As I also say where to find it, the image should help people to get to know about the properties that I am certain it has."

NOTES

1. Jesús M. Carrillo Castillo, *Naturaleza e imperio: La representación del mundo natural en la "Historia general y natural de las Indias" de Gonzalo Fernández de Oviedo* (Madrid, 2004).

2. Francesco Patrizi, *Dieci dialoghi sulla historia* (Venice, 1560), dialogue 2, folio 8v.

3. Gonzalo Fernández de Oviedo, *De la natural hystoria de las Indias* (Toledo, 1526).

4. See Jesús Carrillo Castillo, "Cultura cortesana e imperio: El 'Libro del blasón,' de Gonzalo Fernández de Oviedo," in *Locus amoenus* 4 (1998–1999): 137–154.

5. Gonzalo Fernández de Oviedo, *Libro del muy esforzado cavallero de la Fortuna propiamente llamado Claribalte* (Valencia, 1519).

6. Gonzalo Fernández de Oviedo, *La Historia general y natural de las Indias con privilegio imperial* (Seville, 1535).

7. Clive Griffin, *The Crombergers of Seville: The History of a Printing and Merchant Dynasty* (Oxford, 1988), 139–140.

8. Gonzalo Fernández de Oviedo, *Libro de la Cámara del Príncipe don Juan,* ed. José María Escudero Peña (Madrid, 1870).

9. Gonzalo Fernández de Oviedo, *Las quinquágenas de los generosos e illustres y no menos famosos reyes, príncipes, duques . . .* , 3 vols. (1555–1556), Biblioteca Nacional de Madrid, MS 2217–2219.

10. Gonzalo Fernández de Oviedo, *Libro secondo delle Indie Occidentali* (Venice, 1534); *Sommario dell'Indie Occidentali,* in Giovanni Battista Ramusio, *Navigatione et viaggi,* vol. 3 (Venice, 1556).

11. *L'Histoire naturelle et généralle des Indes: Isles, et terre ferme de la grand mer oceane / Traduicte de castillan en françois* (Paris, 1555).

12. This was the papal document of 26 June 1540, by which Paul III granted Oviedo the right to publish his book in the papal territories and the whole Christian world for twenty-five years; as cited in Francesca Cantú, "Ideología e storiografia in Oviedo: Problemi di interpretazione e di edizione," in *Critica Storica* (1976): 207–246. Oviedo referred to this fact in the introduction to the second part of his *Historia.* See *Historia general y natural de las Indias,* ed. Juan Pérez de Tudela (Madrid, 1957), 2:216.

13. See Gonzalo Fernández de Oviedo, "Carta al Emperador," in Pérez de Tudela 1957, book 1, 1:7.

14. *Historia general y natural de las Indias,* in Pérez de Tudela 1957, book 17, chapter 26, 2:167.

15. On this subject, see Gregorio Hinojo Andrés, *Obras históricas de Nebrija: Estudio filológico* (Salamanca, 1991), 26–29; and José María Maestre, "La divinatio in scribenda historia de Lebrija," *Euphrosine* 22 (1995): 141–173.

16. Jesús Carrillo Castillo, "From Mt. Ventoux to Mt. Masaya: The Rise and Fall of Subjectivity in Early Modern Travel Narrative," in *Voyages and Visions: Towards a Cultural History of Travel,* ed. Jás Elsner and Joan-Pau Rubiés (London, 1999), 57–73.

17. In the original: "Su plumaje principal del cuerpo e cola es de muy hermoso e lindo color leonado; e la cola es de hasta diez plumas de cuatro palmos de luengo, e donde son más gruesas (que es en su nascimiento e poco más adelante), son de la grandeza de un alfiler de los gruesos, e de allí hasta el cabo e extremos se van adelgazando que parecen dos hilos, y son leonadas escuras, que vuelven al negro color. E tentadas entre los dedos, son asperísimas, como sierra; e no tienen pelo ninguno, como otras plumas, sino cerca de los nascimientos, e poquito; e toda la otra longitud dellas aspera, e delgada, como digo. . . . La cabeza es tan grande como de un tordo, e la pluma della amarilla que tira a color naranjado; y el papo es verde dorado, de muy extremada color; y un fleco de pluma muy espesa e corta, que paresce poco más alto que un terciopelo y muy negro, de donde nasce el pico, el cual es tan grande como de una picaza, y derecho y avivado. Las alas son lo que no sé discantar ni aún relatar llanamente. . . . En conclusión, yo confieso que no habrá pintor que lo pinte, por lo que he dicho; pero, leído esto a par del pájaro se me figura que he dicho algo; y así lo he escripto mirándole, y dando gracias a Dios que estas aves crió. Para mi, yo la tengo por la más extremada en su plumaje o gentileza de todas las que yo he visto, y de la que más me he admirado." See Pérez de Tudela 1957, book 6, chapter 15, 1:175–176.

18. As quoted from the original text: "No pueden la pintura de mi pluma y palabras dar tan particular razón ni tan al propio el blasón desta fructa, que satisfagan tan total y bastantemente que se pueda particularizar el caso sin el pincel o dibujo, y aún con esto, serían menester los colores, para que más conforme (si no en todo, en parte), se diese mejor a entender que yo lo hago y digo, porque en alguna manera la vista del lector pudiese más participar desta verdad. Non obstante lo cual, pondréla como supiere hacerlo, tan mal dibujada como platicada; pero para los que esta fructa hobieren visto, bastará aquesto, y ellos dirán lo demás." See Pérez de Tudela 1957, book 7, chapter 14, 1:240. The use here of the term *blason* to refer to the physical appearance of a fruit reveals the influence of the language of heraldry in Fernández de Oviedo's descriptions of the natural world. On this subject, see the discussion in Oviedo's treatise on heraldry, below and note 23.

19. This has already been discussed in Jesús Carrillo Castillo, "Taming the Visible: Word and Image in Oviedo's 'Historia general y natural de las Indias,'" *Viator* 31 (2000): 399–431.

20. Oviedo 1535, book 11, chapter 5, folio 98r.

21. Oviedo 1535, book 7, chapter 1, folio 83r.

22. Oviedo 1535, book 13, chapter 3, folio 103v.

23. These images are analyzed in "Cultura cortesana e imperio: El 'Libro del blasón,' de Gonzalo Fernández de Oviedo," *Locus amoenus* 4 (1998–1999): 137–154.

24. See "Discorso di messer Gio. Battista Ramusio sopra il terzo volume delle Navigazioni e Viaggi nella parte del mondo nuovo. All'eccellente messer Ieronimo Fracastoro," in *Giovanni Battista Ramusio, Navigazioni e viaggi,* ed. Marica Milanesi (Turin, 1985), 5:9–10.

25. Jesús Carrillo [Castillo], "The 'Historia general y natural de las Indias' by Gonzalo Fernández de Oviedo," *The Huntington Library Quarterly* 65 (2003): 321–344.

26. As quoted in the original text: "Digo que en la hondura e última parte que yo vi deste pozo había un fuego líquido como agua, o la materia que ello es, estaba más que vivas brasas encendida su color e, si se puede decir, muy más fogosa materia parescía que fuego alguno puede ser; la cual todo el suelo e parte inferior del pozo ocupaba y estaba hirviendo, no en todo, pero en partes, mudándose el hervor de un lugar a otro e resurgía un bullir o borbollar sin cesar de un cabo a otro. Y en aquellas partes donde aquel hervor no había (o cesaba), luego se cubría de una tela o tez de napa encima como horura o requebrada e mostraba por aquellas quebraduras de aquella tela o napa ser todo fuego líquido como agua lo de debajo; e así por todo el circuito del pozo. E de cuando en cuando, toda aquella materia se levantaba para suso con gran ímpetu e lanzaba muchas gotas para arriba, las cuales tornaban a caer en la misma materia o fuego, que a la estimación de mi vista más de un estadio subían." See Pérez de Tudela 1957, book 42, chapter 6, 4:397.

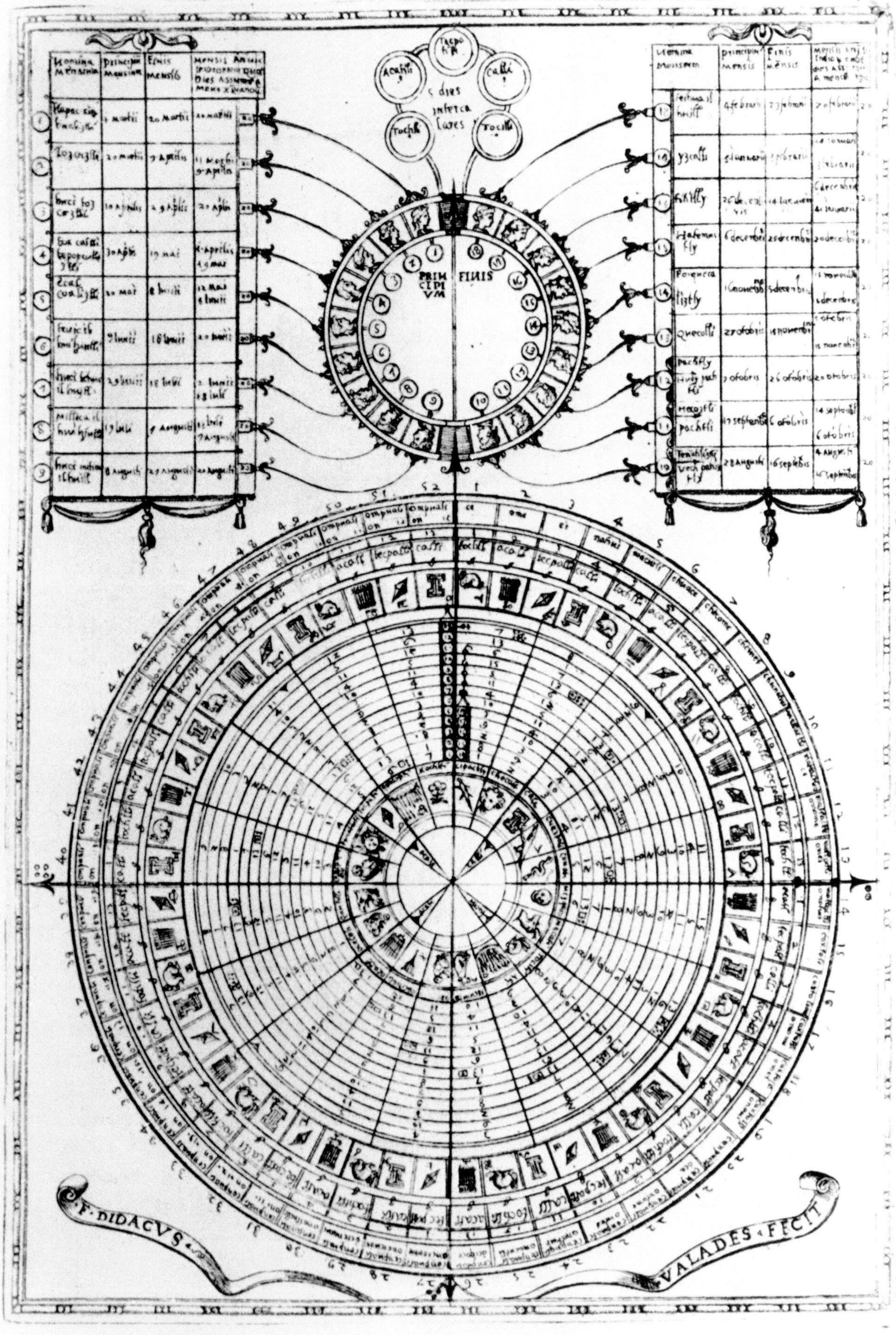
PRIN CIPI VM
FINIS
F. DIDACVS
VALADES FECIT

LINA BOLZONI
Scuola Normale Superiore, Pisa

Mexican Nature in Diego Valadés' Rhetorica christiana *(1579)*

This essay discusses the use of various Mexican plants and precious stones in a treatise of sacred oratory, *Rhetorica christiana* (fig. 1), published in Perugia in 1579 by Diego Valadés (1533–after 1582), a Franciscan friar who was born in Mexico and spent most of his life there.[1] This case is interesting from many points of view, because it is closely linked with other, more general topics; I focus on these aspects first, before discussing Valadés' text, so that readers may appreciate more thoroughly the meaning and importance of what Valadés was attempting to accomplish. The discussion covers four points:

1. The relationship between books, specifically the illustrated book and the art of memory;

2. How images of memory were used in the Christian tradition—by Franciscan and Dominican friars, and later also by Jesuits—in preaching for missionary ends and as a guide to interior elevation;

3. How Valadés interacted with this tradition and, at the same time, how he used this tradition to negotiate a path between the European world and the Mexican world, in order to create new instruments of communication, understanding, and mutual respect; and

4. How his use of various elements drawn from nature in the New World—such as plants and precious gems—forms an important part of his program and an emblem of his unique endeavor.

There is a striking paradox in the European cultural experience of the sixteenth century: the tradition of the art of memory reached its moment of greatest diffusion and complexity during an age—the Cinquecento—that also witnessed the birth and diffusion of the printed book. Thus the techniques of memory reached their greatest development in a world in which their meaning and importance were gradually being eroded. At the same time, one can see that the techniques of memory began to interact, often in productive and creative ways, with the new possibilities created by the printing press. In order to understand how this happened, we must shift from the traditional perspective of scholarly research on the Renaissance art of memory. The classic works by Paolo Rossi and Frances Yates were primarily concerned with theories of memory, treatises on the art of memory, and their relationship with the philosophical, religious, and scientific European tradition.[2] A different perspective, which I propose to take, is one based on the concept that treatises on the art of memory represent merely the suggestion of a much larger phenomenon and that their often dry and repetitive rules are merely the backdrop to a cultural drama that developed on many

Aztec Calendar, from Diego Valadés, *Rhetorica christiana* (Perugia, 1579), unpaginated, located between pages 100 and 101, engraving
Private collection

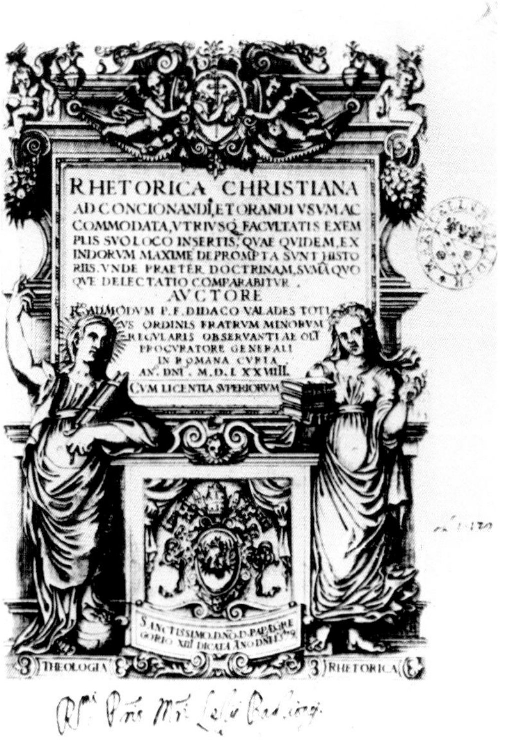

1. Frontispiece from Diego Valadés, *Rhetorica christiana* (Perugia, 1579), engraving
Private collection

levels.[3] It is possible to use treatises on memory as aids to remap the territory in which the techniques of memory interact with different experiences, for example: with the translation of words into images and images into words; with experiments involving the imagination or, as in the present case, involving sacred oratory; with the experience of missionaries; and with the fascinating and complex problem of establishing communication and understanding between different cultures.

From the very outset, the art of memory used *loci* and *imagines* (places and images) that could be purely mental; thus the master of memory would have been very familiar with both the construction of mental buildings and the control of the interior spaces of the mind. At the same time, he would also have been adept at the construction of mental pictures—*imagines agentes*—powerful images, rich in emotional impact and capable of creating and recalling associations with words and ideas. Thus there existed a *memoria verborum* (memory of words) and a *memoria rerum* (memory of things). These images and places could also be real, perceptible to the senses. The interplay between physical and mental reality, between inner experience and the external world, is one of the most important components of the art of memory, one of its "secrets."

The printed book had a complex impact on this tradition, with far-reaching consequences for the loci, the *ordo* (order), and the images. In the mid-sixteenth century, method became one of the new aspects of the art of memory (through the work of Rudolph Agricola and Pierre de la Ramée, for example). Great faith was placed in the possibility of formulating a method that would strictly regulate both knowledge and the means of communicating and recalling it.[4] Diagrams, tables, and large, tree-form schemas were designed to visualize the logical path to be taken. All of the information was presented to the eye reordered and reorganized in a clear, effective fashion that would be easy to remember. The new direction taken by logic and dialectics also interacted fruitfully with the new possibilities created by the book format and by the ordered and reproducible space of the printed page. An apt expression describing this new moment of the art of memory can be found in one of the texts that Paolo Manuzio published for the Accademia Veneziana: with this kind of book, it was possible to "make knowledge visible."[5]

In Valadés' text we find a table (fig. 2) that reduces the entire structure of rhetoric to a tree, accompanied by a note on how memory, an absolutely essential talent, is cultivated through reading, toil, and meditation, "but perfected also through the positioning and arrangement of the things that we wish to remember" (sed etiam perficiatur collocatione dispositioneque eorum, quae in memoria habere cupimus).[6] We see that with printed books, places and images become visible, more precisely fixed, more clear, but also less complex, less closely linked to individual experimentation, creation, and variation. From a certain point of view, it is possible to think of the relationship between the printed book and the systems of the art of memory in opposing terms, but upon further consideration, it soon becomes clear that the reality is somewhat more complex. In some cases the book—and above all the illustrated book—can be perceived as a

2. *Diagram of Rhetoric,* from Diego Valadés, *Rhetorica christiana* (Perugia, 1579), page 49, engraving
Private collection

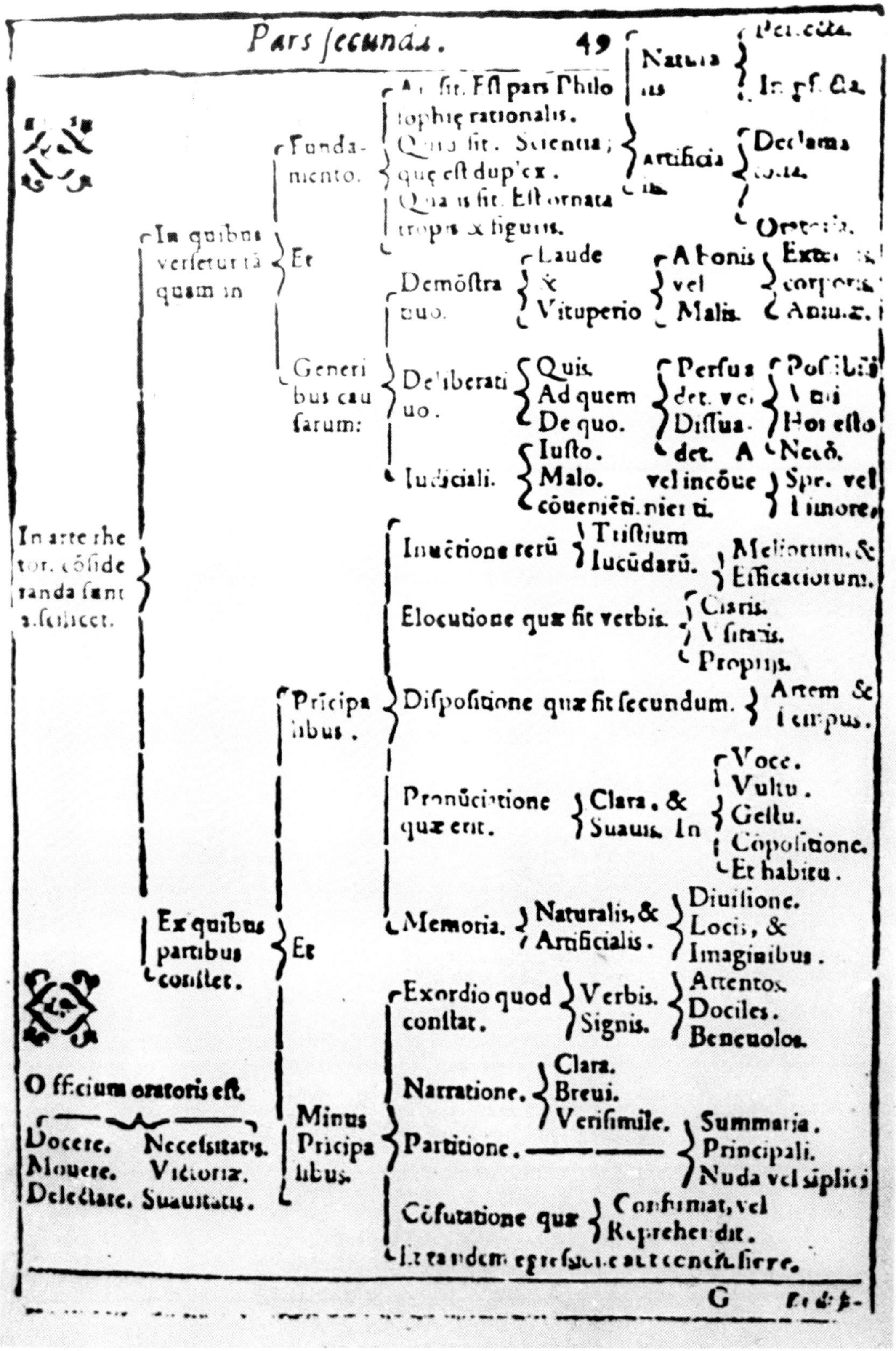

visual translation of a mnemonic system. There is, for example, an interesting passage in Lodovico Dolce's treatise *Dialogo nel quale si ragiona del modo di accrescer et conservar la memoria* (Venice, 1562), in which the images of great painters and the images of the book are described as being very useful for the art of memory:

> Those who wish to remember the story of Europa could use the example of the painting by Titian, and they can do the same for the story of Adonis, and any fictional story, profane or sacred, by choosing especially those figures that delight and excite the memory. Illustrated books, so commonly printed today, are useful for this, like most of those that issue from the press of the most accurate [Gabriele] Giolito.[7]

Like the images created by painters, the illustrated book, Dolce informs us, offers models and material for images of memory.[8] In this way a paradoxical balance is achieved between the art of memory and the printing press. (It was this very collaboration between ancient techniques and modern technology that would ultimately render the former meaningless.) In any case, it is important to note that Dolce offers an interpretive key that we can apply to some of his more successful works, such as his Italian translation of Ovid's *Metamorphoses.*

Dolce, of course, was not the only one to associate illustrated books with systems of memory. Treatises focusing on the religious dimension of memory offer evidence of just how widespread this association and usage actually were. In his *Thesaurus artificiosae memoriae,* for example, the Florentine mnemologist Cosma Rosselli (d. 1578) recommended drawing inspiration from illustrations in the Bible and other books of devotion: "You will find varied and beautiful figures in the illustrated editions of the sacred books" (varia autem et pulcherrima schemata in Bibliis sacris et figuratis invenies).[9]

In the *Gazophylacium artis memoriae* (Treasure House of the Art of Memory, 1610), the Belgian memory expert Lambertus Schenkel (1547–c. 1630) affirmed that it is an excellent thing to form images of memory based on the Bible, and "for this purpose,

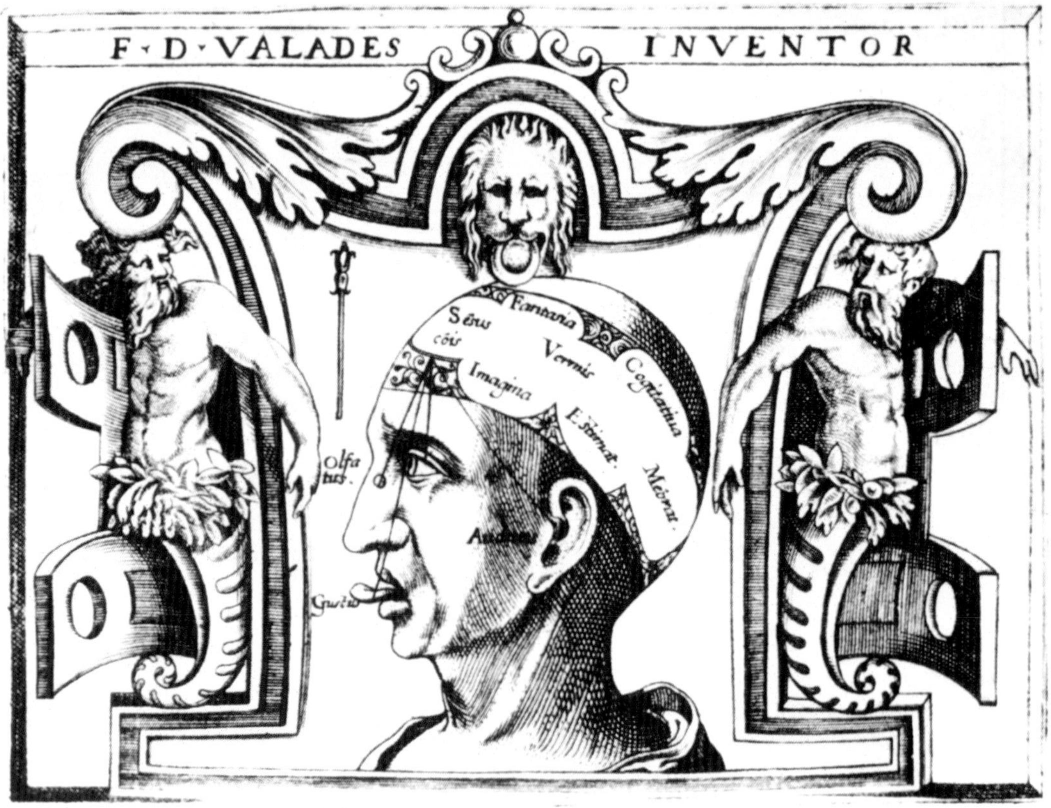

3. *Map of the Brain,* from Diego Valadés, *Rhetorica christiana* (Perugia, 1579), page 88, engraving
Private collection

the illustrated Bible printed in folio is highly useful" (ad quod valde iuvabimur editione Bibliorum per imagines in folio). Schenkel also recommended frequently reviewing the set of places and images mentally. To do this, he wrote, very little time is needed:

> In fact, using our eyes, we can rapidly retrace and review all of a Bible illustrated with material images in the course of an hour; it will not take long for us to turn the pages and see with a glance what it is about, on the condition, naturally, that we are already familiar with it, [and] so it must be for the things that we have positioned in the mnemonic places. How much more can one do with the eyes of the mind, which operate with much greater speed![10]

The position of our author, Valadés, is even more radical: according to him, it is not only possible to *use* an illustrated book as a system of memory; it is also possible to *create* an illustrated book in order to use it as a mnemonic system of the contents. To this end, he equipped the text of *Rhetorica christiana* with a series of images which he himself had conceived: indeed, in his treatise the panel above the representation of the map of the brain is inscribed *F. D. VALADÉS / INVENTOR* (fig. 3). These images, as he explained in the text in great detail, help one to remember the text itself. Valadés therefore, furnished both the text and its own system of memory. In this sense, what we have in this treatise is a more radical interpretation of the mnemonic tradition, because the function of memory is present from the very beginning in the book itself.

An analogous process takes place in Valadés' interpretation of a fresco cycle. In the tradition of the art of memory, a cycle of frescoes can be used as a system of memory.[11] When Valadés mentioned the frescoes that he saw in the Vatican Palace, he declared that they were created as an encyclopedic system of memory:

> What in fact do you think is the meaning of those images painted in such wonderful fashion in the most august and holy palace in Saint Peter's, in this divine and illustrious city of Rome? It is a sort of art of memory, used to help us perceive and understand marvelous concepts, both in themselves and in all of their detail, in part through the mediation of the images, in part thanks to faith. To whoever looks, it places before his eyes the idea of every thing, in a more precise way, so that he can remember it better. With marvelous artifice theology, the science of law both sacred and civil, logic, philosophy, rhetoric, religion, peace, war, and in short all knowledge are arranged there in such a way that one may truly say that the things are adapted to the place, and the place to the things arranged therein.[12]

A rich scholarly tradition sprang up around the use of mnemonic images—for different purposes—by the Roman Catholic Church, from the Middle Ages to the early modern age. One of these was in preaching, in order to impress religious teachings on the minds of the listeners.[13] A second was in techniques of meditation, to attain interior elevation and arrive at mystical experiences (for example, in *Spiritual Exercises* by Ignatius of Loyola and, going back to the Quattrocento, in *Giardino de oratione* and in Jacopone da Todi's mystical poems).[14] A third was in the missionary experience. The most famous example is that of the Jesuit Matteo Ricci, who served in China at the end of the sixteenth century. It is important to note that Ricci used the tradition of the art of memory not only as an instrument for his missionary purposes, but also as a tool to break through the wall that separates two different worlds, two different languages, and two different ways of writing and thinking—European and Chinese. Thus he attempted to create a parallel between Chinese ideograms and European images of memory.[15]

Valadés' attempt to create such parallels was similar but was conducted in a very different situation. Valadés adopted the idea of a universal code, expressed through images, in order to demonstrate and defend the human and cultural values of the Mexican peoples against the European idea of their inarticulate and "bestial nature." For example, in 1533 the Dominican friar Domingo de Betanzos, speaking to the Council of the Indies, said straight out that Indians are bestial in nature and do not have the capacity to receive the Christian faith.[16] Arguments on this topic reached a climax in 1550–1551 during the debate between Bartolomé de Las Casas and Juan de Sepúlveda.

It is true, Valadés admitted, that native populations do not know the European technique of writing, but they do use images to communicate and to remember, exactly as the ancient Egyptians did, and as Europeans do today when they practice the art of memory. To those who denied the human dignity of these indigenous peoples, Valadés offered a reinterpretation, a new cultural translation: he presented the mysterious images of pre-Columbian civilization in Mexico in a framework that made them accessible to European culture by associating them with a cultural artifact from remote antiquity that Western scholars regarded with reverence—the hieroglyphics of ancient Egypt—as well as with a contemporary and widely recognized component of European culture, the art of memory.

For this reason, his attempt was similar to that of Matteo Ricci, but Valadés' argument unfolded in a more dramatic context: for Valadés, it was a matter of defending those who had been subjected to complete domination by outside forces. It is significant to note that his attempt at cultural translation was addressed more to the conquistadors than to the indigenous peoples who were to be evangelized.

Valadés probably experienced firsthand the problematic relationship between two very different worlds: apparently he was a mestizo, born of the union between a conquistador and a Native American noblewoman. It is likely for this reason that there is no sixteenth-century record of his association with the Franciscan order; instead his name first appears much later, in seventeenth-century archives. Yet Valadés had once held a very high position: in 1575, during a trip to Europe, he was nominated procurator general for the Franciscan order. However, one must remember that by the second half of the sixteenth century, Valadés' experience as a mestizo, and his prospects for a religious career in particular, had radically changed. In fact, a synod held in Mexico in 1555 explicitly prohibited the ordination of blacks, mestizos, and indigenous peoples, and the Inquisition set a policy of examining all works written by missionaries in indigenous languages in order to detect any possible contamination by indigenous religions.[17] In 1577, two years after Valadés' appointment as procurator general, the Spanish king, Philip II, removed him from office and ordered his departure from Rome. As a result, Valadés went to Perugia, where he published *Rhetorica christiana.*

We can therefore understand the special significance underlying the care with which Valadés conceived the images that would illustrate his treatise on sacred oratory and entrust it to the memory of the reader. Those images constitute, in fact, a bridge between two cultures. They became the very tools of communication and memory by which the Old World and the New could meet on common ground.

Let us now examine more closely just how Valadés used images in his treatise and how he described the use of images by the Franciscan missionaries in Mexico. In his text, he presented the Franciscan practice not only as an efficient tool (in the European tradition),[18] but also as a way of adapting a similar cultural practice already deeply rooted among native peoples. In these images, we can recognize different levels and different modes of communication and interaction between the European and the Mexican experience.

Valadés presented examples of images in which the European model stood alone, with no native counterpart, such as the representation of the Crucifixion (fig. 4). In this image, a Franciscan friar points with his staff to the figure of Christ, while on the right is a group of *indios.* As Mario Sartor wrote, the models for this image of the Crucifixion can be traced back to a number of European painters and printmakers, including Martin Schongauer, Albrecht Dürer, Titian, and Marten de Vos. Here, the native *indios* are represented as ancient Romans.[19]

On the other hand, Valadés also offered an image in which the indigenous model is dominant: the Aztec calendar (see essay frontispiece, page 128). This image, which appears in the middle of the section dedicated to the art of memory, is linked to a reinterpretation of the Aztec calendar in terms of a mnemonic system with its *loci* and *imagines.* The accurate reproduction of an Aztec calendar was certainly not common in the sixteenth century: indeed, European missionaries systematically destroyed indigenous ritual calendar manuscripts.[20] But in this context, it is pertinent to recall that *Rhetorica christiana* was dedicated to

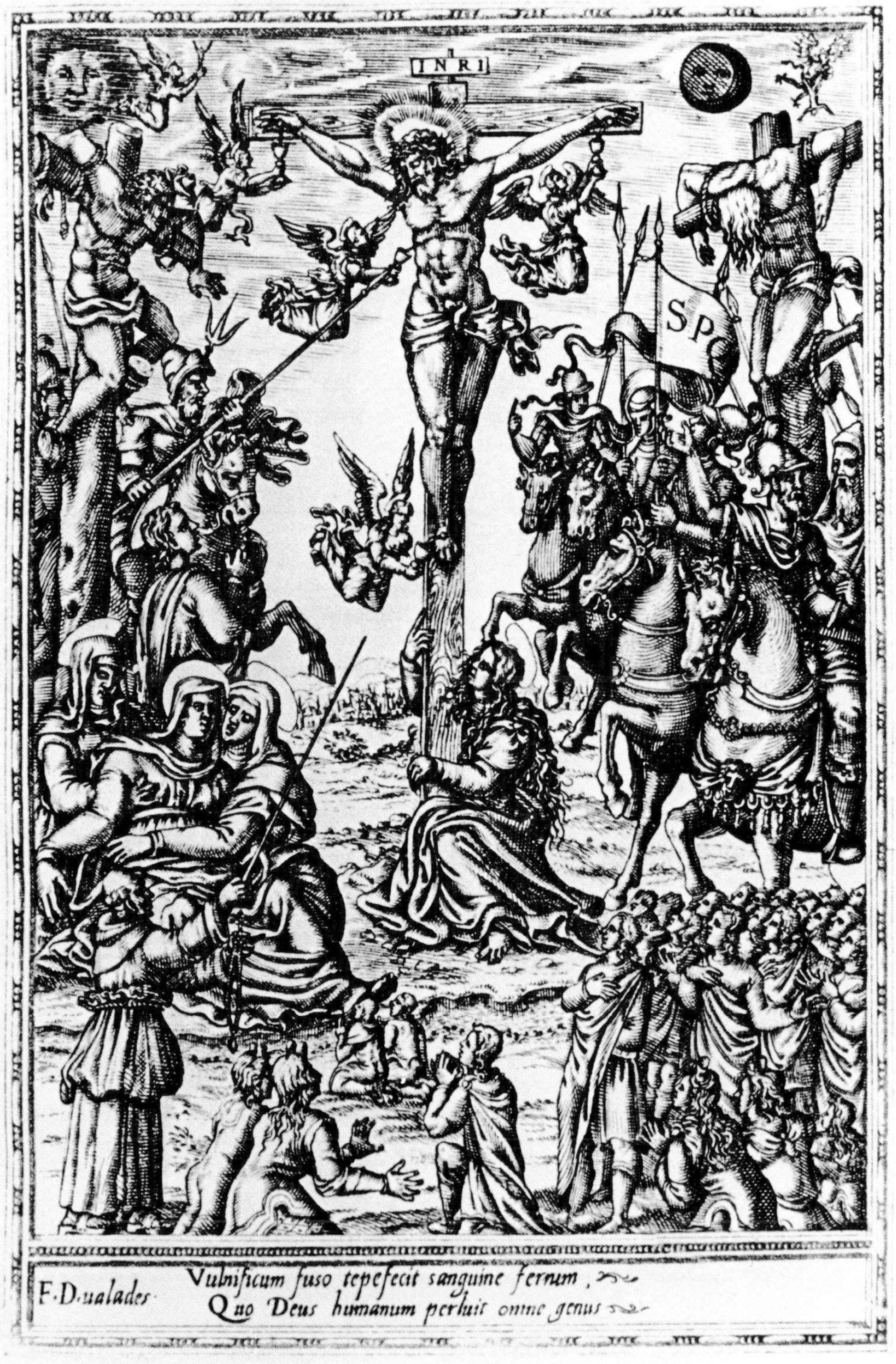

4. *Crucifixion,* from Diego Valadés, *Rhetorica christiana* (Perugia, 1579), page 220, engraving
Private collection

Pope Gregory XIII, who was interested in calendar reform. In any case, it is significant that Valadés cited the construction of this extremely precise and efficient calendar to demonstrate the intelligence of these native peoples.[21]

Between these two opposite examples are various images that occupy a middle ground, where the reality of the New World has carved out a niche for itself in the old, traditional schemas of European mnemonic and devotional devices. A first—and highly interesting—example is the attempt to use a classical mnemonic device, the alphabet, in which the shapes of the letters are taught through their visual similarity to a given natural or artificial object. "Using this ancient mnemonic system," Pauline Moffitt Watts writes, Valadés constructed "a phonetic alphabet correlating letters with indigenous images that are consonant." In the text that accompanies the image of the alphabet (fig. 5), "Valadés also suggests that the sounds of letters can be taught by associations with proper names familiar to the Indians: the sound of *A* through Antonius, *B* through Bartholomeus, and so on."[22] In this way, a traditional mnemonic device becomes an instrument of communication and translation, "a technique for breaking through the 'wall' that separated the written language of the missionaries' script from the ideographic language of Indian hieroglyphics."[23]

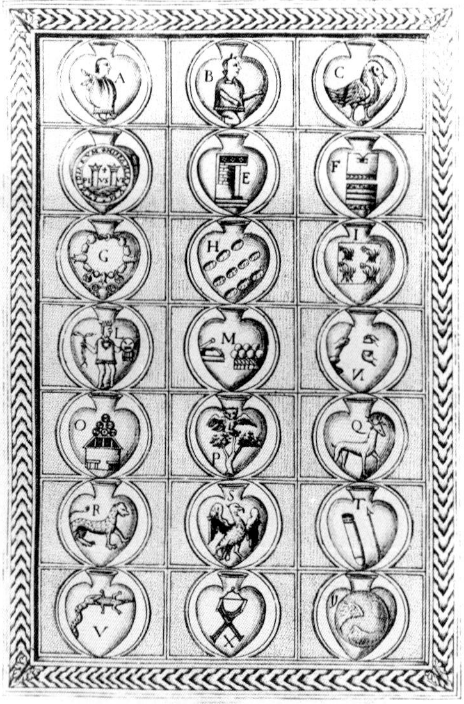

5. *A Figurative Alphabet*, from Diego Valadés, *Rhetorica christiana* (Perugia, 1579), unpaginated, located between pages 100 and 101, engraving
Private collection

Valadés asserted that the Franciscan missionaries carried to Mexico the traditional miscellaneous repertoire of images that had been used in Europe since the Middle Ages:

> Hence in the sacred assemblies they hold among the Indians, the religious men use unusual and amazing figures in order to instill in them divine doctrine more perfectly and more manifestly. To this end they have tapestries into which are woven the principles of the Christian religion, the Apostolic Creed, the Ten Commandments, the seven mortal sins with their numerous offshoots and details, the seven works of mercy, and the seven sacraments, all arranged in a most ingenious manner. It is beyond others an elegant and memorable invention indeed.[24]

This passage offers a very precise description of a set of visual schemas that can be found in many medieval manuscripts. Thanks to the mnemonic structure of the schema, all the essential texts could be condensed into a few pages. In this way, the cycle of schemas became a sort of highly compact library, easy to carry and to use.

Let us examine, for example, the cycle of visual schemas, probably dating from the late fourteenth century, conserved in a manuscript in the Laurentian Library in Florence: the Pluteo codex 30.24. Here one finds a visualization of the topic mentioned by Valadés. At folio 1v, a cuspidated schema links the twelve articles of faith recited in the Credo with twelve prophets and twelve apostles. At folio 4v, an analogous schema draws a visual correspondence between the Ten Commandments, the Ten Plagues of Egypt, and the Ten Errors of the Populace. At folio 3v, the tree of virtues (fig. 6) grows from the root of humility; its branches, leaves, and fruit point upward, toward the sky; its trunk represents the way of life; and in the tree's crown are the luxuriant fruits of the spirit. At folio 4r, the tree of vices (fig. 7) has its roots in pride; the trunk indicates the way of death; and its fruits, which are the fruits of the flesh, hang down toward the earth, as do its branches and leaves. The placement of

6. *The Tree of Virtues,* miniature, c. late fourteenth century, ink and tempera
Biblioteca Medicea Laurenziana, Florence, cod. Pluteo 30.24, folio 3v; photograph Ministero per i Beni e le Attività Culturali della Repubblica Italiana, Florence

7. *The Tree of Vices,* miniature, c. late fourteenth century, ink and tempera
Biblioteca Medicea Laurenziana, Florence, cod. Pluteo 30.24, folio 4r; photograph Ministero per i Beni e le Attività Culturali della Repubblica Italiana, Florence

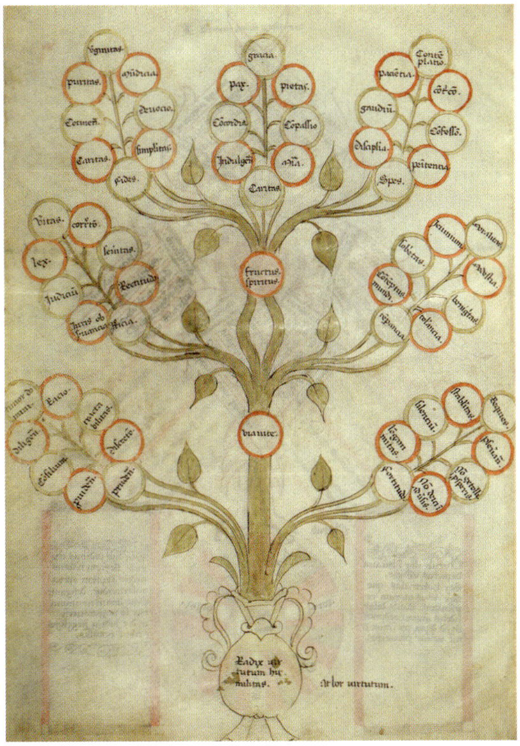

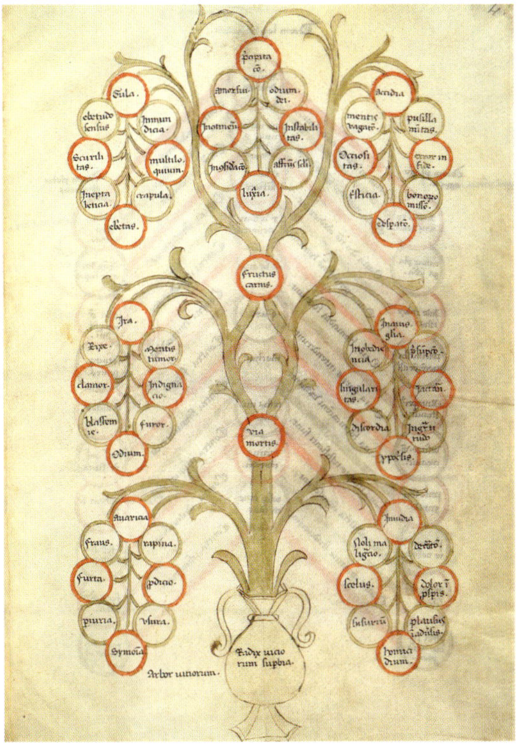

the two trees next to each other enables the eye to take in both the articulation of each one and the correspondences and oppositions between them. On folio 5v is a wheel formed of seven concentric circles, each circle divided into seven segments. The wheel visualizes a whole series of seven-based correspondences: the sacraments, the gifts of the Holy Spirit, the spiritual weapons of the virtues, the works of mercy, the principle virtues, the deadly sins, the spiritual and physical endowments, and so on. It is important to underscore that this repertoire had been established by the end of the thirteenth century and would prove to be enduring.

It is interesting to see how Valadés reinterpreted this tradition of visual, mnemonic schemas in relation to the experience of the New World. In the tree of political hierarchy (fig. 8), he presents a generic schema based on an ornamental version of the tree diagram. The only reference to the Mexican situation is the mention of "vice-rex" (an oblique reference to the Council of the Indies). This image was intended to show that the political and social order is a part of a divine order: beneath the tree, outside of its structured order, is Hell.

A more intriguing image is Valadés' representation of the creation of the world (fig. 9). On the left are the five seals of the Council of the Indies, and on the fifth tier, close to the figure of Adam, Valadés depicts indigenous people under the guidance of a Franciscan friar. Some elements of Mexican flora and fauna are also present, as Mario Sartor noted: these are plants and animals typical of the New World that were also described by Oviedo, Las Casas, Martín de la Cruz, and, later, Francisco Hernández and Bernardino de Sahagún. Among the birds are the quetzal, or snake with feathers, and the *guajolote,* or turkey; among the animals depicted is the Peruvian llama.[25] Many typical New World plants are also represented: the prickly pear, Indian corn, the pineapple, the cocoa plant, and the banana tree. In this way, nature and the peoples of the New World are represented as various parts of the divine order, as essential components of the *vinculum,* the chain that extends from God to unify the different levels of the cosmos.[26]

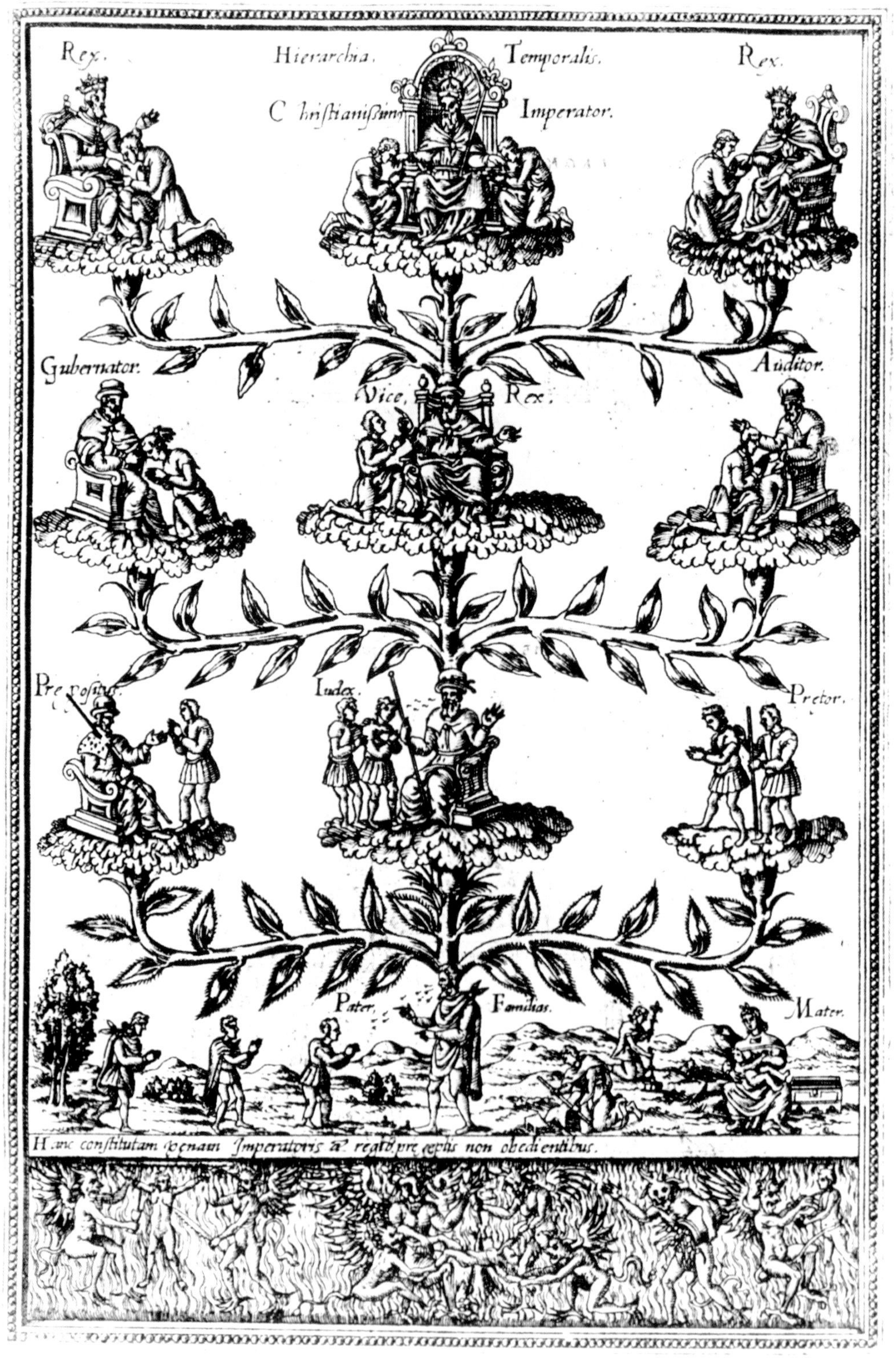

8. *The Tree of Political Hierarchy*, from Diego Valadés, *Rhetorica christiana* (Perugia, 1579), page 173, engraving
Private collection

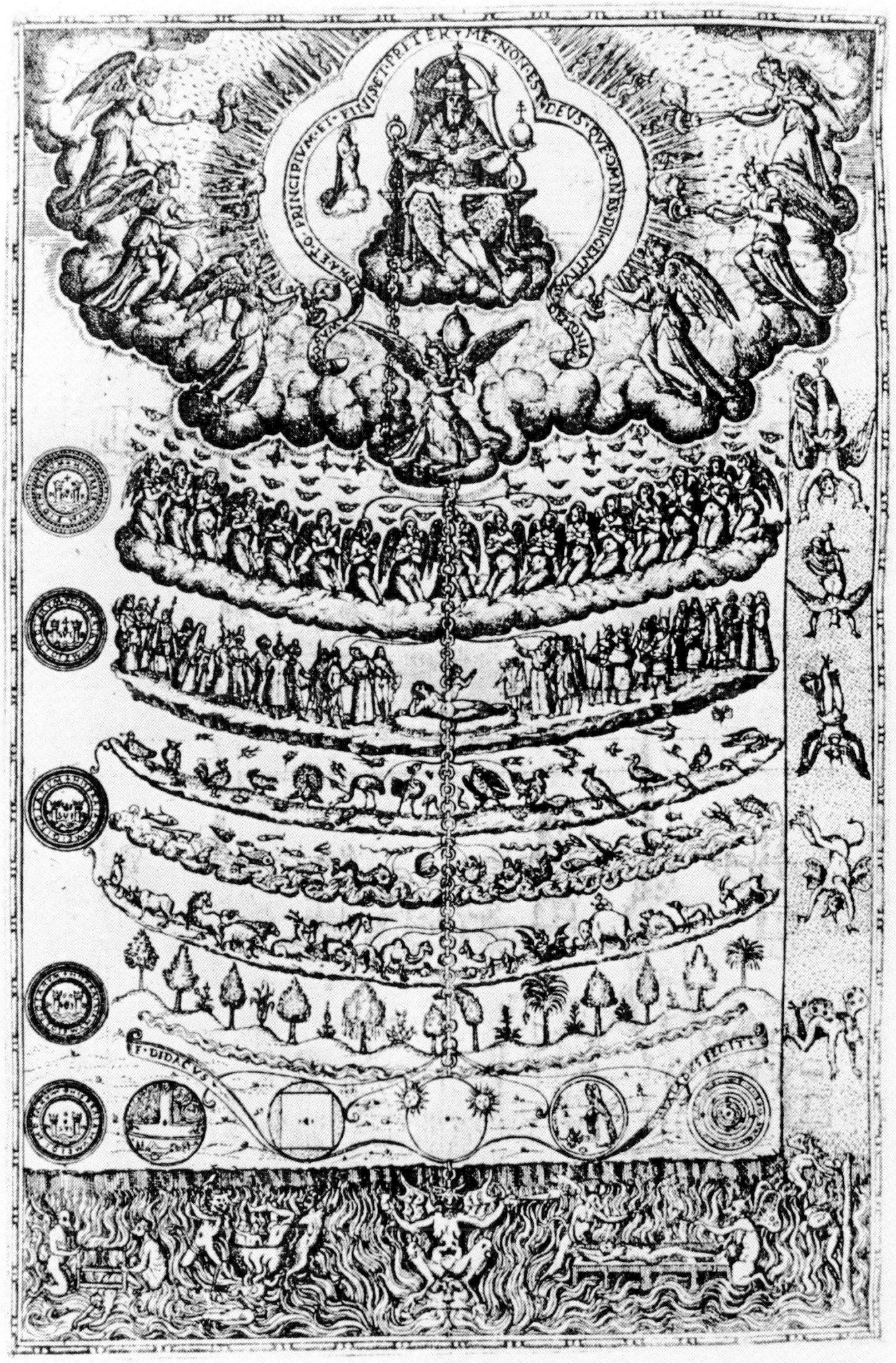

9. *The Creation of the World,* from Diego Valadés, *Rhetorica christiana* (Perugia, 1579), located between pages 220 and 221, engraving
Private collection

Some specifically Mexican plants also appear in an image that has no roots in the European tradition because it specifically illustrates the Mexican world (fig. 10). In the center of this image is a very unrealistic representation of a religious sacrifice, but at the bottom actual Mexican plants are depicted, their names indicated by inscriptions: *cacao, pinha* (pineapple), *tuna* (the fruit of *Opuntia,* or prickly pear cactus), liquid amber (a gum substance mixed with tobacco, from *Liquidambar,* or the sweet-gum tree), *maguei* (or agave, American aloe), and *plátano* (plantain). The illustration also shows an oversized stalk of Indian corn, although the inscription, *flor,* is generic.

Mario Sartor suggests that the snakes in one of Valadés' images dedicated to the torments of Hell have a specific Mexican connotation.[27] Here the snakes are used to represent the devils' hair and tails. The traditional European association between the devil and the snake has a much stronger, more specific meaning in the Mexican context, in which Quetzalcóatl (the snake with feathers) plays a central role in religious rituals and in the indigenous peoples' conception of the cosmos. With this representation, however, the snake is relegated to the demonic world.

We have seen how Mexican nature succeeds in carving a niche for itself in the images of memory that are used to accompany the text. Valadés went even further, however, for Mexican nature can be found at the heart of one mental image, an image that he created solely through words. At the end of the second part of the treatise is a very complex image that has been constructed by combining buildings described in the Bible: the ark of the tabernacle, Noah's ark, and the mystical body of Christ (the church). Its components are drawn from an ancient patristic and medieval tradition. This image was intended to provide *loci* sufficient to store all of the sacred scriptures and to recall all the books of the Bible. Valadés began by creating a mental image of the tabernacle that God ordered Moses to construct. The biblical description is embellished with many details that generate new images, such as the pillars decorated with precious gems. Alongside gems known to the European world (their names and properties being associated with a long literary tradition) are new precious stones of the West Indies, as underscored by Valadés:

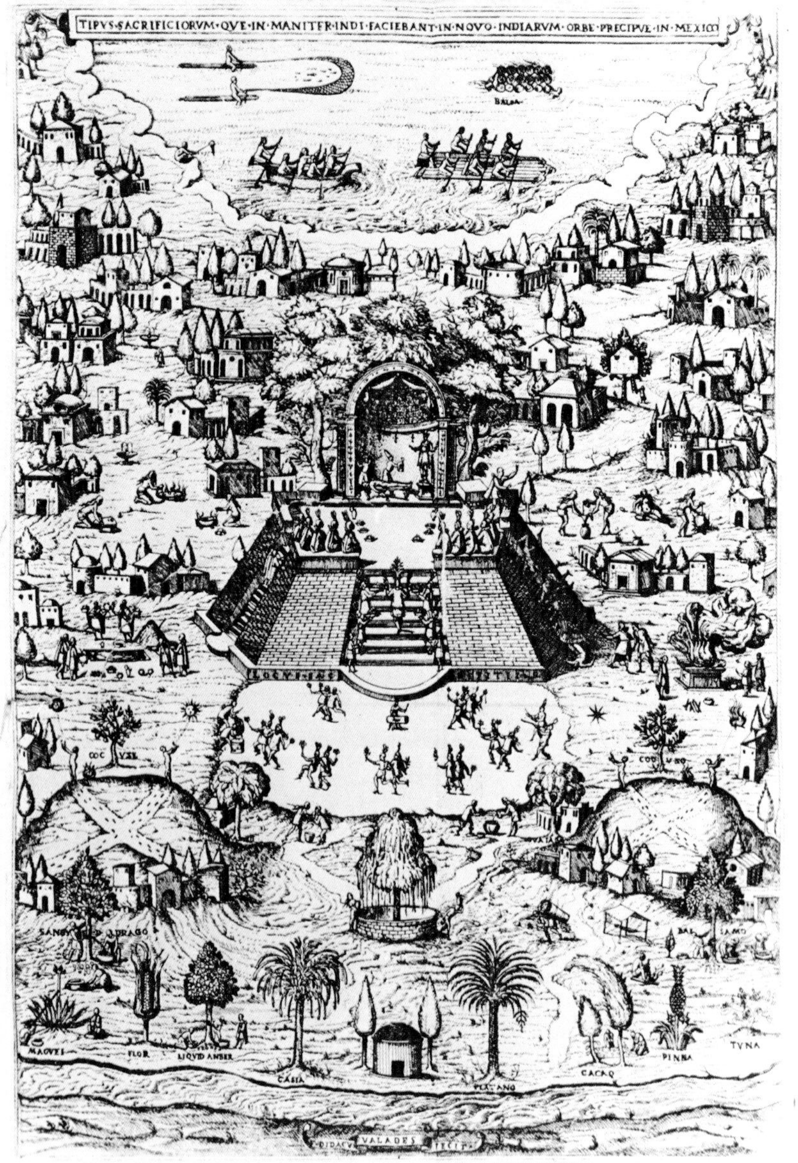

10. *The Mexican World,* from Diego Valadés, *Rhetorica christiana* (Perugia, 1579), located between pages 164 and 165, engraving
Private collection

> On the fifth column—which we would like to imagine is made of *nephriticus,* since this very stone which comes from New Spain, that is from our Western Indies, a stone which has marvelous properties, as we will discuss in more detail later, would look very well here—we will arrange Deuteronomy, whose image will be a throne.[28]

This final, complex image is above all a mental image: it was meant to be used by preachers,

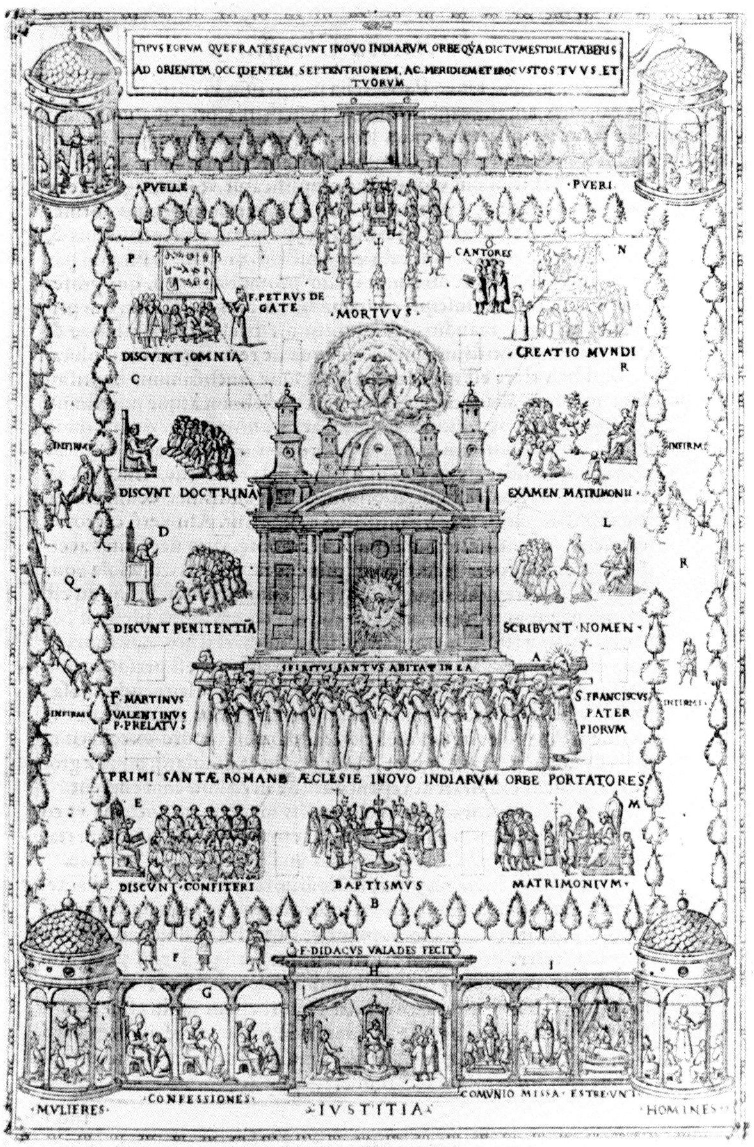

11. *Ideal Representation of the Franciscan Mission in Mexico*, from Diego Valadés, *Rhetorica christiana* (Perugia, 1579), page 207, engraving
Private collection

specifically by Franciscan missionaries, for didactic purposes, for preaching, and for meditation, in situations where it would have been very difficult to use books. But we also have a visual representation of this synthesized image in Valadés' description of Franciscan missionary activities in Mexico. In a procession led by Saint Francis and Friar Martin of Valencia, head of the twelve first Franciscan missionaries, monks are depicted carrying this very building—the ark-tabernacle, "the container and the content of their consciousness"[29]—to the Indians of the New World (fig. 11).

The fact that precious gems from the New World have also been used to construct this building carried by the missionaries—the spiritual building that contains within itself the entire Bible, in other words, everything necessary for spiritual salvation—provides us with a powerful emblem of Valadés' work, of his attempt to use images of memory to create the conditions of translatability, not only between words and images, between the invisible and the visible, but also between two cultures and two worlds that have suddenly and dramatically come into contact with one another.[30]

NOTES

1. Diego Valadés, *Rhetorica christiana ad concionandi et orandi usum accommodata, utriusque facultatis exemplis suo loco insertis, quae quidem ex Indorum maxime deprompta sunt historiis unde praeter doctrinam, sum[m]a quoque delectatio comparabitur* (Perugia, 1579). See Adriano Prosperi, "Intorno a un catechismo figurato del tardo '500," *Quaderni di Palazzo Te* 2 (1985): 45–53; René Taylor, *El arte de la memoria en el nuevo mundo* (Madrid, 1987); Pauline Moffit Watts, "Hieroglyphs of Conversion: Alien Discourses in Diego Valadés' *Rhetorica christiana," Memorie domenicane* 22 (1991): 405–433; Mario Sartor, *Ars dicendi et excudendi: Diego Valadés incisore messicano in Italia* (Padova, 1992); Claudio Finzi and Adolfo Morganti, eds., *Un francescano tra gli Indios: Diego Valadès e la 'rhetorica christiana'* (Rimini, 1995).

2. Paolo Rossi, *Clavis universalis: Arti della memoria e logica combinatoria da Lullo a Leibniz* (Naples, 1960) (English translation: *Logic and the Art of Memory: The Quest for a Universal Language* [London, 2000]); Frances Yates, *The Art of Memory* (London, 1966).

3. See Lina Bolzoni, *La stanza della memoria: Modelli letterari e iconografici nell'età della stampa* (Turin, 1995) (English translation: *The Gallery of Memory: Literary and Iconographic Models in the Age of the Printing Press* [Toronto, 2001]); Jörg Jochen Berns and Wolfgang Neuber, eds., *Seelen Maschinen: Gattungstraditionen, Funktionen, und Leistungsgrenzen der Mnemotechniken vom späten Mittelalter bis zum Beginn der Moderne* (Vienna, 2000).

4. For more on this topic, see two books by Walter J. Ong, *Ramus: Method and the Decay of Dialogue, from the Art of Discourse to the Art of Reason* (Cambridge, Mass., 1958); and *Interfaces of the Word: Studies in the Evolution of Consciousness and Culture* (Ithaca, N.Y., 1977). See also Cesare Vasoli, *Dialettica e retorica dell'Umanesimo* (Milan, 1968); and Anthony Grafton and Lisa Jardine, *From Humanism to the Humanities: Education and the Liberal Arts in Fifteenth- and Sixteenth-Century Europe* (Cambridge, Mass., 1986), chapters 6–7.

5. On the Accademia Veneziana, see Bolzoni 2001, 3–22. The quotation "acciò che visibilmente s'intendano le cose, che sono dall'intelletto solo comprese" can be found in "Supplica ai Procuratori di San Marco, 12 luglio 1560," in Paul L. Rose, "The Accademia Venetiana: Science and Culture in Renaissance Venice," *Studi veneziani* 11 (1969): 231.

6. Valadés 1579, 48.

7. Lodovico Dolce, *Dialogo del modo di accrescere e conservar la memoria*, ed. Andrea Torre (Pisa, 2001), 146–147: "Come chi volesse raccordarsi della favola di Europa, potrebbe valersi dell'esempio della pittura di Tiziano; et altretanto di Adone e di qual si voglia altra favola, o historia profana o sacra, eleggendo specialmente quelle figure che dilettano e quindi sogliono la memoria eccitare. A che sono di utile i libri con figure, come per lo più hoggidì si sogliono stampare, nella guisa che si possono vedere nella maggior parte di quelli che escono dalle stampe dell'accuratissimo Giolito."

8. See Michel Plaisance, ed., *Le Livre illustré italien au XVIe siècle: Texte/image* (Paris, 1999).

9. Cosma Rosselli, *Thesaurus artificiosae memoriae, concionatoribus, philosophis, medicis, juristis, oratoribus, procuratoribus, caeterisq[ue] bonarum litterarum amatoribus* (Venice, 1579), 31.

10. Lambertus Schenkel, *Gazophylacium artis memoriae* (Strasbourg, 1610), 141 and 122: "Nam si oculis externis una hora pervolare et repetere possumus tota Biblia imaginibus expressa materialibus, tam cito enim atque folia vertimus, uno intuitu videmus quid in iis agatur, si videlicet omnia ante nobis cognita sint, ut debent quae per artem locis afficimus: Quanto magis mentis oculis, qui multo citius operantur."

11. Bolzoni 2001, chapter 5.

12. Valadés 1579, 92: "Quid enim putas significare miras illas figuras tam admirabili modo depictas in Augustissimo Sanctissimi Palatio in S. Petro in hac alma egregiaque urbe Romana? Nisi memoria quaedam localis, ut res mirabiles tam in se quam etiam in suis accidentibus, aliqua ipsis mediantibus, aliqua fide, sentire intelligereque possimus ut uniuscuiusque rei typus poenitius rem inspicientibus subiicit ac ministrat, atque ut ea firmius memoriae habeamus? Ibi enim miro artificio theologia, utriusque iuris scientia, logica, philosophia, retorica, religio, pax, bellum et tandem omnia ita sunt ut merito dici queat res esse proportionatas loco et locus locatis."

13. I discuss many examples, from the relationship between the frescoes of *Triumph of Death* in Pisa and Dominican preachers in the Trecento to the sermon cycles of Bernardino of Siena. See Lina Bolzoni, *La rete delle immagini: Predicazione in volgare dalle origini a Bernadino da Siena* (Turin, 2002); English edition: *The Web of Images: Vernacular Preaching from Its Origins to St. Bernardino of Siena* (London, 2003).

14. On *Giardino de oratione*, see Carlo Ginzburg, "Folklore, magia, religione," in *Storia d'Italia* (Turin, 1972), 1:603–676 (esp. 631–632), and Michael Baxandall, *Painting and Experience in Fifteenth-Century Italy: A Primer in the Social History of Pictorial Style* (1972; 2nd edition, Oxford, 1985), 45–50. On Jacopone, see Bolzoni 2002, 121–138. On the medieval tradition, see two important books by Mary Carruthers, *The Book of Memory: A Study of Memory in Medieval Culture* (Cambridge, 1990) and *The Craft of Thought: Meditation, Rhetoric, and the Making of Images, 400–1200* (Cambridge, 1998).

15. See Jonathan D. Spence, *The Memory Palace of Matteo Ricci* (New York, 1983).

16. Watts 1991, 406; and Lewis Hanke, *All Mankind in One: A Study of the Disputation between Bartolomé de Las Casas and Juan Ginés de Sepúlveda on the Religious and Intellectual Capacity of the American Indians* (De Kalb, Ill., 1974). Pauline Moffitt Watts quotes the letter written to Emperor Charles V from Mexico on 6 May 1532 by the Franciscan provincial, Jacopo da Testera. He "spoke of the 'wall' that separated the Indians of New Spain from himself and his fellow missionaries. It was a wall which could be penetrated only by the arduous task of learning native languages. Without such fluency, Testera asserted, the door through which Europeans could enter and 'contemplate the secrets of this people . . . the sentiments of their souls' would remain closed. And, he charged, those Europeans who did not undertake this task would remain 'sleeping witnesses of that which they will never see,' like the soldiers who guarded Christ's tomb at the Resurrection" (Watts 1991, 407).

17. Watts 1991, 407–413, 432–433.

18. Louis Gougaud, "Muta praedicatio," *Revue benedictine* 42 (1930): 168–171; Lucien Febvre, "Iconographie et évangelisation chrétienne," in *Pour une histoire à part entière* (Paris, 1962), 795–819; Michael Camille, "Seeing and Reading: Some Visual Implications of Medieval Literacy and Illiteracy," *Art History* 8 (1985): 26–49; Jérôme Baschet and Jean Claude Schmitt, eds., *L'Image: Fonctions et usages des images dans l'Occident médiéval* (Paris, 1996).

19. Sartor 1992, 57.

20. Watts 1991, 421–422.

21. Valadés 1579, 94.

22. Watts 1991, 426.

23. Watts 1991, 416.

24. Valadés 1579, 95: "Hinc viri religiosi in sacris concionibus quas apud indigetes habent, ad instillandum illis perfectius et manifestius doctrinam divinam, utuntur inauditis et stupendis figuris, in eumque finem habent aulea quibus intexta sunt capita religionis Christianae, ut sunt symbolum Apostolorum, Decalogus, septem peccata mortalia cum sua numerosa sobole et circumstantiis, septena opera misericordiae, et septem sacramenta, via et ordine artificiosissima, quod quidem inventum praeter caetera elegans est et memorabile." I use here the translation in Watts 1991, 423.

25. Sartor 1992, 37–39.

26. See Arthur O. Lovejoy, *The Great Chain of Being: A Study of the History of an Idea* (New York, 1965).

27. Valadés 1579, 217. Sartor 1992, 49: "Alla iconologia 'classica' europea, che aveva sempre coniugato il serpente al demonio su base biblica, si aggiunge in questo caso una trasposizione iconografica piuttosto precisa dal mondo indigeno. Probabilmente ciò avviene non a caso, ovvero come eredità inconsapevole dell'incisore, ma come precisa volontà di coniugare il culto di Quetzalcóatl, divinità positiva ed eroe culturale messicano, ma pur sempre un dio pagano, alle espressioni demoniache da respingere definitivamente." (To the "classical" European iconography that has always associated the serpent with the devil on biblical grounds, one may add in this case a very exact iconographic transposition from the world of the native Indians. This is very probably occurring not by chance, nor as a result of the unconscious artistic patrimony of the engraver, but represents a deliberate attempt to link the cult of Quetzalcóatl, a positive divinity and hero of Mexican culture, but still a pagan god, to demonic aspects [of religion] that must be firmly rejected.)

28. Valadés 1579, 105: "In quinta columna, quam ex Nephritico erigere cupimus, quoniam hic lapis maxime laudatur, ac ex Nova Hispania, hoc est de nostris Indis occidentalibus defertur: cuius operationes mire laudantur ut latius postea tractabimus, Deuteronomium . . . collocabimus, cuius demonstrativum signum erit tronum."

29. Watts 1991, 431.

30. See also the attempts to use the new points of view offered by the revival of the art of memory to interpret non-Western cultures: Mary Nooter Roberts and Allen F. Roberts, eds., *Memory: Luba Art and the Making of History* (New York, 1996); Susanne Küchler and Walter Melion, eds., *Images of Memory: On Remembering and Representation* (Washington, D.C., and London, 1991); Carlo Severi, "Scritture figurate e arti della memoria nel nuovo mondo: Valadés, Schoolcraft, Löwy," in Lina Bolzoni, Vittorio Erlindo, and Marcello Morelli, eds., *Memoria e memorie* (Florence, 1998), 29–66.

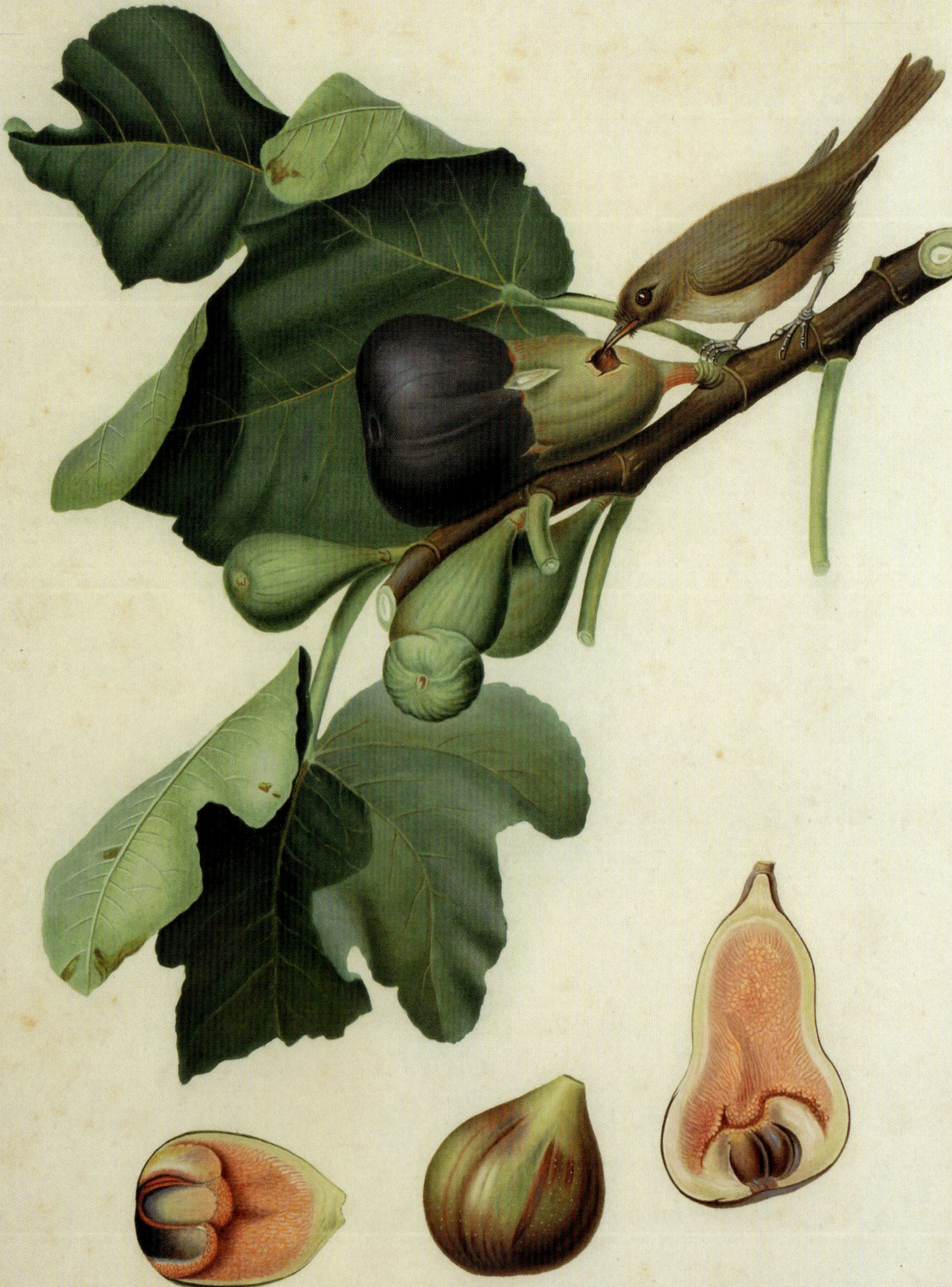

Fico Tetifero, o Fico dell' Osso

Domenico Del Pino disegnò in Finale 1826

Giuseppe Pera inc. in Firenze

ALESSANDRO TOSI
Università di Pisa

From Florence to Philadelphia: Naturalistic Illustration in Tuscany, c. 1760–c. 1840

Some of the most beautiful and important volumes on natural history produced in the second half of the eighteenth century and the first half of the nineteenth were published in Tuscany. As documents of the development of scientific thought in relation to new horizons in research and taxonomy, these undertakings also reflect a figurative culture ready for significant technical and linguistic progress.

In the 1700s, during the reign of the Lorena, cultural ferment in Tuscany gave rise to considerable developments in scientific studies, especially in the area of natural history. In the second half of the century, intense academic debate flourished in Florence at the Società Botanica and the Accademia dei Georgofili (the firstagricultural academy in Europe, founded in 1753), as well as at the Società Botanica in Cortona and the Accademia dei Fisiocritici in Siena. This newfound vitality is also reflected in the work of enlightened intellectuals such as the naturalist and mineralogist Giovanni Targioni Tozzetti (1712–1783); the physician and director of the Orto Botanico of Florence, Saverio Manetti (1723–1785); and the physiologist and natural scientist Felice Fontana (1730–1805); and in the expansion of the universities of Pisa and Florence. These are only a few examples of a Tuscany eager to embrace Enlightenment Europe's desire for renewal.[1]

In this fertile context, the splendid *Storia naturale degli uccelli, trattata con metodo ed adornata di figure intagliate in rame e miniate al naturale* was published in Florence in five volumes from 1767 to 1776 (fig. 1). Sponsored by Carlo Gerini (1733–1796, nephew of Giovanni and Andrea Gerini, two of the most influential of Florentine art patrons) and with an explanatory text by Saverio Manetti, this edition contains more than six hundred plates portraying the ornithological collections of the marquis and senator Giovanni Gerini. Most of the plates were prepared from preliminary tempera paintings by the abbot Lorenzo Lorenzi of Volterra (active second half of eighteenth century). Andrea Gerini commissioned Lorenzi to execute these paintings because of his uncommon ability in the "tocco in penna" (touch of the pen). The paintings were then translated into engravings and were hand-colored by Violante Vanni (act. c. 1760–1770) and Abbot Lorenzi himself (fig. 2).

Every figure was "drawn from life," although the artists certainly also had recourse to stuffed specimens, as well as reuse of plates drawn by other artists such as the Vallombrosan monk and botanist Bruno Tozzi (1656–1743).[2] Thus, scientific inquiry was united with the desire to realize a work of remarkable visual impact, attested by the involvement of the celebrated painters Giovanni Domenico Ferretti (1692–1768) and Giuseppe Zocchi (1715–1767).

Ferretti, described at the opening of the work as "our expert and famed painter,"

Domenico Del Pino and Giuseppe Pera, *Fico fetifero, o fico dall'osso*, from Giorgio Gallesio, *Pomona italiana* (1817–1839), hand-colored engraving
Dumbarton Oaks Research Library and Collection, Washington, D.C.

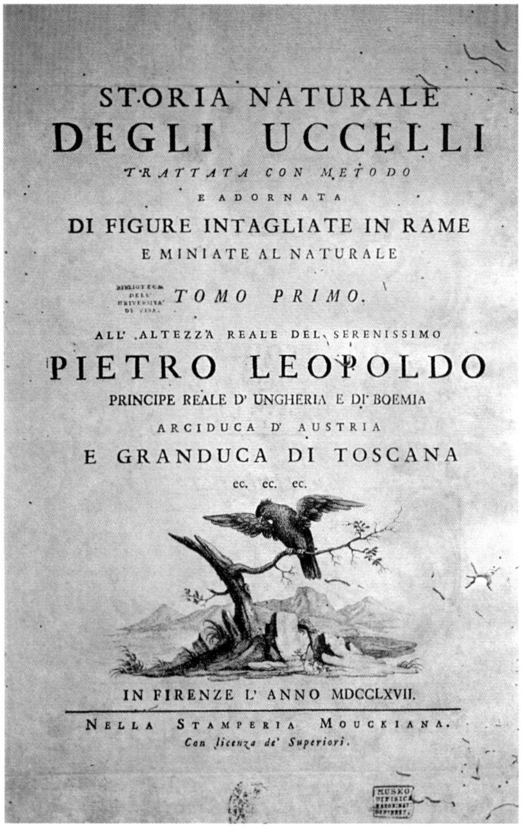
STORIA NATURALE
DEGLI UCCELLI
TRATTATA CON METODO
E ADORNATA
DI FIGURE INTAGLIATE IN RAME
E MINIATE AL NATURALE
TOMO PRIMO.
ALL' ALTEZZA REALE DEL SERENISSIMO
PIETRO LEOPOLDO
PRINCIPE REALE D' UNGHERIA E DI BOEMIA
ARCIDUCA D' AUSTRIA
E GRANDUCA DI TOSCANA
ec. ec. ec.
IN FIRENZE L' ANNO MDCCLXVII.
NELLA STAMPERIA MOUCKIANA.
Con licenza de' Superiori.

1. Title page from Saverio Manetti, *Storia naturale degli uccelli, trattata con metodo ed adornata di figure intagliate in rame e miniate al naturale* (Florence, 1767)
Biblioteca Universitaria, Pisa; author photograph

was the author of several "original figures for this work," demonstrating "not only all of the naturalness and beauty innate in the individual figures portrayed, but also all of the beauty and skill of art."[3] The work also contained several illustrations by Giuseppe Zocchi, a painter sponsored by Andrea Gerini, who had promoted the artist's brilliant career as a *vedutista* who worked in and around Florence, Venice, and Rome, and for whom Zocchi had drawn the well-known series of views of the city of Florence and of Tuscan villas printed in 1744.[4] Zocchi executed the drawings for the book's elegant frontispiece, which shows a refined *rocaille* taste unusual for Florence (fig. 3); for the vignette of the dedication to the new Grand Duke Pietro Leopoldo, signed by Manetti, Lorenzi, and Vanni; and for the portrait of Giovanni Gerini included in the first volume (fig. 4).

The *Storia naturale degli uccelli,* one of the most important publishing events of its day, was a monument to ornithology and to natural history collecting. Moreover, it was a testimonial to patronage; it had not been long since the 1759 publication of the *Raccolta di stampe rappresentanti i quadri più scelti de' signori marchesi Gerini,* sponsored by Andrea Gerini, which reproduced various works from the Gerini family's extraordinary painting collection and in which Lorenzi also participated. The *Storia naturale degli uccelli* also emphasized the importance of the Florentine school of painting—represented by its leading exponents—as well as the genre of naturalistic illustration, which boasted a noble and ancient tradition in Tuscany. Zocchi and Ferretti were the only Florentines mentioned by the writer and social commentator Giuseppe Baretti (1719–1789) in his *Account of the Manners and Customs of Italy; with Observations on the Mistakes of Some Travellers, with Regard to That Country.* The two artists were named in a short list of the best Italian artists of the century, which also included Canaletto (1697–1768) and Pompeo Batoni (1708–1787).[5]

Other events underline the close ties between art and science, publishing, and Enlightenment culture. Foremost among these was the appearance of the *Gazzettiere americano,* the translation of the popular and successful *American Gazetteer,* published in London in 1762. This Italian version was printed in three volumes in Livorno in 1763 by Marco Coltellini, who also printed Cesare Beccaria's *Dei delitti e delle pene* and who—as librettist for the works of composers Christoph Willibald Gluck, Wolfgang Amadeus Mozart, and Antonio Salieri—would enjoy ample success in Vienna at the court of the Austrian archduchess Maria Teresa.[6]

Reflecting Livorno's lively and cosmopolitan cultural environment, the *Gazzettiere americano* presented a refined iconographic display with plates engraved by Giuseppe Maria Terreni, Carlo Faucci, Antonio and Ferdinando Gregori, Giovanni Ottaviani, Giuseppe Pazzi, the Vallombrosan Domenico Veremondo Rossi, Andrea Scacciati, and Violante Vanni. The illustrations not only document American flora, fauna, and commercial products, but also reproduce various plates of the first edition of the *Encyclopédie ou Dictionnaire raisonné des sciences, des arts et des mètiers,* published in Paris in 1751–1772 (fig. 5).[7]

2. *Pappagallo minore verde, e cremisi d'America,* from *Storia naturale degli uccelli* (Florence, 1767–1776), hand-colored engraving
Biblioteca Universitaria, Pisa; author photograph

3. Giuseppe Zocchi, frontispiece from *Storia naturale degli uccelli* (Florence, 1767–1776)
Biblioteca Universitaria, Pisa; author photograph

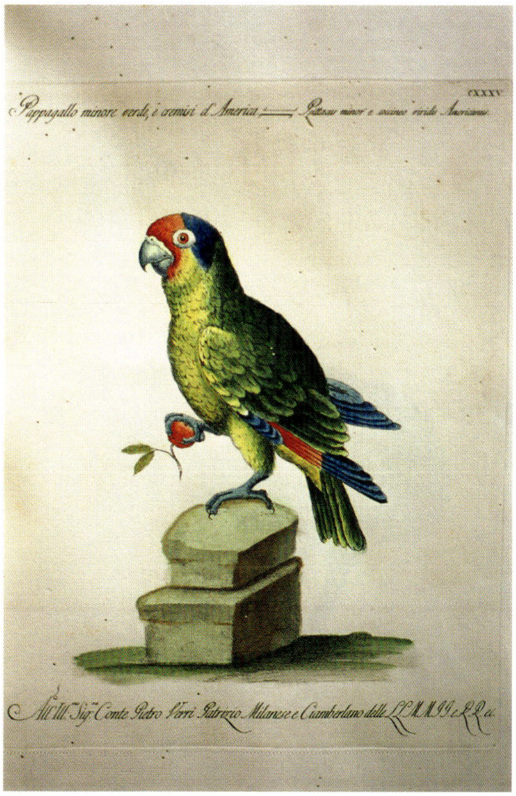

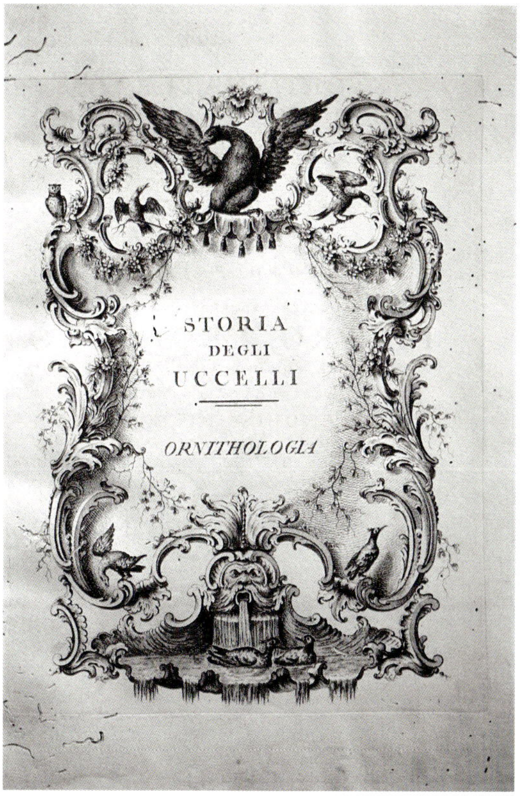

The inclusion of these plates was particularly relevant since Marco Coltellini's initiative anticipates the Livorno edition of the *Encyclopédie* (1771–1779), which—along with the one published in Lucca in 1758–1776—proved to be a milestone in the field of eighteenth-century Italian engraving. But the Livorno edition was not the only one planned in Tuscany under Pietro Leopoldo. Another edition of the *Encyclopédie* was planned in Florence in 1769, when the printers Allegrini and Pisoni envisioned a pocket-sized format "for the convenience of those who, even while strolling, may wish to pass some time in useful meditation" (per comodo di chi fino al passeggio impiegar volesse il tempo in utili meditazioni) and whose plates "at exorbitant cost, will be newly engraved by a master hand" (con spesa esorbitante saranno tutte incise di nuovo da mano maestra).[8]

In 1777, two years after the Museo di Fisica e Storia Naturale, directed by Felice Fontana, opened to the public in Florence, the printer Giuseppe Allegrini conceived the *Pomona toscana,* subtitled *Raccolta delle frutte stampate in rame, ed illustrate colla loro coltura ed uso* (Collection of fruits in copperplates, with explanations as to their cultivation and use). Modeled on other horticultural works published in Europe—such as the *Pomologia,* by Johann Hermann Knoop (1758); the *Traité des arbres fruitiers,* by Henri-Louis Duhamel du Monceau (1768); and the *Pomona franconica,* which Johann Prokop Mayer had begun to publish in Nuremberg in 1777[9]—the *Pomona toscana* attempted to illustrate "the principal varieties of Tuscan fruits":

In the vast and pleasant study of all creation, there is nothing more lovely and seductive than the fruits that are the object of this science, and increasingly appreciated by man, who makes use of their products, so that this alone has given rise to the origin and progress of this part of Natural History. Fruit trees therefore enjoy for good reason their preference more than any other family of plants, not only for their utility, which they provide by their healthful fruits, but still more for the general ornament they bring to Gardens and even more to the countryside. Our Tuscany has always been generous with this type of tree, both for its mild and temperate climate, and

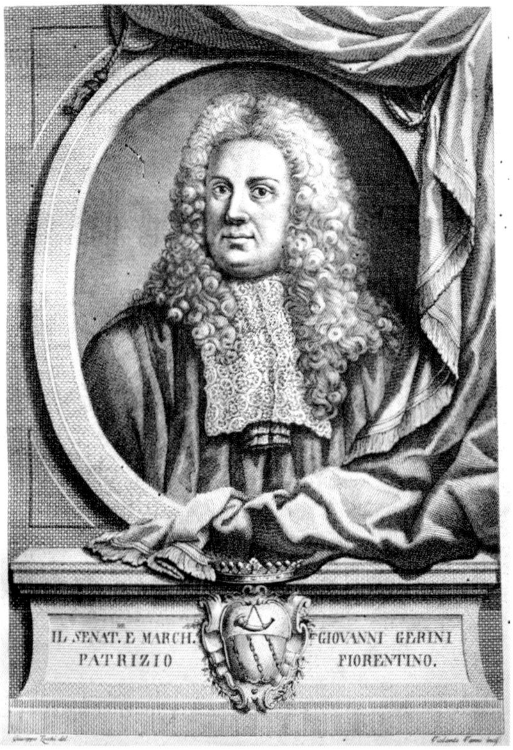

4. Giuseppe Zocchi and Violante Vanni, dedication page, with portrait of Giovanni Gerini, from *Storia naturale degli uccelli* (Florence, 1767–1776)
Biblioteca Universitaria, Pisa; author photograph

> the industrious work of its Farmers, and for the favor of its Rulers, who have always tried to enrich this part of Italy with new kinds of fruit, sparing no expense to obtain them from foreign countries.[10]

The figures would have illustrated "the principal varieties of Tuscan fruits," each with "a detailed description of its tree, the best methods for its cultivation, for the propagation of the species, and finally its nutritional and sometimes technical uses." The engravings, to be executed "in three different manners, that is printed in black, hand-colored, and printed in colors as natural as possible," were entrusted to the "clever artists" directed by Carlo Faucci, "our famed engraver" and pupil of Carlo Gregori. Faucci, who studied advanced engraving techniques in London, had also designed the splendid introductory plate of the Livorno edition of the *Encyclopédie,* which featured a medallion displaying the portrait of the Grand Duke Pietro Leopoldo held up by Fame and surrounded by various allegorical figures representing Livorno, Justice, Commerce, and Father Time, with a smiling cherub holding open an *Encyclopédie* in the foreground (fig. 6).[11]

We do not know why the *Pomona toscana* was never realized. But if on the one hand its ambitious planning expressed the desire to bring Tuscan culture up to date, on the other hand its failure heralded the new problems attending the relationship between art and science in the early modern age, which relegated Italy to the outskirts of the international renewal of scientific knowledge that was occurring so enthusiastically across Europe.

It is true that there would be still other notable examples of scientific publishing, such as *Fauna etrusca,* by Pietro Rossi, printed in Livorno in 1790, or *Materia medica vegetabile toscana,* by Gaetano Savi, printed in Florence by Molini and Landi in 1805, with plates reflecting the influence of Pierre Joseph Redouté's stylistic experimentation. But what was lacking in this decisive moment in the passage to a more modern professional definition of the painter-naturalist was institutional—not intellectual—awareness of the roles of and relationships between artist and scientist.

The extraordinary development of naturalistic studies in France, England, and Germany around the end of the eighteenth and the beginning of the nineteenth century was accompanied by typographic enterprises made possible by a generation of artists educated in natural history museums and botanical gardens. The Jardin des plantes and the Muséum d'histoire naturelle in Paris supported the activity of Gerard van Spaendonck, Pierre Joseph and Henri-Joseph Redouté, Charles-Alexandre Lesueur, Jacques Milbert, Pierre-Antoine Poiteau, and Pierre-Jean-François Turpin, artists who mastered innovations in reproduction techniques along with a more modern and convincing figurative language. Such institutions were lacking in Italy, however, and naturalistic illustration was not considered a suitable subject for debate or a means of bringing academic information up to date.

This was clear to Giorgio Gallesio (1772–1839), the Ligurian naturalist who, around 1816, decided to publish *Pomona italiana,* the first census of Italian horticultural species. Referring to the difficulties encountered in this monumental editorial project, Gallesio would write:

> My position differed from that of past pomologists. Inhabitants of metropolitan cities representing their nations could know beforehand the number of varieties they had to describe and draw, these having already been gathered there by gardeners, or belonging to nearby and familiar dilettanti; they were able to obtain examples without leaving their studios; they had the convenience of studying without voyages or inconvenience; they had the advantage of the aid of the foremost agronomists and botanists of the time; and, finally, they had at their disposal at all times and seasons specialized artists for executing the figures, appropriate engravers hired to transfer them to copper with the appropriate methods, and colorists trained in this particular trade. Instead, I lacked all of this. Italy has no central city containing all these resources: abounding with artists of the classical kind, it had none in this secondary genre, heretofore neglected and regarded as exclusive to foreigners.[12]

Still, the choice of Tuscany for Gallesio's project was no accident. Here, thriving Pisan typographers and Florentine draftsmen joined a noble tradition embodied by extraordinary painters such as Jacopo Ligozzi and Bartolomeo Bimbi and by naturalists such as Pier Antonio Micheli and Giovanni Targioni Tozzetti.[13] Following in that tradition, together with a desire to bring it up to date in response to modern needs, Gallesio would create one of the masterpieces of nineteenth-century publishing, the *Pomona italiana, ossia trattato degli alberi fruttiferi* (fig. 7). The *Pomona italiana* was printed in Pisa in 1817–1839, set in the typeface named Niccolò Capurro, created by the scholar and writer on art Giovanni Rosini (1776–1855). The book contains 160 color plates executed by Florentine engravers using a variety of techniques, including stipple engraving, aquatint, and mezzotint.[14] A monumental publishing enterprise, modeled on French, German, and English counterparts published from 1806 to 1818—such as the *Pomona* (1808) of Peter Henderson, the *Pomona Herefordiensis* (1811) of Thomas Andrew Knight, the *Pomona Britannica* (1812) of George Brookshaw, and the *Pomona Londinensis* (1818) of William Hooker—Gallesio's

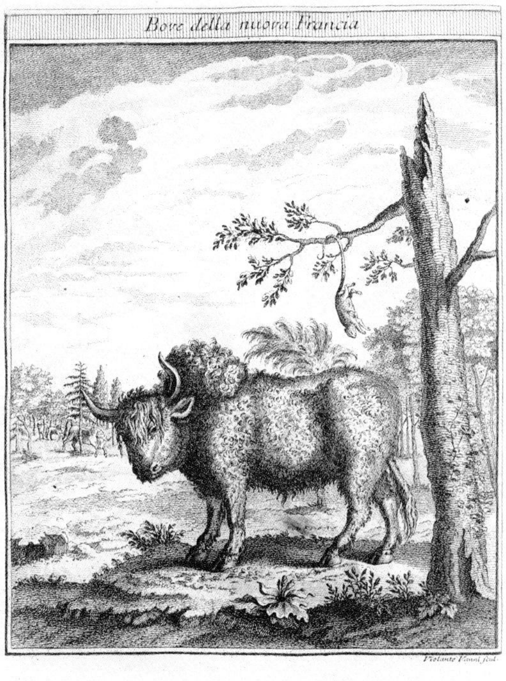

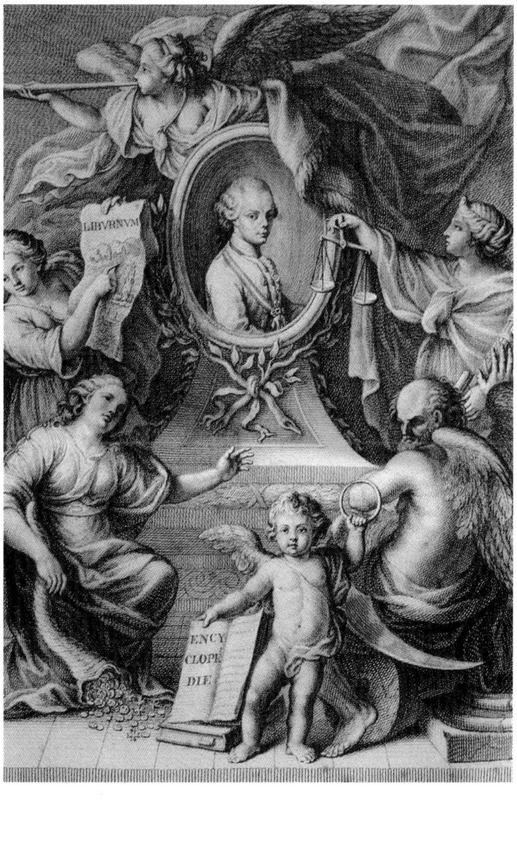

5. Violante Vanni, *Bove della nuova Francia,* from *Il Gazzettiere americano, contenente un distinto ragguaglio di tutte le parti del nuovo mondo* (Livorno, 1763), engraving
Biblioteca di Storia delle Arti, Pisa; author photograph

6. Filippo Bracci and Carlo Faucci, frontispiece, with portrait of Pietro Leopoldo, from *Encyclopédie ou Dictionnaire raisonné des sciences, des arts et des mètiers* (Livorno, 1770), engraving
Biblioteca Universitaria, Pisa; author photograph

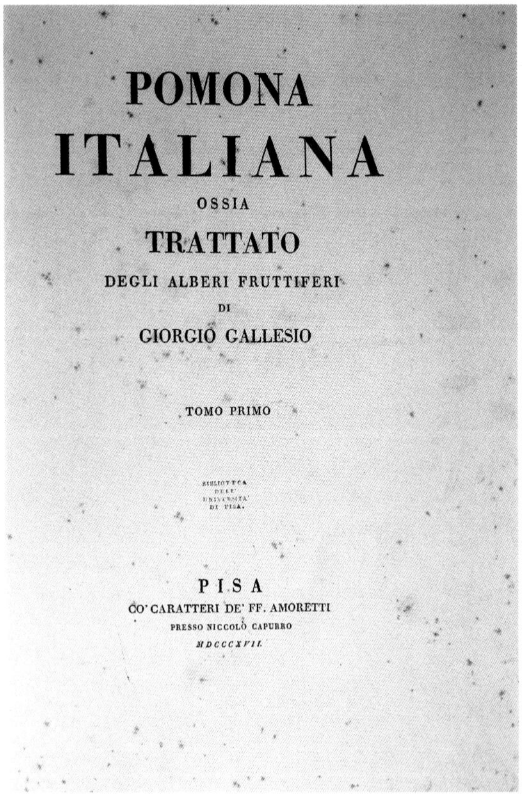
POMONA
ITALIANA
OSSIA
TRATTATO
DEGLI ALBERI FRUTTIFERI
DI
GIORGIO GALLESIO

TOMO PRIMO

PISA
CO' CARATTERI DE' FF. AMORETTI
PRESSO NICCOLÒ CAPURRO
MDCCCXVII.

7. Title page from Giorgio Gallesio, *Pomona italiana, ossia trattato degli alberi fruttiferi* (Pisa, 1817)
Biblioteca Universitaria, Pisa; author photograph

work was a notable achievement precisely because of the new dimension and dignity of naturalistic illustration (the "artistic part" being "perhaps more important than the literary," wrote Gallesio).[15]

The first artist entrusted with the plates of Gallesio's work was Antonio Serantoni (1780–1837), trained at the Accademia di Belle Arti in Florence. Serantoni was an anatomical draftsman who had executed many of the plates of the so-called *Grande anatomia* and the *Anatomia pittorica,* by Paolo Mascagni, and was the creator of much-admired anatomical wax sculptures, such as a "Venere decomponibile," that is, a Venus or "statue of the female sex composed of many detachable parts for observation of the four cavities," commissioned in 1822 by Dr. Edward Nolbroh of South Carolina for use in his anatomy class.[16]

Gallesio's observations on the tempera paintings executed by Serantoni from 1816 emphasize the naturalist's close collaboration with "his" painter: either carried out "with a naturalist's precision" or to be retouched because "the colour of the fruit was untrue," each image seeks a perfect correspondence between scientific description and iconographic representation (figs. 8, 9). Serantoni's images so surpassed the results of most other renowned contemporary painter-botanists that several works on vellum by Pierre-Jean-François Turpin (1775–1840), "possibly the greatest natural genius of all the French botanical painters of his day," and Turpin's collaborator, Pierre-Antoine Poiteau (1766–1854), ordered specially from Paris, proved inferior to Serantoni's plates.[17]

For Gallesio, this direct collaboration with the artist, leading to the recognition of the new profession of "naturalistic illustrator," was a fundamental moment in scientific research. Therefore in September 1816, at the time of the announcement of the publication of the *Pomona italiana,* while he was busily engaged in painting fruit for Gallesio's work, Serantoni petitioned the grand duke of Tuscany, requesting that his profession of anatomical and botanical draftsman and engraver be recognized, "since the progress of Natural History is certainly owed to the exact figures of plants and animals in natural colors." Likewise, following the example of Gerard van Spaendonck and his teaching of *peinture des fleurs* (flower painting) at the Jardin des plantes (because, "like Spaendonck, he draws by hand and engraves with the greatest skill and truthfulness various fruits for the *Pomona Italica* [*sic*] which is to be published by Count Gallesio"), Serantoni proposed setting up a similar class at the hospital of Santa Maria Novella, at the Museo di Fisica e Storia Naturale, or "at another establishment of public instruction" in Florence.[18]

The artist's request was not granted but was clearly emblematic of the desire to overcome the organizational and ideological lapses noted at the same time and with the same clarity of vision by the naturalist Gallesio. Furthermore, this episode permits us to comprehend better the enthusiasm for naturalistic illustration that swept Florence and Tuscany during the first half of the nineteenth century. For example, in 1817 at the Accademia dei Georgofili, the gardener Antonio Piccioli presented the first volume of a collection of watercolors of various plants and flowers drawn "from nature" and "with particular exactitude, and

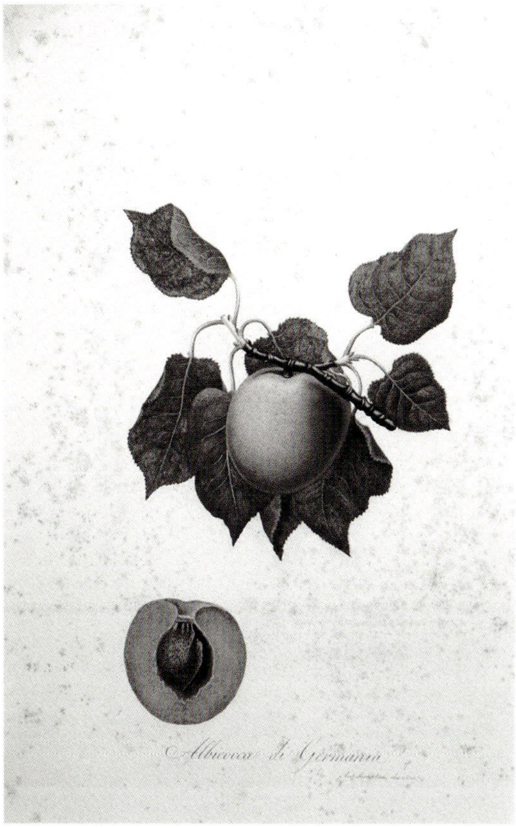

8. Antonio Serantoni, *Albicocca di Germania,* from Giorgio Gallesio, *Pomona italiana* (Pisa, 1817–1839), hand-colored engraving
Biblioteca Universitaria, Pisa; author photograph

the greatest of good taste," from the garden of the Museo di Fisica e Storia Naturale in Florence. Later, Piccioli produced a collection of plaster fruits entitled Pomona Toscana, following the illustrations of Gallesio and Serantoni but perhaps also modeled on the failed project conceived by the printer Giuseppe Allegrini in 1777.[19]

The immediate success of *Pomona italiana* encouraged other naturalists to publish valuable botanical and horticultural collections that were also related to the rapid spread of the taste for landscape gardening. Thus, for the *Flora italiana ossia raccolta delle piante più belle che si coltivano nei giardini d'Italia,* by Gaetano Savi (1769–1844), printed in Pisa in 1818–1824, Antonio Serantoni, "with extreme diligence," executed plates "drawn and colored from life" and depicting many plants from the botanical garden in Pisa (fig. 10). Likewise, the *Raccolta dei fiori, frutti ed agrumi più ricercati per l'adornamento dei giardini disegnati al naturale da vari artisti,* published in Florence in 1825 by Antonio Targioni Tozzetti "to satisfy the desire of many dilettante gardeners," involved the collaboration of the artists who worked on the *Pomona italiana* as well as the talents of several "dilettanti" such as Giovanni Geri, gardener of the Giardino di Boboli (fig. 11).[20]

Meanwhile painters, engravers, and colorists, as well as gardeners and dilettanti, were working on the *Pomona italiana,* and their illustrations arrived from every part of Italy. In September 1824, Gallesio visited the Galleria degli Uffizi in Florence to study the splendid temperas of Jacopo Ligozzi (1547–1627) and to compare them to the plates of his closest collaborators, especially the skilled Ligurian painter Domenico Del Pino (dates unknown). He invited artists to follow the "anatomical exactitude" and especially that "certain glow which I have never been able to obtain from our colorists, and which is found in nature," qualities that characterized several of Ligozzi's masterpieces—for instance, *Fig Branch with Exotic Finches* (fig. 12), the model for a plate from *Pomona italiana* executed by Domenico Del Pino (fig. 13). With these, Gallesio pointed out a fundamental requirement of scientific observation and artistic inquiry, calling

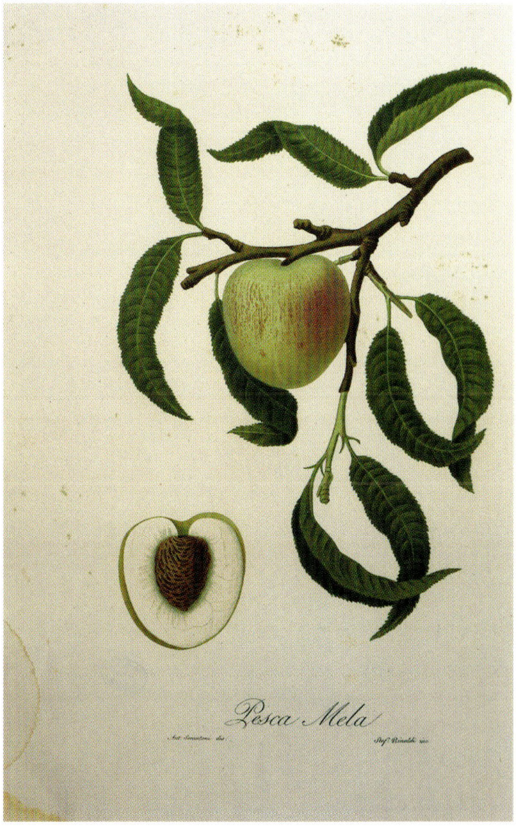

9. Antonio Serantoni and Stefano Rinaldi, *Pesca mela,* from Giorgio Gallesio, *Pomona italiana* (Pisa, 1817–1839), hand-colored engraving
Biblioteca Universitaria, Pisa; author photograph

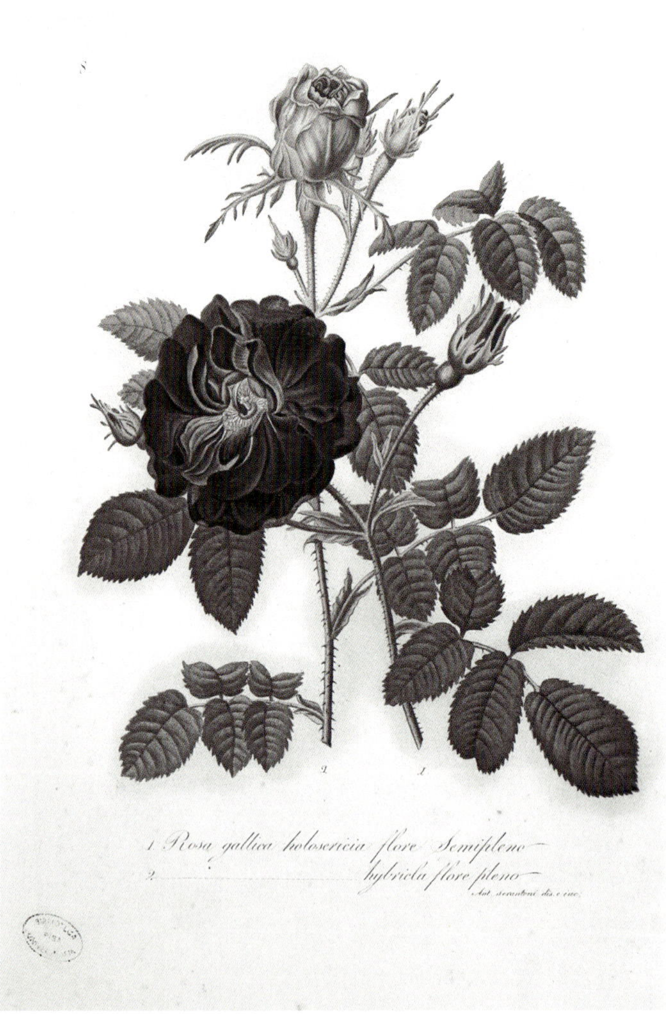

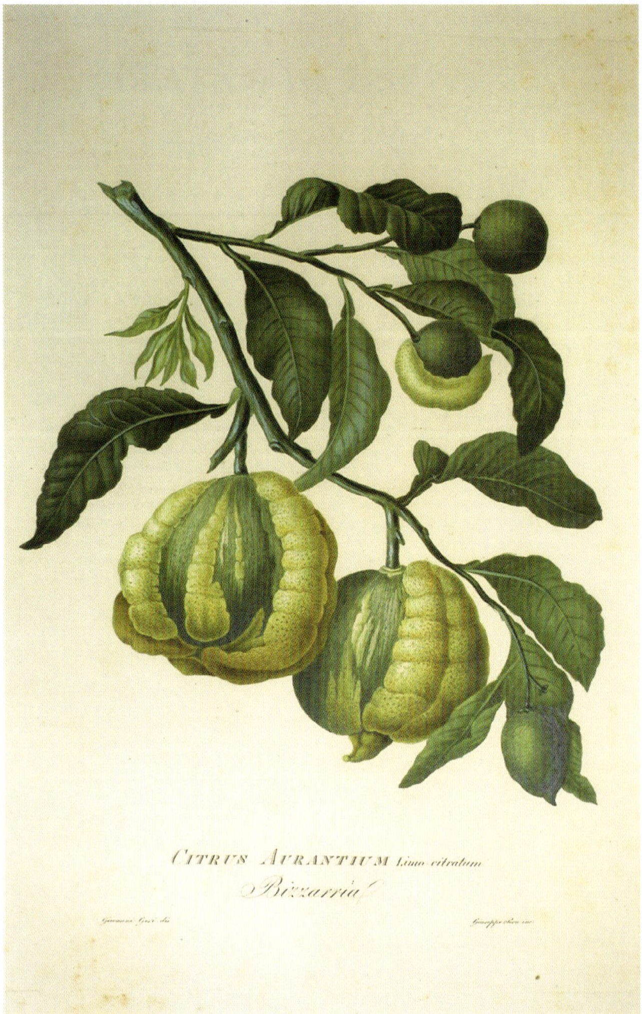

upon the highest tradition of naturalistic illustration.[21]

Another incident would emphasize and amplify the importance of this episode. At this very moment, in the spring of 1824, young Prince Charles-Lucien Bonaparte (1803–1857), nephew of Napoleon Bonaparte and a naturalist with a solid artistic background, met the naturalist and illustrator John James Audubon in Philadelphia. He was so astonished by the extraordinary talent of that "zealous observer of nature and skilful artist" that he invited him to participate in continuing Alexander Wilson's ambitious editorial project, *American Ornithology*.[22] And as we know, the plate illustrating the "great crow–blackbird," or purple grackle, despite criticism of its supposed lack of "correctness," would be published in the first volume of Bonaparte's *American Ornithology,* printed in Philadelphia in 1825, next to plates executed with, as the text described it, "the happy pencil" of Titian Peale, colored "from nature" by Alexander Rider, and engraved with "accuracy and ability" by Alexander Lawson, "the foremost ornithological engraver of our age" (fig. 14).[23]

The incident was truly of great significance. From Florence to Philadelphia, widespread conviction grew that the naturalist's task of describing and naming new species was inseparable from a precise visual description provided by means of botanical and zoological illustration. Starting from the shared analytical observation of natural data by artist and scientist, this new definition arrived at a final confirmation and comparison with either tradition (the legacy of Jacopo Ligozzi) or modernity (the lesson

10. Antonio Serantoni, *Rosa gallica,* from Gaetano Savi, *Flora italiana ossia raccolta delle piante più belle che si coltivano nei giardini d'Italia* (Pisa, 1818–1824), hand-colored engraving
Biblioteca Universitaria, Pisa; author photograph

11. Giovanni Geri and Giuseppe Pera, *Citrus aurantium, Limo citratum, Bizzarria,* from Antonio Targioni Tozzetti, *Raccolta dei fiori, frutti ed agrumi più ricercati per l'adornamento dei giardini disegnati al naturale da vari artisti* (Florence, 1825), hand-colored engraving
Biblioteca Universitaria, Pisa; author photograph

12. Jacopo Ligozzi, *Fig Branch with Exotic Finches,* c. 1580–1590, gouache on paper
Gabinetto Disegni e Stampe degli Uffizi, Florence, 1958 Orn.

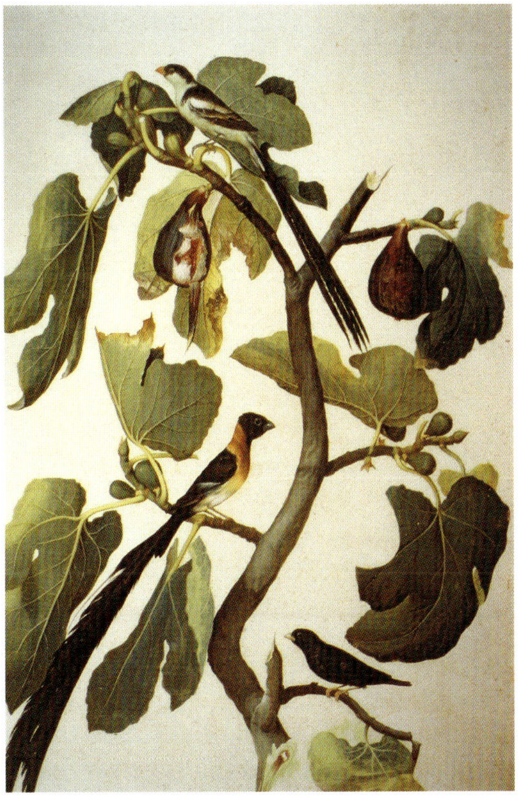

13. Domenico Del Pino and Giuseppe Pera, *Fico fetifero, o fico dall'osso,* from Giorgio Gallesio, *Pomona italiana* (1817–1839), hand-colored engraving
Biblioteca Universitaria, Pisa; author photograph

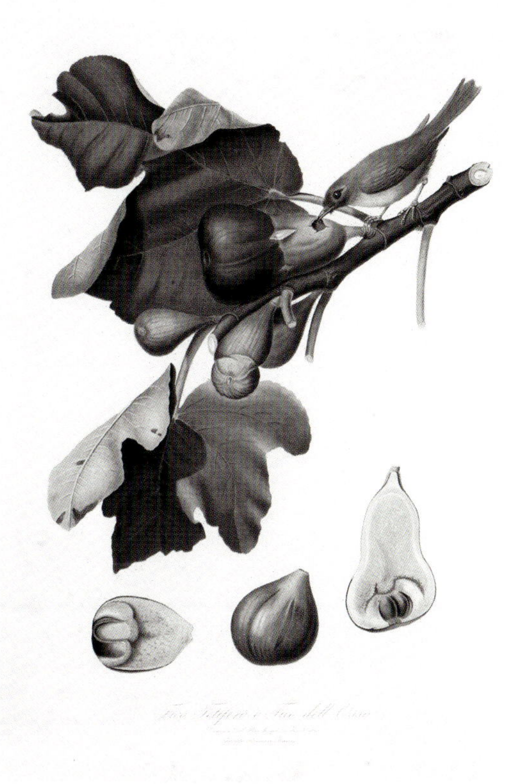

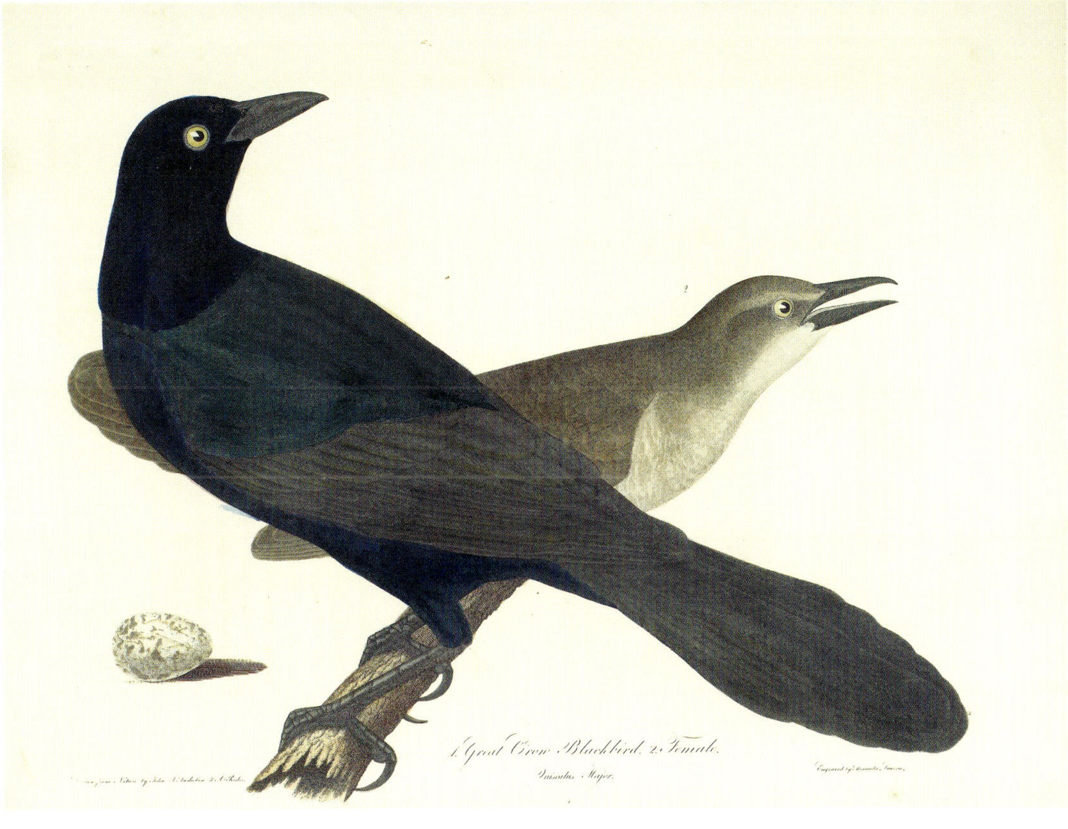

14. John James Audubon and Alexander Lawson, *Great Crow-Blackbird,* from Charles-Lucien Bonaparte, *American Ornithology, or, The Natural History of Birds Inhabiting the United States* (Philadelphia, 1825–1833), hand-colored engraving
University of Pennsylvania, Annenberg Rare Book and Manuscript Library

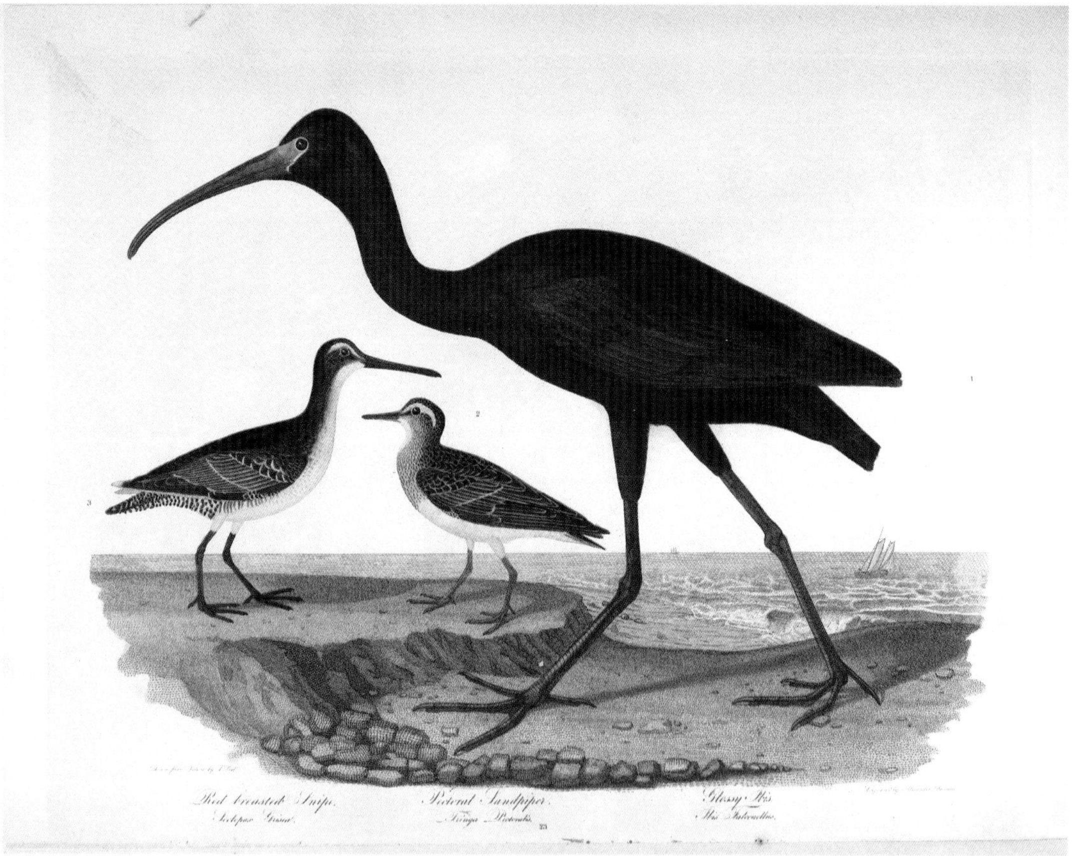

15. Titian Peale, Alexander Rider, and Alexander Lawson, *Red Crested Snipe, Pectoral Sandpiper, Glossy Ibis,* from Charles-Lucien Bonaparte, *American Ornithology, or, The Natural History of Birds Inhabiting the United States* (Philadelphia, 1825–1833), hand-colored engraving
University of Pennsylvania, Annenberg Rare Book and Manuscript Library

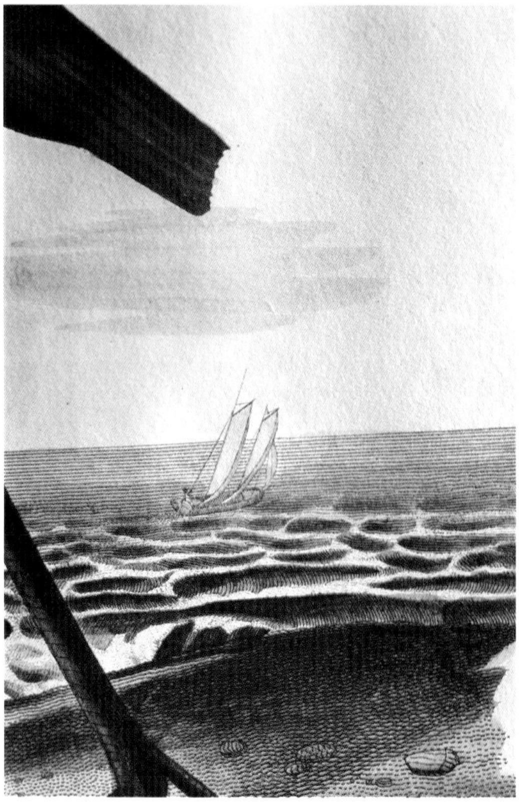

16. Detail of figure 15, showing a double-masted sailboat on the water in the background

of Alexander Wilson). From there spread a figurative culture, characterized by technical experimentation and thoughtful stylistic research, ranging from the elegant layout of the plates of the *Pomona italiana* to a greater attention to landscape, evident in the plates of *American Ornithology* (figs. 15, 16). The practice of drawing from nature, now fully accepted in the scientific field, was the visual development of the eighteenth-century naturalists' desire to "catch nature in the act" (cogliere la natura sul fatto); and this had to be strongly connected not only to an expanding search for precise description, but most of all to the emergence of new meditations on landscape.

In the first volume of his *American Ornithology,* Bonaparte notes that "the chief defect of various magnificent European works" was "the unnatural stiffness faithfully copied from stuffed skins."[24] But the "accuracy" and "correctness" sought after by Bonaparte—as well as by Lawson and Audubon (who accused Peale "of drawing birds as if they were seated for their portraits")[25]—

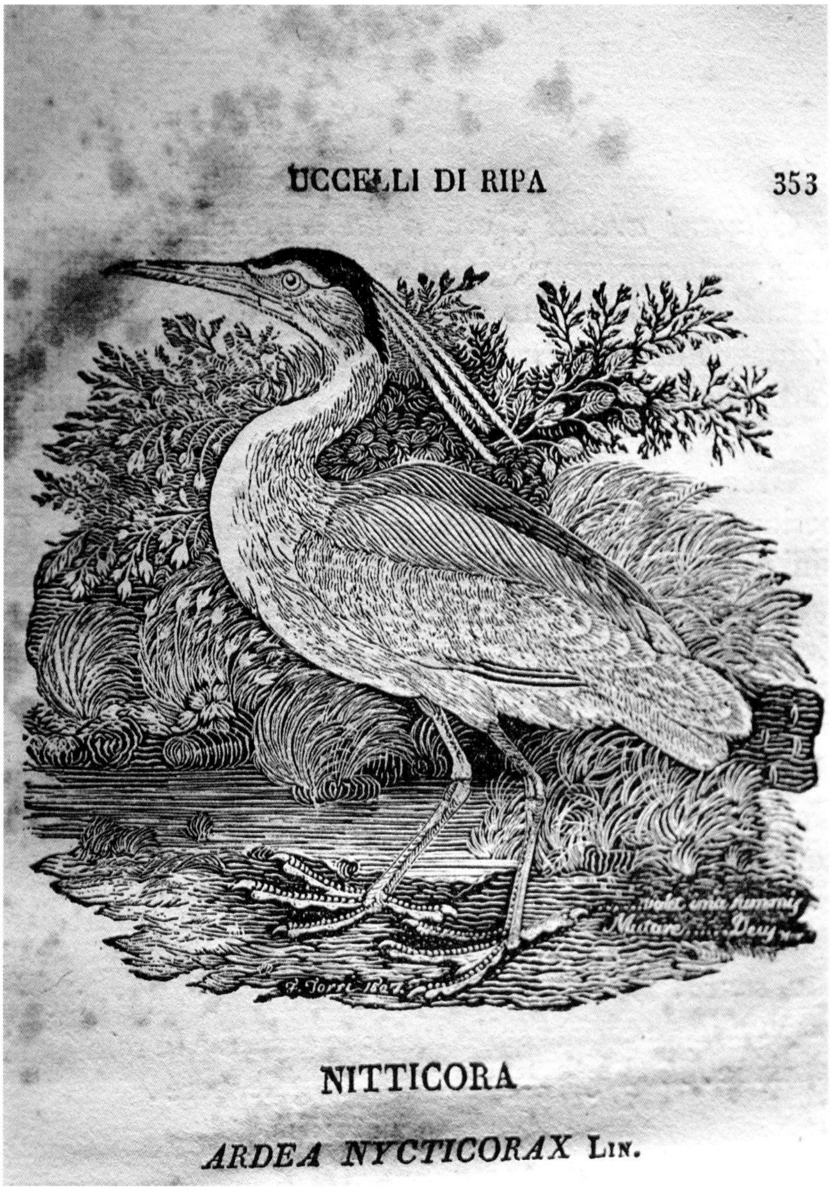

17. Francesco Torri, *Nitticora: Ardea nycticorax* Lin., from Paolo Savi, *Ornitologia toscana* (Pisa, 1827–1831), wood engraving
Biblioteca di Scienze naturali e ambientali, Pisa; author photograph

was a natural consequence of working from nature based on a relationship with the landscape, perceived as fundamental to figurative meditations. An example of this process can be seen in the experience—truly exemplary—of Titian Peale on the Stephen Harriman Long expedition, or in Bonaparte's enthusiasm for the art of a painter-naturalist such as Audubon, "who has devoted twenty years of this life to studying nature in the forests of the West."[26]

If this convergence came to characterize the entire course of American natural history illustration, no less significant is the criticism directed to the plates of the *Storia naturale degli uccelli* by Paolo Savi (1798–1871; son of Gaetano Savi, director of the Museo di Storia Naturale in Pisa, and a correspondent of Charles-Lucien Bonaparte). In his introduction to the first volume of *Ornitologia toscana,* printed in Pisa in 1827, Savi stated, "The biggest defect of those figures is in their strange and unnatural positions. It is clearly evident that the artist had never observed the natural attitudes of the birds he has represented, and has merely let himself be guided by picturesque fantasy."[27]

In order to avoid the incorrectness of "picturesque fantasy," Savi included several wood engravings by Francesco Torri inspired by the "most beautiful" *History of British Birds* by Thomas Bewick, and insisted especially on the value of direct observation of the terrain, which was fundamental to his important zoological and geological studies (fig. 17).[28]

It was hardly by chance that, again in 1824, the very first invitation to plein air painting was addressed to Tuscan artists. During the Florentine exhibition of the Accademia di Belle Arti, the reviewer of *Antologia,* the magazine founded and edited in Florence by Giovan Pietro Vieusseux, recommended to the artist "to be not indolent and reluctant to abandon his own hearth, but instead to climb the hills, descend into the valleys, pass over the mountains, tarry on the plains; and always sketching, and perhaps painting along the way, to gather many surprising and varying views of nature. Working thus, one would never be subject to the accusation of having painted such and such a view, without ever having left his own room."[29] It was not until the middle of the nineteenth century that artists in general would heed this message, but the role of naturalistic illustration was fundamental to the growth of a new pictorial sensibility. In this sense, Bonaparte's experience in America was quite important, and was destined to have great repercussions on Italian culture. Returning to Italy in 1828, and inspired by his stay in Philadelphia, Bonaparte published the *Iconografia della fauna italica per le quattro classi degli animali vertebrati,* the first national zoological census, printed in Rome in 1832–1841.[30]

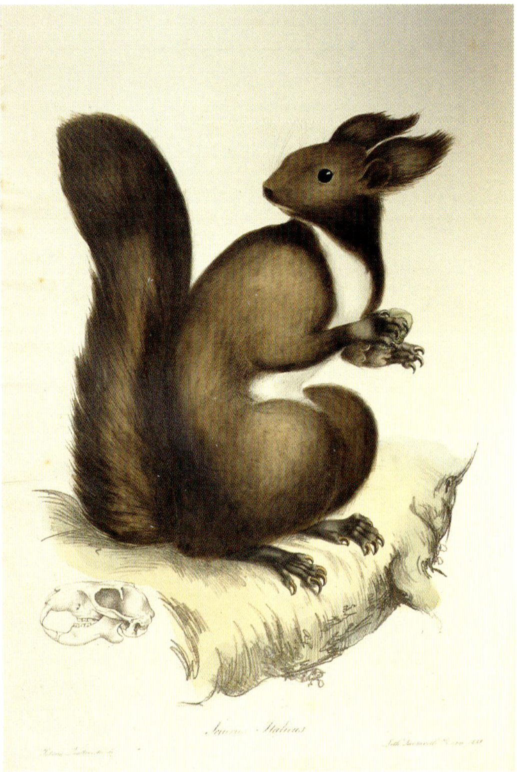

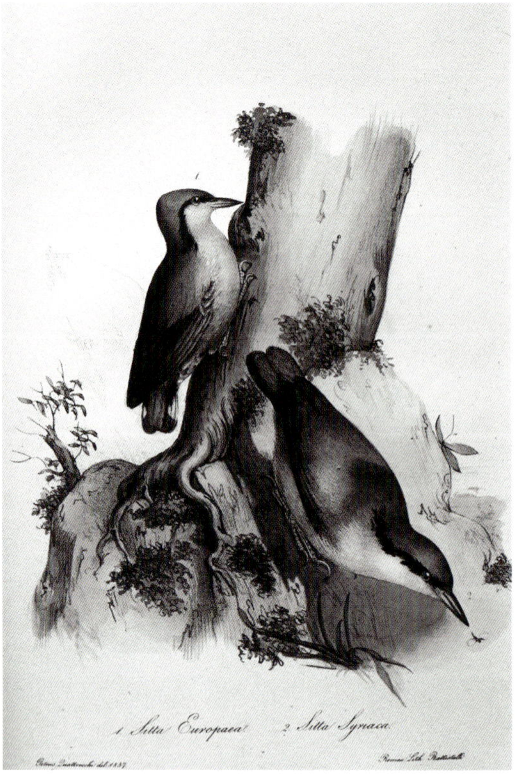

18. Pietro Quattrocchi, *Sciurus italicus,* from Carlo Luciano (Charles-Lucien) Bonaparte, *Iconografia della fauna italica per le quattro classi degli animali vertebrati* (Rome, 1832–1841), lithograph
Biblioteca Universitaria, Pisa; author photograph

19. Pietro Quattrocchi, *Eurasian Nuthatch, White-Breasted Nuthatch,* from Carlo Luciano (Charles-Lucien) Bonaparte, *Iconografia della fauna italica per le quattro classi degli animali vertebrati* (Rome, 1832–1841), lithograph
Biblioteca Universitaria, Pisa; author photograph

Lacking a great "painter-naturalist" such as Audubon, or the "happy pencil" of Titian Peale and the "accuracy and ability" of Alexander Lawson, the naturalist prince sought artists of solid background such as Carlo Ruspi (dates unknown), who was trained in rendering ancient artifacts; the young Alessandro Capalti (1810–1868), a pupil of Tommaso Minardi and an excellent fresco and portrait painter; Costantino Squanquerillo (dates unknown), an anatomical draftsman; and Pietro Quattrocchi (dates unknown). The plates, "all drawn and colored originally from the living specimens, under the eyes of the author," were printed in Rome by the lithographers Battistelli, Rosi, and Santerelli (figs. 18, 19). The choice of the relatively new medium of lithography reflects Charles-Lucien Bonaparte's awareness of the most famous contemporary examples favoring a fresher approach (such as *Birds of Europe* [1837] by John Gould), and his formal instructions to the artists were quite precise: "Each animal is painted from life, and the artist, in order not to depart from natural forms, should avoid unnatural foreshortening, shadows, and every other artistic nicety."[31]

Aside from criticism of its sometimes unconvincing illustrations,[32] the *Iconografia della fauna italica* acquired a wider programmatic and political significance. In fact, Bonaparte was committed to promoting scientific conferences in Italy, and in 1839, the date of the first such meeting in Pisa, he dedicated his work to Grand Duke Leopold II, "who welcomed many Scientists to happy Tuscany and sponsored their first reunion in Italy."[33] And on this historical occasion marking the beginning of modern scientific organization in Italy—not merely by chance in Tuscany—zoologists and botanists greatly admired the plates of the *Iconografia della fauna italica* as well as those of the *Pomona italiana.* To mark the occasion, Bonaparte's portrait was among those sketched at the congress in Pisa by Francesco Boggi (fig. 20).

In conclusion, although the works of Bonaparte and Gallesio reflect a traditional conception of the naturalist's work based on a vast taxonomic experience—not always linked to the latest developments

20. Francesco Boggi, *Charles-Lucien Bonaparte,* from *Album di 57 ritratti di scienziati intervenuti al Primo Congresso in Pisa nell'ottobre 1839 disegnati da F. Boggi* (Pisa, 1839), lithograph
Biblioteca Universitaria, Pisa; author photograph

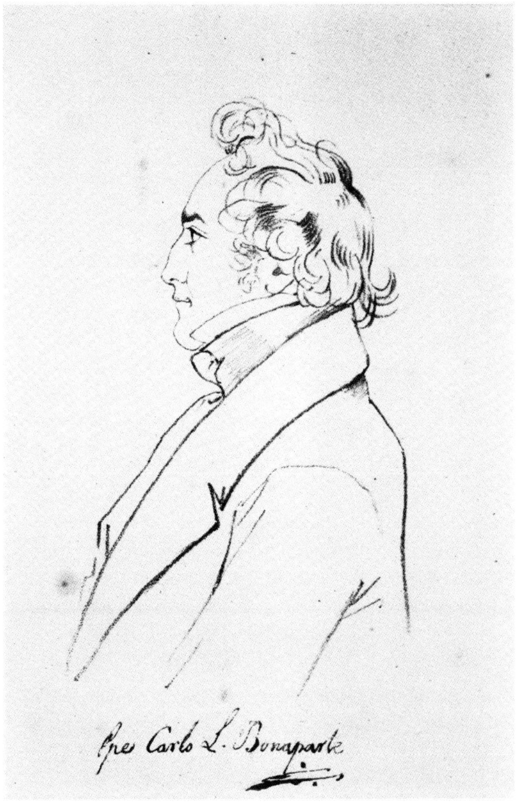

in biology and a more comprehensive and modern systematic vision—the importance they accorded to the naturalistic image is a key moment in descriptive investigation. These "romantic dreams" permitted Italian science to pursue its own identity and to lay the groundwork for its own organization; at the same time, they had an undeniable impact on artistic culture. This impact can be measured in terms of technical knowledge and didactic aspirations, of linguistic progress—here Bonaparte's American experience was truly crucial—and of theoretical reflections on the significance of drawing "from nature," in the light of a new consciousness of nature on which the modern system of images would be founded.

NOTES

1. See Eric W. Cochrane, *Tradition and Enlightenment in the Tuscan Academies, 1690–1800* (Rome, 1961) and *Florence in the Forgotten Centuries, 1527–1800: A History of Florence and the Florentines in the Age of the Grand Dukes* (Chicago and London, 1973); Alessandro Tosi, "Biblioteche della natura: collezioni naturalistiche nella Toscana del primo Settecento," *Annali della Scuola Normale Superiore di Pisa: classe di lettere e filosofia,* series 3, vol. 19, no. 3 (1989): 1027–1064; Lucia Tongiorgi Tomasi, "Naturalistic Illustration and Collections in Tuscany in the Eighteenth Century," in *Science and Visual Image in the Enlightenment,* ed. William R. Shea (Canton, Mass., 2000), 111–136.

2. Tongiorgi Tomasi 2000, 131.

3. *Storia naturale degli uccelli, trattata con metodo e adornata di figure intagliate in rame e miniate al naturale;* text by Saverio Manetti; hand-colored engravings by Violante Vanni and Abbot Lorenzo Lorenzi, after original paintings by Lorenzi (Florence, 1767–1776), 1: "le quali riguardo alle varie positure, ed alla vivacità, e simetrie delle medesime dimostrano non solo tutta la naturalezza, e bellezza propria degl'individui effigiati, ma tutta ancora la bellezza, e perizia dell'arte." Also quoted by Tongiorgi Tomasi 2000, 130.

4. I refer to Zocchi's *Scelta di XXIV vedute delle principali contrade, piazze, chiese e palazzi della Città di Firenze* and to *Vedute delle ville, e d'altri luoghi della Toscana,* the two series of views sponsored by Andrea Gerini and printed by Giuseppe Allegrini in 1744, the preparatory drawings for which are at the Pierpont Morgan Library, New York. See Elaine Evans Dee, *Views of Florence and Tuscany by Giuseppe Zocchi, 1711–1767: Seventy-seven Drawings from the Collection of the Pierpont Morgan Library* (New York, 1968); and especially Alessandro Tosi, *Inventare la realtà: Giuseppe Zocchi e la Toscana del Settecento* (Florence, 1997).

5. Giuseppe Baretti, *An Account of the Manners and Customs of Italy; with Observations on the Mistakes of Some Travellers, with Regard to That Country* (London, 1768), 275.

6. *Il gazzettiere americano contenente un distinto ragguaglio di tutte le parti del nuovo mondo* (Livorno, 1763). For more about Coltellini and the *Gazzettiere americano,* see Susanna Corrieri, "Marco Coltellini e la sua stamperia nella Toscana del Settecento," *Nuovi Studi Livornesi* 1 (1993): 165; see also Tongiorgi Tomasi 2000, 134.

7. The plates include illustrations of the processing of indigo and cotton, a sugar plantation, and a sugar mill, all engraved by Giuseppe Maria Terreni of Livorno. On the plates of the Tuscan editions of the *Encyclopédie,* see Alessandro Tosi, "Le planches nelle edizioni toscane dell'*Encyclopédie,*" *Studi settecenteschi* 16 (1996): 221–248.

8. "Prospetto di un'edizione fiorentina dell'Enciclopedia," in *Novelle letterarie* (1769): 69–74, 117–119, and 162–164.

9. See Sandra Raphael, *An Oak Spring Pomona: A Selection of the Rare Books on Fruit in the Oak Spring Garden Library* (Upperville, Va., 1990).

10. *Novelle letterarie* (1777): 210–212: "Nel vasto ed ameno studio delle create cose, non vi è sicuramente parte più bella e più seducente dei vegetabili, che formano l'oggetto di tale scienza, quegli appunto più si apprezzano dagli uomini, che con i loro prodotti arrecano ai medesimi qualche vantaggio, giacché questo solo ha dato motivo all'origine e ai progressi di questa bella parte d'Istoria Naturale. Gli alberi fruttiferi pertanto godono con tutta ragione la preferenza sopra ogn'altra famiglia di piante, non solo per l'utilità grande che essi ci arrecano, mediante la salubrità dei loro frutti, quanto ancora per il vago ornamento che essi apportano ai Giardini, ed alle campagne ancora. La nostra Toscana è stata in ogni tempo doviziosa di tal genere di alberi, sì per il dolce e temperato suo clima, quanto ancora per la massima industria degli Agricoltori, e per il favore dei Sovrani, i quali hanno sempre procurato di arricchire questa parte d'Italia di nuovi generi di frutti, non avendo risparmiato cosa alcuna per procurarseli da stranieri paesi." See Alessandro Tosi, "Tra flora e pomona: arte, natura e scienza in Toscana nella prima metà dell'800," in *Flora e pomona: L'orticoltura nei disegni e nelle incisioni dei secoli XVI–XIX,* ed. Lucia Tongiorgi Tomasi and Alessandro Tosi [exh. cat., Gabinetto Disegni e Stampe degli Uffizi] (Florence, 1990), 34–35. For discussion of other images of the fruits of Tuscany, see the essay by Hans W. Hubert in this volume on the paintings of Bartolomeo Bimbi for the Casino della Topaia.

11. Tosi 1996, 229.

12. "Avvertimento agli associati," in Giorgio Gallesio, *Pomona italiana* (Pisa, 1817–1839), vol. 4: "La mia posizione era diversa da quella dei Pomologi che mi hanno preceduto. Abitanti tutti di città metropoli ove era concentrata la Nazione che rappresentavano, essi potevano conoscere anticipatamente il numero delle Varietà che avevano a descrivere e figurare, perché si trovavano già per la maggior parte raccolte dai pepinieristi e dai giardinieri, o possedute da dilettanti vicini e conosciuti; essi avevano modo di procurarsene i campioni senz'uscire dal loro gabinetto; avevano la facilità di poterle studiare senza viaggi e senza dissesti; avevano il vantaggio di potersi aiutare dai lumi degli Agronomi che le coltivavano e dei Botanici più distinti del tempo; avevano finalmente a disposizione in tutti i tempi e in tutte le stagioni artisti speciali per eseguirne le figure, incisori appropriati onde portarle sul rame con metodi adattati; e coloritori iniziati in questo genere particolare di lavori. Io, invece, mancava di tutto questo. L'Italia non ha una città centrale ove si riconoscano tante risorse: Ricca di Artisti nei generi classici, essa non ne aveva in questo genere secondario, trascurato sino a quell'epoca e riguardato come esclusivo degli Oltramontani." See also Tosi 1990, 40–41.

13. *Manifesto della Pomona italiana*, 1816: "In tutti i paesi i Naturalisti si sono occupati di questo studio; né vi è più Nazione, che non abbia la sua Pomona. L'Italia è stata la prima a concepire l'idea di un lavoro di questa natura, e le collezioni dei Gran-Duchi Medici, di cui esistono le figure nel Museo di Firenze, e i manoscritti del Micheli, che si conservano presso uno dei suoi degni successori, il Sig. Targioni Tozzetti, ne sono la prova." (In all countries, naturalists are preoccupied with this study; there is no country that does not have its Pomona. Italy established the first idea of a work of this nature, and the collections of the Grand Dukes of the Medici, of which exist the figures in the Museum of Florence, and the manuscripts of Micheli, which are preserved with one of his successors, signor Targioni Tozzetti, are living proof.) The "figures" mentioned by Gallesio are the paintings by Bartolomeo Bimbi, for which see Lucia Tongiorgi Tomasi, "The Flowering of Florence: Botanical Art for the Medici," in Lucia Tongiorgi Tomasi and Gretchen A. Hirschauer, eds., *The Flowering of Florence: Botanical Art for the Medici* [exh. cat., National Gallery of Art] (Washington, 2002), 90–101. See also the essay on Bimbi's paintings by Hans W. Hubert in this volume.

14. On the *Pomona italiana,* see Enrico Baldini and Alessandro Tosi, *Scienza e arte nella Pomona italiana di Giorgio Gallesio* (Florence, 1994).

15. Giorgio Gallesio to N. Palmerini, 4 April 1822, in "Giornale di agricoltura e viaggi," unpublished manuscript, Biblioteca dell'Accademia delle Scienze di Torino, vol. 5; quoted in Tosi 1990, 35.

16. For more information about Serantoni, see P. Vannoni, *Biografia di Antonio Serantoni disegnatore, incisore e lavoratore di cere anatomiche* (Florence, 1838); Tosi 1990, 35–36; and Alessandro Tosi, "Pittori e incisori per la *Pomona italiana,*" in Baldini and Tosi 1994, 77. Aside from many anatomical waxes based on the plates for Mascagni's treatises and admired in Florence and abroad, Serantoni also planned a series of wax fruits based on the plates of the *Pomona italiana.*

17. See Tosi 1990, 36–37; and for Turpin, see Wilfred Blunt and William T. Stearn, *The Art of Botanical Illustration* (Woodbridge, Suffolk, U.K., 1994), 205–209.

18. "Supplica di Ferdinando Alliata," 28 September 1816, Biblioteca dell'Accademia di Belle Arti di Firenze, 1816–1844.

19. See Tosi 1990, 47; Lucia Tongiorgi Tomasi, *An Oak Spring Flora: Flower Illustration from the Fifteenth Century to the Present Time* (Upperville, Va., 1997), 367–369. Piccioli was the author of *L'Antotrofia ossia la coltivazione de' fiori* (Florence, 1834).

20. See Tosi 1990, 44–50; see also Alessandro Tosi, "Fruit and Flower Gardens from the Neoclassical and Romantic Periods in Tuscany," in *The Italian Garden: Art, Design and Culture,* ed. John Dixon Hunt (Cambridge, 1996), 202–221.

21. Tosi 1990, 42–43; Tongiorgi Tomasi 2002, 49–51.

22. Charles-Lucien Bonaparte, *American Ornithology, or the Natural History of Birds Inhabiting the United States, Not Given by Wilson, with Figures Drawn, Engraved and Coloured from Nature* (Philadelphia, 1825–1833), 1:36. See Joseph Kastner, *A Species of Eternity* (New York, 1977), 221–222; Charlotte M. Porter, *The Eagle's Nest: Natural History and American Ideas, 1812–1842* (Tuscaloosa, Ala., 1986), 62–64; Ann Shelby Blum, *Picturing Nature: American Nineteenth-Century Zoological Illustration* (Princeton, 1993), 89–92; and Patricia Tyson Stroud, *The Emperor of Nature: Charles-Lucien Bonaparte and His World* (Philadelphia, 2000), 55–56. For Charles-Lucien Bonaparte, see also G. Petit, s.n., in *Dictionary of Scientific Biography* (New York, 1981), 1:281–283; M. Cappelletti Alippi, s.n., in *Dizionario biografico degli italiani* (Rome, 1969), 11:549–556; Alessandro Tosi, "Nuove frontiere dell'arte e della scienza: l'illustrazione naturalistica nel XIX secolo," in *Natura-cultura. L'interpretazione del mondo fisico nei testi e nelle immagini,* ed. Giuseppe Olmi, Lucia Tongiorgi Tomasi, and Attilio Zanca (Florence, 1999), 345–362; and Carlo Luciano [Charles-Lucien] Bonaparte, *Iconografia della fauna italica per le quattro classi degli animali vertebrati,* ed. Alessandro Minelli and Augusto Vigna Taglianti (Rome, 1832–1841; repr. Treviso, 2003; subsequently cited as Bonaparte 1832–1841 or Bonaparte 2003).

23. Bonaparte 1825–1833, 1:iv–v. For the drawings of the purple grackle see also Linda Dugan Partridge, "Audubon and the Tradition of Ornithological Illustration," in *Art and Science in America. Issues of Representation,* ed. Amy R. W. Meyers (San Marino, Calif., 1994), 102–103; John James Audubon, *Writings and Drawings* (New York, 1999), 211–217.

24. Bonaparte 1825–1833, 1:iv–v; also quoted in Porter 1986, 64.

25. Blum 1993, 92.

26. Bonaparte 1825–1833, 2:v. "Mr. J. J. Audubon, painter-naturalist, who has devoted twenty years of his life to studying nature in the forests of the West, has gratified us with the sight of several drawings of new species which will appear among the plates he is now engaged in publishing. It is greatly to be wished, for the advancement of American ornithology, that while his work, so magnificent, but necessarily so slow in coming forth, is preparing, a scientific abstract of his discoveries should be drawn up without delay." For the relationship between natural history illustration and landscape see Amy R. W. Meyers, *Sketches from the Wilderness: Changing Conceptions of Nature in American Natural History Illustration* (New Haven, 1985), especially the chapter on Titian Ramsey Peale, 194–258; and Meyers 1994. See also the essay by Kenneth Hartman on the Long expedition in *The Culture of Nature: Art and Science in Philadelphia, 1740–1840* (New Haven, 2008).

27. Paolo Savi, *Ornitologia toscana,* 3 vols. (Pisa, 1827–1831), 1:v: "il difetto più grande di quelle figure, è nella loro positura strana e forzata. Vedesi chiaramente che il disegnatore non aveva mai osservato le attitudini naturali degli uccelli da lui rappresentati, e che solo lasciavasi guidare da una fantasia pittoresca." For Bonaparte and Savi, see Stroud 2000, 81, 86, 168.

28. Savi 1827–1831, 1:xiii: "E siccome trovai nel sig. Francesco Torri studente in Legge, la necessaria capacità per simili lavori, ad esso feci imitare le incisioni che in questo libro riporto." For geological research by Paolo Savi, see Pietro Corsi, "The Pisa School of Geology of the Nineteenth Century: An Exercise in Interpretation," *Palaeontographia italica* 82 (1995).

29. *Antologia* 47 (1824): 94–109: "Di non essere neghittoso e restio ad abbandonare i propri focolari, ma bensì di salire i colli, d'internarsi nelle valli, di varcare i monti, di arrestarsi ai piani; e sempre delineando, e fors'anche dipingendo per via, farsi corredo di tanti sorprendenti e variati quadri della natura. In simil guisa operando non andrebbe mai incontro al rimprovero di avere dipinta tale e tal altra veduta, senza uscire dalla propria stanza." See Alessandro Tosi, "Sguardi sull'Ottocento. Dal fondo iconografico del Gabinetto Vieusseux," in *Antologia Vieusseux* 1 (1995): 76–77.

30. Bonaparte 1832–1841. See Iacopella Manfredini, "Il principe naturalista: Carlo Luciano Bonaparte e la sua Iconografia della fauna italica," *Rara volumina* 2 (1996): 35–54; Tosi 1999, 357–359; Stroud 2000, 124; and Alessandro Tosi, "Il Principe e gli artisti: le tavole dell' *Iconografia della fauna italica,*" in Bonaparte 2003, 69–79.

31. Bonaparte 1832–1841, 1:i: "Ogni animale è ritratto dal vivo; e l'artefice per non dipartirsi dalle naturali forme ha dovuto evitare gli scorci, le ombre, ed ogni altro artistico accorgimento."

32. For criticism by George Ord, see Stroud 2000, 193.

33. For Bonaparte's role in Italian scientific conferences, see Giuliano Pancaldi, "Un linguaggio per la zoologia," in Giuliano Pancaldi, ed., *I congressi degli scienziati italiani nell'età del positivismo* (Bologna, 1983), 135–152.

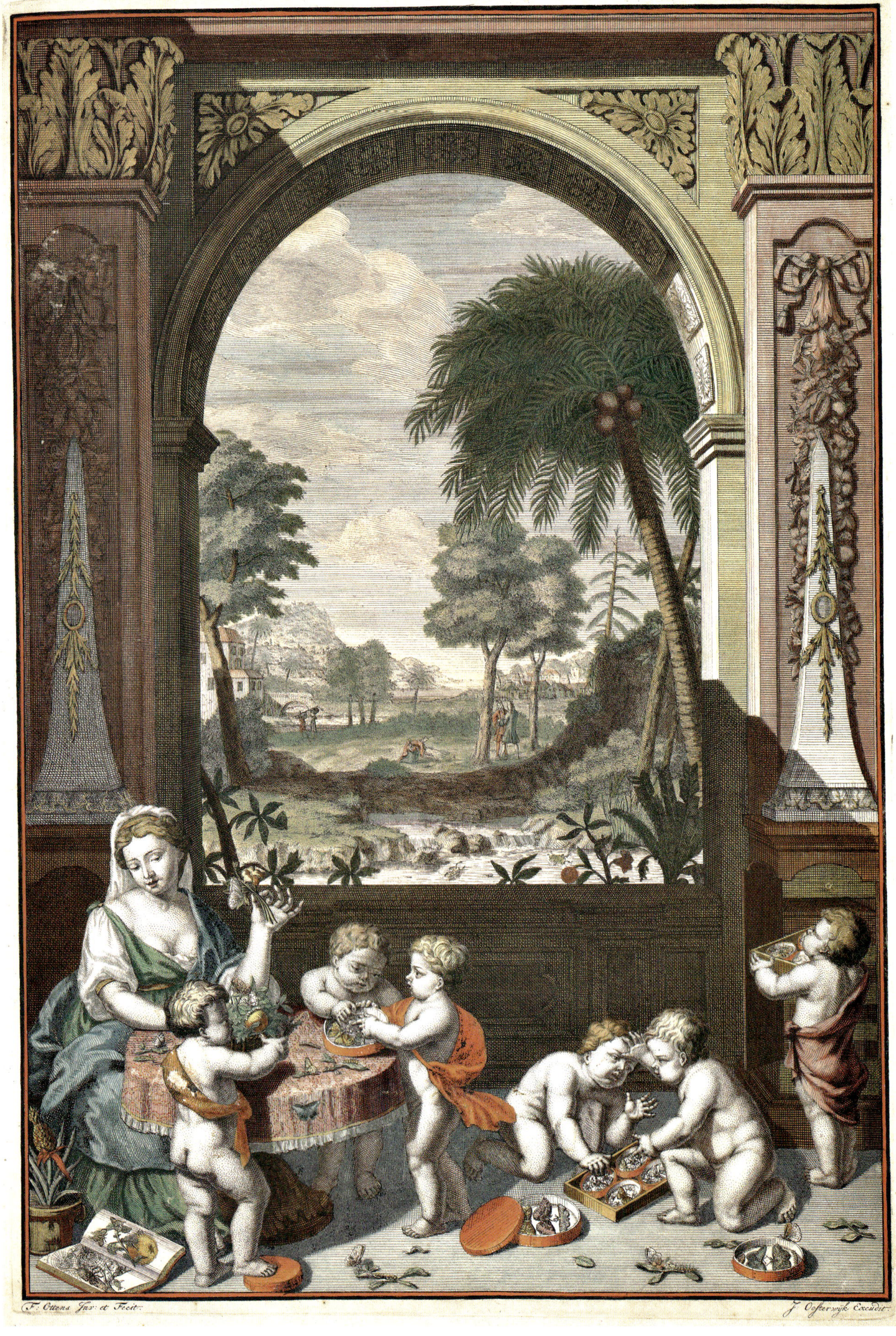
F: Ottens Inv: et Fecit:
J: Oosterwyk Excudit:

LUCIA TONGIORGI TOMASI
Università di Pisa

"La femminil pazienza": Women Painters and Natural History in the Seventeenth and Early Eighteenth Centuries

In a provocative essay that generated considerable discussion when it was published in 1971, Linda Nochlin affirmed:

> Thus, the so-called woman question, far from being a minor, peripheral and laughably provincial sub-issue grafted on to a serious, established discipline, can become a catalyst, an intellectual instrument, probing basic and "natural" assumptions, providing a paradigm for other kinds of internal questioning, and in turn providing links with paradigms established by radical approaches in other fields. Even a simple question like "Why have there been no great women artists?" can, if answered adequately, create a sort of chain reaction, expanding not merely to encompass the accepted assumptions of the single field, but outward to embrace history and the social sciences, or even psychology and literature, and thereby, from the outset, can challenge the assumption that the traditional divisions of intellectual inquiry are still adequate to deal with the meaningful questions of our time, rather than the merely convenient or self-generated ones.[1]

Following Nochlin's lead, more than a generation of feminist art historians have investigated the historical, sociological, and anthropological reasons why women were perennially relegated to "the margins," as Natalie Zemon Davis so aptly described it,[2] and hindered from developing their own artistic identities, succeeding in emerging from the private sphere only in rare cases during the period that extended from the sixteenth century to about the middle of the eighteenth century. Women were usually found working in a semi-clandestine manner, hidden from sight in family-based ateliers, although it may be noted that "family" and "convent"—the two social poles which defined the existence of women in that period—were in fact much more open than one might tend to think today.[3]

Few indeed were the exceptions, among whom may be singled out the outstanding Roman artist Artemisia Gentileschi (1593–1651/1653). Elizabeth Cropper, who for many years has studied women and their roles in the panorama of sixteenth- and seventeenth-century art, raised many interesting issues about this painter. In an essay she wrote for an exhibition shown in 2002, Cropper pointed out how Gentileschi's life and works "became a magnet for the exercise of every new form of art history."[4] Indeed, all research focusing on art produced by women has the potential to generate "magnets" for similar exercises, leading to new methodological approaches and reflections of a much broader nature. Our perspective on their work can only open further each time that, freed of the rigid schematicism of the past, we focus on actual personages, their historical and cultural backgrounds, and specific works of art, challenging traditional orientations and the

Maria Sibylla Merian, frontispiece from *Dissertatio de generatione et metamorphosibus insectorum Surinamensium* (Amsterdam, 1705)
Collection of Mrs. Paul Mellon, Oak Spring Garden Library, Upperville, Virginia

numerous topoi or clichés that persist in the study of art history. These might include the myth of the genius or that of the artist as a completely free and independent creator, not to mention the value judgments implicit in many of the conventional classifications, including the dichotomies "man-woman," "nature-culture," and "reason-intuition."[5]

"Feminine" Genres

One of the preconceived notions that has proved most difficult to uproot concerns the artistic genres traditionally associated with women artists over the course of the centuries. Although it is now agreed that they regularly worked in the studios of their fathers, husbands, or other male relatives—where they may even have participated in such prestigious projects as historical, mythological, and religious paintings—women received public recognition only for a limited number of "low-style" subjects conventionally reserved for them. These were genres in which neither the gift of *inventio* nor any great technical or stylistic virtuosity—both of which were considered the prerogative of male artists—was believed to be necessary. These themes, imposed by long tradition, were regarded as the only ones truly suitable for the *donnesca mano* (womanly hand) and *femminil pazienza* (feminine patience), characteristics that even today tend to be associated with the so-called fair sex.[6]

The prevailing conviction, therefore, was that embroidery, the *sermo humilis* (humble language) of flowers, insects and birds; the still life; and in certain cases domestic scenes and portraits were peculiarly adapted to women artists. Techniques requiring the precise application of colors or a delicate touch, such as the painting of miniatures and works in gouache or watercolor, were considered appropriate to the feminine hand, just as the painting of frescoes and large-scale canvases in oils, not to mention sculpting in stone—which demanded physical strength and, above all, were sustained by the power of *genio*—were regarded as the domain of men. It is certainly not fortuitous that most historical sources describe those *virtuose donne* (talented women) who dedicated themselves to the arts as *miniatrici* (miniaturists) or *intagliatrici* (engravers) rather than as genuine *pittrici* (painters), thus underlining the distinction between the higher and lower forms of artistic endeavor.[7]

Another art form in which women distinguished themselves at a very early stage, but on which much research remains to be done, was that of naturalistic illustration. The aim of this genre, whose emergence coincided with the birth of the modern sciences, was to provide an accurate portrayal of plants and animals, a faithful copy of the natural world. Between the end of the sixteenth and the middle of the eighteenth century, in the stimulating atmosphere created by the Scientific Revolution and the Enlightenment, Europe saw an outpouring of artistic works marked by particular attention to naturalistic phenomena. During this period, in which new criteria and new instruments for scientific study (such as the microscope) were introduced, naturalistic illustration achieved the status of an artistic genre. Although generally commissioned by men of science as an aid to their studies, such works soon won the appreciation of sovereigns and *curiosi,* who vied with one another to add examples to their collections. The genre quickly attracted many women artists, some of whom achieved both wealth and fame, although they have until now received little attention from scholars in comparison to women from more recent epochs, such as the Victorian age or the early twentieth century.[8]

Since antiquity, women have been considered the privileged repository of knowledge regarding herbs and their curative properties, and they were often credited with magical powers: the figure of Circe in Homer's *Odyssey* is a classic example. However, the activities of these women were usually relegated to the level of popular medicine and superstition, women herbalists being considered little more than witches and held in contempt by the officially recognized practitioners of medicine—physicians and apothecaries. Notable exceptions did exist, however, such as the German abbess Saint Hildegard von Bingen (1098–1179). In addition to her musical compositions, she is distinguished for her treatises on natural history and the medicinal uses of plants, animals, and minerals. Between 1150 and 1160, she compiled a list, in both Latin and German, of no fewer than 213 different medicinal herbs.

Artists or Scientists?

The role of women in the sciences has been studied in considerable detail, and it may be said that here they played an even more sporadic and marginal role than in the arts, at least until the Age of Enlightenment. This happened in part because the sciences were considered a highly elite and intellectually exacting pursuit, while the arts, after all, included a component of manual dexterity and craftsmanship which women were admitted to possess.[9] Furthermore, as scientific discoveries and information were disseminated through the medium of the printed word, the authors of scientific texts were invariably men, for the pen was considered to be an instrument even less suited to the female hand than the brush.

As the sciences became institutionalized during the course of the seventeenth century, they barred women far more unconditionally than the arts. Women were not admitted to the scientific academies of Europe (the first of which, the Accademia dei Lincei, was established in Rome in 1603) until well into the eighteenth century—even though, ironically, these academies were often symbolized by female figures.[10] In contrast, during the seventeenth century, some art academies were already accepting women as members, although they were never granted official duties or allowed to participate in meetings.[11]

Denied an active role in the theoretical and experimental sciences, many women with an inclination for such subjects turned instead to scientific illustration, an art form which provided an outlet for their innate curiosity and sensitivity to the myriad aspects of the natural world. With their attributes of patience and minute accuracy, women were acknowledged to excel in this art form. Furthermore, the favorable climate that developed beginning in the late sixteenth century allowed some women to separate themselves from the paternal atelier system and to strike out on their own, placing their talents at the disposition of scientists or, in a few cases, affirming themselves in autonomous careers.

In addition, advances in printing technology during the sixteenth and seventeenth centuries led to the publication of ever more lavishly illustrated texts, whose popularity increased the demand for specialized artists and engravers, among whom could be counted many women. For example, it is known that at the end of the sixteenth century the renowned bookbinder and printer Christophe Plantin of Antwerp (1520–1589), who specialized in the publication of scientific texts, retained three widows on his staff to hand-color the plates of important botanical texts. The coloring was done not only to make the books more attractive to potential buyers but also in the interest of scientific accuracy, as such work was always carried out under the direct supervision of scientists.[12] A limited number of copies intended for presentation to illustrious personages were prepared with particular care and attention to detail, so that their plates approached the quality of miniature paintings. Plantin's employment of women confirms that they had been found to be particularly skilled at this painstaking task.

Before long, however, a handful of women painters began to dedicate themselves to the representation of plants and animals as fully autonomous artists, achieving impressive results not only on an aesthetic level but also on a scientific one. Botanical, floral, and zoological (in particular ornithological) subjects traditionally associated with embroidery, the feminine art par excellence, became the preferred subjects for these artists. Once women had advanced beyond the stage of copying the masters, they did not limit themselves to the mere documentation of naturalistic specimens but sought instead to capture and portray their subjects' very essence, in the process developing a highly personal style.

This period also saw the affirmation of the still life, which came to share some of the same themes as naturalistic illustration. As reflected by the various terms devised for this new genre in the many languages of Europe—still life, *nature morte, natura morta, Stilleben*—its aim was to depict "immobile" nature, as opposed to the "living" model of the human figure. In the hierarchy of painting, the still life was relegated to a subordinate position. Initially it was considered a "low" genre, and indeed the Roman polymath and art collector Vincenzo Giustiniani (1564–1637), around 1620 placed no higher than fifth in his classification of the *gradi* (grades) of painting "il saper ritrarre fiori ed altre cose minute" (the ability to portray flowers and other small things).[13] This low rank was assigned to flower painting in spite of the judgment of the revolutionary painter Caravaggio (1571–1610), as reported by Giustiniani himself, that "tanta manifattura gli era a fare un quadro buono di fiori come di figure" (there was as much effort involved in executing a fine painting of flowers as one of human figures).

The still life was obviously as fitting to the female hand as was scientific illustration, and many women—especially in northern Europe—specialized in this genre,[14] including the French artist Louise Moillon (1610–1696), the Italian painter Fede Galizia (1578–1630), the Dutch painter Clara Peeters (1594–c. 1657), and—perhaps the most famous of them all—Rachel Ruysch (1664–1750), the daughter of Frederik Ruysch (1638–1731), illustrious professor of anatomy at the university in Amsterdam and owner of an exceptional collection of scientific specimens.

Inevitable and immediate were the ties and exchanges that sprang up between the still life and naturalistic illustration, a fascinating theme of which many aspects remain to be investigated. While artists focused on and polished the "naturalistic citations" contained in their works, scientific illustrations tended "irresistibilmente a farsi quadri" (irresistibly to turn into paintings). Transcending their specialized status as a form of scientific documentation, still-life paintings soon acquired a decorative dimension and the configuration of genuine works of art.[15] Indeed, it is sometimes impossible to decide whether to assign these works to the category of still life or that of scientific illustration. Many of the florilegia, manuscripts of miniature paintings, and printed books illustrated with engravings of flowers (usually exotic species or horticultural varieties) which were being produced all over Europe during this period constituted a veritable marriage between botanical illustration and the floral still life. These works in turn came to be used as sources for designs in the applied arts.[16]

If one examines the careers of those women who dedicated themselves to naturalistic art throughout the seventeenth and in the first decades of the eighteenth centuries, linking their body of work, where possible, to the salient events in their personal lives, it may be possible to understand better the contributions they made to this new genre in which art and science fused, with intriguing consequences. The close links that developed in this period between these two traditionally separate cultures have formed the subject of study by many scholars.[17] Ultimately, the recomposition of these two branches of human endeavor, illuminated by an examination of one of their most pleasing manifestations—scientific illustration—could provide points for further reflection and unexpected insights into various issues relating to the role of women in art.

The Seventeenth Century: Three Women Artists in Rome

During the reign of Pope Urban VIII (Maffeo Barberini; 1623–1644), Rome was one of the most open and vivacious cultural capitals of Europe. A number of women artists managed to establish careers for themselves in cultural circles among the gifted and the powerful, sharing a passion for the sciences and the arts. In this period, the nobleman Federico Cesi (1585–1630) founded the illustrious Accademia dei Lincei, of which the astronomer Galileo Galilei was a member. Various members of the papal family, in particular Cardinal Francesco, the nephew of Pope Urban VIII, cultivated an interest in the sciences, especially botany, and constructed magnificent gardens in which exotic plants and rare cultivars were grown.[18]

In this fertile climate, where the activities of artists and scientists were so closely intertwined, many painters chose to specialize in the genre of botanical painting, among them the Frenchman Nicolas Robert (1614–1685), whose work would later win great favor at the court of Louis XIV and the approval of Nicolas Guillaume (c. 1600–1663) of Lorraine, also known as Nicolas de La Fleur and "Monsù Fiore" for his great skill as a botanical artist. Both Robert and de La Fleur came to Rome early in their careers, and it was there that they painted the leaves of their first florilegia. Indeed, for many artists a sojourn in the Italian capital allowed them to gain solid training and establish a reputation that would later help them to obtain prestigious commissions from wealthy patrons in other Italian and European states.[19]

Rome in this period also saw the publication of books on the subjects of botany and horticulture, which met with immediate success and became models of their kind because of the importance of their scientific content and the remarkable quality of their illustrations. In 1616 the printer Jacopo Mascardi published *Minus cognitarum rariorumque nostro coelo orientium stirpium ekphrasis,* an encyclopedic work on the flora of Europe by the Linceian academician Fabio Colonna of Naples (1566–1640), which included a particularly fine set of etched plates. In 1625 the same printer published *Exactissima descriptio rariorum quarundam plantarum . . . in Horto Farnesiano,* in which the alchemist and herbalist Tobias Aldini (dates unknown), formerly curator of the "gabinetto di cose rare del cardinal Barberini" (collection of rare objects [owned by] Cardinal Barberini), in collaboration with the celebrated physician and botanist Pietro Castelli (c. 1575–1657), described the rare plants growing in the garden of Cardinal Odoardo Farnese on the Palatine Hill. This text was accompanied by extraordinary plates engraved by Luca Ciamaberlano (c. 1580–c. 1645).[20]

Botanical illustration in this period therefore became not only a well-established and respected genre, but also a vastly popular one which attracted such women of talent as Isabella Cattani Parasole (or Parasoli; act. 1597–1616), Anna Maria Vaiani (act. 1627; d. c. 1655), and Giovanna Garzoni (1600–1670). Parasole was Roman, while Vaiani was from Florence and Garzoni was born in Ascoli. All three came from families of artists and artisans but were irresistibly drawn to Rome, where they eventually gravitated to the scientific circles of the Accademia dei Lincei and the entourage of the Barberini family.

Isabella Cattani Parasole

Isabella Cattani Parasole, artist and engraver, won a certain renown for her volume of designs for embroidery and lacework, *Teatro delle nobili e virtuose donne,* published in 1616 with plates drawn and engraved by the artist herself. Earlier in her career, her talent had attracted the attention of Castore Durante (1529–1590), who commissioned her to prepare the illustrations for his treatise, *Herbario nuovo.* Isabella's husband, Leonardo Parasole (c. 1570–c. 1630, also known as Leonardo Norsino) engraved the woodblocks to print her designs. Durante, a native of Umbria, was chief physician to the pope and professor of materia medica at the university in Rome. After the *Herbario nuovo* was first published in 1585, it was so well received that it immediately went through several reprintings.[21]

Although the title of the work alludes to "figure che rappresentavano piante vive" (pictures representing live plants), Cattani Parasole's illustrations were in large part drawn—perhaps on the advice of Durante himself—from illustrations published in the most authoritative botanical treatises of the period, such as the many editions of *I discorsi* by Pietro Andrea Mattioli and *De historia stirpium commentarii insignes* (Basel, 1542) by Leonhard Fuchs. Despite Isabella's use of prototypes, however, her drawings, and the engravings into which they were translated, exhibit a noteworthy originality that is clearly discernible, even in the reduced dimensions of the illustrations accompanying Durante's text. While the artist took care to provide an exact, if somewhat abridged, portrayal of each plant, including its leaves, roots, flowers, and fruit, she also sought to add her own personal touches, departing from the traditional canons of botanical illustration and showing a considerable freedom of expression and style. In some of the drawings the plant is shown in a landscape illustrating its natural habitat, with perhaps the addition of a human figure or two. In others, the plant is depicted in a garden setting, for example,

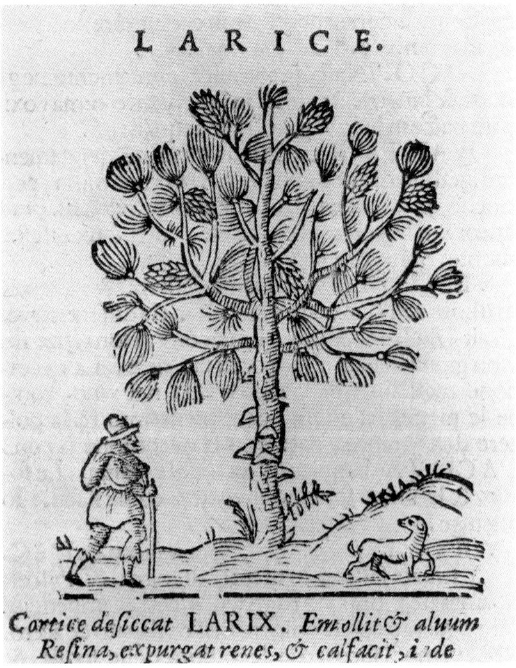

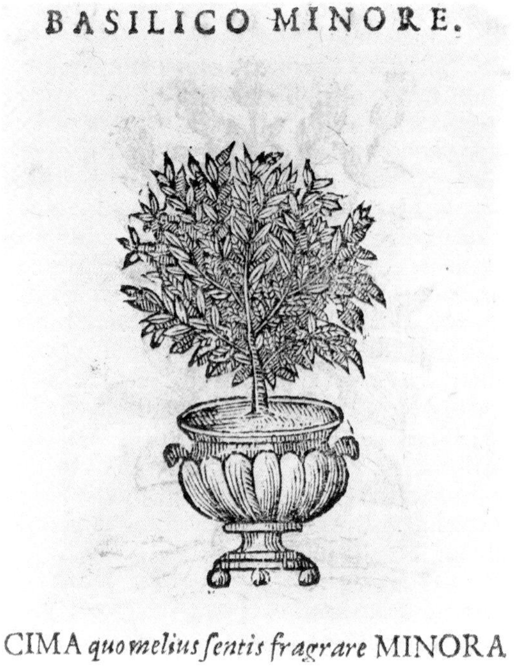

1. Isabella Cattani Parasole, *Larice* (larch), from Castore Durante, *Herbario nuovo* (Rome, 1636), wood engraving
National Library of Medicine, Bethesda, Maryland

2. Isabella Cattani Parasole, *Basilico minore* (basil), from Castore Durante, *Herbario nuovo* (Rome, 1636), wood engraving
National Library of Medicine, Bethesda, Maryland

growing in an ornamental pot such as those "tienesi . . . su le logge" (kept on loggias), in a charming allusion to the everyday details of gardening (figs. 1 and 2).[22]

After the successful completion of this project, Cattani Parasole was invited to work with one of the most illustrious and demanding patrons in Rome, Prince Federico Cesi. Giovanni Baglione, a painter and writer on the arts, describes Cattani Parasole's "intagli nel Libro dell'herbe del Principe Cesi d'Acquasparta, letteratissimo signore" (engravings in the book of plants of Prince Cesi of Acquasparta, most learned gentleman),[23] some of which would later be reproduced in *Tesoro messicano* (officially titled *Rerum medicarum Novae Hispaniae thesaurus)*, the monumental work on the natural history of the "Nuove Indie" (New Indies) begun by Cesi and finished in 1651, many years after his death, by members of his scientific circle. Some of the engravings in *Tesoro messicano* were clearly adapted from the *Herbario nuovo,* although they have been shorn of the added details that made the illustrations to the original work by the Umbrian physician so captivating.

Anna Maria Vaiani

Little is known of Anna Maria Vaiani, a native of Florence like the Barberini pope and daughter of the painter and engraver Alessandro Vaiani, from whom she learned her art.[24] A portrait of her, executed in Rome by the French artist and engraver Claude Mellan (1598–1688), has been preserved.[25]

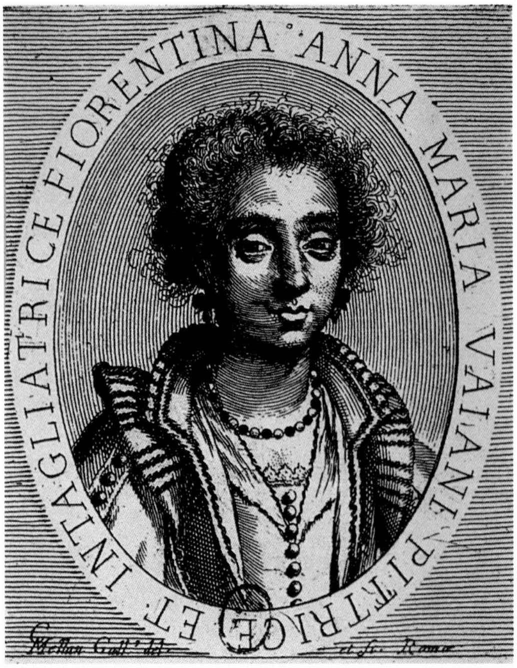

3. Claude Mellan, *Anna Maria Vaiani,* c. 1626, etching
Istituto Nazionale per la Grafica, Rome

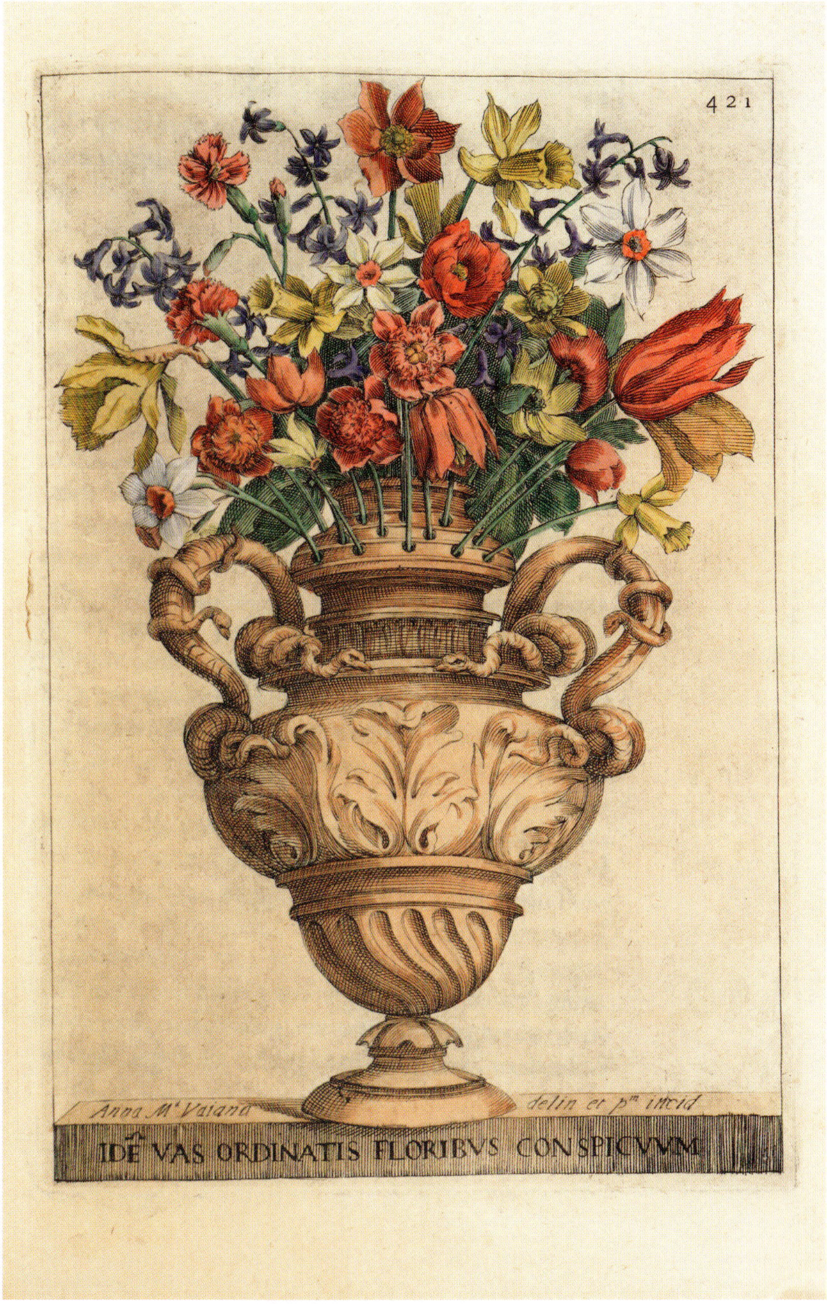

4. Anna Maria Vaiani, *Vase of Flowers,* from Giovan Battista Ferrari, *De florum cultura* (Rome, 1633), hand-colored etching
Collection of Mrs. Paul Mellon, Oak Spring Garden Library, Upperville, Virginia

The engraving and the original drawing show the artist in elegant attire and wearing jewels, to all appearances a lady of rank, with a cartouche that identifies her as "pittrice et intagliatrice fiorentina" (fig. 3). Although her name is mentioned by various sources, however, the woman behind the portrait—as Elizabeth Cropper has observed—remains a tantalizingly elusive figure.[26]

Vaiani is cited, for example, in admiring terms in several letters by Galileo Galilei, who, describing her as a "fanciulla di grandissimo talento" (young lady of the greatest talent), did his best to introduce her into influential cultural circles in Rome.[27] Galileo, who dedicated one of his letters on Copernicus to Christine of Lorraine, grand duchess of Tuscany, appreciated the friendship of gifted women and the talent of women artists, as is demonstrated by his close ties with Artemisia Gentileschi.[28] Vaiani soon showed herself to be not only a skilled engraver, translating drawings of antiquities into elegant plates, but also a fine floral painter whose works became quite sought after.

Although the floral paintings that she executed for Cardinal Francesco have been lost, an impressive work by Vaiani's hand may be found in the treatise on gardening and horticulture published in Rome in 1633 by the Jesuit Giovan Battista Ferrari (1584–1655). *De florum cultura* (1633) was one of the most prestigious books to be produced in seventeenth-century Rome, and many well-known artists collaborated on its illustration, including Pietro da Cortona, Andrea Sacchi, Giovanni Lanfranco, Guido Reni, Johann Friederich Greuter, and Claude Mellan. Vaiani herself designed and engraved a plate showing an elaborate bouquet of some of the rarest flowers—double anemones, a tulip, and other prized bulbous species—then being cultivated in the most famous gardens in Rome (fig. 4). Although this is the only signed engraving by the artist in *De florum cultura,* it is not unreasonable to hypothesize a closer and more prolific collaboration with Monsù Fiore, who prepared the botanical drawings that were reproduced in the treatise.[29]

Other works allow us to appreciate the breadth of Vaiani's oeuvre. In the short period from 1632 to 1633, for example, she not only participated in the illustration of an exceptional work dedicated to the antiquities preserved in the Roman Galleria Giustiniana;[30] she also painted the coats of arms in the *cappella segreta papale* in the Vatican and was an active member of the atelier of the eclectic connoisseur and collector Cassiano dal Pozzo (1588–1657).[31] Anna Maria Vaiani probably also had ties with the Accademia di San Luca, the artists' academy of Rome, as her name is cited in some of the institution's later documents.

A few sources mention her possible (but brief) marriage to the well-known painter of battle scenes Jacques Courtois (1621–1676), and later a trip to England (on which she was duly accompanied by her mother). She was ostensibly invited to paint floral pictures for Aletheia Talbot, Countess of Arundel (d. 1654), the wife of Thomas Howard, second Earl of Arundel, but the actual reason for Vaiani's presence was apparently to add luster to the reputation of George Conn, the pope's envoy to the Court of Saint James'. Serving officially as artistic adviser to Charles I, Conn was at the same time seeking (unofficially) to convert him to Catholicism.[32]

Giovanna Garzoni

While the information we have regarding Isabella Cattani Parasole and Anna Maria Vaiana is disappointingly meager, much more is known about the life and career of Giovanna Garzoni, the *miniatora* (miniaturist), as she was often referred to by contemporary sources. Indeed, her personality stands out distinctly against the panorama of the seventeenth-century art world.[33] Garzoni was born in Ascoli, in the Marches, in 1600 to a family of goldsmiths and artisans originally from Venice. Boldly striking out on her own at an early age, she eventually established a brilliant career for herself in many of the capitals of Europe. In 1648 she would be cited by the seventeenth-century art historian Carlo Ridolfi in *Le meraviglie dell'arte* as demonstrating "a quale segno arrivi la perspicaccia donnesca" (to what great heights feminine perspicacity may reach), a most laudatory judgment indeed, especially since his contemporary, the historian Emanuele Tesauro, described this very quality of *perspicaccia* as enabling the possessor "penetrare le più minute e lontane circostanze di ogni soggetto" (to penetrate the most minute and subtle causes underlying every subject).[34]

Garzoni's portraits and her copies of famous paintings, superbly executed in the miniaturist style, were greatly admired, but she was above all known for her luminous still-life paintings—beautiful floral compositions and luxuriant vessels of fruits and vegetables painted in an unmistakable style based on the application of minute dots and strokes of color using the finest brushes. She was showered with commissions from members of the court of Savoy in Turin and the Medici family in Florence, as well as numerous virtuosi in Naples, Venice, Florence, and Rome, who were ready to pay considerable sums to possess one of her works.

The artist arrived at her mastery of the principles of still life, however, only after a long period of training in botanical illustration during the early seventeenth century in Rome. Testifying to this is a series of botanical studies Garzoni executed at age sixteen and preserved in a remarkable miniated herbal at Dumbarton Oaks in Washington, D.C.[35] As we learn from an unedited manuscript note by the eighteenth-century Tuscan naturalist Giovanni Targioni Tozzetti (1712–1783), in 1616 the pharmacist Enrico Corvino (Hendrik de Raeff, dates unknown) commissioned the young artist (who had just completed a fine painting of the Holy Family for a church in her native city) to paint an herbal for him. Corvino was born in Delft and moved to Rome, where he married the sister of the botanist Pietro Castelli. He owned the celebrated pharmacy "at the sign of the Imperial Eagle" on Mount Giordano, which became a meeting place for many of the naturalists who lived in Rome, including members of the Accademia dei Lincei. Like other pharmacists of the period, Corvino assembled a fine collection of naturalia and cultivated a large garden of medicinal plants. Corvino was also considered a capable botanical artist in his own right, and two of his children became miniature painters, Francesco specializing in insects and butterflies and Maddalena in religious subjects.

The *Erbario miniato* (which Targioni Tozzetti tells us was later acquired by Monsignor Leone Strozzi, as confirmed by the owner's stamp on the frontispiece), had until recently been thought to be a fairly late work because of the inclusion of a presumed self-portrait showing the artist as a fully mature woman (fig. 5). This sheet was actually inserted at a later date, however, and is not by Giovanna Garzoni herself, but rather by an anonymous

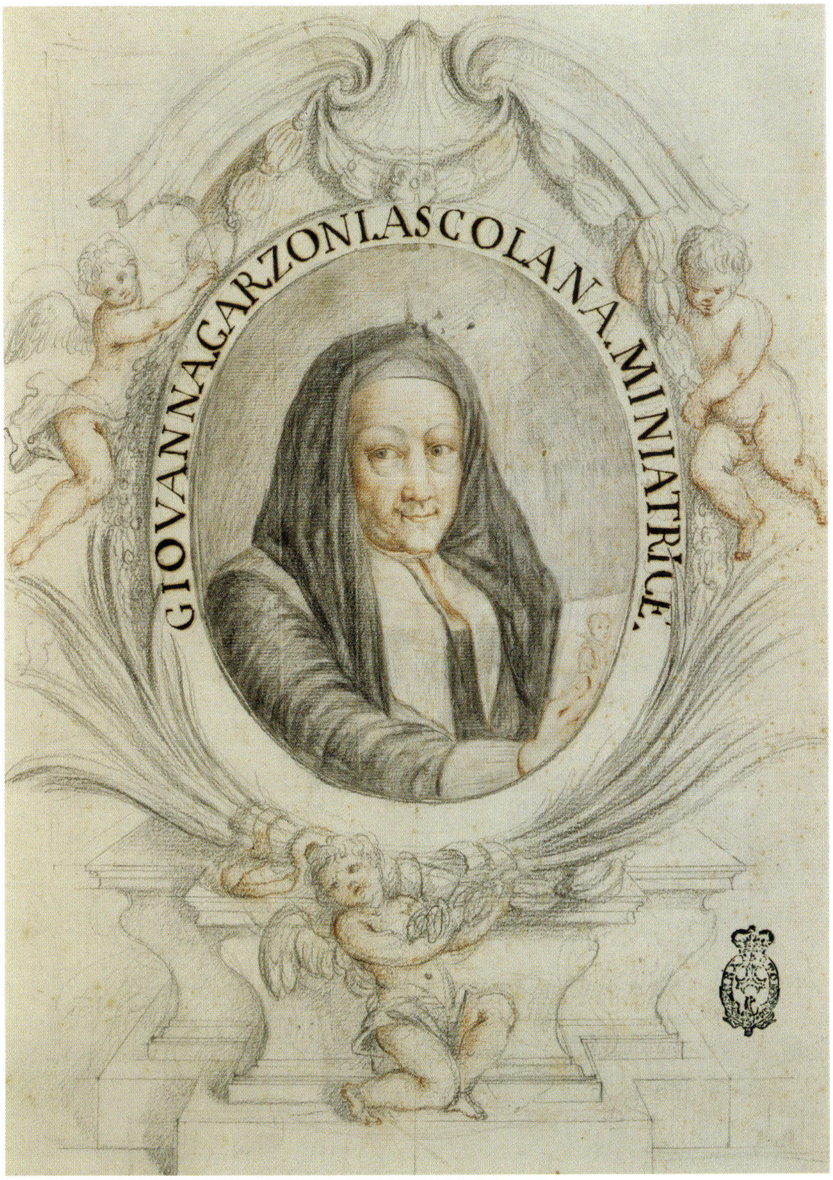

5. Anonymous, *Giovanna Garzoni*, from Giovanna Garzoni, *Erbario miniato*, undated, pencil, ink, and watercolor
Dumbarton Oaks Research Library and Collection, Washington, D.C.

artist who based his work upon portraits of her executed either by Carlo Maratta or by the artist's compatriot, Giuseppe Ghezzi. One of these portraits is preserved at the Accademia di San Luca and the other adorns Giovanna Garzoni's tomb in the church of Santi Luca e Martina in Rome.

The Dumbarton Oaks herbal actually consists of works painted by Garzoni in different periods of her life. At least some of the pages are clearly by the hand of an immature artist, for they were not executed from life but instead are based on illustrations from one of the most celebrated botanical works of the sixteenth century, Pietro Andrea Mattioli's translation, with commentary, of Dioscorides' *De materia medica.* Perhaps Corvino gave her a copy to use as a model, just as Castore Durante seems to have done with Isabella Cattani Parasole. The manuscript also contains many later paintings in which the artist provides ample demonstration of her ability to draw from life and, in the case of the illustration of *Lathrea clandestina* (purple toothwort), of her close ties with Federico Cesi and his academy, the prince having made a special study of this parasitic plant.[36]

It may be presumed that the artist's connection with the botanist and antiquarian Cassiano dal Pozzo also dates from this early period in Rome. Cassiano, who arrived in the papal city in 1612, became an enthusiastic advocate of Garzoni's work, and his support helped her to obtain many prestigious commissions, as revealed by the letters of gratitude she sent him.[37]

It is known that Giovanna Garzoni lived for some time in Venice, where in 1622 she entered into a brief marriage to the portrait artist Tiberio Tinelli. Perhaps because she had taken a vow of chastity when she was a young woman, her experience of matrimony was less than happy, but it may, in fact, have been her husband who convinced her to turn her exceptional talent as a miniaturist to the art of portraiture.[38] Garzoni, perhaps realizing that she was wedded exclusively to her art, spent the next three decades of her life working for a series of enlightened patrons, moving from one Italian court to another, always accompanied by her devoted brother Mattio.

The artist's name was by this time familiar to connoisseurs all over Italy and abroad. After a sojourn in Naples from 1626 to 1631, where she painted portraits and religious scenes for the Duke of Acalà, viceroy of Naples, Garzoni transferred to the Savoy court in Turin. She remained there until 1637, painting fruit, flowers, and animals on vellum for Christine of Lorraine and executing an extraordinary series of portraits for other members of the family.

Evidence has recently come to light confirming her trip to Paris: in 1640, Garzoni received warm accolades and expressions of esteem from no less than the distinguished Italian artist and engraver Stefano della Bella

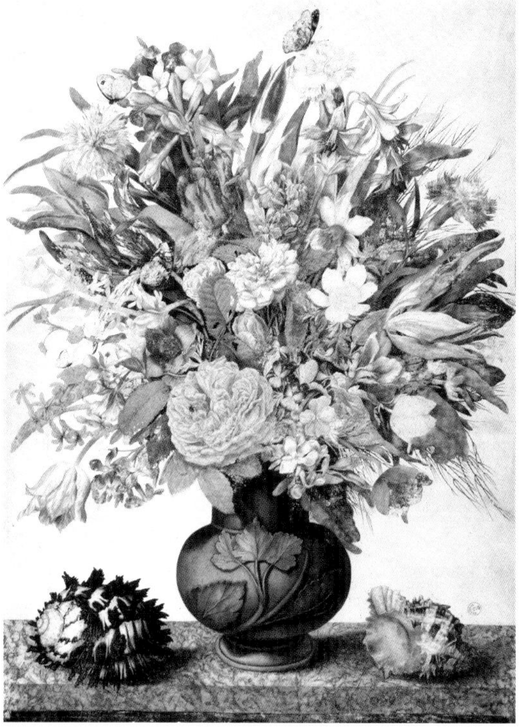

6. Giovanna Garzoni, *Vase of Flowers*, undated, gouache on vellum
Gabinetto Disegni e Stampe, Galleria degli Uffizi, Florence; photograph Scala / Art Resource, New York

(1610–1664), who was then living in the French capital. Still, his flattering reception did not lessen her desire to return to Italy as soon as possible.[39] Garzoni's stay in Paris, brief as it was, must have played a crucial role in her artistic formation. It was there that she had the opportunity to meet many well-established still-life painters and, inspired by their example, to develop the themes that would characterize her later work and win her lasting fame. She seems to have been particularly impressed by the opulent floral arrangements she saw in Paris, especially by the dishes overflowing with fruit and vegetables painted by Jacques Linard, François Garnier, and Louise Moillon (Garnier's daughter-in-law), and by the Flemish artist Jean Michel Picart.[40] Evidence of her close ties with Picart can be found in a comparison of his painting *Bouquet of Flowers in a Vase* (De Boer Collection, Amsterdam) with a work in gouache on vellum by Garzoni, today conserved in the Gabinetto Disegni e Stampe at the Uffizi.[41] Indeed, the very same vase, embellished with a design of leaves in low relief, appears in paintings by both artists (figs. 6 and 7). It may be hypothesized that Garzoni executed her work in Paris and took it with her to Florence in 1642, where it was later acquired by Cardinal Leopoldo of the Medici, as revealed in an inventory of his collections.

It may also be observed that Giovanna Garzoni produced many works during her extended stay in the Tuscan capital. These consist primarily of vases of flowers or vessels piled high with fruit or vegetables, harmoniously arranged on tables or on stone shelves, with such added details as insects or flowering vines, themes that she had apparently mastered in Paris (fig. 8). All the members of the grand duke's court were great admirers of her art, as evidenced by the many paintings on vellum by her hand to be found today in the museums of Florence.[42]

The Medici family esteemed Garzoni's talent as a miniaturist so highly that, long after she left the Tuscan city in 1651, she continued to receive commissions for still lifes, *carte di semplici* (drawings of plants), naturalistic subjects, copies of famous paintings, and decorative work (painted fans, designs for *pietre dure* mosaics, and the like). The artistic tradition so generously fostered by the Medici, which had featured the naturalistic image since the time of the artist Jacopo Ligozzi (1547–1627), as well as the more recent interest of many members

7. Jean Michel Picart, *Vase of Flowers*, undated, oil
Portland Art Museum, Oregon; museum purchase, funds provided by the Reserve Fund and the Marybeth Branaman and Georgia Stoetzel Branaman Bequest

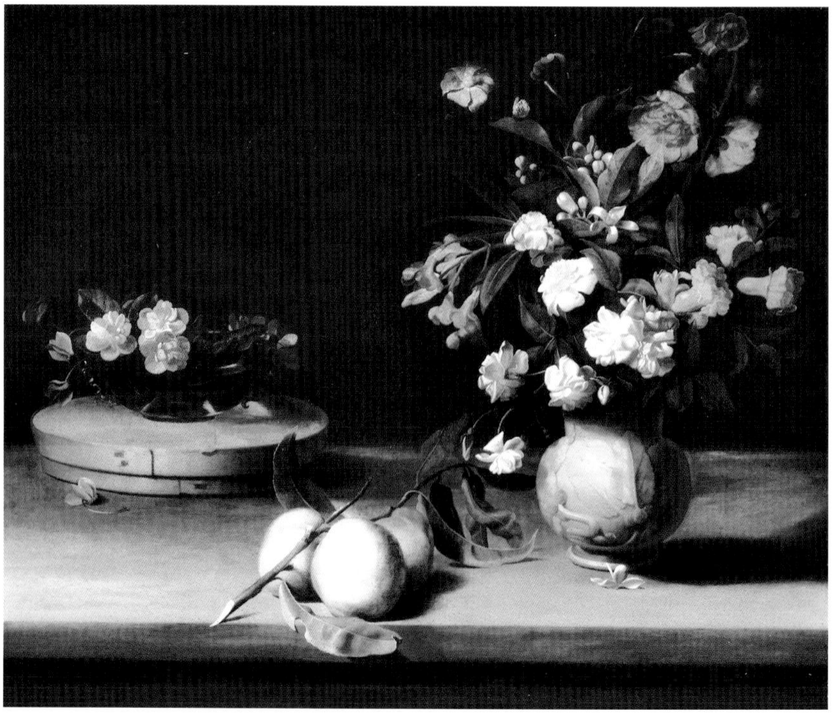

8. Giovanna Garzoni, *Chinese Plate with Cherries and Bean Pods,* tempera on vellum, c. 1620
Private collection

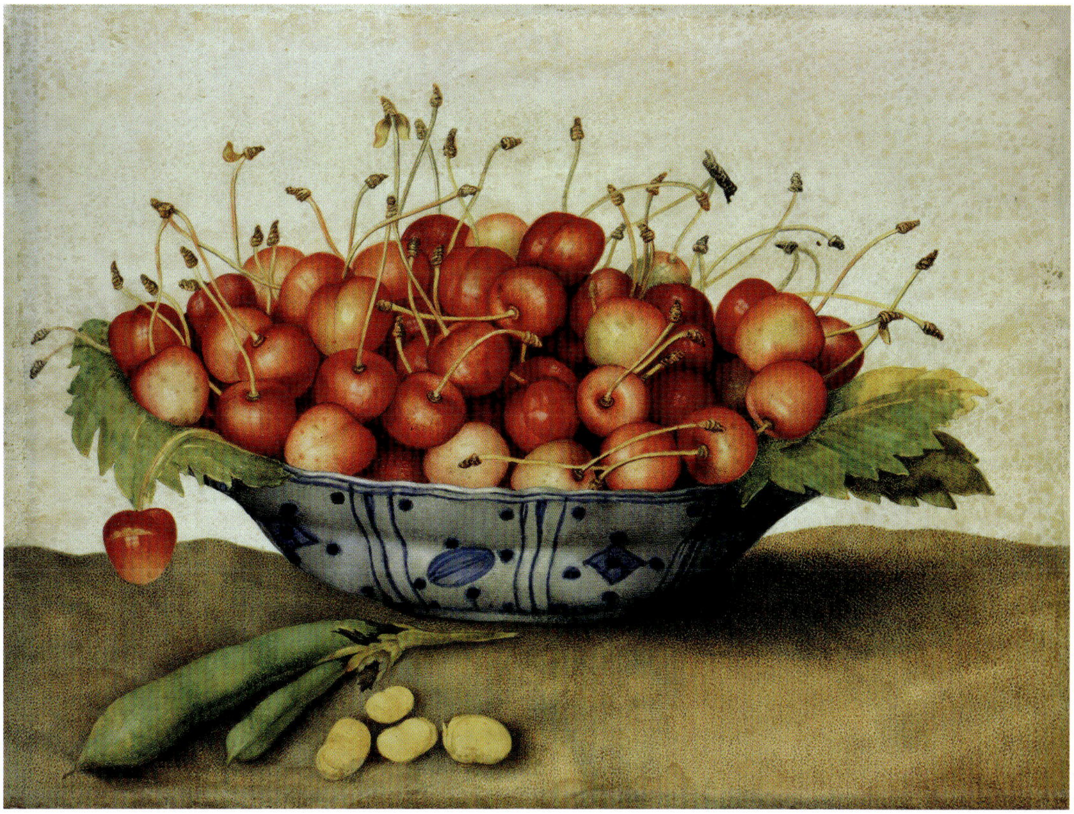

of the family in still life, fit admirably with Garzoni's chosen themes and refined style.

When Garzoni returned to her beloved Rome in 1651 she was a wealthy and celebrated artist. There she became closely associated with, although never an official member of, the Accademia di San Luca, to which she bequeathed many works, some of her papers, and her personal fortune upon her death in 1670. Today Garzoni's name is associated with a large body of work consisting of elegant floral paintings and glowing "portraits" of fruits and vegetables. To her known oeuvre, most of which is preserved in Florence, may be added some previously unpublished works presented in the exhibition *The Flowering of Florence* (2002)[43] and another still life (private collection) depicting a platter of asparagus, several carnations, and a lively green grasshopper (fig. 9). These are all further testimony to the pictorial and naturalistic sensibility, the fruit of an infinite *femminil pazienza,* of this exceptional artist.

Catherine Perrot, Painter and Teacher

Giovanna Garzoni may never have been allowed to become a full-fledged member of the Accademia di San Luca, but her slightly younger French colleague, Catherine Perrot, was officially received into the Académie royale de peinture et de sculpture of Paris in 1682. She was one of the few women ever to obtain this honor. Unfortunately, little information about Perrot has come down to us, and no paintings can be attributed with certainty to her hand. We do know that she studied under the French painter Charles Le Brun (1619–1690) and that she learned to paint miniatures of naturalistic subjects from the great botanical artist Nicolas Robert. Returning to Paris after his trip to Rome, Robert was appointed *premier peintre du roy en miniature,* a prestigious but demanding position. His appointment required him not only to document the king's botanical collections in the Jardins royals, producing the superb *vélins* (paintings on vellum) in which he so excelled, but also to spend considerable time teaching his art to others.[44]

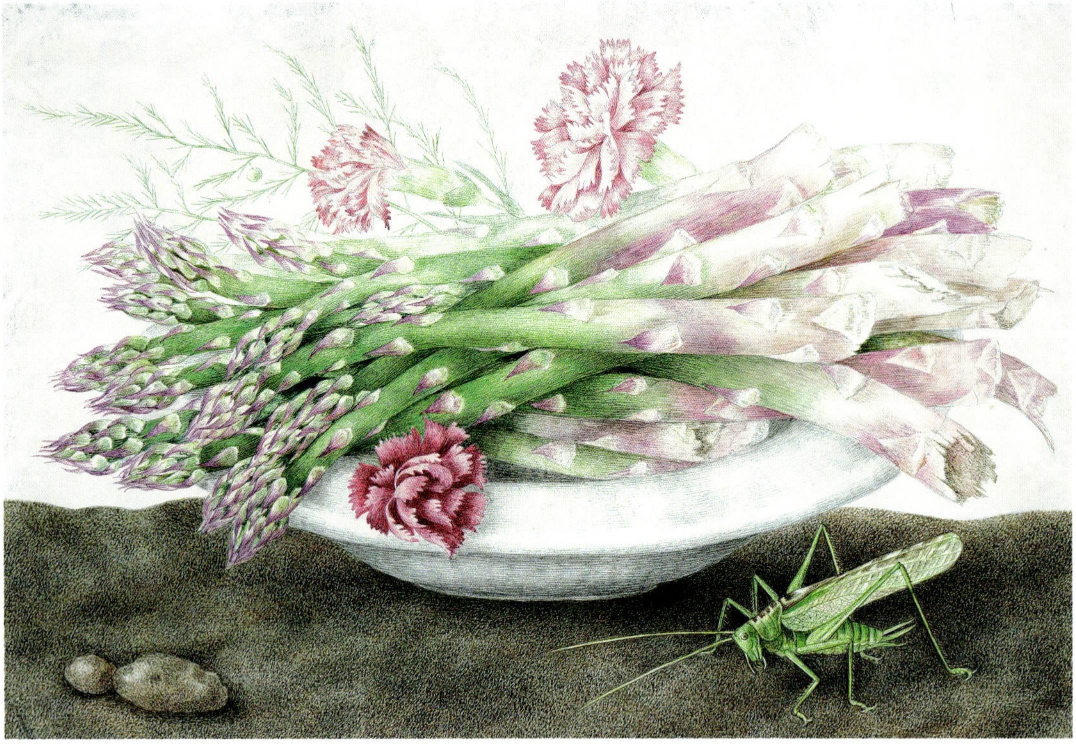

9. Giovanna Garzoni, *Plate of Asparagus with Carnations and a Grasshopper,* undated, gouache on vellum
Private collection, Italy, reproduced by permission of the owner; photograph Trinity Fine Art, London

Catherine Perrot married Claude Horry, notary to the archbishop of Paris, and was assigned to teach the art of miniature painting to the younger members of the royal family, including Gabrielle de Savoie, the future wife of Philip V, king of Spain. Perrot was thus one of the first of her gender to become a teacher of naturalistic painting, an occupation that was destined to become a specialty for women artists. This unusual appointment also presaged the emergence, in the eighteenth century, of the predominantly female figure of the amateur painter.

Perrot was the author of two modest works intended for students of the art of flower painting that were published in Paris in 1686 and 1693.[45] The first, *Les Leçons royales, ou la manière de peindre en mignature les fleurs et les oyseaux,* was dedicated to "Madame la Dauphine" and explained, step by step, how to paint flowers and birds on vellum, imparting some of the "secret" techniques that had become an essential component of the tradition of naturalistic painting (fig. 10). This work gives us an idea of Perrot's teaching methods, which consisted primarily in having her students copy the works of the great master Nicolas Robert. "To paint flowers . . . ," she wrote, "and to succeed at it, you cannot find better examples than the prints of Nicolas Robert . . . the most excellent flower-painter who has ever lived, until now."[46] She also furnished a list of the necessary pigments, together with their prices, and noted that these materials could be purchased at the shop "À la Cornemuse," run by the widow Foubert, in the parish of Saint Nicolas Deschamps.

Perrot borrowed thirty engravings from Robert's florilegium, *Multiformes florum species,* to serve as models and taught her students how to make careful copies of these as the first step to becoming true flower painters. They had to begin by tracing the outline of an engraving, either in silverpoint or by piercing a series of tiny holes with a needle onto a piece of high-quality vellum fixed to a wooden board. The colors, diluted with gum arabic, could then be applied within the outlines, taking care to employ only the most delicate tints where there were no shadows. For each engraving Perrot specified the exact pigments to be used and their gradations, following the example of the original by Robert.

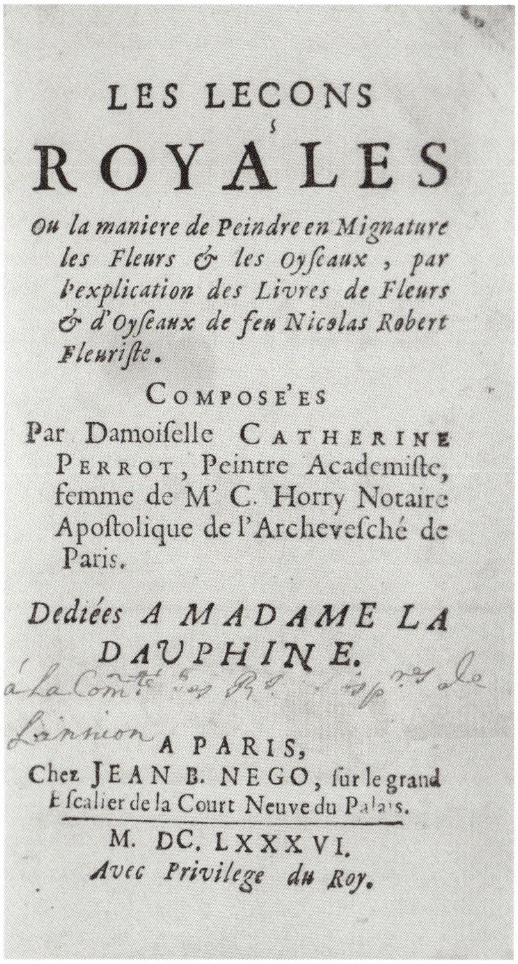

LES LECONS ROYALES

Ou la maniere de Peindre en Mignature les Fleurs & les Oyseaux, par l'explication des Livres de Fleurs & d'Oyseaux de feu Nicolas Robert Fleuriste.

COMPOSE'ES

Par Damoiselle CATHERINE PERROT, Peintre Academiste, femme de Mr C. Horry Notaire Apostolique de l'Archevesché de Paris.

Dediées A MADAME LA DAUPHINE.

A PARIS,

Chez JEAN B. NEGO, sur le grand Escalier de la Court Neuve du Palais.

M. DC. LXXXVI.

Avec Privilege du Roy.

10. Catherine Perrot, title page from *Les Leçons royales, ou la manière de peindre en mignature les fleurs et les oyseaux* (Paris, 1686)
Collection of Mrs. Paul Mellon, Oak Spring Garden Library, Upperville, Virginia

Les Leçons royales, which could be purchased at the printing house of François Poilly, was not, as the introduction explained, addressed exclusively to "personnes de qualité." Rather, it was intended to serve as a do-it-yourself manual for anyone interested in learning the art of flower painting. The simple yet effective format invented by Catherine Perrot proved highly successful and would later be adopted by other authors of painting treatises.

Maria Sibylla Merian

One of the greatest painters in the history of scientific illustration was Maria Sibylla Merian (1647–1717) (fig. 11), an artist whose life and work have formed the subject of numerous studies, some of them quite recent. Researchers have focused, in turn, on her unique contributions to the art of botanical painting, her impressive scientific achievements, and lately on an analysis of her career from the perspective of gender studies.

Merian was born into a family of artists; her father, Matthaüs Merian the Elder (1593–1650), was a well-known engraver from Basel who eventually settled in Frankfurt. His daughter's natural talents were brought out by early training, so that she became proficient in drawing, painting, and embroidery, as well as in the use of various printmaking techniques—etching, engraving, drypoint, and autography (in which a dampened sheet of paper or vellum was pressed onto a freshly prepared plate and both were passed through a press). After the death of her father, Maria Sibylla learned the art of flower painting from her stepfather, the still-life painter and art dealer Jacob Marrel (1614–1681). She also studied under two artists from Marrel's atelier, Abraham Mignon, a painter specializing in still lifes with insects, and the German Johann Andreas Graff (1637–1701), whom she married in 1665.[47]

Even as a child, Merian had always been fascinated by insects and their various metamorphoses; it is known that she collected silkworms and fed them leaves, carefully observing and recording their subsequent transformations.[48] After her marriage she continued her studies, depicting the life cycles of the moths and butterflies to be

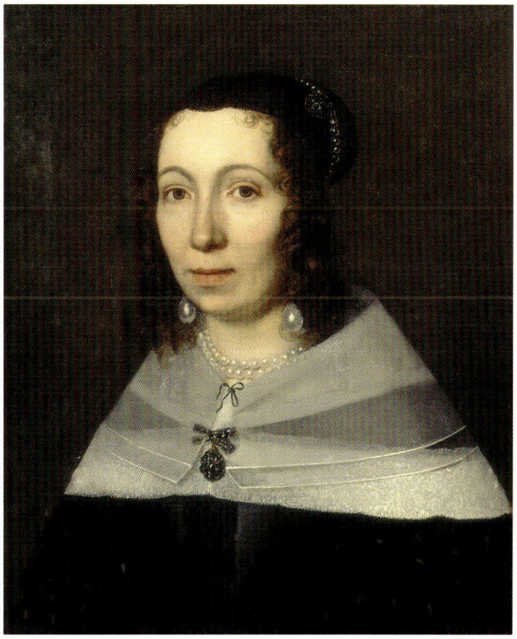

11. Netherlandish artist, *Maria Sibylla Merian,* 1679, oil
Kunstmuseum, Basel

found around her new home in Nuremberg. Although entomology was her main interest, plants and flowers also figured prominently in her art, for she usually portrayed her insects together with a branch of the plant on which they fed. In addition to her highly original insect studies, she also produced more traditional works in the genre of botanical painting, such as a volume of floral motifs for embroiderers, *Neues Blumenbuch* (1680), and a magnificent florilegium in three volumes for the nobleman Vincent Moller, a prominent figure in the political, economic, social, and religious affairs of the city of Hamburg.

Merian's first important work, *Der rupsen begin, voedzel en wonderbaare verandering,* published originally in German as *Der Raupen wunderbare Verwandelung und sonderbare Blumennahrung,* was devoted to caterpillars and their transformation into butterflies. It was originally divided into three volumes, the first appearing in 1679 and the second in 1683, both published by her husband. The fifty plates in each volume were accompanied by a text prepared by the artist and based on her own observations. The third volume was published posthumously in Amsterdam, together with a Dutch translation of the first two volumes, under the supervision of the artist's elder daughter and close collaborator, Dorothea Maria Henrietta. In keeping with the time-honored tradition of the family atelier, the artist generally signed her works with her maiden name, Maria Sibylla Merian, but on the title page of the second volume of *Der rupsen begin* appear the words "Maria Sibylla Gräffin Sculpsit."

In this work we find early evidence of a rare characteristic of the artist, one that has not received much attention from scholars. Most of her illustrations were accompanied by descriptions that reveal not only the faculty for observation of the born scientist, but also a genuine literary talent. She conceived her studies as an enthralling experience of discovery of the marvels of creation, and her words contain an irrepressible note of wonder and joy, sometimes verging on the lyrical tones of poetry, in their descriptions of even the most humble plants and insects, couched in a style which could have not been more different from the works of her male colleagues.

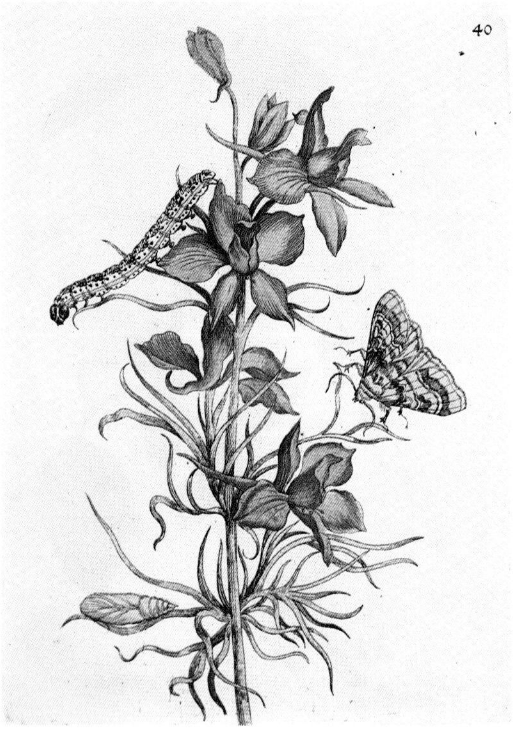

12. Maria Sibylla Merian, *Meadow Larkspur and Pease Blossom Moth,* undated, gouache on parchment
Collection of Mrs. Paul Mellon, Oak Spring Garden Library, Upperville, Virginia

For example, the description that accompanies Maria Sibylla's drawing of the moth *Periphanes delphinii* not only reflects the passionate interest and boundless patience which she brought to her studies; word and image work together to provide the most precise idea possible of the phenomenon under investigation (fig. 12). She writes:

> I have often seen hovering over the light blue flowers of the *Consolida regalis* the enchanting little moth that I depict here: so well known is it for its beauty and unusual coloring that I found myself wondering more than once from what caterpillar it might spring. I therefore pursued my research until I found the caterpillars I was looking for on the flowers of this very plant, to which they cause great damage since they not only like to feed upon them, but often devour the leaves and flowers with such voracity that they leave the stem completely bare.

The text continues with an account of the magnificent colors which the insect assumes in its various mutations, and concludes:

> I have portrayed one of these small moths in the center of the picture, poised on two green

leaves, the more to delight the eye of the nature lover the more attentive and acute that eye is, and to lend luster to this tiny work of art of indefatigable Nature.[49]

On the subject of the humble plantain, the artist observes:

I have often found these endearingly pretty green caterpillars on the greater plantain and, ascertaining that this weed is so much to their taste that they will eat it with the greatest voracity, I was extremely glad to be able to include in my forthcoming treatise this salutary little plant with its changeable colors, which many thoughtlessly tread upon as they walk through meadows or along roads: whereas an artist might perhaps make good use of it because of its unusual and lovely form, arranging it in the foreground of a landscape or similar painting.[50]

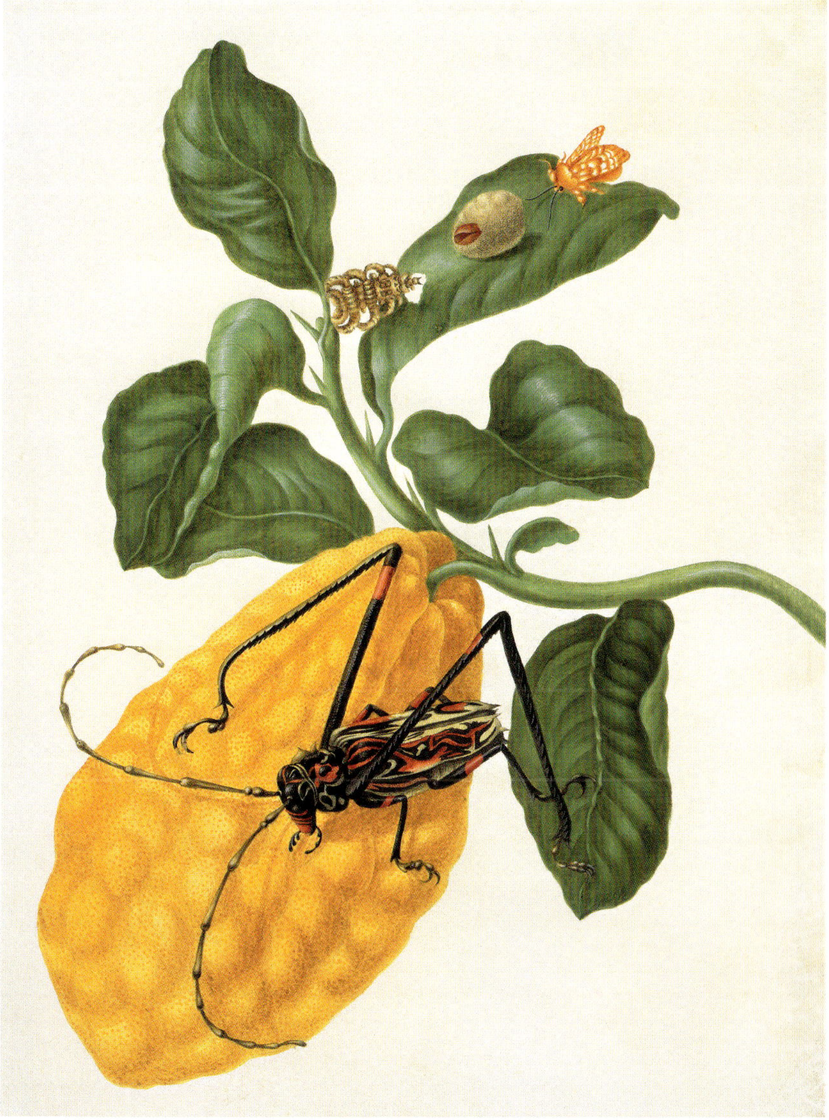

13. Maria Sibylla Merian, *Citron with a Moth and a Harlequin Beetle,* c. 1701–1702, watercolor and body color over pencil and etching on vellum
The Royal Collection © 2005, Her Majesty Queen Elizabeth II

In 1685, at the age of thirty-eight, Merian was persuaded by her beloved half-brother Caspar to abandon husband and home and, together with her two daughters and her mother, to join the Labadists, a pious sect founded by Jean de Labadie (1610–1674). This ex-Jesuit priest called for a return to the virtues of the early church and to this end established a religious community at the isolated castle of Waltha, near Leeuwarden, in Frisia. Taking advantage of the sect's belief in absolute equality between the sexes, the artist devoted all her time to scientific study. At Waltha, she discovered a unique collection of insects from Surinam that had been acquired by a member of the Sommelsdijk family, proprietors of the castle. The color, form, and dimensions of these tropical insects, so different from the European species with which she was familiar, must have been a revelation to her.

Merian left the Labadist community in 1691 and moved to Amsterdam, where she earned her living by painting floral still lifes for wealthy art collectors. There she also began working on the third volume of *Der rupsen begin,* her treatise on caterpillars and butterflies. At the same time she had the opportunity to associate with such intellectuals as Nicolaas Witsen, the enlightened burgomaster of Amsterdam, who was a member of the Dutch East India Company and owned a magnificent insect collection. She also met two illustrious scientists, the anatomist Frederik Ruysch and the botanist Caspar Commelijn, director of the city's botanical garden.

In 1699, when she was fifty-two years of age, with the encouragement and financial assistance of these friends, Merian was able to realize her dream of setting off for Surinam to study in situ the insects that she had hitherto seen only as inanimate specimens. From this extraordinary two-year expedition, which she undertook with her daughter Dorothea Maria Henrietta but "without a man's protection,"[51] Merian produced *Dissertatio de generatione et metamorphosibus insectorum Surinamensium,* the crowning achievement of a life dedicated to art and to intellectual and spiritual inquiry. Both mother and daughter suffered great hardships during their stay in the South American colony. As the artist wrote in 1702 upon her return to Amsterdam, "In that country there reigns a torrid heat, and every task becomes an enormous effort; I myself risked paying with my very life and for this reason could not prolong my stay." Indeed, she had contracted malarial fever, and thus in September 1701 she and her daughter set sail for home.

Merian spent the rest of her life completing the record of the entomological observations that she had made in Surinam, editing the scientific data with the help of the botanist Commelijn and arranging and retouching the paintings which she had executed on site. The extraordinarily refined technique of these works, which today can be seen in the British Museum, the Royal Collection at Windsor Castle, and the Russian Academy of Sciences in Saint Petersburg (fig. 13), reflects the artist's early training as an engraver. Beneath the tempera colors one can trace the marks of her original pencil drawing, with strokes that are reminiscent of the fine lines incised by the engraver's tool. In fact, these pencil marks must have been of invaluable help to the artists charged with the task of transferring her drawings to metal plates. Merian herself engraved perhaps twenty of the plates for the book she was preparing, while the remaining forty plates were executed by Joseph Mulder (1659/1660–after 1718) and Jan Pieter Sluyter (1675–1713). The artist insisted, however, that only she and her daughter, who had studied the insects from life, should be entrusted with the final task of coloring the illustrations in the printed copies of the book.

The first edition of *Dissertatio de generatione et metamorphosibus insectorum Surinamensium* was published in Latin and Dutch in 1705, with sixty plates. The second edition, which appeared in 1719, two years after the artist's death, contained an additional twelve plates, some the work of Merian and others perhaps by the hand of her younger daughter, Johanna Helena. The sheer magnificence of its engravings and its unusual subject matter make the *Dissertatio* one of the most important scientific books ever to be produced in Europe, a true masterpiece of the printer's art.

The work opens with a frontispiece showing the artist examining specimens presented to her by six putti, while behind her a spacious arch frames a tropical landscape (fig. 14). The foreword is full of factual information and personal reflections, the author describing in detail her "longinquum et sumptuosum" (venturesome and costly) voyage, her working methods, and the help that she received from the natives of Surinam. She explains that each insect

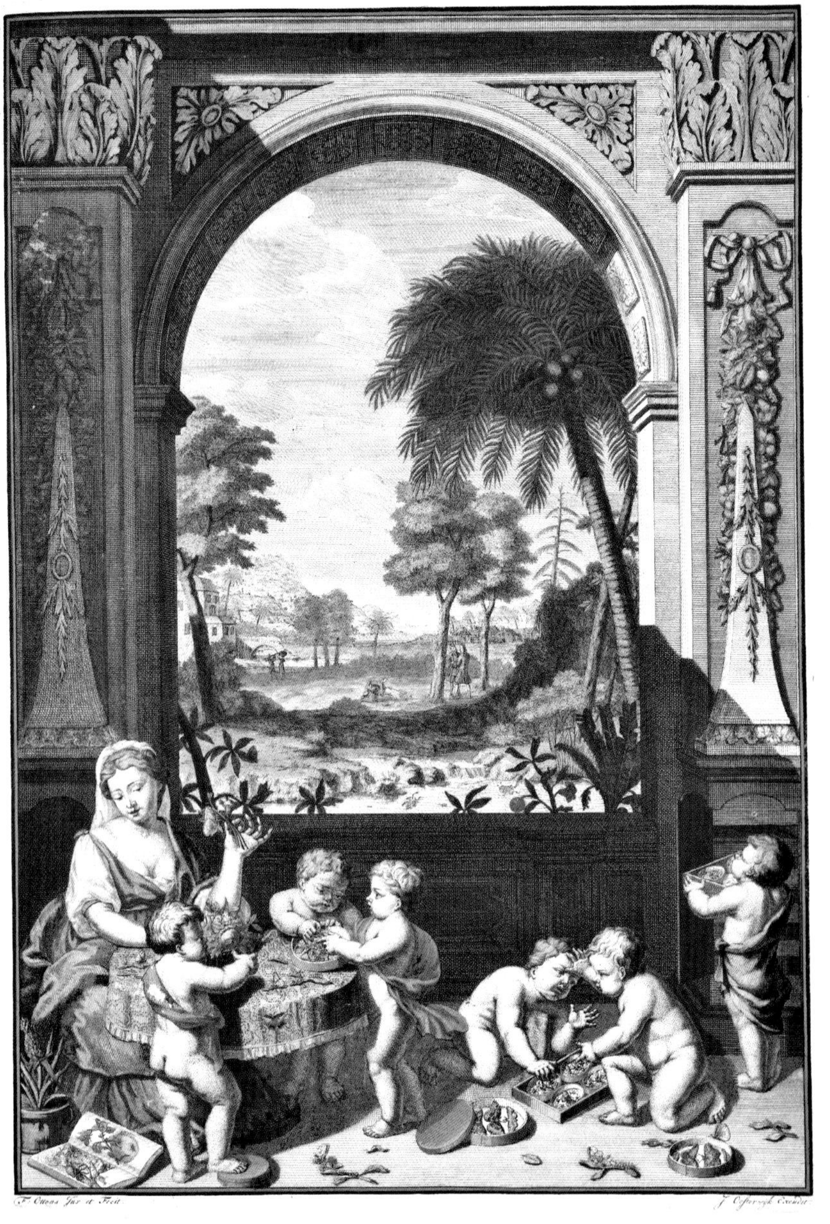

14. Maria Sibylla Merian, frontispiece from *Dissertatio de generatione et metamorphosibus insectorum Surinamensium* (Amsterdam, 1705)
Collection of Mrs. Paul Mellon, Oak Spring Garden Library, Upperville, Virginia

has been carefully studied in the various phases of its life cycle—often with the aid of a microscope—and is depicted life-sized together with the plant, flower, or fruit that made up its usual diet. Each entry begins with such botanical data, thus furnishing additional information on how the artist composed her pictures.

As we have already seen in *Der rupsen begin,* here the artist's images are reinforced by her words. A description of the pineapple (fig. 15), which opens the *Dissertatio,* is characteristic and illustrates the artist's approach to her subject matter:

> The pineapple being the most important fruit [in the Americas] among those which are edible, it is only fair that it occupy first place in this work, just as it has in my research. On the page I have painted it at the moment of its flowering. The tiny colored leaves that lie just beneath the fruit are like satin spotted with yellow. The lateral shoots continue to grow; when the mature fruit is gathered, the longer leaves are sea green on the outside and grass green on their inner sides, while the borders are slightly reddish and armed with sharp spines. Of all the insects in America, the best known are the cockroaches, primarily because of the heavy damage that they cause to the plants they feed upon. They have a special predilection for the pineapple, because they are attracted to all that is sweet. They deposit their eggs very close to one another, with a protective covering that is round in form to defend them from the numerous spiders. On the other part of the fruit is a different species of cockroach, which carries its eggs in a brown abdominal sac. If the creature is attacked it immediately drops the covering, from which the little ones can emerge. This defensive action allows them to escape more easily.[52]

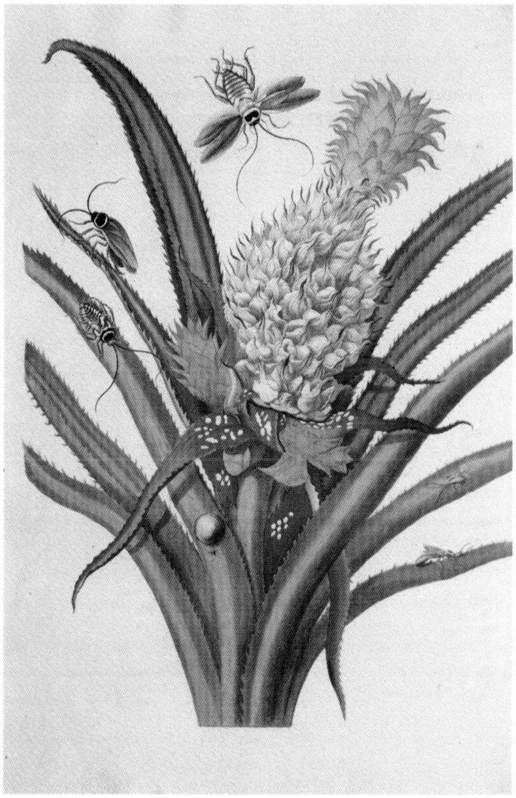

15. Maria Sibylla Merian, *Pineapple* (detail), plate I from *Dissertatio de generatione et metamorphosibus insectorum Surinamensium* (Amsterdam, 1705), hand-colored etching
Collection of Mrs. Paul Mellon, Oak Spring Garden Library, Upperville, Virginia

Merian's sensitivity to the most minute aspects of the natural world and her rich visual vocabulary, the fruit of a lifetime of study and practice, are reflected in every detail of the *Dissertatio.* Unlike the authors of other botanical and entomological treatises of the period, our artist-naturalist provides a perspective that has been defined as "ecological"[53] rather than systematic, for she always studied her animals in their natural setting rather than in the sterile environment of the laboratory. Moreover, although she made ample use of the microscope to examine the external details and form of her insects, Merian never dissected them in order to analyze their anatomy as her colleagues generally did, from Francesco Redi and Marcello Malpighi to Jan Swammerdam, a Dutch naturalist whose work was well known to Merian and whom she cites in her preface. As the nineteenth-century historian Jules Michelet observed, the artist's unique approach amounted to a veritable "theology of the insect." In his work *L'Insecte,* published in 1858 in Paris, Michelet discusses the complex and sometimes contradictory aspects of Merian's personality; while on the one hand he admits her "noble vigor, strong simplicity, and virile seriousness," he concludes by seeing in her work "the hand of a woman, conscientious and tender, that has reached out to touch [her subjects] with the most loving respect."[54]

Merian's vision of Surinam was far from that of an idyllic tropical paradise, and in many of her paintings she shows next to the insect its natural predator. As the French physiologist and chemist Louis Figuier (1819–1894) wrote: "Every one of her paintings depicts a drama in miniature. Next to the frightened and suspicious insect one can see an avid lizard waiting patiently, or a ferocious spider weaving its sinister snare."[55] Nor do the implacable laws of nature spare the splendid tropical flowers, often depicted by the artist as falling prey to ravening insects.

Among all the countries of Europe, Holland was perhaps the only one where the career of such an exceptional figure as Maria Sibylla Merian could have been possible. During the late seventeenth century, there was a passionate interest in the exotic plants and other natural history specimens that were arriving, many of them through the Dutch East India Company. In addition, an exemplary rigor was exercised in the approach to scientific research established by the great universities, such as that of Leiden, and by the botanical gardens of Leiden and Amsterdam. For example, an extensive and systematic effort to introduce exotic plants to meet the demands of the public botanical gardens and the private gardens of wealthy collectors was begun around 1665 and reached its apex around 1682.

One notable figure among many in this stimulating ambience was Agnes Block (1629–1704), known as the Lady of Flines, a noblewoman with an interest in the natural sciences who won the admiration of botanists in Holland and abroad, including Lelio Trionfetti of Bologna (1656–c. 1708), with whom she conducted an interesting correspondence. In 1687, on the estate of Vijverhov, near Loenen, along the Vecht River, her gardeners succeeded in cultivating a pineapple to fruition. Agnes had this exceptional plant depicted, along with many other exotic specimens from her garden, by botanical artists then active in Amsterdam, including Jan Moninckx and his sister Maria; Alida Withous, the daughter of Matthias Graf and the sister of Jan and Johanna Helena Graf; and Dorothea Maria Henrietta, the daughter of Maria Sibylla Merian.[56]

Behind the Convent Walls: Artist-Nuns and Botanical Painting

During this period, many members of religious orders, both male and female, found an outlet for their energy and talents in the arts. Embroidery and miniature painting constituted regular activities in the convent, and Tuscany, especially Florence, could boast a long tradition of nuns who were also artists, the most celebrated being Sister Plautilla Nelli (1523–1588). A student of Fra Paolino and a follower of Fra Bartolomeo, Sister Plautilla entered the Florentine convent of Santa Caterina da Siena, which was renowned for its highly productive atelier specializing in religious paintings. The names of other artists active there between the end of the sixteenth century and the middle of the eighteenth century—such as Sister Prudenza Cambi, Sister Felicia Lumicini, and Sister Maria Angelica Razzi—have come down to us, although their lives and their oeuvre still remain to be reconstructed.[57]

Even less is known about the nuns who devoted themselves to naturalistic illustration, although the close ties between this genre and the art of embroidery, with its bird and flower motifs, allow us to hypothesize that within the cloister many of the artistically inclined must have been attracted to the painting of subjects of this kind.

Teresa Berenice Vitelli, or Sister Veronica (act. 1706–1729), was one such figure. Probably born sometime during the last decades of the seventeenth century, she took her vows as a nun of the Benedictine order and joined the convent of Sant'Apollonia in Florence. It is known that her work was admired by Violante of Bavaria, wife of Grand Duke Ferdinando de' Medici, for whom she executed copies in pastels of works by Rembrandt van Rijn, Francesco Trevisani, and Livio Mehus, a Flemish painter who died in Florence in 1691. She is also mentioned as one of the participants in an exhibition organized by the Accademia dell'Arti del Disegno and held in the cloister of the church of the Santissima Annunziata.

The first known painting by Sister Veronica is a lively composition of birds and lizards on vellum, signed and dated 1706, to which other works on similar subjects have recently been added. Her realistically portrayed birds and other animals are arranged in charming little scenes against domestic backgrounds.[58] Sister Veronica was clearly familiar with the oeuvre of Giovanna Garzoni, however, for an elegant painting in gouache on vellum of two tulips, a rose, and a delphinium, bound together with a flowing blue ribbon, contains distinct echoes of the latter's style. This particular work allows us to ascribe to Sister Veronica other unattributed paintings on floral subjects contained in the fourth volume of *Ornithologiae vivis coloribus expressae,* a miscellaneous collection of paintings assembled by the Vallombrosan monk Bruno Tozzi (1656–1743) from the end of the seveteenth century into the first decades of the eighteenth. Tozzi was a naturalist and botanical artist working at the abbey of Vallombrosa, near Florence, a center for botanical studies that was renowned beyond the Alps and as far as away England.[59]

The fourth volume of *Ornithologiae* contains gouache and watercolor paintings by various hands on a variety of subjects. Some of the paintings at the beginning of the manuscript, in particular several small bouquets bound with fluttering ribbons, vases of flowers, and paintings of single blossoms, can in my opinion be attributed to Sister Veronica because of their stylistic affinities with the works mentioned above (fig. 16). It should also be noted that since the convent of Sant'Apollonia and the abbey of Vallombrosa belonged to the same religious order, there must certainly have been close ties and regular exchanges between them.

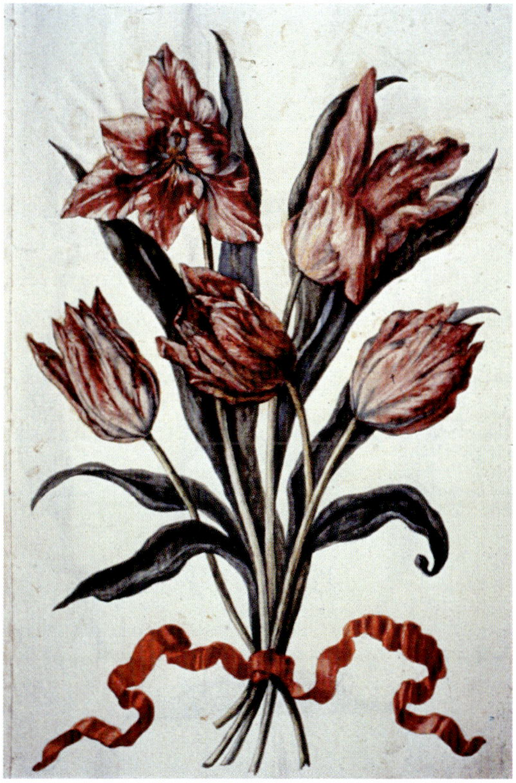

16. Teresa Berenice Vitelli (Sister Veronica), *Bouquet of Tulips,* from Bruno Tozzi, *Ornithologiae vivis coloribus expressae,* undated, gouache on paper
Ministero per i Beni e le Attività Culturali della Repubblica Italiana / Biblioteca Nazionale Centrale di Firenze, MS Conventi Soppressi, A.I.830, volume 3

The Eighteenth Century: Women Artists and Students of Flower Painting

During the eighteenth century, as the popularity of naturalistic illustration spread beyond the ruling classes, men of science, and virtuosi to an emerging class of wealthy bourgeois, the genre began to assume new dimensions and greater complexity. As the number of women artists in Europe multiplied, they also began to turn their hands to subjects traditionally reserved for men, although the still life would always retain an important place in their oeuvre.

Among the artists who gained particular recognition in this genre was Rachel Ruysch, daughter of the anatomist Frederik Ruysch and the most talented student of Maria Sibylla Merian, whose trip to Surinam was partly organized and financed by Rachel's father. Rachel Ruysch was best known for her large paintings of floral bouquets, marked by brilliant plastic and chromatic effects. She did not forget her early training, however, and—in the rigorous tradition of scientific illustration—focused with microscopic precision on the individual elements in her compositions.

Because the eighteenth century was also the age of *femmes savantes,* the important role that women played in the study of the exact and natural sciences has already been thoroughly examined. What has received considerably less attention is an analogous phenomenon in the arts: women skilled in botanical and zoological illustration began to play a dominant role in the family atelier. The unique contributions of a few such artists deserve brief examination here.

Barbara Regina Dietzsch

A German painter of botanical and zoological subjects, Barbara Regina Dietzsch (1706–1783) provides a typical example of the emancipated woman who enjoyed complete autonomy in the practice of her art. She was the eldest of nine children, born into a family that counted many painters, engravers, and musicians among its members; her father was the well-known landscape painter Johann Israel Dietzsch (1681–1754). Barbara Regina grew up in Nuremberg, a city with an illustrious artistic and scientific tradition,

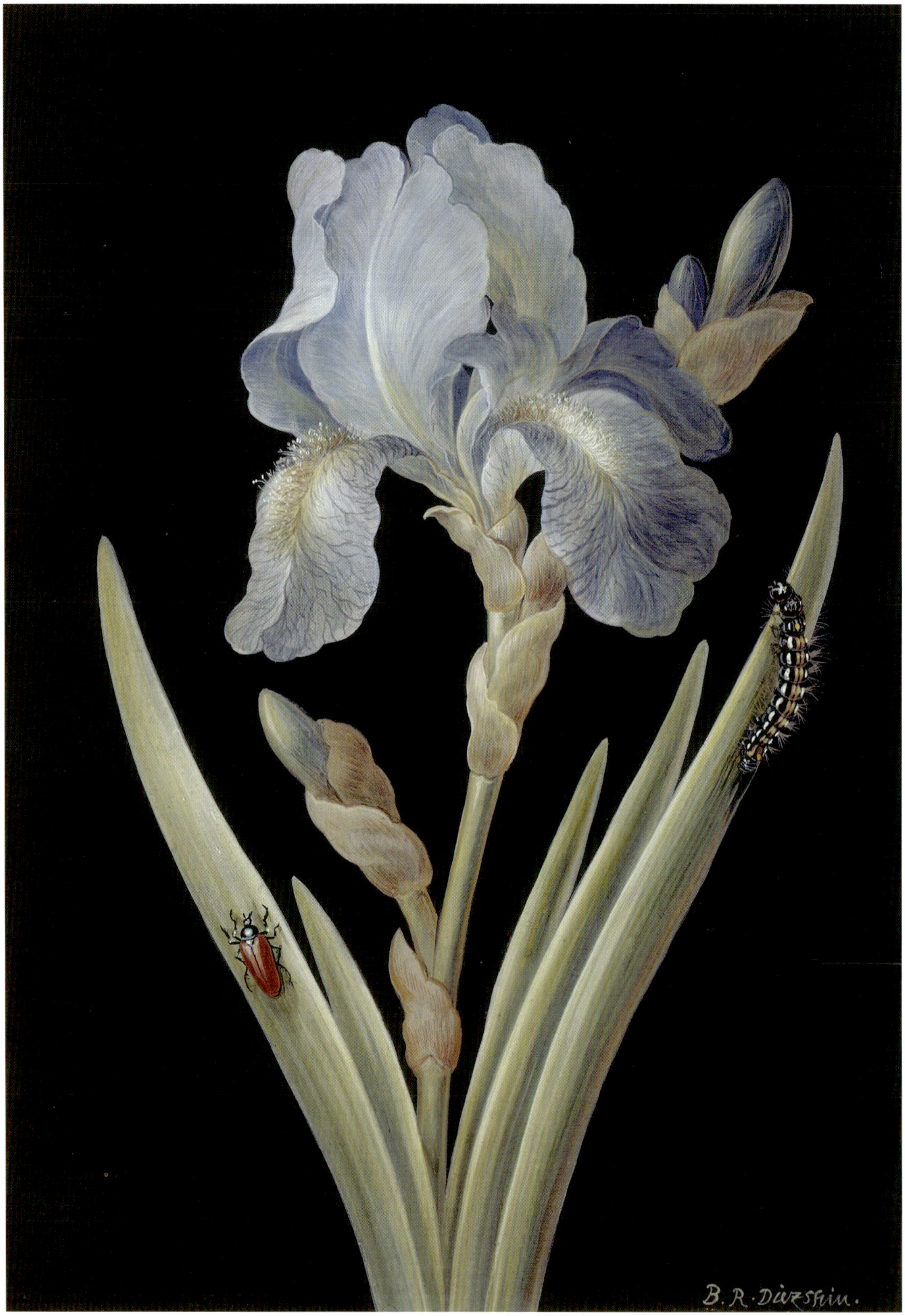

17. Barbara Regina Dietzsch, *Iris germanica*, undated, gouache on vellum
Fitzwilliam Museum, University of Cambridge

18. Elizabeth Blackwell, *Paeonia femina* (female peony), from *A Curious Herbal, Containing Five Hundred Cuts of the Most Useful Plants* (London, 1737–1739), hand-colored etching

Collection of Mrs. Paul Mellon, Oak Spring Garden Library, Upperville, Virginia

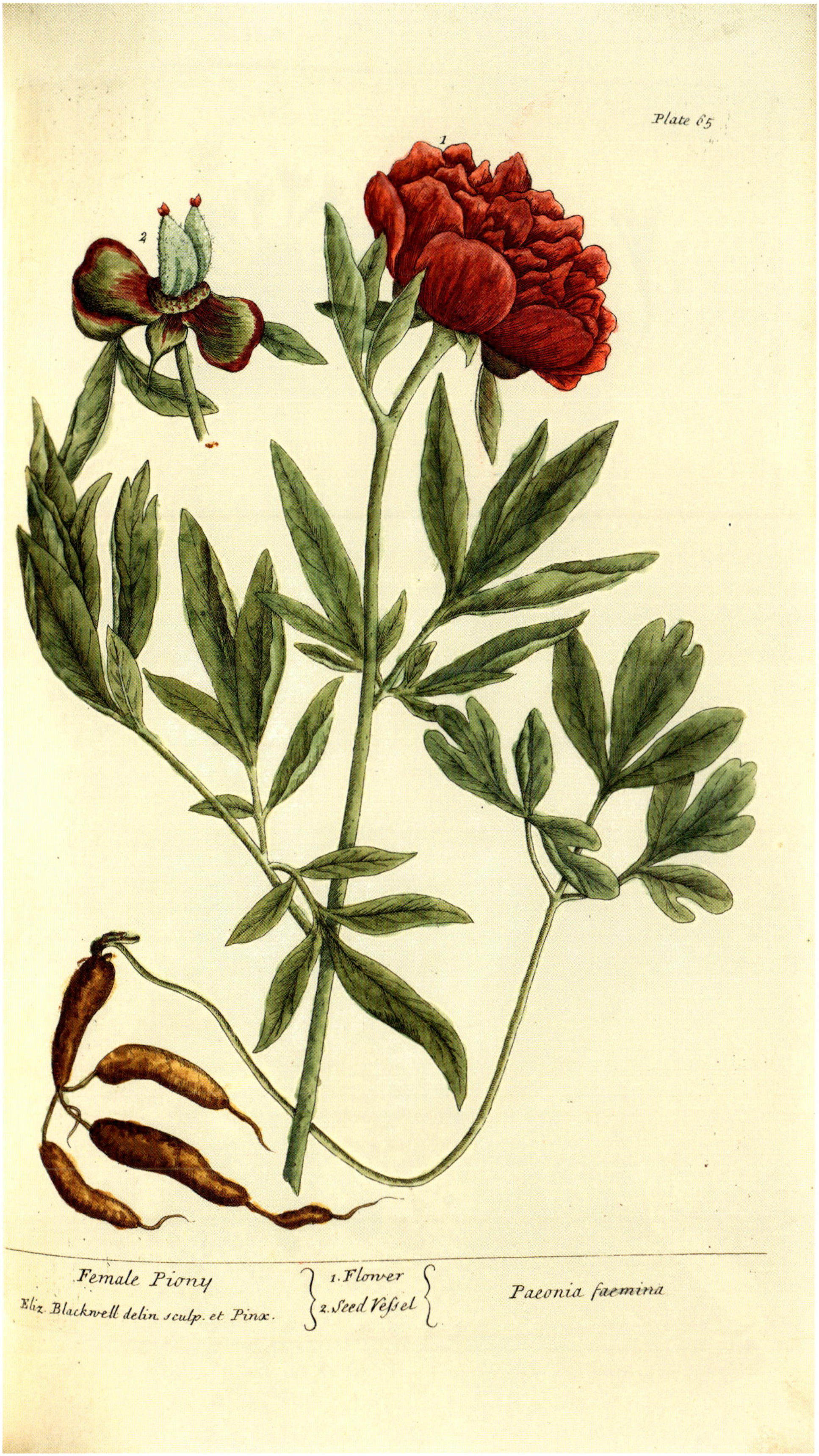

during the same period in which the eminent botanist and physician Christoph Jacob Trew (1695–1769) wrote many of his lavishly illustrated encyclopedic treatises.

Barbara Regina Dietzsch specialized in painting flowers, birds, and insects, while her sister Margaretha Barbara concentrated primarily on botanical subjects and provided some of the illustrations for Trew's works. Their brother Johann Christoph Dietzsch (1710–1769) chose instead to continue the tradition of landscape painting established by their father, but few works on scientific subjects by his hand have come down to us. Although Barbara Regina was undoubtedly the most gifted of the three siblings, it is not always easy to determine the exact authorship of the botanical and floral compositions in gouache on vellum that came out of the highly productive Dietzsch workshop, compositions that were often enlivened by the addition of insects, lizards, and other small animals, posed against velvety dark backgrounds (fig. 17).[60]

Elizabeth Blackwell

Elizabeth Blackwell (1700?–1758) was not an artist by profession and a somewhat involved personal story lies behind the creation of her work, *A Curious Herbal, Containing Five Hundred Cuts of the Most Useful Plants* (London, 1737–1739). Elizabeth's husband, Alexander Blackwell, was a wealthy Scottish intellectual and self-educated physician who, when his credentials were challenged, was forced to give up his medical practice in Aberdeen. The couple then moved to London, where he set up a successful printing house. He again ignored professional licensing requirements, however, and his violation of trade regulations resulted in a series of heavy fines. He was forced to close his shop and, being unable to pay his mounting debts, he was eventually arrested and sent to debtor's prison.[61]

Elizabeth began casting about for a means to restore the family's fortunes, and when she learned from Sir Hans Sloane, president of the Royal Society, that English physicians and botanists lacked an up-to-date illustrated herbal of medicinal plants, she decided to remedy this lacuna. Renting a house close to the Chelsea Physic Garden (the only botanical garden in London at the time), she began in 1715 to make botanical drawings of the garden's collection of medicinal herbs. Her husband, still in prison, wrote the text describing their uses and properties, basing his work in large part on the *Botanicum officinale* published by Joseph Miller in 1722.

A Curious Herbal, featuring illustrations of five hundred flowering plants engraved and hand-colored by the artist herself, was published in several parts from 1737 to 1739 and then reissued in two volumes by the London publisher Samuel Harding (fig. 18). The work came to the attention of the naturalist Dr. Christoph Jacob Trew, who undertook the publication of a new edition of the work, with the text translated into Latin and German and the plates re-engraved by N. F. Eisenberger (who added another hundred illustrations of his own). This expanded version of the herbal, with various supplements, appeared as *Herbarium Blackwellianum* in Nuremberg over the period 1750–1773.

The simplicity of the drawings in *A Curious Herbal* betrays the fact that Elizabeth Blackwell was not a professional botanical artist, but they are nonetheless commendably accurate (each plant being drawn in careful outline, colored, and neatly arranged on the page, together with details of its flower and seeds) and possess an undeniable charm of their own. *A Curious Herbal* enjoyed a modest success, and Elizabeth was able to obtain her husband's release from prison and regulate his affairs so that he could once again take up his practice.

The story of the Blackwells and their herbal underscores the fact that in the middle of the eighteenth century there was a growing demand for botanical illustrations. With resourcefulness and determination, the wife of the eccentric physician was able to take advantage of this opportunity.[62]

A Genteel Pastime

While Elizabeth Blackwell took up her brush in the face of economic necessity—choosing the genre of botanical illustration because it was popular and marketable and using her native talent to the best of her ability—many other women began painting flowers simply because social convention deemed it to be a suitable pastime for them. Thus, as had occurred in France in the time of Catherine Perrot a century earlier, a vogue for naturalistic painting sprang up among young ladies in eighteenth-century England. Although it was conceded that members of the fair sex were particularly sensitive to the beauties of nature, it was also believed that their natural propensity needed to be carefully directed and trained. The popularity of the flower-painting genre was therefore reflected in the large number of female pupils who clamored to study under the most celebrated botanical painters of the day.

One of these was Georg Dionysius Ehret (1710–1770), who had settled in England after collaborating with the Swedish botanist Carl Linnaeus (1707–1778) on some of his groundbreaking studies. Ehret was in such great demand as a teacher that for many years he directed an informal and remarkably successful school of botanical painting for well-born young ladies,[63] in which he sought to teach botanical painting in accordance with the highest standards of the art. Furthermore, he codified the rules of botanical illustration, which gave the genre a solid professional dimension based on technical training and a careful grounding in the natural sciences. Ehret taught his students not only how to paint flowers in accordance with the strict rules of scientific illustration, but also how to conserve their works; he even emphasized the importance of recording on each sheet the scientific names of the species portrayed using the system of Linnaeus. In his memoirs, the artist enumerated with pride the names of the many ladies of aristocratic birth who frequented his classes, including Catherine, Duchess of Norfolk, and Mary, Duchess of Leeds. Some of his students, such as Lady Frances Howard and Lady Ann Hamilton, later gained a modest fame of their own as botanical artists, although they rarely surpassed the level of a conscientious reinterpretation of the master's style.[64]

The panorama described here, embracing such different personalities, all passionately engaged in the art of naturalistic illustration, gives an idea not only of the richness and complexity of this art form, but also of a uniquely feminine approach to the natural world and scientific phenomena. In this perspective, the *femminil pazienza* that was otherwise considered a lesser gift traditionally attributed to women artists acquired extraordinary potential. Maria Sibylla Merian was fully aware of this fact when she wrote in June of 1704 to the physician Johann Georg Volckamer of Nuremberg: "patience is a very beneficial little herb."[65]

NOTES

I thank Rachel Lambert Mellon for permission to reproduce a number of images from books and manuscripts in her collection, Oak Spring Garden Library, Upperville, Virginia.

The text was translated from the Italian by Lisa Chien.

1. Linda L. Nochlin, "Why Have There Been No Great Women Artists?" in *Women, Art, and Power and Other Essays* (London, 1989), 146–147.

2. Natalie Zemon Davis, *Women on the Margins: Three Seventeenth-Century Lives* (Cambridge, Mass., 1995); Italian translation: *Donne ai margini: tre vite del XVII secolo* (Bari, 1996).

3. Giulia Calvi, "Introduzione," in *Barocco al femminile*, ed. Giulia Calvi (Rome and Bari, 1992), vii–xxvii.

4. Elizabeth Cropper, "Life on the Edge: Artemisia Gentileschi, Famous Woman Painter," in *Orazio e Artemisia Gentileschi*, ed. Keith Christiansen and Judith W. Mann [exh. cat., Palazzo Venezia and the Metropolitan Museum of Art] (Rome and New York, 2002), 263–281. By the same author, see "Artemisia Gentilesca, la 'pittora,'" in Calvi 1992, 191–218.

5. Gaia Salvatori, "La sindrome delle Muse: motivi, problemi e orientamenti della critica sulle donne artiste del Seicento Europeo," in *Donne, filosofia e cultura nel Seicento*, ed. Pina Totaro (Rome, 1999), 393–396.

6. Luigi Lanzi discussed the concept of "femminil pazienza," with reference to the artist Lavinia Fontana, in *Storia pittorica della Italia dal risorgimento delle belle arti fin presso al fine del XVIII secolo*, 6 vols. (Milan, 1823), 6:57. A vast bibliography exists on the subject of women artists, of which I will mention just a few works here: Ann Sutherland Harris and Linda Nochlin, *Women Artists: 1550–1950* [exh. cat., Los Angeles County Museum of Art] (New York, 1976); Rozsika Parker and Griselda Pollock, *Old Mistresses: Woman, Art and Ideology* (London 1981); Nancy G. Heller, *Women Artists: An Illustrated History* (New York, 1987); Whitney Chadwick, *Women, Art, and Society* (London, 1990); Claudio Strinati and Jordana Pomeroy, *Italian Women Artists from Renaissance to Baroque* [exh. cat., National Museum of Women in the Arts] (Washington, D.C., 2007).

7. Artemisia Gentileschi was instead always referred to as "pittora." On the concept of *virtù*, see Fredrika H. Jacobs, *Defining the Renaissance Virtuosa: Women Artists and the Language of Art History and Criticism* (Cambridge, 1997), 71–72.

8. Sandra Raphael, ed., *Women and Natural History* [exh. cat., Bodleian Library] (Oxford, 1996); Julia Lindkvist, "Women Illustrators of Natural History in the Enlightenment," in *Science and the Visual Image in the Enlightenment*, ed. William Shea (Sagamore Beach, Mass., 2000), 185–212.

9. Carolyn Merchant, *The Death of Nature: Woman, Ecology and the Scientific Revolution* (San Francisco, 1982); Italian translation, *La morte della natura: le donne, l'ecologia e la rivoluzione scientifica* (Milan, 1988); Margaret Alic, *Hypatia's Heritage: A History of Women in Science from Antiquity through the Nineteenth Century* (Boston, 1986); Londa Schiebinger, *The Mind Has No Sex? Women in the Origins of Modern Science* (Cambridge, Mass., 1989).

10. See, for example, the title page to Richard Waller, *Essayes of Natural Experiments Made in the Academie del Cimento, under the Protection of the Most Serene Prince Leopold of Tuscany* (London, 1684), where both the Royal Society and the Accademia del Cimento are represented allegorically as female figures.

11. It may be noted that the Accademia di San Luca, the most celebrated artists' academy in Rome, did not even consider the question of allowing women artists as members until well into the eighteenth century. Nevertheless, an ample respect for their talent emerges from an examination of the institution's documents. The academy also provided assistance to elderly female artists, such as Giovanna Garzoni; see Angela Cipriani, "Giovanna Garzoni miniatrice," in *Ricerche di storia dell'arte* 1–2 (1976): 240–245. The academy's policy toward women, however, may actually have been more nuanced than previously thought; Melchior Missirini, in his work *Memorie per servire alla storia della romana Accademia di San Luca* (Rome, 1823) includes Isabella Parasole, Anna Maria Vaiana, and Giovanna Garzoni in his list of the members of the academy. In a small bound manuscript (vol. 29) dating from the middle of the eighteenth century and conserved in the academy's archives, the names of Vaiani and Garzoni appear among the "Nomi delle Signore Pittrici Accademiche" (Names of the Academy's Lady Painters). I am grateful to Angela Cipriani for furnishing this information.

12. Agnes Arber, *Herbals, Their Origin and Evolution: A Chapter in the History of Botany, 1470–1670* (1912; reprint, Cambridge, 1986), 315. On the practice of coloring the engravings in scientific texts by hand, see Lucia Tongiorgi Tomasi, "L'illustrazione naturalistica: tecnica e invenzione," in *Natura-Cultura: L'interpretazione del mondo fisico nei testi e nelle immagini*, ed. Giuseppe Olmi, Lucia Tongiorgi Tomasi, and Attilio Zanca (Florence, 2000), 145–147.

13. Vincenzo Giustiniani, "Lettera al signor Teodoro Amideni," in *Raccolta di lettere sulla pittura, scultura ed architettura*, ed. Giovanni Bottari and Stefano Ticozzi (Milan, 1822–1825), 6:122.

14. On the links between still-life painting and women artists, see Norman Bryson, *Looking at the Overlooked: Four Essays on Still Life Painting* (Cambridge, Mass., 1990), 136–178.

15. Giuseppe Olmi, *L'inventario del mondo: Catalogazione della natura e luoghi del sapere nella prima età moderna* (Bologna, 1992), 126; Lucia Tongiorgi Tomasi, "Naturaleza muerta y ciencia de la naturaleza en Italia durante los siglos XVI y XVII," in *El bodegón* (Madrid, 2000): 201–224.

16. Tongiorgi Tomasi 2000, 149.

17. See, for example, C. P. Snow, *The Two Cultures and the Scientific Revolution* (New York, 1959); Italian translation, *Le due culture* (Milan, 1964).

18. Many scholarly studies have been published on the cultural and intellectual ambience in Rome during the first decades of the seventeenth century. I will limit myself here to citing Elisabeth MacDougall, "A Cardinal's Bulb Garden: A 'Giardino Segreto' at the Palazzo Barberini in Rome," in *Fountains, Statues, and Flowers: Studies in Italian Gardens of the Sixteenth and Seventeenth Centuries* (Washington, D.C., 1994), 219–347; Francesco Solinas, ed., *Cassiano dal Pozzo: Atti del Seminario Internazionale di Studi* (Rome, 1989); Francesco Solinas, ed., *I segreti di un collezionista: Le straordinarie raccolte di Cassiano dal Pozzo, 1588–1657* [exh. cat., Palazzo Barberini] (Rome, 2000); Francesco Solinas, *L'Uccelliera: Un libro di arte e di scienza nella Roma dei primi Lincei* (Florence, 2000); Giovan Battista Ferrari, *Flora overo cultura di fiori*, facsimile edition, ed. Lucia Tongiorgi Tomasi (Florence, 2001).

19. Nicolas Robert, *Fiori diversi novame[n]te posti in luce* (Rome, 1640) and *Variae ac multiformes forum species* (Paris, c. 1660). In 1638 Nicolas de La Fleur engraved the copper plates for a flower book, which was later printed in Paris by Mariette.

20. Lucia Tongiorgi Tomasi, *An Oak Spring Flora: Flower Illustration from the Fifteenth Century to the Present Time. A Selection of the Rare Books, Manuscripts and Works of Art in the Collection of Rachel Lambert Mellon* (Upperville, Va., 1997), 113–116.

21. The work was reprinted in a great number of editions. See in particular the facsimile of the 1717 Venice edition, ed. Giuliana Forneris and Annalaura Pistarino (Ivrea, 2000).

22. See Castore Durante, *Herbario nuovo* (Venice, 1636), 23.

23. Giovanni Baglione, *Le vite de' pittori, scultori e architetti: dal pontificato di Gregorio XIII del 1572 fino a' tempi di papa Urbano ottavo nel 1642* (Rome, 1642), 394–395. On Isabella Cattani Parasole, see Jacobs 1997, 105–106; Washington 2007, 195–197.

24. Ulrich Thieme and Felix Becker, eds., *Allgemeines Lexikon der bildenden Künstler* (Leipzig, 1940), 34:40–41. On the artist, see Lucia Tongiorgi Tomasi and Angela Ferrari, "Botanica barocca," in *Gazzetta del bibliofilo* 17 (1986): 6–9.

25. Solinas 2000, fig. 32. The preparatory drawing by Mellan, which dates from 1626, was included in the exhibition *Claude Mellan, gli anni romani: un incisore tra Vouet e Bernini*, ed. Luigi Ficacci [exh. cat., Istituto Nazionale per la Grafica, Rome] (Rome, 1989), 259.

26. Elizabeth Cropper, "Vincenzo Giustiniani's 'Galleria': The Pygmalion Effect," in Jennifer Montagu et al., *Cassiano dal Pozzo's Paper Museum, Quaderni puteani*, vol. 3 (Milan, 1992), 102.

27. Galileo Galilei, *Le opere di Galileo Galilei*, ed. Antonio Favaro, 20 vols. (Florence, 1890–1909), 14:110, 115, 134, 148, 158; 15:66; 16:538; 17:250, 258, 351.

28. Mary Garrard, *Artemisia Gentileschi: The Image of the Female Hero in Italian Baroque Art* (Princeton, 1989), 383–384.

29. Tongiorgi Tomasi 2001, xii.

30. *Galleria Giustiniana del Marchese Vincenzo Giustiniani*, 2 vols. (Rome, 1640), 1:3; 2:33, 52; Cropper in Montagu et al. 1992; and *Caravaggio e i Giustiniani: Toccar con mano una collezione del Seicento*, ed. Silvia Danesi Squarzina [exh. cat., Palazzo Giustiniani, Rome] (Milan, 2001), 362–365, 374.

31. Francesco Solinas, "L'Erbario Miniato e altri fogli di iconografia botanica appartenuti a Cassiano dal Pozzo," in Francis Haskell et al., *Il museo cartaceo di Cassiano dal Pozzo: Cassiano naturalista, Quaderni puteani*, vol. 1 (Milan, 1989), 60; and Francesco Solinas, "Sull'atelier di Cassiano dal Pozzo: Metodi di ricerca e documenti inediti," in Montagu et al. 1992, 64.

32. See Solinas 2000, 38.

33. Angela Cipriani, "Giovanna Garzoni, miniatrice," *Ricerche di storia dell'arte* 1–2 (1976): 241–254; Silvia Meloni, "Giovanna Garzoni 'miniatora medicea,'" in *FMR* (1983): 15, 77–96; Lucia Tongiorgi Tomasi, "L'immagine naturalistica a Firenze tra XVI e XVII secolo: Contributo al rapporto 'Arte-Natura' tra manierismo e prima età barocca," in *Immagini anatomiche e naturalistiche nei disegni degli Uffizi: secc. XVI e XVII*, ed. Paolo Roberto Ciardi and Lucia Tongiorgi Tomasi [exh. cat., Uffizi, Gabinetto Disegni e Stampe] (Florence, 1984), 37–67; Gerardo Casale, *Giovanna Garzoni "insigne miniatrice" 1600–1670* (Milan and Rome, 1991); Gerardo Casale, ed., *Gli incanti dell'iride: Giovanna Garzoni pittrice nel Seicento* [exh. cat., San Severino Marche, Fondazione Salimbeni] (Milan, 1996); Marco Chiarini, ed., *Il giardino del granduca: Natura morta nelle collezioni medicee* (Turin, 1997), 90–103; Silvia Meloni Trkulja and Elena Fumagalli, *Nature morte: Giovanna Garzoni* (Paris, 2000); Washington 2007, 220–238.

34. Carlo Ridolfi, *Le meraviglie dell'arte, overo le vite de gl'illustri pittori veneti, e dello stato* (Venice, 1648), 71; Emanuele Tesauro, *Il cannocchiale aristotelico* (Venice, 1669), 64; Agnes Mongan, "A Fête of Flowers: Women Artists' Contribution to Botanical Illustration," *Apollo* 119 (1984): 264–267.

35. Giacinto Cantalamessa Carboni, *Memorie intorno i letterati e gli artisti della città di Ascoli nel Piceno* (Ascoli, 1830), 202–204. The manuscript by Giovanna Garzoni is mentioned by Giovanni Targioni Tozzetti in *Selve,* MS Targioni Tozzetti 189, XIII, Biblioteca Nazionale Centrale, Florence, f. 89v; Lucia Tongiorgi Tomasi, "Fiori, giardini, giardinieri, naturalisti e artisti a Roma nella prima metà del Seicento," in *Studi in onore di Corrado Maltese,* ed. Stefano Marconi (Rome, 1997), 183–188.

36. Two plates illustrating *Lathrea clandestina* may be found in manuscripts 974 and 976 of *Plantae et flores* by Federico Cesi (Institut de France, Paris), folios 235 and 276, respectively. See also Paola Lanzara, "Un problematico erbario figurato a Dumbarton Oaks," in Casale 1991, 34–47.

37. Stefano Bottari, *Raccolta di lettere sulla pittura, scultura e architettura scritte da' più celebri professori che in dette arti fiorirono dal sec. XV al XVII* (Rome, 1754), 1:238–242; and Casale 1991, 205–206.

38. Tiberio Tinelli (1586–1638) was a portrait artist whose name is mentioned in the papers of Marco Boschini and Carlo Ridolfi (who praised his "naturalism"). See Diane DeGrazia and Eric Garberson, eds., *Italian Paintings of the Seventeenth and Eighteenth Centuries* [systematic catalogue, National Gallery of Art, Washington] (New York, 1996), 317.

39. Elena Fumagalli, "Giovanna Garzoni a Parigi," in Meloni Trkulja and Fumagalli 2000, 11–12.

40. Michel Faré, *Le Grand Siècle de la nature morte en France: le XVIIe siècle* (Paris-Fribourg, 1974), 93.

41. Giovanna Garzoni, *Vaso contenente diversi fiori posto su un piano di marmo con ai lati due conghiglie,* Florence, Uffizi, Gabinetto Disegni e Stampe, 2152 Orn. This work is mentioned in Guardaroba Medicea 992, Archivio di Stato di Firenze, f 32v.

42. Silvia Meloni, "Giovanna Garzoni 'miniatora medicea,'" *FMR* (1983): 15, 77–96.

43. Lucia Tongiorgi Tomasi and Gretchen A. Hirschauer, *The Flowering of Florence: Botanical Art for the Medici* [exh. cat., National Gallery of Art] (Washington, 2002).

44. Tongiorgi Tomasi 1997, 161–169.

45. Catherine Perrot, *Les Leçons royales, ou la manière de peindre en mignature les fleurs et les oyseaux* (Paris, 1686) and *Traité de la mignature . . .* (Paris, 1693).

46. "Pour peindre les fleurs . . . Et pour bien réussir, vous ne pouvez prendre de meilleurs Exemples que les estampes de Nicolas Robert . . . le plus excellent fleuriste qui ait paru jusqu'au present."

47. There is an extensive bibliography on the artist. See in particular Kurt Wettengl, ed., *Maria Sibylla Merian, 1647–1717: Artist and Naturalist* [exh. cat., Historisches Museum, Frankfurt am Main] (Ostfildern, 1998); Maria Sibylla Merian, *La meravigliosa metamorfosi dei bruchi,* trans. Maria Gregorio and Daniela Majerna (Turin, 1993); Davis 1995, 140–202; Tongiorgi Tomasi 1997, 312–319 and 380–384; Schiebinger 1989, 66–79; and Lindkvist 2000, 184–212.

48. William T. Stearn, *The Wondrous Transformation of Caterpillars: Fifty Engravings Selected from "Erucarum Ortus" (1718)* (London, 1978), 14.

49. Maria Sibylla Merian, *Der rupsen begin, voedzel en wonderbaare verandering* (Amsterdam, 1679), 60.

50. Merian 1679, 75.

51. Viktoria Schmidt-Linsenhoff, "Metamorphoses of Perspective: Merian as a Subject of Feminist Discourse," in Frankfurt am Main 1998, 203.

52. Maria Sibylla Merian, *Dissertatio de generatione et metamorphosibus insectorum Surinamensium* (Amsterdam, 1719), 5.

53. Davis 1995, 151.

54. In Michelet's words, "noble vigueur," "simplicité forte," "gravité virile . . . la main de femme, consciencieuse, tendre, qui n'a touché à tout cela qu'avec un respect plein d'amour." See Jules Michelet, *L'Insecte* (Paris, 1858), 393–401.

55. Louis Figuier, *La Vie et les moeurs des animaux: les insectes* (Paris, 1867), consulted in the Italian edition, *Vita e costume degli animali, gli insetti* (Milan, 1893), 192. See also Lucia Tongiorgi Tomasi and Paolo Tongiorgi, "Tradizione e novità nella ricerca e nell'iconografia zoologica del tardo '600," in *Immagine e natura: l'immagine naturalistica nei codici e libri a stampa delle Biblioteche Estense e Universitaria, Secoli XV–XVII,* ed. Maria Gioia Tavoni, Lucia Tongiorgi Tomasi, and Paolo Tongiorgi [exh. cat., Biblioteca Estense] (Modena, 1984), 210–211.

56. Erik de Jong, "Flora Batava: Agnes Block op haar buiten Vijverhof," *Kunstlicht* 5 (1984–1985): 22–28; D. Onno Wijnands, *The Botany of the Commelins* (Rotterdam, 1983), 21–22; D. Onno Wijnands, "Hortus Auriaci: The Gardens of Orange and Their Place in Late Seventeenth-Century Botany and Horticulture," in *The Anglo-Dutch Garden in the Age of William and Mary,* ed. John Dixon Hunt and Erik de Jong, *Journal of Garden History,* special issue, 8 (1988): 2–3, 74.

57. See Jacobs 1997, 165–168.

58. On Sister Veronica, see Marilena Mosco and Silvia Meloni Trkulja, eds., *Natura viva in Casa Medici* [exh. cat., Palazzo Pitti] (Florence, 1985), 142; Marilena Mosco and Milena Rizzotto, eds., *Floralia: florilegio delle collezioni fiorentine del Sei–Settecento* [exh. cat., Palazzo Pitti] (Florence, 1988), 122; and Milan 1996, 120–125.

59. Silvia Meloni Trkulja and Lucia Tongiorgi Tomasi, eds., *Bartolomeo Bimbi: un pittore di piante e animali alla corte dei Medici* (Florence, 1998), 32–35.

60. Tongiorgi Tomasi 1997, 322–324; Heidrun Ludwig, *Nürnberger naturgeschichtliche Malerei im 17. und 18. Jahrundert* (Marburg, 1988), 111–138.

61. British Library Online Gallery, Landmarks in Printing, "Elizabeth Blackwell's 'A Curious Herbal,'" http://www.bl.uk/onlinegallery/themes/landmarks/blackwells.html (accessed 4 January 2006).

62. Tongiorgi Tomasi 1997, 320–322.

63. Gerta Calmann, *Ehret: Flower Painter Extraordinary* (Oxford, 1977); Tongiorgi Tomasi 1997, 185–192.

64. Sandra Raphael, ed., *An Oak Spring Garland: Illustrated Books, Prints, and Drawings from the Oak Spring Garden Library, Upperville, Virginia* [exh. cat., Milberg Gallery, Princeton] (Upperville, Va., 1989), 22; Tongiorgi Tomasi 1997, 207–211.

65. Frankfurt am Main 1998, 267.

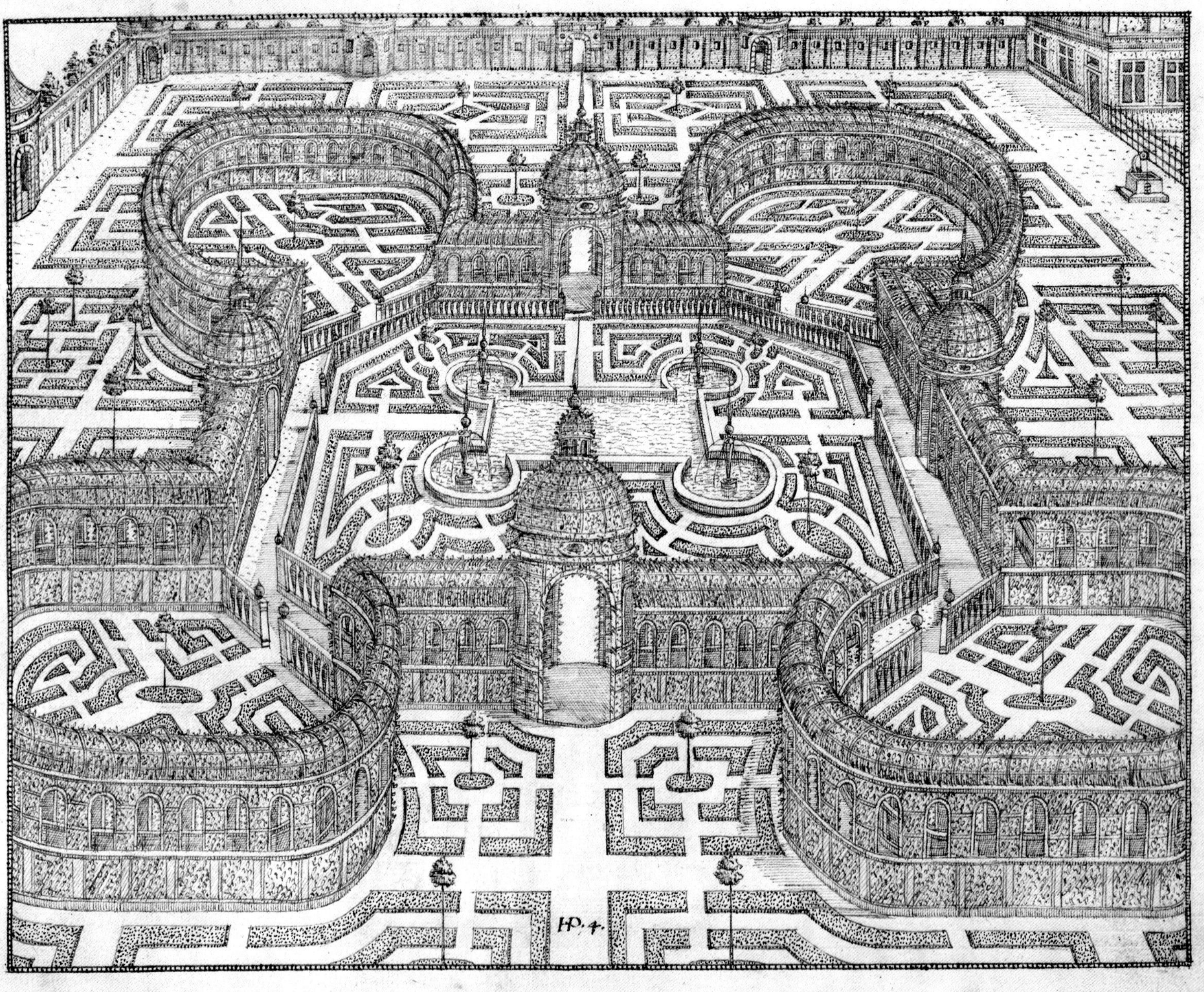
HP.4

ERIK A. DE JONG
Wageningen University and Research Center

A Garden Book Made for Emperor Rudolf II in 1593: Hans Puechfeldner's Nützliches Khünstbüech der Gartnereij

The collection of the Garden Library at Dumbarton Oaks in Washington, D.C., contains a unique document for the study of gardens in the second half of the sixteenth century. It is an album dated 1593, containing three pages of text and fifty-six designs for architecture and gardens. In the catalogue of the Dumbarton Oaks Garden Library it is attributed with a question mark to the Dutch architect Jan or Johannes or Hans Vredeman de Vries (1526–c. 1606) or to a member of his family. In this essay, I will suggest a new attribution for the Dumbarton Oaks album and demonstrate its centrality to understanding the role of gardens and gardening at the court of Rudolf II in Prague and, indeed, art and culture in the Northern Renaissance.

The album's first two folios contain two texts in German script on royal herb gardens and royal orchards. These texts function as the opening to a dedication page and forty-four drawings of perspectives of gardens and architecture. These drawings elaborate on distinct themes, including centralized buildings and berceaux (fig. 1), different kinds of parterres (fig. 2), plantation designs (fig. 3), and gardens with fountains (fig. 4). A third text, on royal orchards, introduces twelve drawings of perspectives and plans for plantations with fountains. Almost no other existing documentation so clearly illustrates an early architectural concept of the garden for the Northern Renaissance.

1. Folio 6 from Hans Puechfeldner, *Nützliches Khünstbüech der Gartnereij*, 1593, pen and bister on white laid paper
Dumbarton Oaks Research Library and Collection, Washington, D.C.

One of the extra attractions of the album is that the garden designs are supposedly by Vredeman de Vries, a well-known Dutch architect, painter, and designer of architecture and fortifications, who lived in Holland in the sixteenth century and who studied in Antwerp and Flanders. He is known mostly for his large oeuvre of engraved designs for architecture, gardens, and ornaments, besides being a designer in situ. His engravings for the *Hortorum viridariorumque elegantes & multiplicis formae, ad architectonicae artis normam affabrè*, published in Antwerp in 1583, are widely known among garden historians and are used, and often misused, as some of the few illustrations of late sixteenth-century Northern Renaissance gardens (fig. 5).[1]

A first look into the album convinced me that, indeed, there must exist an interdependency between the drawings it contains and designs by Vredeman de Vries, but also that the album could not be his work but must be that of a pupil of some kind. This observation makes the Dumbarton Oaks album all the more intriguing, since it could perhaps illuminate the spread and influence of Vredeman de Vries' Flemish-Dutch style in northern Europe and shed light on the role and function of garden manuscripts at the end of the sixteenth century. It seems, therefore, important to look closely at the album as a physical object: its binding, its paper, and its composition.

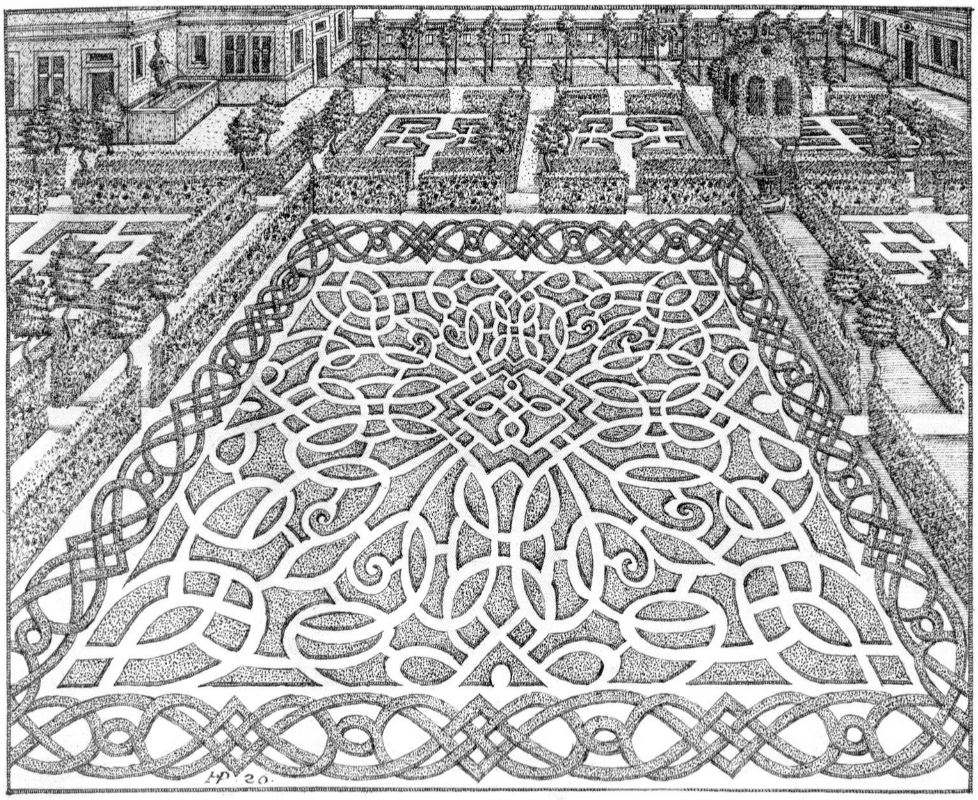

2. Folio 28 from Hans Puechfeldner, *Nützliches Khünstbüech der Gartnereij,* 1593, pen and bister on white laid paper
Dumbarton Oaks Research Library and Collection, Washington, D.C.

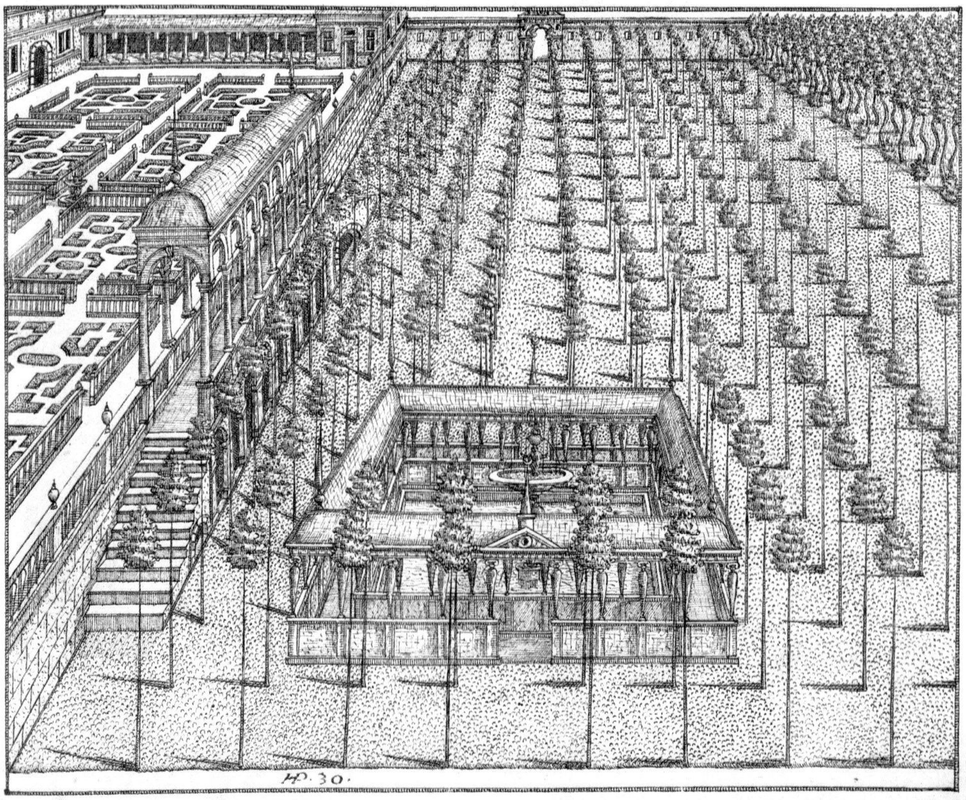

3. Folio 32 from Hans Puechfeldner, *Nützliches Khünstbüech der Gartnereij,* 1593, pen and bister on white laid paper
Dumbarton Oaks Research Library and Collection, Washington, D.C.

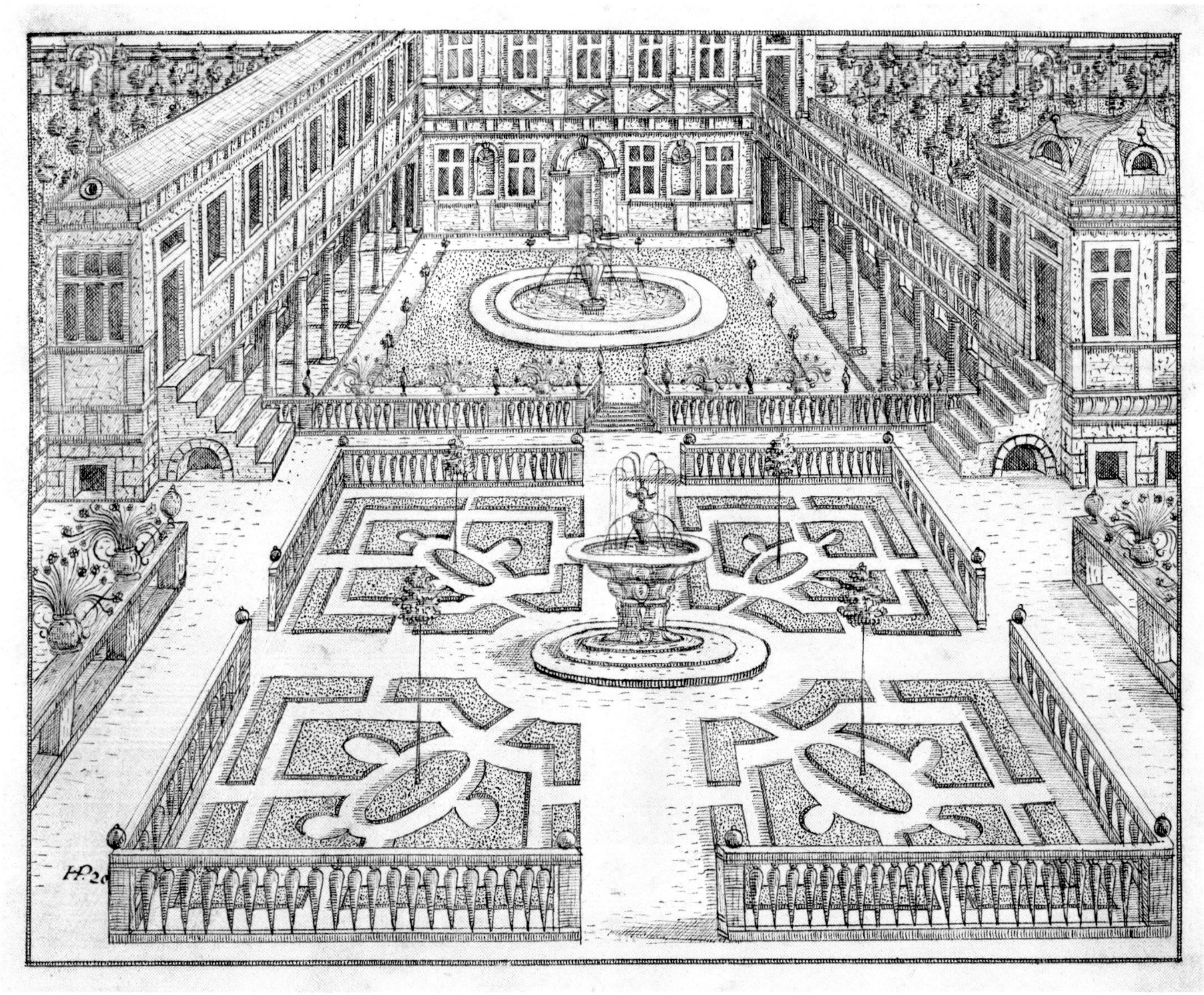

4. Folio 22 from Hans Puechfeldner, *Nützliches Khünstbüech der Gartnereij*, 1593, pen and bister on white laid paper
Dumbarton Oaks Research Library and Collection, Washington, D.C.

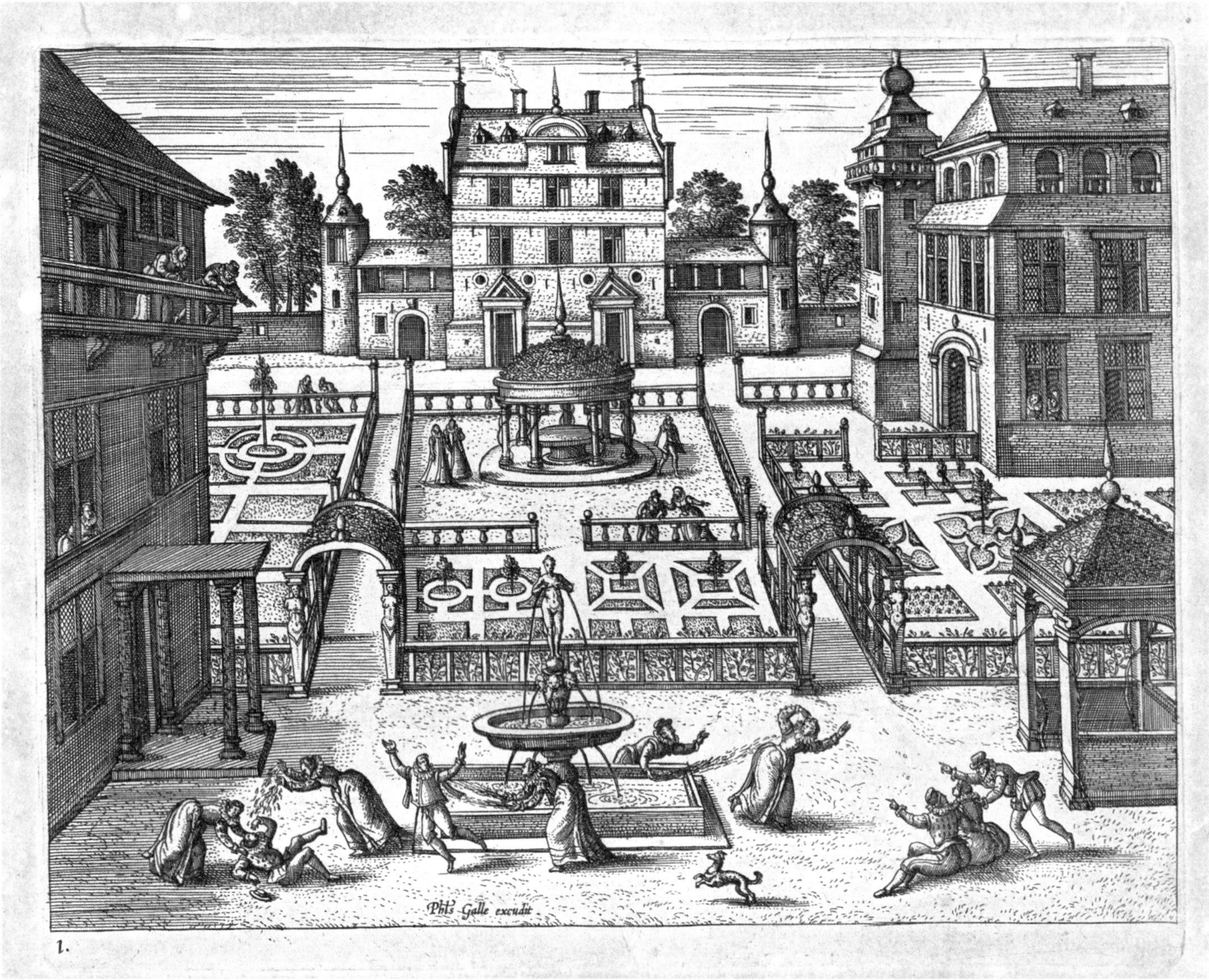

5. Garden scene from Hans Vredeman de Vries, *Hortorum viridariorumque elegantes & multiplicis formae, ad architectonicae artis normam affabrè*, 1583, engraving
Dumbarton Oaks Research Library and Collection, Washington, D.C.

The binding is made of full, straight-grain, red morocco leather over pasteboards with gold tooling on the boards and spine (fig. 6).[2] A gold and arabesque vignette in the center and two borders ornament the cover on both front and back. Within the second border on the front, "1593" is printed in gold. On the upper right we find in ink the number 64, probably referring to an original inventory number. The inside of the cover bears remnants of an inscription in ink of the word *bibliotheca*. Any identification of whose *bibliotheca* it was has, unfortunately, been erased. We also find two different bookplates of the Blisses, the owners of Dumbarton Oaks (fig. 7). The support consists of fifty-nine folios of fine white laid paper. The watermarks on all pages show a scrolled shield within a cross on a three-lobed mountain (fig. 8), identifying the paper as a type made in Prague between 1577 and 1594, a period corresponding closely to the date on the cover. This date range for the compilation of the album is further confirmed by the inscription of 1592 on architectural elements depicted in four folios (fig. 9), which leads me to conclude that the project was begun in that year and finished in 1593. All drawings are made with pen and bister and bear marks of blind ruling used to construct a perspective. The handwriting is a German script. From the album we may glean some further knowledge of the artist. From the initials *HP* throughout (fig. 10) and a

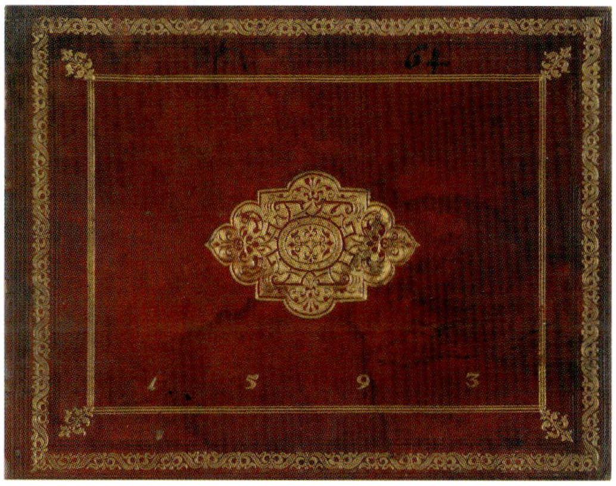

6. Hans Puechfeldner, *Nützliches Khünstbüech der Gartnereij*, 1593, binding, gold-tooled red morocco leather
Dumbarton Oaks Research Library and Collection, Washington, D.C.

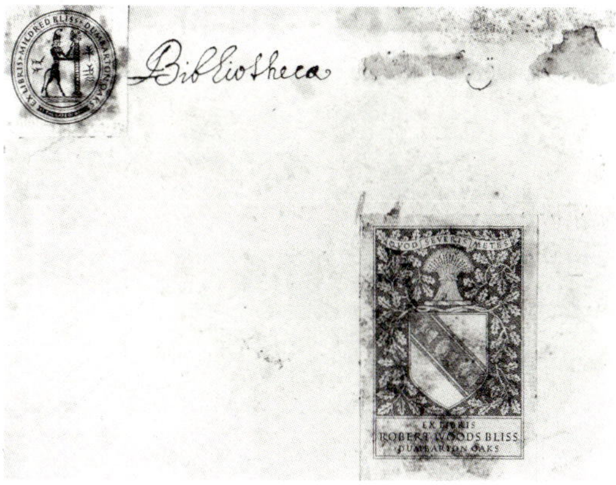

7. Bookplates of the Bliss family, from Hans Puechfeldner, *Nützliches Khünstbüech der Gartnereij*, 1593
Dumbarton Oaks Research Library and Collection, Washington, D.C.

corresponding "Hans Puech" on folio 34, we might conclude that he was the maker of these drawings. It remains to be seen in what way he was related to Hans Vredeman de Vries or if his designs were done after Vredeman de Vries' work.

Apart from the paper, the monograms, and the name Puech, the key to the provenance and the identity of the artist resides on the dedication page (fig. 11). Set in a berceau in place of an ornamental parterre is the inscription *RVDOLLFVS DER ANNDER*. This is, of course, the Holy Roman emperor and renowned collector Rudolf II, who ascended to the throne in 1576 and died in 1612. At his court in Prague in Bohemia, Rudolf's activities as a patron of the arts and sciences provide a context in which to view this album and begin to understand its origins. The many archival documents related to Rudolf II are valuable resources for the study of the album at Dumbarton Oaks. The court accounts of 14 February 1597 refer to "drei garten buchs" (three garden books) made by the gardener Hans Puechfeldner in 1593 for the emperor. Puechfeldner is identified in the document as "his majesty's gardener here in Prague, [who] is to be paid one hundred gilders for these and for seven models or designs of a more recent date."

The "Hans Puech" of the Dumbarton Oaks album may now be identified as Rudolf II's gardener Hans Puechfeldner. Can the album also be identified as one of the three books made by him and mentioned in the court accounts? The provenance of the Dumbarton Oaks album is enhanced by two comparable works, dated 1591 (fig. 12) and 1594 (fig. 13), by the same artist, now in the Österreichische Nationalbibliothek in Vienna.[3] Although the Viennese albums contain no text and have less elaborate bindings, the paper, tooling, and drawings in both are identical with those of the Dumbarton Oaks copy. Both are signed in full by Hans Puechfeldner, gardener to his Roman Imperial Majesty, Rudolf II, and bear the titles *Ein Nützliches Khünstbüech der Gartnereij* and *Ein Nützliches Künstbüech der Lüstigen Gardtnereij*. These titles are a valuable hint because the two principal concepts, art and nature (represented by the garden), were of vital importance at the Prague court. Since the Viennese albums are respectively

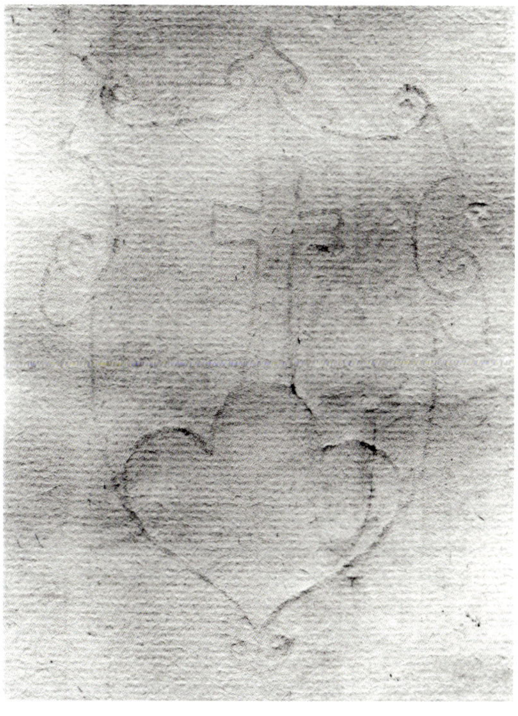

8. Watermark from Hans Puechfeldner, *Nützliches Khünstbüech der Gartnereij*, 1593, white laid paper
Dumbarton Oaks Research Library and Collection, Washington, D.C.

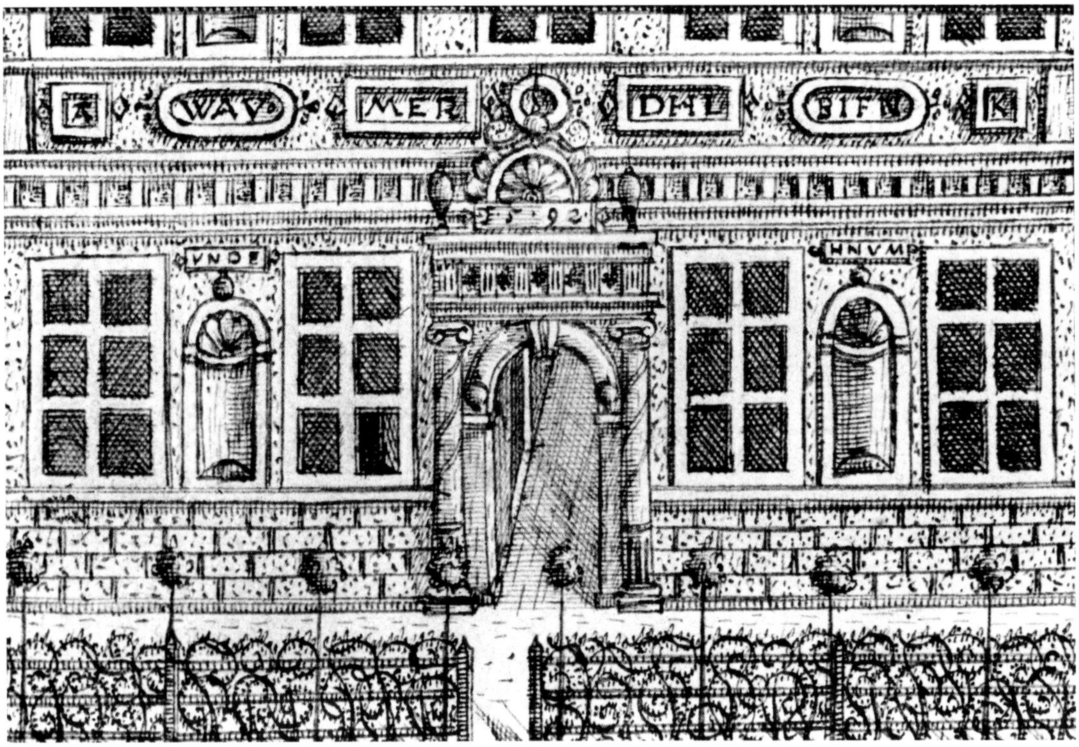

9. Detail of date from Hans Puechfeldner, *Nützliches Khünstbüech der Gartnereij*, 1593
Dumbarton Oaks Research Library and Collection, Washington, D.C.

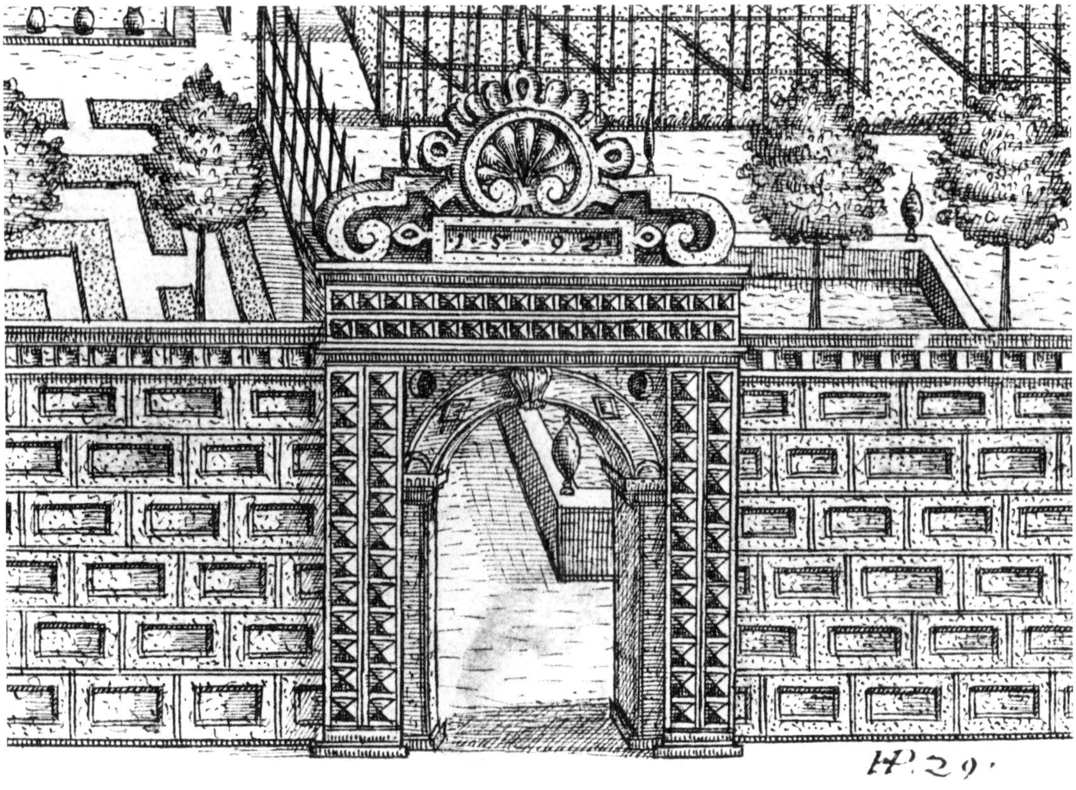

10. Detail of artist's signature, folio 31, from Hans Puechfeldner, *Nützliches Khünstbüech der Gartnereij*
Dumbarton Oaks Research Library and Collection, Washington, D.C.

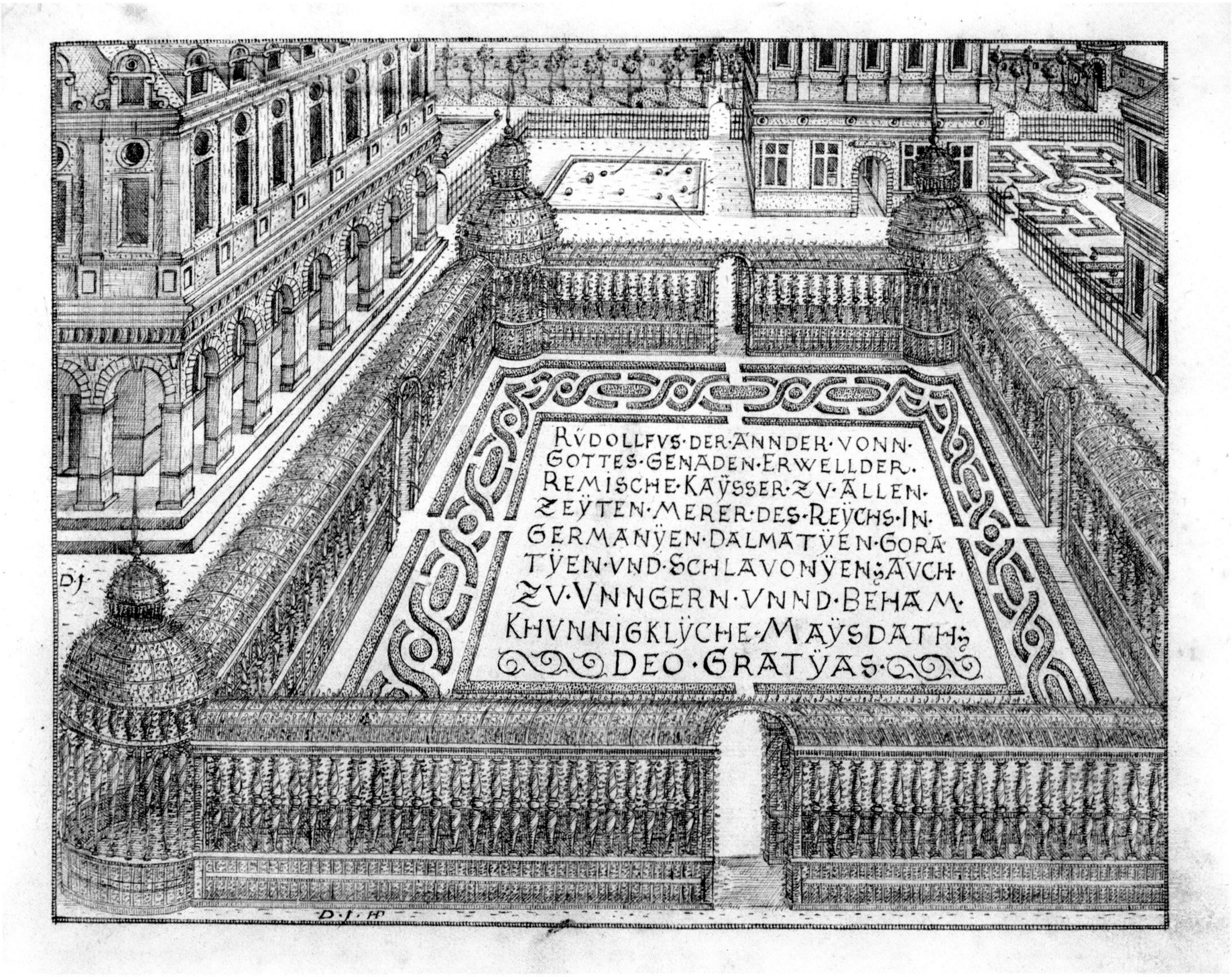

11. Dedication page from Hans Puechfeldner, *Nützliches Khünstbüech der Gartnereij*, 1593, pen and bister on white laid paper
Dumbarton Oaks Research Library and Collection, Washington, D.C.

dated 1591 and 1594 on their covers, the Dumbarton Oaks album functions as the second in the series. Its missing title page would have borne a like title, so we may consider this album also as a *Nützliches Khünstbüech der Gartnereij*. Thus we may establish that the three books were designed as presentation copies for Rudolf II, 1591, 1592–1593, and 1594. I suggest that the title *Khünstbüech* (art book) may mean, among other things, that they were to be stored in Rudolf's famous *Kunstkammer*, where so many other objects and manuscripts pertaining to the world of nature and art were collected.[4]

To say something about the content of the texts of the album at Dumbarton Oaks, they were written not in Czech but in German, the preferred language of Rudolf II. First, they identify gardens as fit for kings and other great lords who used the garden to renew their spirits after conducting business of state, a topos that dates to classical antiquity. Puechfeldner advises on the importance of careful siting of the garden where an eastern morning breeze has access and a freshwater spring runs in different places. This advice comes from classical writers such as Pliny and Varro on siting gardens and villas.[5] Typical is his advice to wall the garden and to build to the north a palace where one may escape melancholy and seek joyful renewal of the spirit. Apparently, architecture and gardens function here as an antidote for melancholy. Important are

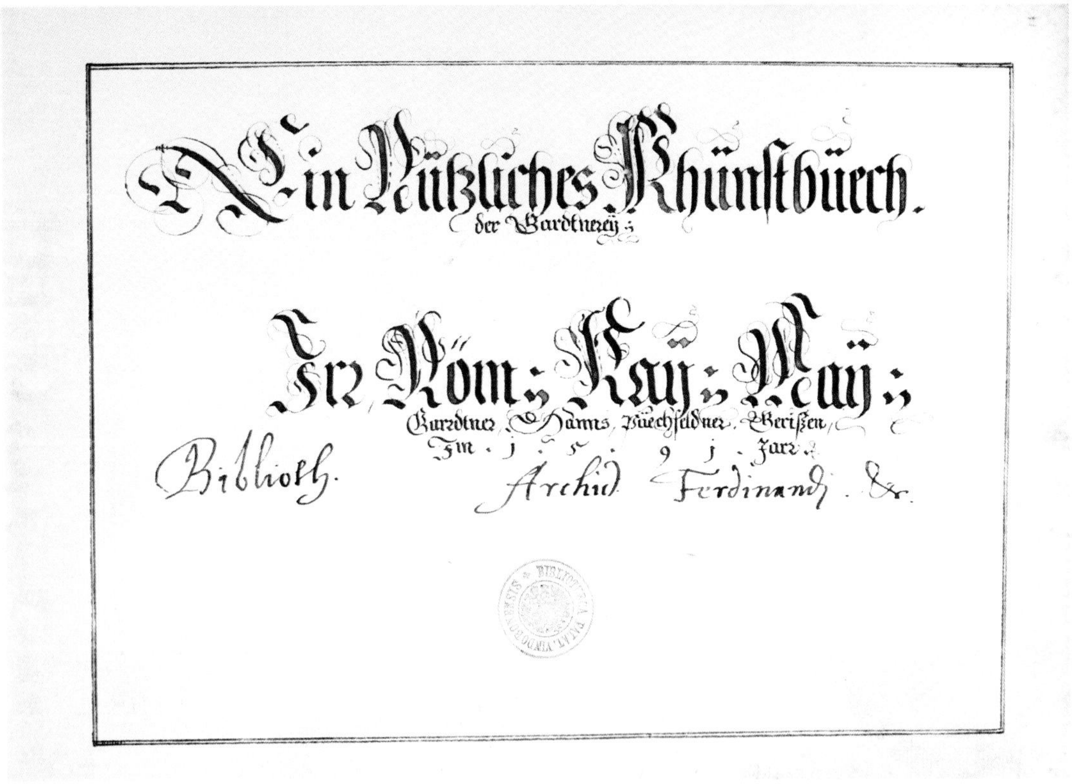

12. Title page from Hans Puechfeldner, *Ein Nützliches Khünstbüech der Gartnereij,* 1591
Österreichische Nationalbibliothek, Vienna

the gardener's repeated remarks on order in the garden. Through artful construction, carried out with method and skill, variety diverts the wandering emperor without exposing him to the intrusion of others. Furthermore, Puechfeldner writes that the garden and plantations should be full of rare trees, flowers and herbs, animals such as deer, and caged birds such as nightingales, goldfinches, linnets, and turtledoves. The garden is a collection of the natural world that brings delight and consolation. As the Dumbarton Oaks album says, "For we are not born for fun or play but for earnestness and strain to act justly and seriously. Play and fun may be enjoyed together when enough serious and necessary business has been conducted."

Closer inspection reveals that these texts are inspired by passages in the German edition of Pietro de Crescenzi's famous and popular *Opus ruralium commodorum,* originally dating from 1304–1309 with ten German editions between 1490 and 1583.[6] Thus, the Dumbarton Oaks album contains a more or less traditional text accompanying modern designs for gardens, which themselves are an architectural transformation of the medieval *hortus conclusus* and other royal imperial garden motifs from the late medieval period. For his presentation of gardens, Puechfeldner used—not without difficulty, it must be said—the laws of perspective to create a geometrical and architectonic garden space, imposing order and control, physically and optically, on nature. In this he must have been heavily influenced by Vredeman de Vries. However, the latter did not arrive in Prague until 1596, the year he designed fountains for the emperor and painted large garden perspectives for him. This is three years later than the date of the Dumbarton Oaks album, yet Puechfeldner composed his first work five years earlier, a testimony that Vredeman de Vries already had a solid reputation in Habsburg culture.

We know from inventories that Vredeman de Vries' works were avidly collected by Habsburg princes in Innsbrück, Graz, Vienna, and Prague. Many designs in Puechfeldner's first work from 1591 are, in fact, copies of Vredeman de Vries' designs. However, the drawings by Puechfeldner are both much more detailed and much livelier than

13. Title page from Hans Puechfeldner, *Ein Nützliches Künstbüech der Lüstigen Gardtnereij,* 1594
Österreichische Nationalbibliothek, Vienna

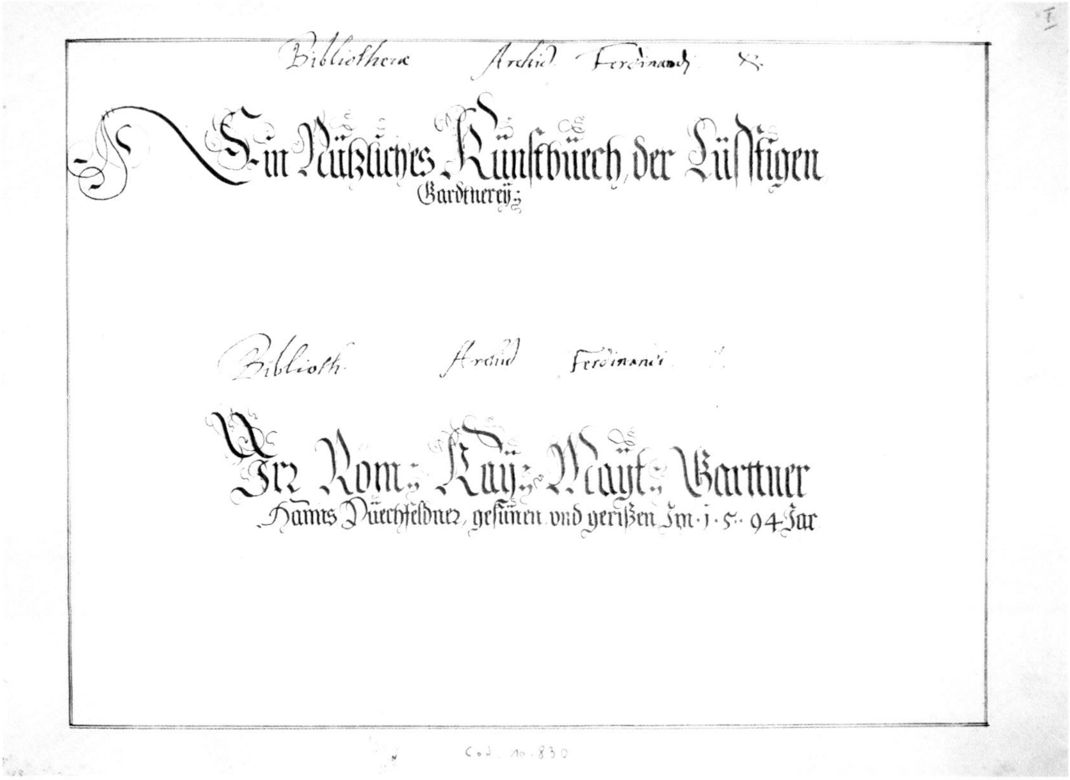

Vredeman de Vries' rather dry engravings. But from these designs, we may conclude that Puechfeldner learned his craft by way of *translatio* and *imitatio,* since all other designs in the second and third volumes are his own.

The designs in the two later books were imbued with Puechfeldner's own *inventio,* inspired also by local tastes, seen in the many pavilions that figure in his own drawings (fig. 14), which are a reflection of the Renaissance Anna Pavilion, already in existence in the mid-sixteenth century in the Prague Castle gardens. These designs also reflect the influence of other decorative prints published in Antwerp, for instance, in knotted and interlaced moresque patterns in the parterres. But this constitutes a problem that I shall not address here.

Hans Vredeman de Vries was a talented man, born in the secluded north of Holland. He learned new Renaissance theories of architecture from books during his stay in Antwerp, and I must emphasize that it was through the highly developed state of Flemish gardens and garden design as well as of Flemish botany and horticulture, in which late medieval traditions were fused with Renaissance ideas, that he was able to fuse Flemish garden practice with new Renaissance theories. This can be seen at the gardens of the royal palace in Brussels (fig. 15) as well as in the gardens at the Casa del Campo in Madrid, made by Flemish gardeners for Philip II.[7] Since these Flemish gardeners spread all over the Habsburg empire and worked in Spain at exactly the time when the young Rudolf II was a member of the household of Phillip II in Madrid, his interest in gardens may indeed have been aroused in that period.

For his garden designs, Vredeman de Vries used a method introduced in the Renaissance by architects such as Sebastiano Serlio to represent space by means of perspective construction. It is above all this lesson that Puechfeldner, as a practical gardener, must have learned from Vredeman de Vries' published engravings. This is demonstrated in the Dumbarton Oaks album by the blind ruling on his drawings, used to make the perspective as correct as he could and his construction of garden space convincing. The order Puechfeldner stresses so

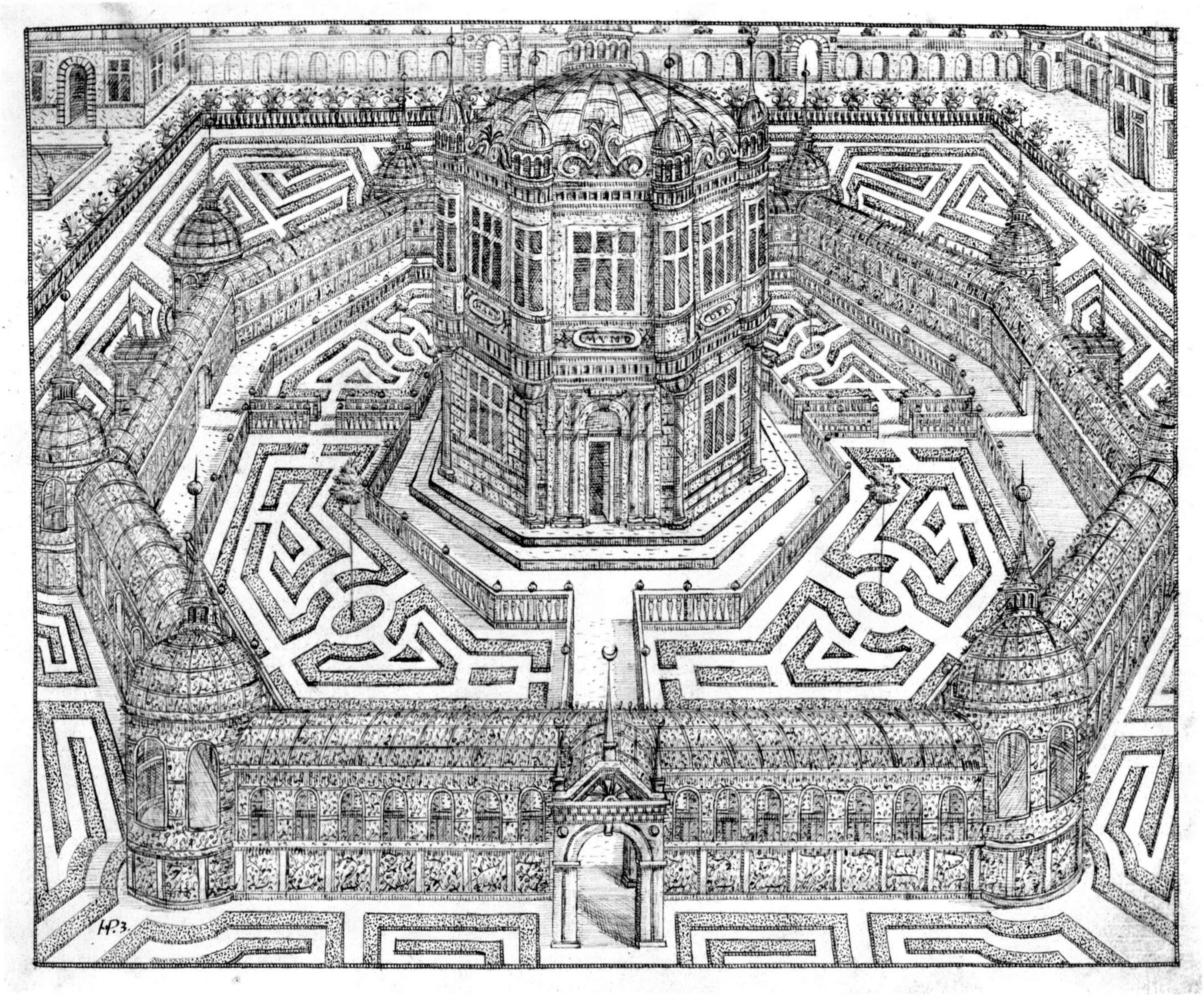

14. Folio 5 from Hans Puechfeldner, *Nützliches Khünstbüech der Gartnereij*, 1593, pen and bister on white laid paper
Dumbarton Oaks Research Library and Collection, Washington, D.C.

clearly in his text held a special place in the architectural treatises used in the sixteenth century because order was related to dignity, beauty, and harmony—meaning symmetry and balance in a composition.

Taking his lessons from the study of nature to heart, the architect used the circle, the square, and the octagon as perfect forms, exact in number, arrangement, and form. Beauty was seen as a form of sympathy and consonance of the parts within a body according to definite number, outline, and position. Within this system, the column was considered the principal element within a vocabulary of architectural ornament that expressed meanings of order and worthiness appropriate for every occasion and building type. Vredeman de Vries applied Serlio's expressive system of the three orders, Doric, Ionic, and Corinthian, to garden architecture in his *Hortorum viridariorumque* to distinguish garden designs by a gradated complexity of ornament. Gardens were called Doric (fig. 16), Ionic (fig. 17), or Corinthian (fig. 18), following the division of the architectural orders. Parterres were thus given the imprint of ornament in a classical order, ensuring the quality of design, expression, and character. It is here that the garden indeed becomes architecture.

Following the example of the architectural orders, fountains, berceaux, and parterres

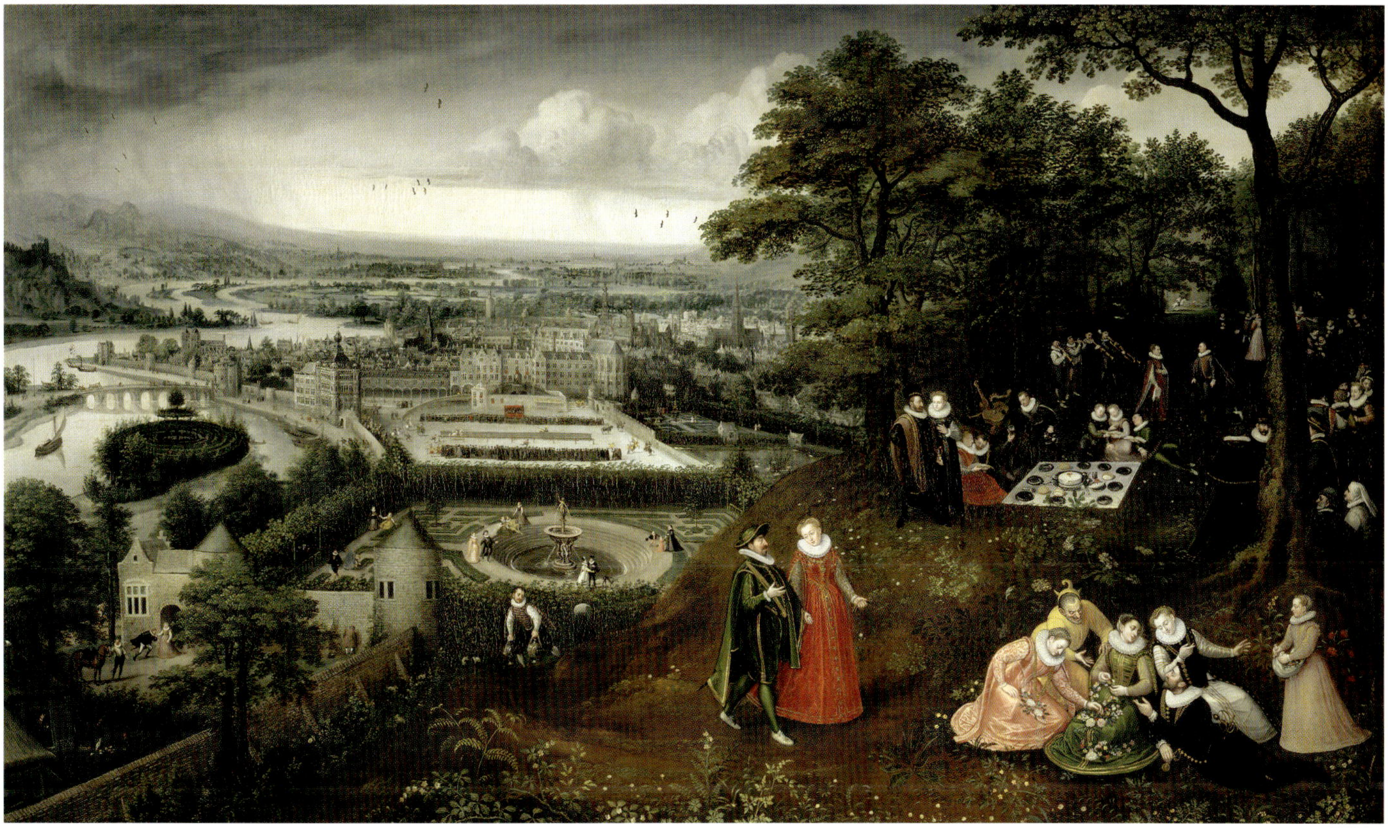

15. Lucas I van Valckenborch, *Allegory of May with the Gardens of the Royal Palace in Brussels*, 1587, oil
Kunsthistorisches Museum, Vienna

helped to infuse both Vredeman de Vries' and Puechfeldner's designs with the necessary classical dignity, character, and beauty. It is here that Puechfeldner was a diligent student of Vredeman de Vries. His designs abound in essays on the circle, the octagon, and the square, and many may be classified, though not systematically, according to Vredeman de Vries' principles of Doric, Ionic, and Corinthian. If we add to these drawings the texts referring to authorities such as Pliny, Varro, and de Crescenzi, Puechfeldner's album becomes imbued with classical authority, form, and meaning, making it fit for the ruler of the Holy Roman Empire. The variety of designs that Puechfeldner offers may strike us as superfluous, yet as a follower of Vredeman de Vries he understood, perhaps very well, the concept of invention (*inventie* in Dutch), that is, design as a product of a fertile mind and of an intellectual ability to master mathematics and perspective as a new science.

Puechfeldner's album is perhaps also a proud proof of the talent and emancipation of a new class of German gardeners trying to develop themselves from craftsmen and practitioners into designers with a knowledge of theory. Puechfeldner's liberation could perhaps be connected with the great development of surveying in Bohemia at this time and the manufacture of precise instruments to measure space and time, also carried out by German and Bohemian craftsmen. Still in the Prague city library is a famous manuscript by Jost Bürgi, on the mechanical instruments for triangulation. Bürgi's manuscript also includes gardens, suggesting that the Puechfeldner album and this manuscript are perhaps part of a much larger development in measuring time and space.[8]

As a patron of the garden design album, Rudolf II is perhaps best evoked by the portrait Giuseppe Arcimboldo painted circa 1590, which, according to the Italian artist and writer Giovanni Paolo Lomazzo (1538–1600), was eagerly awaited by the emperor himself.[9] Rudolf II is here depicted as the gardener Vertumnus, the god of seasons and change, strongly associated with Jupiter (fig. 19). The portrait is a symbol of imperial universalism and speaks of the

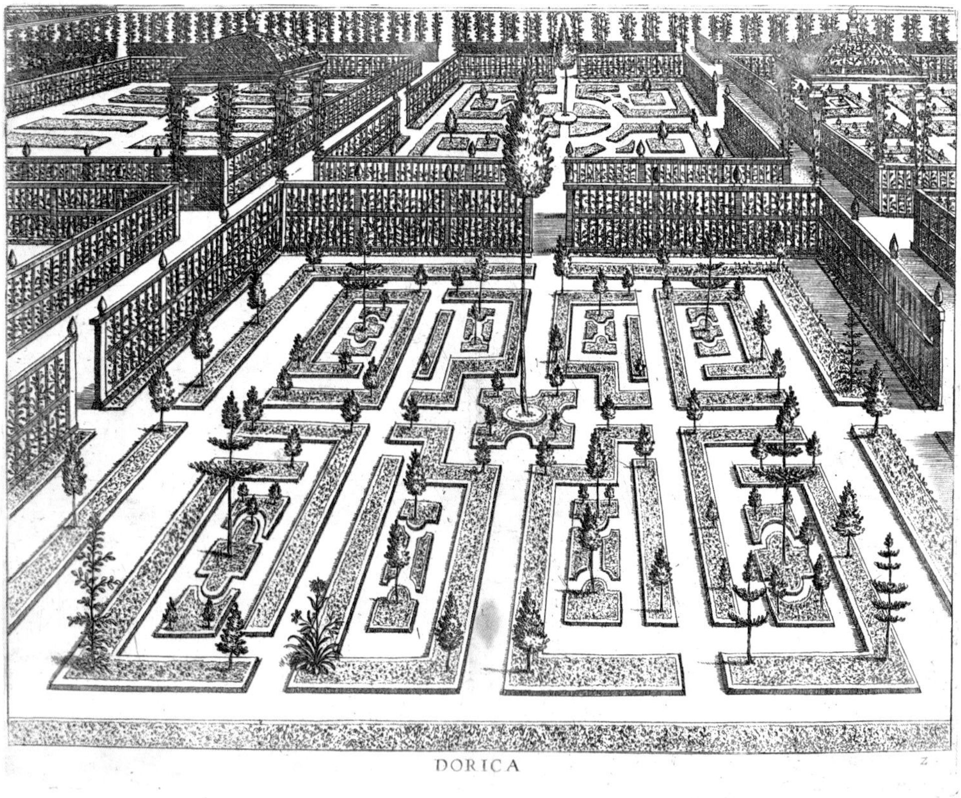

16. *Garden Design in the Doric Order*, from Hans Vredeman de Vries, *Hortorum viridariorumque elegantes & multiplicis formae, ad architectonicae artis normam*, 1583, engraving
Dumbarton Oaks Research Library and Collection, Washington, D.C.

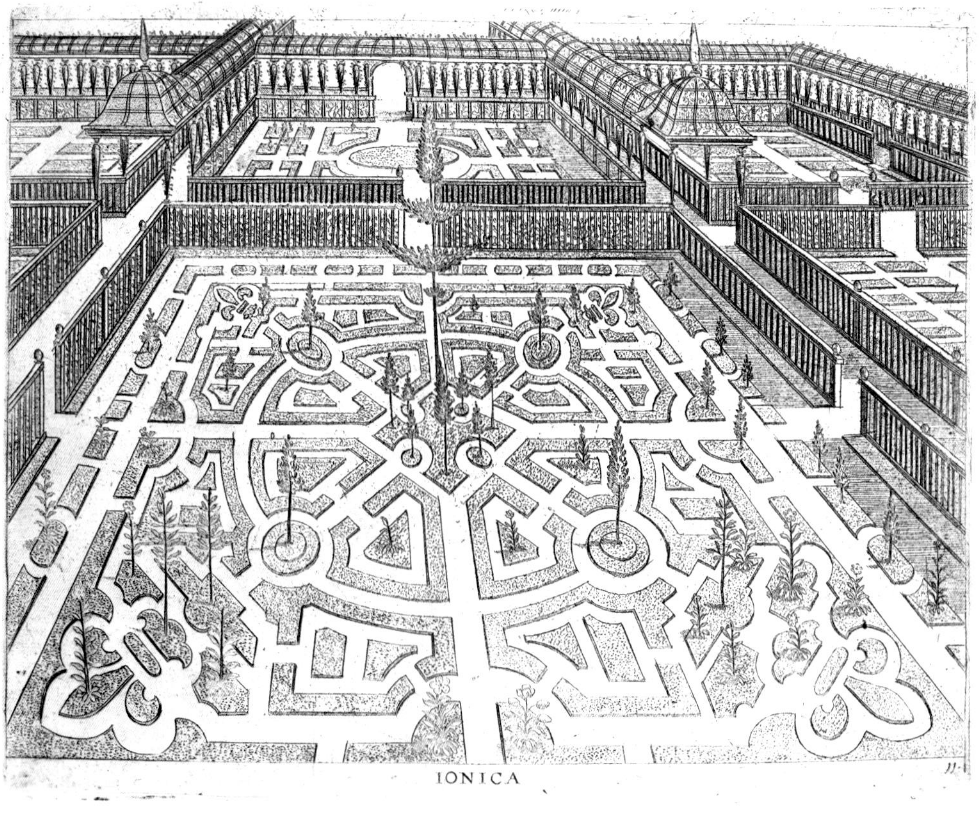

17. *Garden Design in the Ionic Order*, from Hans Vredeman de Vries, *Hortorum viridariorumque elegantes & multiplicis formae, ad architectonicae artis normam*, 1583, engraving
Dumbarton Oaks Research Library and Collection, Washington, D.C.

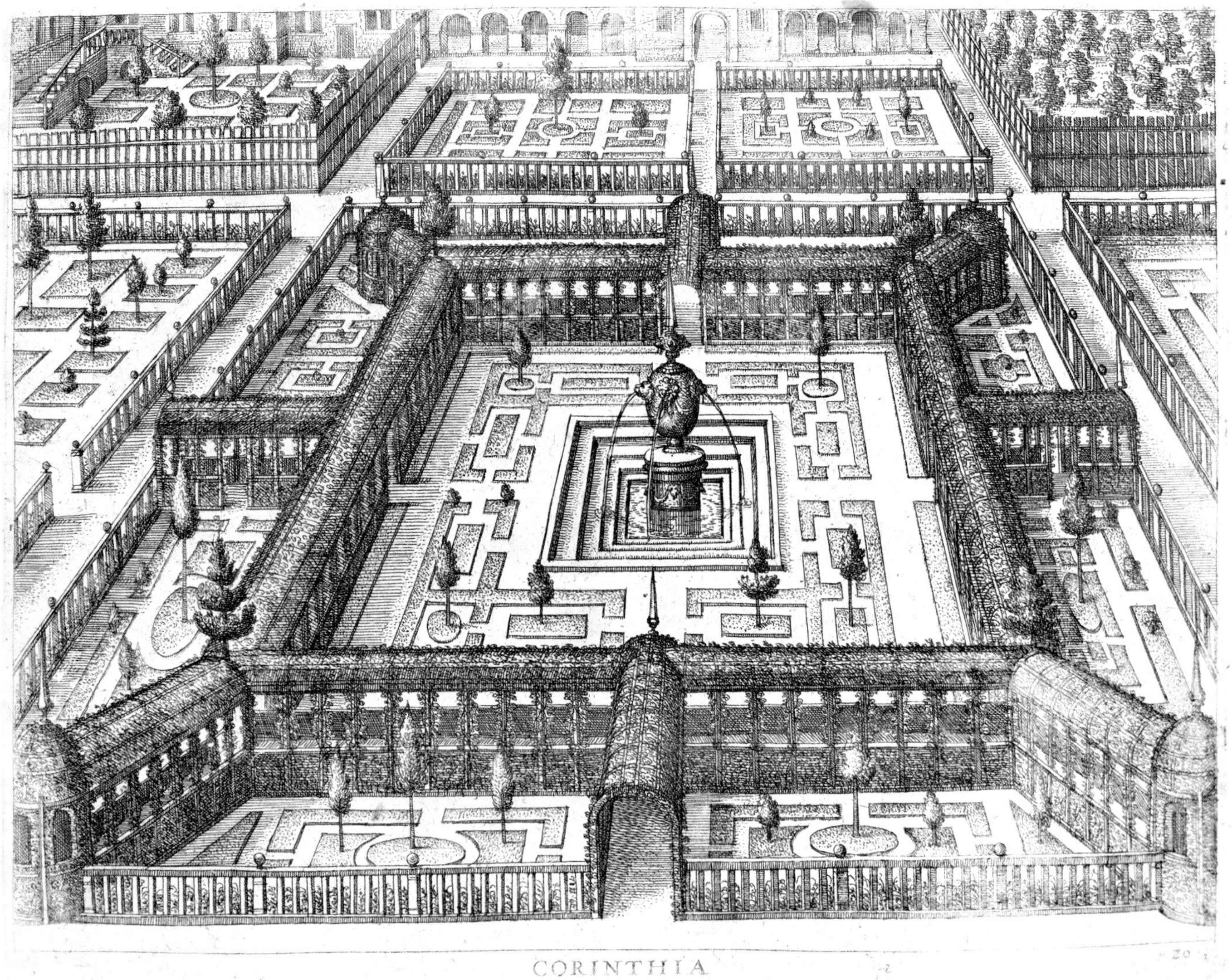

18. *Garden Design in the Corinthian Order,* from Hans Vredeman de Vries, *Hortorum viridariorumque elegantes & multiplicis formae, ad architectonicae artis normam,* 1583, engraving
Dumbarton Oaks Research Library and Collection, Washington, D.C.

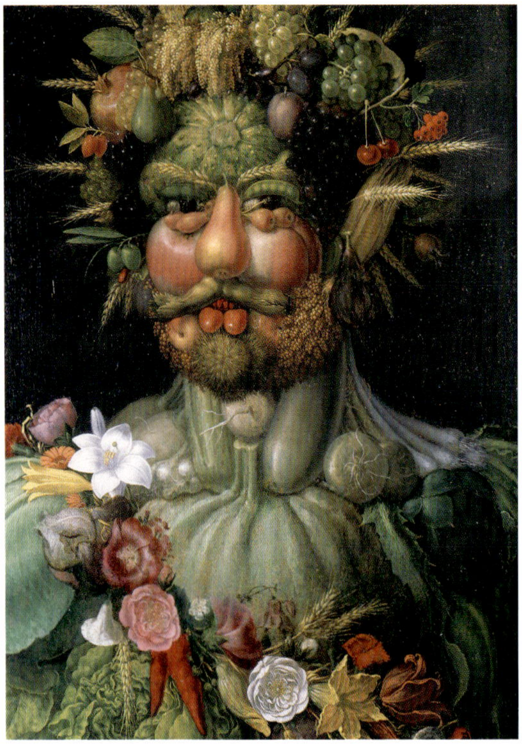

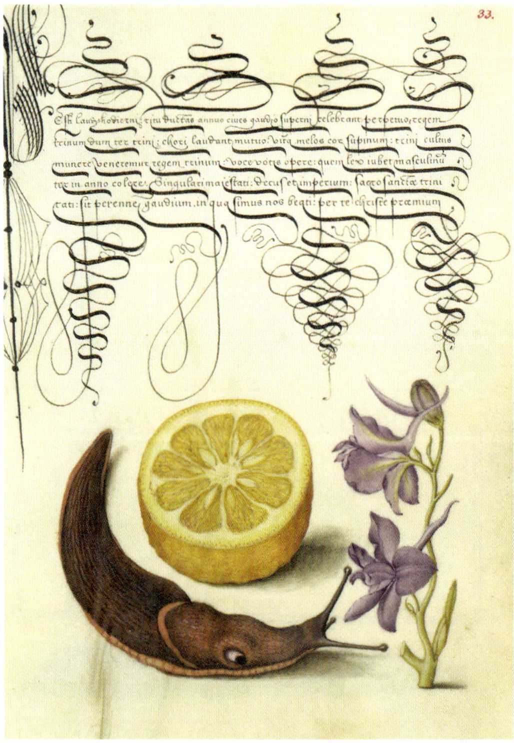

19. Giuseppe Arcimboldo, *Vertumnus*, c. 1590, oil on wood
Skoklosters slott, Bålsta, Sweden; photograph Samuel Uhrdin

20. Joris Hoefnagel, *Sour Orange, Terrestrial Mollusk, and Larkspur*, from *Mira calligraphiae monumenta*, c. 1591–1596, tempera and gold paint on parchment
The J. Paul Getty Museum, Los Angeles

dynastic mystique of the house of Habsburg. In this image, composed of different fruits, flowers, and vegetables that the seasons offer from gardens and orchards, Rudolf II is the gardener par excellence, the master of the universe, bringing order and harmony to the conjunction of the seasons and, by implication, also of the elements and thus of all things visible and invisible. The allegory served one purpose—to introduce the philosophical concept of Rudolf's *Kunstkammer*, since here the variety of manifestations of art and nature were brought together in a museum that served as a key to the understanding and study of the harmony of the creative universe.[10]

Rudolf's collecting was the result of what have been called his pansophic strivings.[11] It was an encyclopedia of the visible world, systematically ordered and ranging from shells, minerals, stuffed animals, eggs, dried fruits and flowers, and seeds to artifacts from the New World, rare and costly objects made from natural materials, scientific instruments, drawings, prints, and books—playing and commenting on the intricate relationships between art and nature, as in two codices by Rudolf's court artist Joris Hoefnagel (1542–1601) (now respectively at the National Gallery of Art and the J. Paul Getty Museum), *Animalia rationalia et insecta* and *Mira calligraphiae monumenta* (fig. 20). At Rudolf's court, nature was seen as a codex whose leaves must be turned and observed to read its meaning. It was described as such in the works of Paracelsus (1493–1541), one of the writers admired by Rudolf II.[12] We know that the emperor used his collection also for private contemplation. We also know from several contemporary statements that Rudolf, disturbed by ailments of melancholy, began to love solitude from about 1586 onward, shutting himself off in his cellars among his collections, neglecting the burdens of state and administration.[13] The garden, so it seems, as an outdoor *Kunstkammer*, served that same function.

It is strange that, in this context, the gardens Rudolf created have not yet played a role of note in modern literature. This can only be explained by the misapprehension that, as one author has explained, scant visual documentation of his gardens exists.[14] But from the archives and many travel descriptions, we now learn of gardening and landscaping activities undertaken around Prague Castle. We read of new plantations;

21. Title page from Carolus Clusius, *Rariorum plantarum historia* (Antwerp, 1601), engraving
Amsterdam University Library (UvA), Special Collections

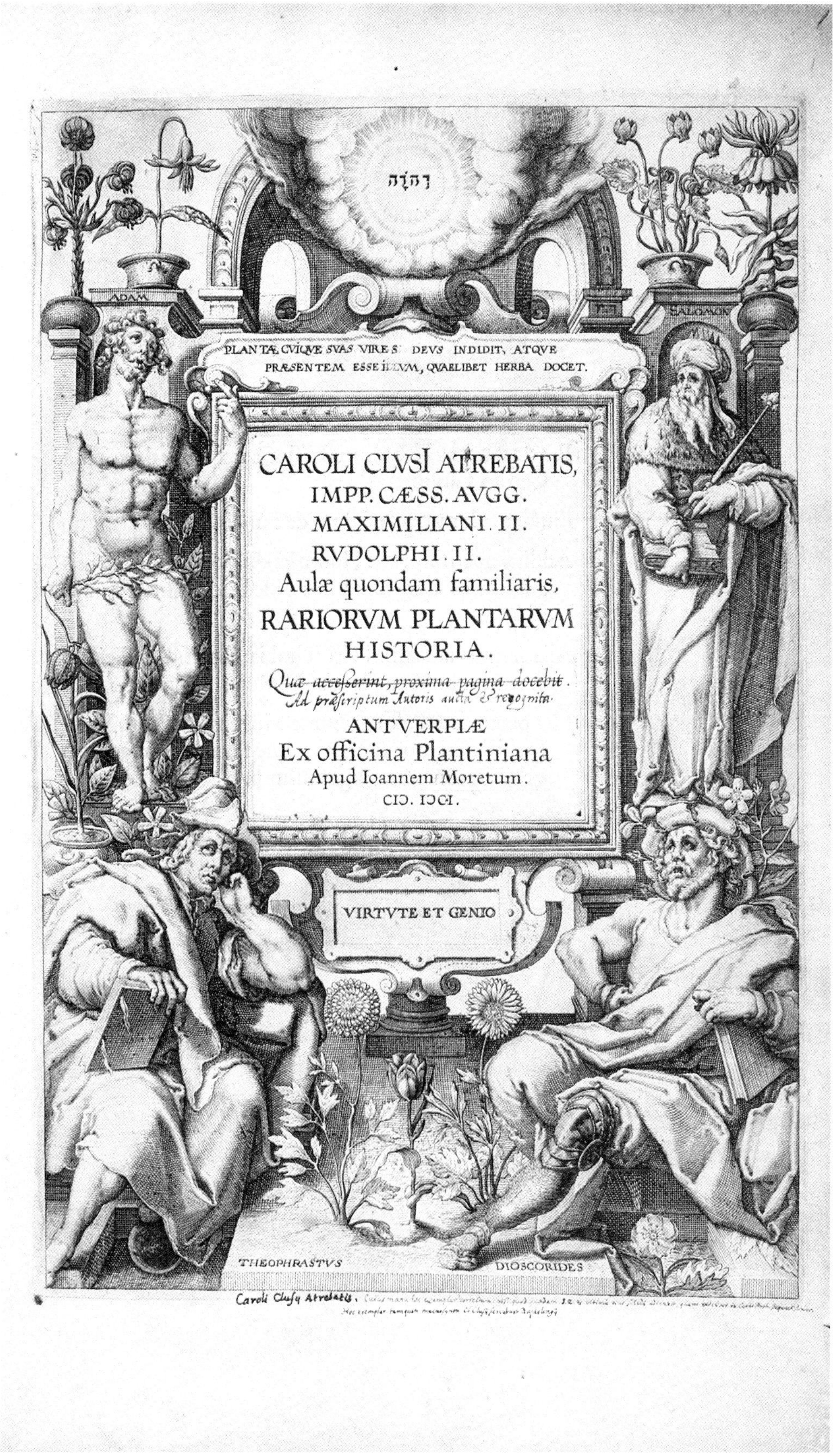

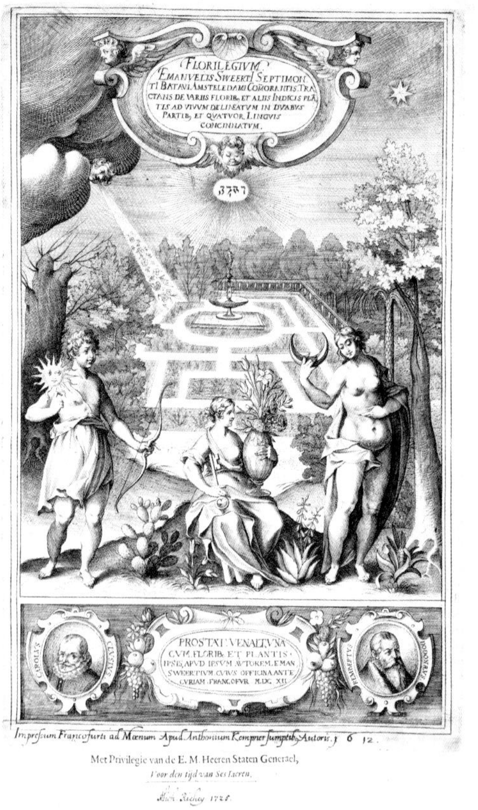

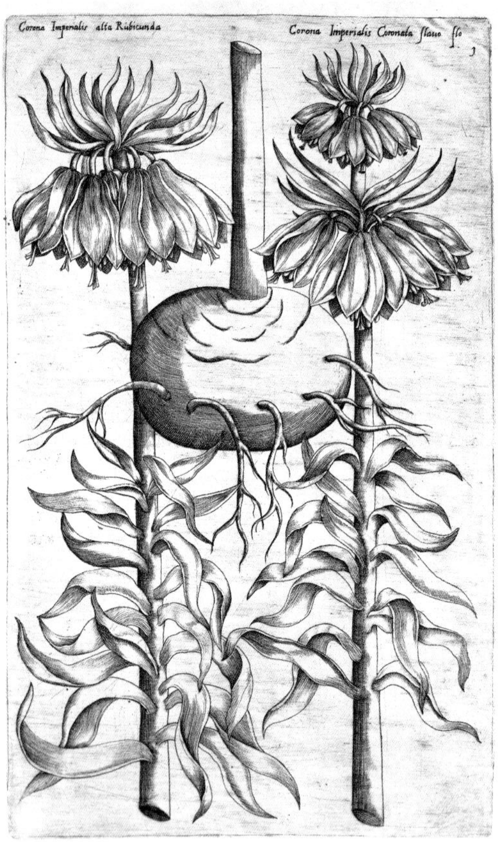

22. Dedication page from Emanuel Sweerts, *Florilegium*, 1612, engraving
Dumbarton Oaks Research Library and Collection, Washington, D.C.

23. *Corona imperialis*, from Emanuel Sweerts, *Florilegium*, 1612, engraving
Dumbarton Oaks Research Library and Collection, Washington, D.C.

basins and aviaries for rare birds; orchards; collections of rare trees and flowers from Peru, North America, and the East Indies; grottoes with minerals; fountains designed by Hans Vredeman de Vries; ponds dug and pavilions constructed. All this affirms Rudolf's concept of nature in the garden as God's second book. In the garden of Prague Castle, visitors could observe how plants and flowers growing through letters cut out in a lead slab made a text in which God's creation of the world could be read. The emperor's love of rare and exotic plants is manifested by the famous Flemish botanist Carolus Clusius (1526–1609), who dedicated his *Rariorum plantarum historia* of 1601 to Rudolf II (fig. 21). The Dutch artist Emanuel Sweerts also dedicated his *Florilegium* (1612) (fig. 22) to Rudolf II, whom he calls *hortorum deus*, the god of gardens, and in whose honor the book opens with a representation of *Corona imperialis* (fig. 23), followed by the flowers Sweerts observed in the imperial gardens in Prague, which are described as the gifts of Vertumnus.

Thus the gardens at the Prague court formed a truly integral part of Rudolf II's collections. They were a microcosm of God's creation and showed the artfulness of nature not as dried specimens but in a living form. Puechfeldner's albums provide us with images of gardens Rudolf II liked to see—ideal objects for personal use and contemplation.

I hope to have demonstrated that the Dumbarton Oaks album stands for a whole world: a world of international connections in the fields of art, gardens, and architecture. It tells the story of the importance of gardens for the Habsburg princes in relation to botany and other studies of nature and of the predominance of Flemish garden models in Habsburg culture. It shows us a world of art and nature, of harmony and order, of collecting and personal use, of form and meaning. In short, it reflects the universal world of Rudolf II, for whom this art book of nature was made in 1593 by his gardener Hans Puechfeldner.

NOTES

This essay was prepared for publication by Nicole Anselona, Anne Nellis, and Elizabeth Pergam from a transcription of the talk presented at the symposium "The Art and History of Botanical Painting and Natural History Treatises," National Gallery of Art, Center for Advanced Study in the Visual Arts, 3 May 2002. The talk was based on an essay previously published by the Sociedad Estatal para la Conmemoración de los Centenarios de Felipe II y Carlos V in *Felipe II, el rey íntimo: Jardin y naturaleza en el siglo XVI* (Madrid, 1998), 173–189.

1. Recent contributions to scholarship on Hans Vredeman de Vries are Petra Sophia Zimmerman, *Die Architectura von Hans Vredeman de Vries: Entwicklung der Renaissancearchitektur in Mitteleuropa* (Munich, 2002), and *Hans Vredeman de Vries und die Renaissance im Norden,* ed. Heiner Borggrefe (Munich, 2002).

2. For further information on bookbinding in this period, see Radim Vondraček, "Bookbinding: Style and Ornament," in *Rudolf II and Prague: The Court and the City,* ed. Eliška Fučíková et al. (Prague and New York, 1997), 340–344.

3. Hans Puechfeldner, *Ein Nützliches Khünstbüech der Gartnereij,* Prague, 1591, Cod. 10.738, Österreichische Nationalbibliothek, Vienna, and Hans Puechfeldner, *Ein Nützliches Künstbüech der Lüstigen Gardtnereij,* Prague, 1594, Cod. 10.830, Österreichische Nationalbibliothek, Vienna.

4. For Rudolph II's *Kunstkammer,* see Eliška Fučíková, "The Collection of Rudolf II at Prague: Cabinet of Curiosities or Scientific Museum?" in *The Origins of Museums: The Cabinet of Curiosities in Sixteenth- and Seventeenth-Century Europe,* ed. Oliver Impey and Arthur MacGregor (Oxford, 1985), 47–53, and Beket Bukovinska, "The *Kunstkammer* of Rudolf II: Where It Was and What It Looked Like," in Vondraček 1997, 199–208. For broader surveys of the milieu of the Bohemian court, see Thomas DaCosta Kaufmann, *School of Prague: Painting at the Court of Rudolf II* (Chicago, 1988), and Robert J. W. Evans, *Rudolf II and His World: A Study in Intellectual History, 1576–1612* (London, 1997).

5. These include Pliny's letters, in which he discussed villas and gardening, and Varro's *Rerum rusticarum libri tres.* See *The Letters of the Younger Pliny,* trans. Betty Radice (Baltimore, 1963), and *Marcus Porcius Cato, On Agriculture; Marcus Terentius Varro, On Agriculture,* trans. William Davis Hooper, rev. Harrison Boyd Ash (Cambridge, Mass., 1936).

6. R. G. Calkins, "Piero de Crescenzi and the Medieval Garden," in *Medieval Gardens,* ed. Elisabeth B. MacDougall (Washington, D.C., 1986), 155–175. See also Johanna Elizabeth Bauman, "Piero de' Crescenzi's *Liber ruralium commodorum:* Unearthing the Origins of the Pleasure Garden" (Ph.D. diss., University of Virginia, 2000).

7. J. Miguel Morán Turina and Fernando Checa Cremades, *Las casas del rey: Casas de campo, cazaderos y jardines: Siglos XVI y XVII* (Madrid, 1986).

8. Jost Bürgi was a clock and instrument maker for Rudolf II. See Bruce T. Moran, "German Prince-Practitioners: Aspects in the Development of Courtly Science, Technology, and Procedures in the Renaissance," *Technology and Culture* 22 (April 1981): 253–274.

9. See Kaufmann 1988, 171.

10. Thomas DaCosta Kaufmann, "Archimboldo's Imperial Allegories," *Zeitschrift für Kunstgeschichte* 39 (1976): 275–296, and, by the same author, "Remarks on the Collection of Rudolf II: The *Kunstkammer* as a Form of Representation," *Art Journal* 38 (1978): 22–38.

11. See Fučíková in Impey and MacGregor 1985. On the role of natural history illustration at Rudolf's courts, see also Lee Hendrix, "Natural History Illustration at the Court of Rudolf II," in Impey and MacGregor 1985, 157–171.

12. See, for example, Paracelsus, *Philosophiae magnae* (Cologne, 1567).

13. Peter Marshall, *The Magic Circle of Rudolf II: Alchemy and Astrology in Renaissance Prague* (New York, 2006).

14. For the gardens at Prague Castle, see Eliška Fučíková, "Prague Castle under Rudolf II, His Predecessor and Successors," in Impey and MacGregor 1985, 3–4, and J. Krčálová, "Die Gärten Rudolfs II," *Leids kunsthistorisch jaarboek* 1 (1982): 149–160.

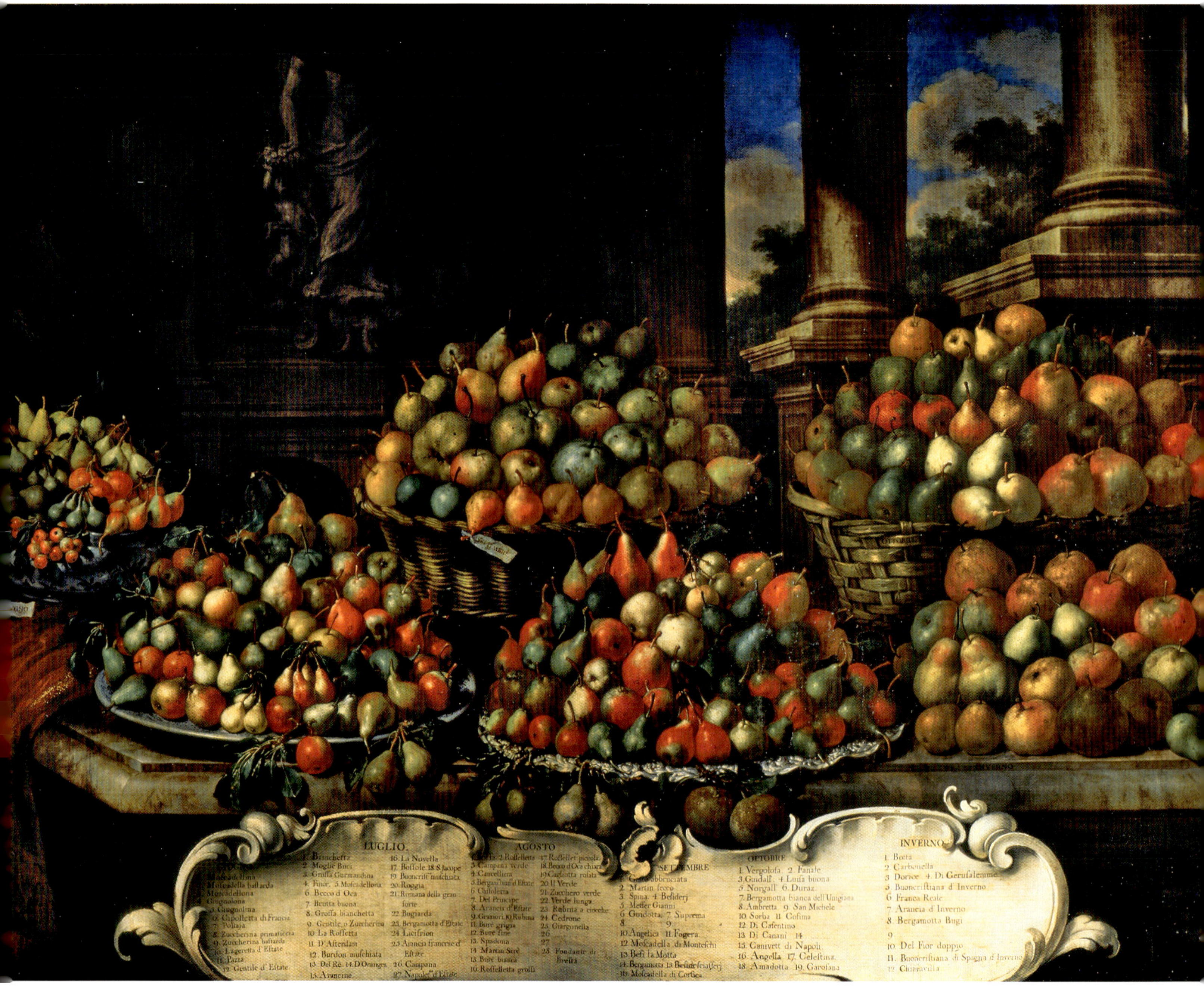
LUGLIO.
Mofcadella baſtarda
Mofcadellona
Giugnolina
6. Cipolletta di Francia
7. Pollaja
8. Zuccherina primaticcia
9. Zuccherina baſtarda
10. Lageretta d'Eſtate
11. Pazza
12. Gentile d'Eſtate
2. Moglie Buci
3. Groſſa Gurmandina
4. Finor. 5. Moſcadellona
6. Becco d'Oca
7. Brutta buona
8. Groſſa bianchetta
9. Gentile, o Zuccherina
10. La Roſſetta
11. D'Aſterdam
12. Burdon muſchiata
13. Del Rè. 14. D'Oranges
15. Arancine
16. La Novella
17. Boſſole 18. S. Jacope
20. Roggia
21. Romana della gran forte
22. Bugiarda
23. Bergamotta d'Eſtate
24. Lucifrion
25. Arancia franceſe d'Eſtate
26. Campana
AGOSTO
4. Cancelliera
6. Caſſoletta
7. Del Principe
8. Arancia d'Eſtate
11. Burè grigia
12. Burè fine
13. Spadona
14. Martin Sirè
15. Burè bianca
16. Roſſelletta groſſa
19. Caghotta roſata
20. Il Verde
21. Zucchero verde
22. Verde lunga
24. Cedrone
25. Giargonella
26
27
28. Fondante di Breſta
SETTEMBRE
2. Martin ſecco
3. Spina. 4. Beſiderj
5. Meſſer Gianni
6. Guidotta. 7. Suprema
8 9
10. Angelica 11. Fogera
12. Moſcadella di Monteſchi
13. Beſi la Motta
16. Moſcadella di Corſica
OTTOBRE
1. Vergoloſa. 2. Fanale
3. Guidall. 4. Luiſa buona
5. Norgall. 6. Duraz.
7. Bergamotta bianca dell'Unigiana
8. Ambretta 9. San Michele
10. Sorba 11. Coſima
12. Di Caſentino
13. Di Canani 14
15. Ganivett di Napoli
16. Angella 17. Celeſtina
18. Amadotta 19. Garofana
INVERNO
1. Botta
2. Carbonella
3. Dorice 4. Di Geruſalemme
5. Buoncriſtiana d'Inverno
6. Franca Reale
7. Arancia d'Inverno
8. Bergamotta Bugi
9
10. Del Fior doppio
11. Buoncriſtiana di Spagna d'Inverno
12. Chiaravilla

HANS W. HUBERT
Albert-Ludwigs-Universität Freiburg

"Cosmic Delight": Bartolomeo Bimbi and the Representation of Nature at the Court of Cosimo III de' Medici

The empirical and experimental study of nature was exercised and favored to an unusual extent at the Medicean court beginning in the sixteenth century. The enormously growing body of knowledge obtained through these studies, however, called into question several assumptions regarding nature: the old conception of the world; the exclusively religious claim to interpretation of this world; and the categories of registration, of order, and of systematic representation of this world.[1] This process of reinterpreting nature, however, was both long lasting and burdened with conflicts. It also had manifold effects on the arts, and on the collecting of art, until its culmination in the eighteenth century. As the scientific process and the concept of nature changed, the keepers of the grand ducal collections in Florence, which contained not only artificialia but also a great variety of naturalia, were confronted with the problem of establishing a new systematic order.

It was Francesco Redi (1626–1698), one of the foremost naturalists of his day, who—supported by Grand Duke Ferdinando II and by his brother Cardinal Leopoldo—turned the observation of nature, anatomical dissection, experimentation, and demonstration into public events at the Medicean court. In so doing, Redi not only contributed to the diffusion of his opinions and his findings but also stimulated others to imitate his experimental methods. His efforts were directed against the erudition of such Jesuit scholars as the German Athanasius Kircher (c. 1601–1680) and the Italian Filippo Bonanni (1638–1725), who had been nourished by an antique and mostly Aristotelian tradition.[2] Redi was also in the confidence of the young Cosimo III de' Medici (1642–1723), who came to power as grand duke of Tuscany in 1671. Cosimo confirmed Redi in his position of court physician (*arciatra*) and appointed him director of the Gabinetto d'Istoria Naturale, founded in 1677.[3] In his treatise *Esperienze intorno a diverse cose naturali e particolarmente a quelle che ci son portate dalle Indie,* published shortly after Cosimo's accession, Redi dissociated himself from the magic-encyclopedic interpretation of nature dominant since the sixteenth century. His work was an expression of the new preference for methodical collecting that should be accompanied by scientific research carried out by naturalists, to interpret what Saint Augustine (354–386) had called the "great book of nature" in a precise and coherent manner.[4] In this way, Redi initiated in Florence the process of departure from the paradigm of the *Kunstkammer* as it had been applied to the distinguished Roman museum established by Kircher. In so doing, Redi established the intellectual premise for Cosimo's collecting of naturalia.[5]

The fine arts stood in a complex relationship to this increasingly scientific approach

Bartolomeo Bimbi, *Pears*, 1699, oil
Villa Medicea, Poggio a Caiano; photograph Studio Fotografico Tosi, Florence

to nature and to collections of naturalia, a relationship that was reciprocal. On the one hand, artists were positively and actively involved in this new process through their faithful representations of nature, while on the other hand they were passive in the face of the increasing loss of meaning of their traditional iconic motifs.[6] Particular phenomena in this context were the representations of plants (fruit, vegetables, flowers) and animals made by the Florentine court painter Bartolomeo Bimbi (1648–1730). These paintings were executed on commission, over a period of three decades, for the villas of Grand Duke Cosimo III de' Medici.[7] The decoration and the significance of one of these ensembles, which was created for the Casino della Topaia, is the subject of this essay (fig. 1).

Ever since some of Bartolomeo Bimbi's paintings emerged from the obscurity of storage in the Florentine museums into the public light, they have been an eye-catching attraction for viewers. The two main reasons are their uncommon subjects and their unusual size. Who had ever seen a giant chard, a grand ducal squash, a monstrous cauliflower, or a horseradish—all painted life size, or even larger? This effect is even more emphatic in Bimbi's large panel paintings of fruit types, especially because the variety and abundance of the depicted fruit is paralleled by the foliage and fruit of the opulent gilt frames made by the Netherlandish wood-carver Vittorio Crosten (fig. 2).[8] Thanks to the sometimes dense, sometimes lavishly layered paint, Bimbi rendered not only the different tones but also the very different surfaces of the various fruits—cherries and lemons, for example (see figs. 7 and 10). By all these means he created an unusual sense of immediacy and intimacy in the presence of the objects depicted. But on closer examination, the viewer will notice that the individual cherries are all numbered, in correspondence with the legend, at the bottom, which includes a complete list of all the different varieties depicted. These inscriptions connect Bimbi's paintings directly to the genre of botanical illustration, and in fact scholars have confirmed the enormous documentary importance of Bimbi's paintings for the reconstruction of the flora of Tuscany of the past.[9] By examining these paintings in their original context, however, we will see that their significance goes much further.

On the basis of recent analysis of the archives of the Guardaroba Medicea in Florence, it has been possible to trace the provenance of many of Bimbi's canvases. This research has shown that many of the

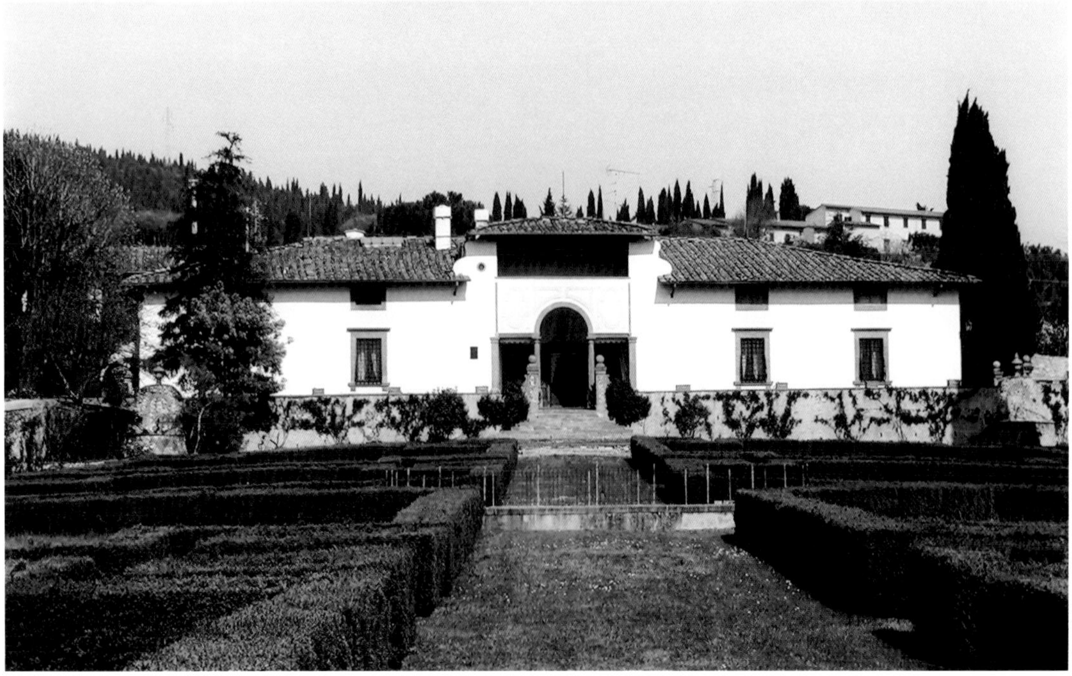

1. Casino della Topaia, Florence, garden facade
Author photograph

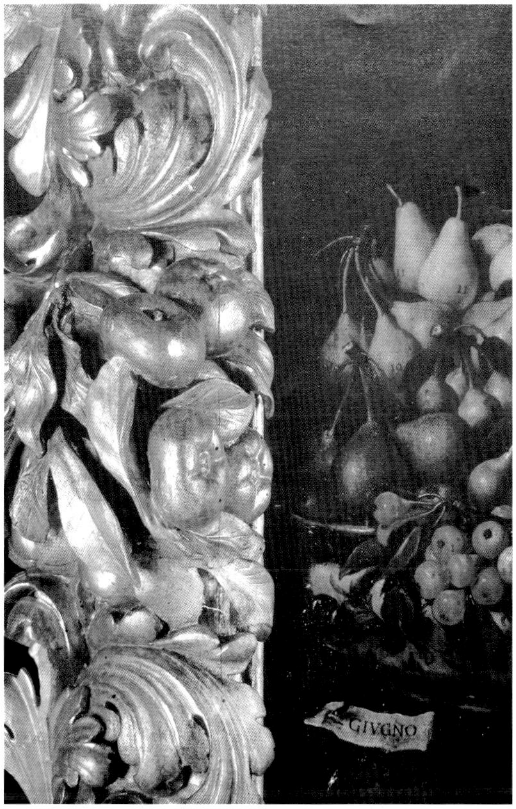

2. Detail of frame with fruit motifs carved by Vittorio Crosten, 1696, carved and gilded wood
Villa Medicea, Poggio a Caiano

paintings of plants and fruit were once in the so-called Casino della Topaia.[10] Located between two famous Medici properties—the Villa Castello and the Villa Petraia—on the hill called Poggio Secco, the Casino della Topaia already existed at the time of Duke Cosimo I in the sixteenth century.[11] In those days it served as guest accommodations for literati such as Scipione Ammirato (1531–1601) and Benedetto Varchi (1503–1565), who wrote their historical works on the city of Florence in this wonderful rural atmosphere.[12] Varchi called this place Cosmiano, in honor of his benefactor. Grand Duke Cosimo III renovated and enlarged the house in the late seventeenth century, probably engaging the Florentine architect and sculptor Giovan Battista Foggini (1652–1725).[13] Foggini laid out a long fruit orchard enclosed by a wall in front of the Casino, giving it the character of a protected *hortus conclusus*, completely hidden from outside observers. After that, the Casino was used mainly by the duke himself, when he came to visit his villas and vineyards and wanted to withdraw from courtly life in search of pious meditation. Once the structure was completed, around 1696, Cosimo began to decorate the rooms with paintings. In the course of the following three decades, Bartolomeo Bimbi painted a group of at least forty-eight canvases, consisting of two types of subjects: on the one hand, twelve large representations of all the different species of one genus of fruit; and on the other, about thirty-six medium and small paintings with single representations of specific fruits and plants. The following is a brief description of the main examples of paintings in this series.

Bimbi painted the large canvases as sets of two or four pieces each, though sometimes the paintings in these sets are of different sizes. The smaller ones are 45 by 61 inches (about 114 by 155 cm) while the larger ones are about 67 by 90 inches (170 by 228.6 cm).[14] Most of them had matching opulent gilt frames, as mentioned above, by Vittorio Crosten, in which the subjects of the paintings, foliage and fruit, were repeated. The earliest pendants were the pictures of figs and apples of 1696 (figs. 3 and 4). In both cases, the composition is conventional: large quantities of fruit are placed on a table in front of a dark background, which is partially covered by a red curtain. No fewer than thirty-three different varieties of figs—divided into groups, the early-maturing *fichi primaticci* and the autumnal *fichi settembrini*—and fifty-two varieties of apples are depicted, on plates, in baskets, or simply lying on the table surface. In the painting of apples, a white-and-blue vase with a branch of roses, a stalk of tuberose, and a flowering aristolochia enriches the well-balanced arrangement. Three years later, in 1699, Bimbi painted two more pairs of pendants (figs. 5 and 6). In the large picture of pears, 115 varieties are grouped and labeled according to the months in which they mature. The background shows the interior of a villa with a niche statue of a male figure, probably Vertumnus (Roman god of the seasons, patron of gardens and fruit trees). Even in the smaller pendant, which depicts apricots and peaches, Bimbi managed to paint nine varieties of apricots and thirty-six kinds of peaches. Their characteristically dull, fuzzy skin and the range of colors—from light yellow to dark red—are rendered very delicately. Most intriguing, perhaps, are the pendants with paintings of

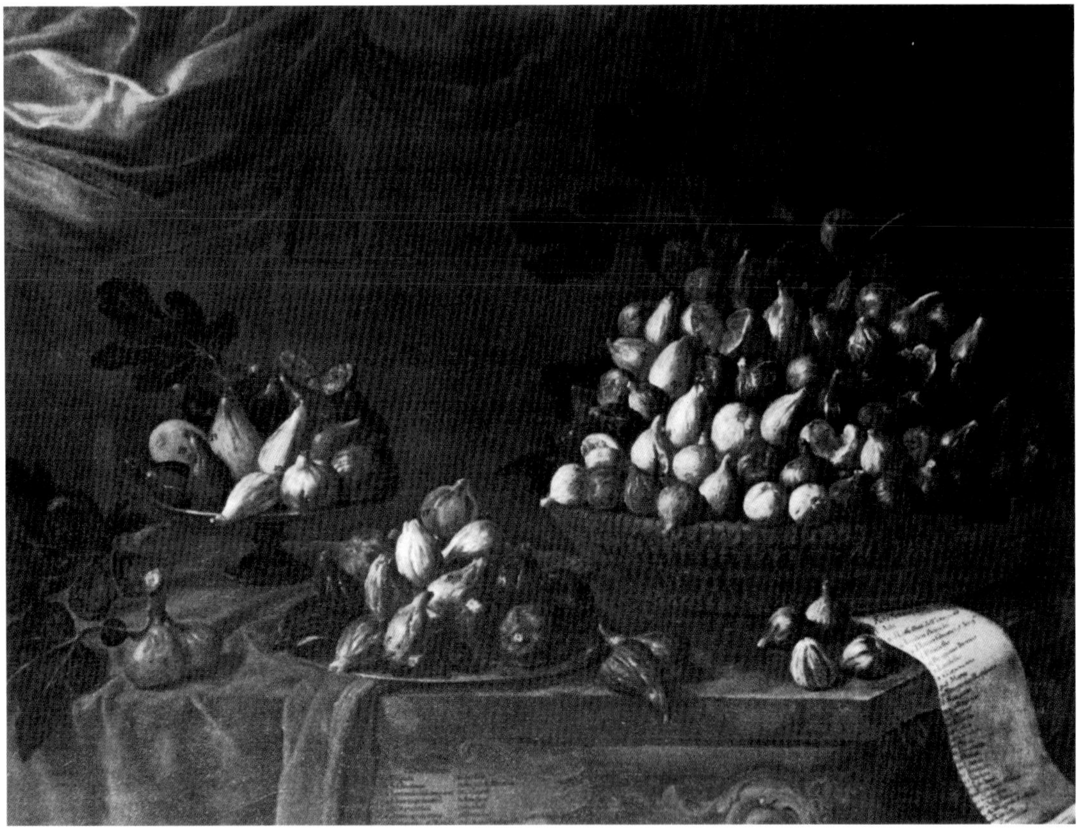

3. Bartolomeo Bimbi, *Figs*, 1696, oil
Villa Medicea, Poggio a Caiano

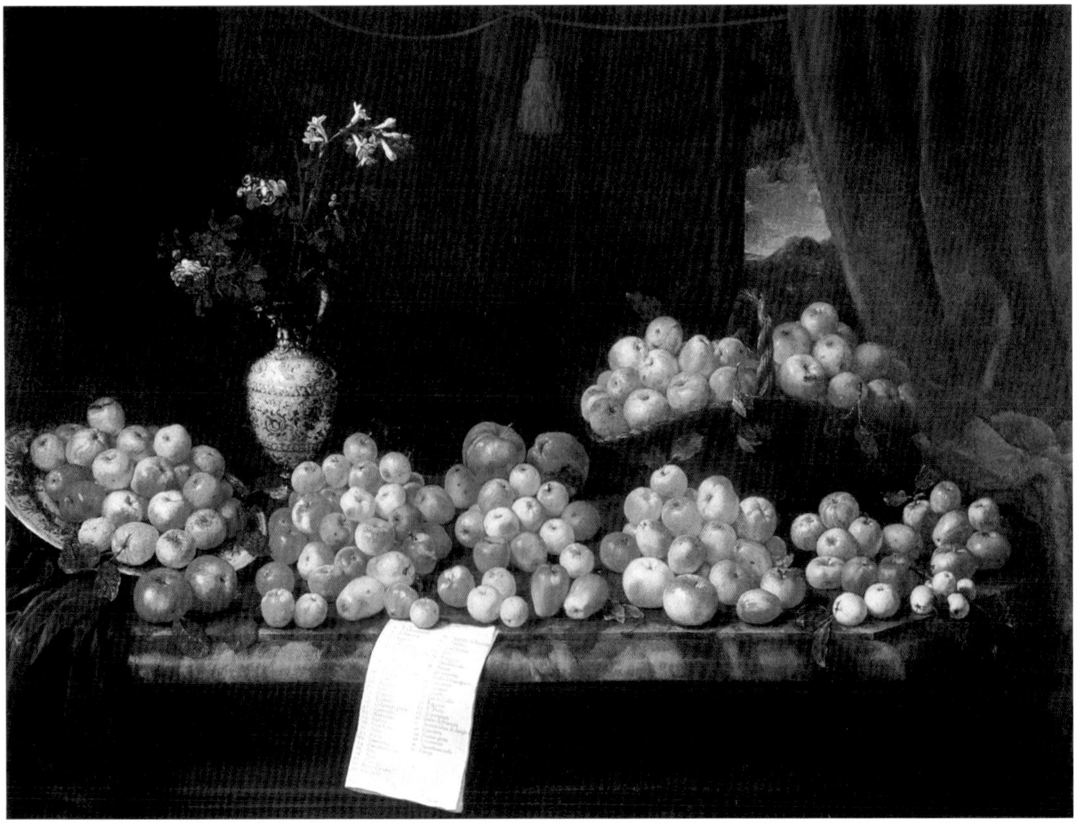

4. Bartolomeo Bimbi, *Apples*, 1696, oil
Villa Medicea, Poggio a Caiano

5. Bartolomeo Bimbi, *Pears,* 1699, oil

Villa Medicea, Poggio a Caiano; photograph Studio Fotografico Tosi, Florence

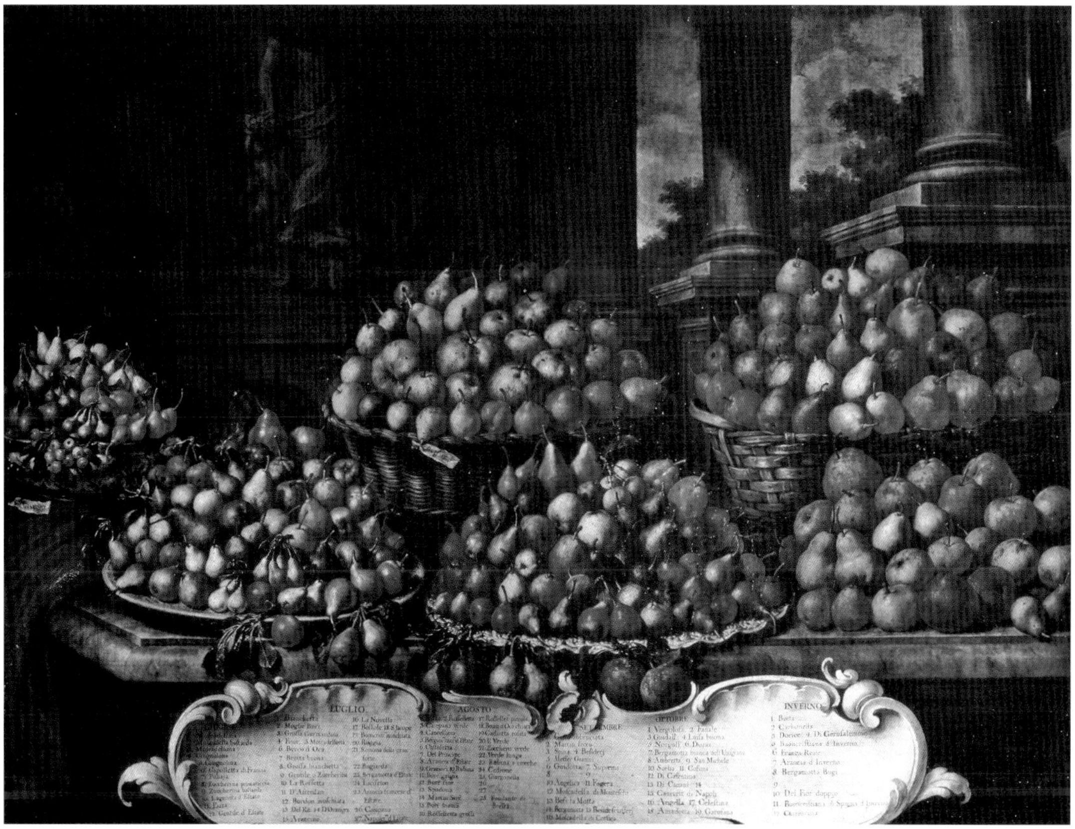

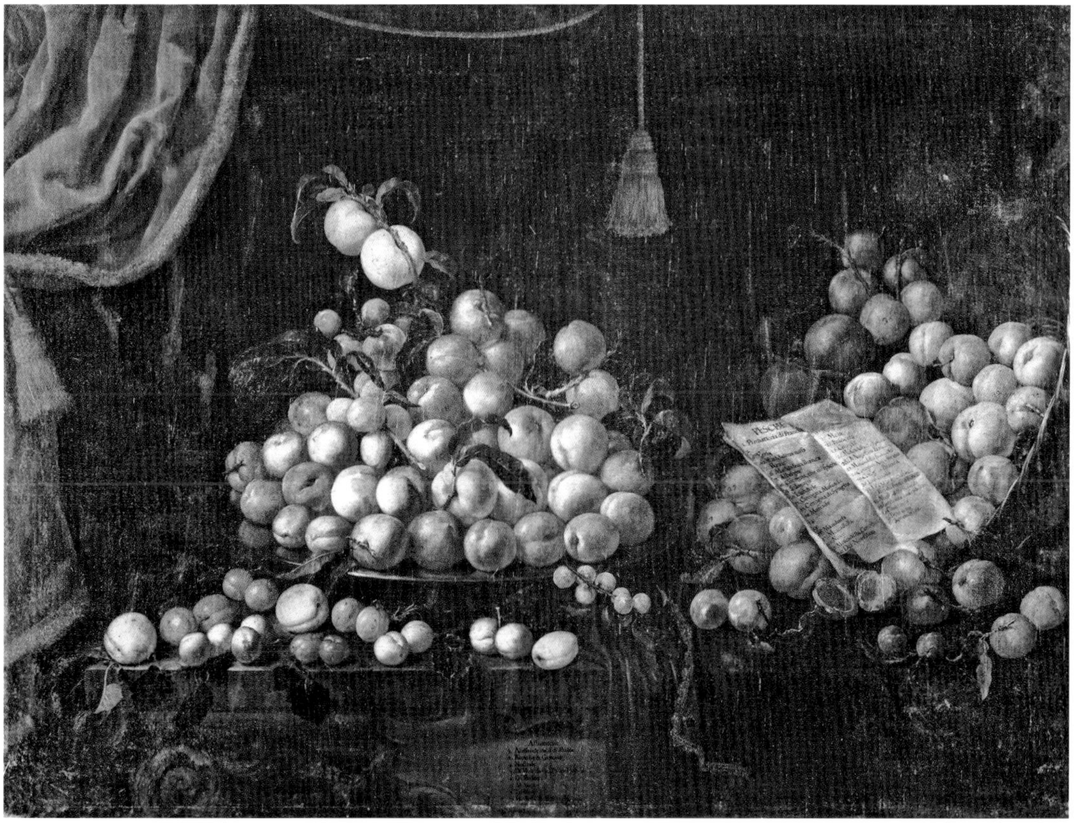

6. Bartolomeo Bimbi, *Apricots and Peaches,* 1699, oil

Villa Medicea, Poggio a Caiano; photograph Gabinetto Fotografico della Soprintendenza alle Gallerie, Florence

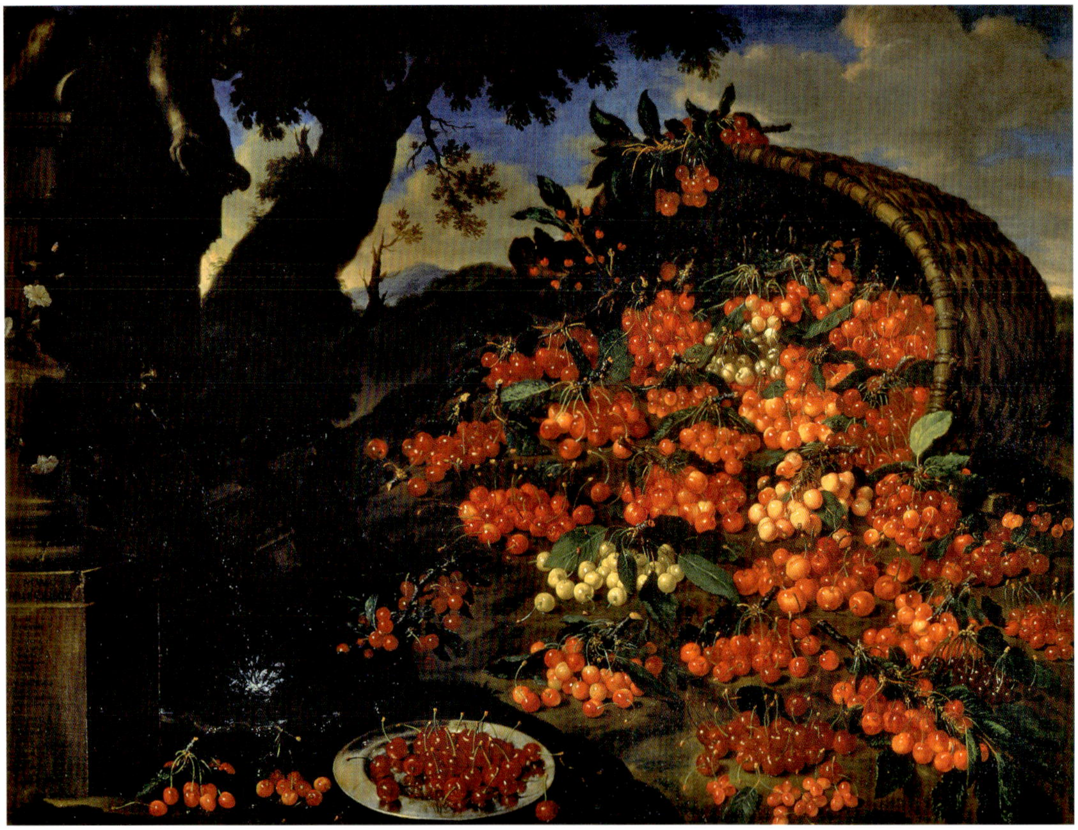

7. Bartolomeo Bimbi, *Cherries*, 1699, oil
Villa Medicea, Poggio a Caiano; photograph Studio Fotografico Tosi, Florence

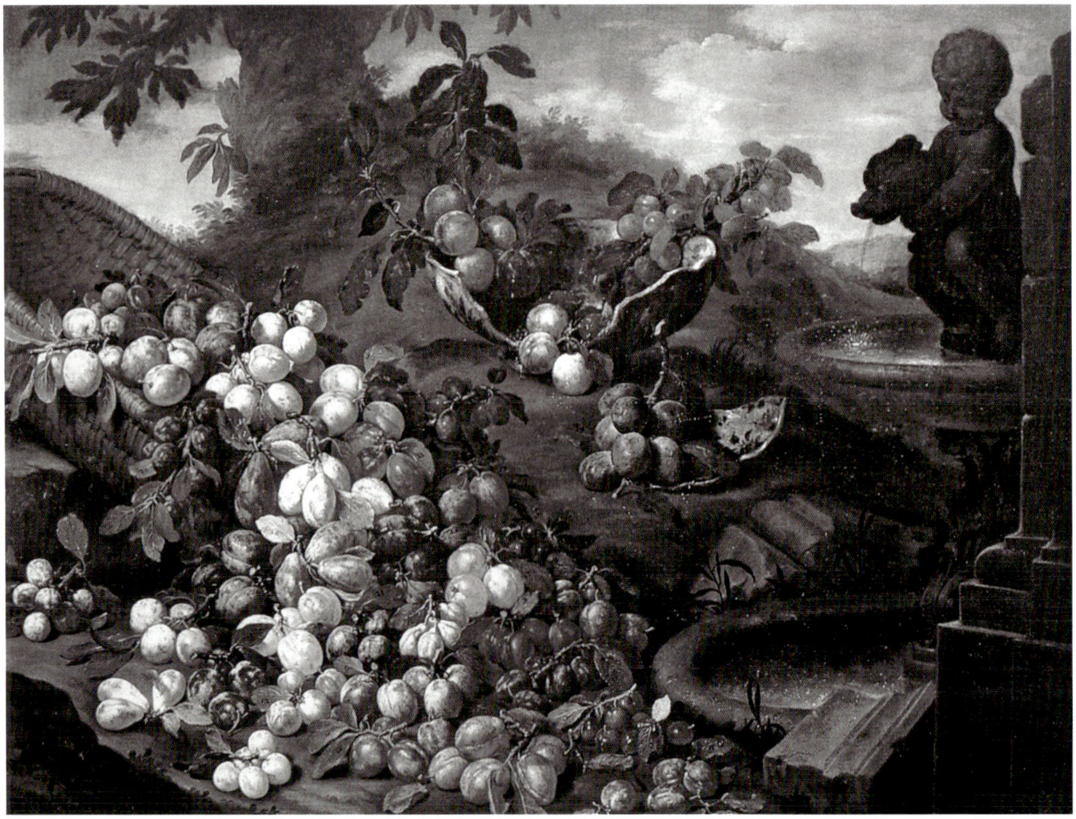

8. Bartolomeo Bimbi, *Plums*, 1699, oil
Villa Medicea, Poggio a Caiano

cherries and plums, which Bimbi finished that same year, 1699 (figs. 7 and 8). In both cases, Bimbi used the symbolism of the cornucopia to demonstrate abundance and variety. In the painting of plums, beyond the horizon of the hills, we see a clouded evening sky. In both paintings, towering trees accentuate the low point of view, and the pictorial space is framed by small garden fountains, each with a putto, on whose pedestals the names of thirty-four varieties of cherries and thirty-nine varieties of plums are inscribed.

In August 1700 two canvases depicting grapes were moved into the Casino della Topaia (fig. 9). Here, Bimbi departed from his earlier compositional schemes by showing the vines growing on an espalier between two stone pillars and by depicting the hanging grapes with hardly any overlapping. Bimbi's strategy of depicting the fruit in its natural surroundings, emphasized by the view onto the hilly vineyards above the parapet, does not work in a botanical sense, however, because all seventy-five different varieties of grapes, labeled on the cartouche below, are shown growing on the same few vines.

Fifteen years later, Bimbi again painted fruit in a similar arrangement. The four large paintings of citrus fruits, all dated 1715, are the last of his series (figs. 10 and 11). Here, Bimbi painted a kind of impenetrable wall of fruits, leaves, and blossoms. But the fruit is depicted artificially, in sets of two or sometimes three. A few specimens are cut in half to reveal the color of the fruit, the number of individual sections, and the thickness of the peel, which were considered important elements in their classification.[15] In this manner, Bimbi documented 114 kinds of citrus fruit—oranges, lemons, citrons, and bergamots—listing their Italian names carefully in the cartouches. But for the first time the corresponding numbers are not simply painted directly on the fruit but instead appear on small white tags that are tied to the plants, following the practice of labeling plants in botanical gardens.

For a reference source on citrus fruit, there existed an extensive and systematic

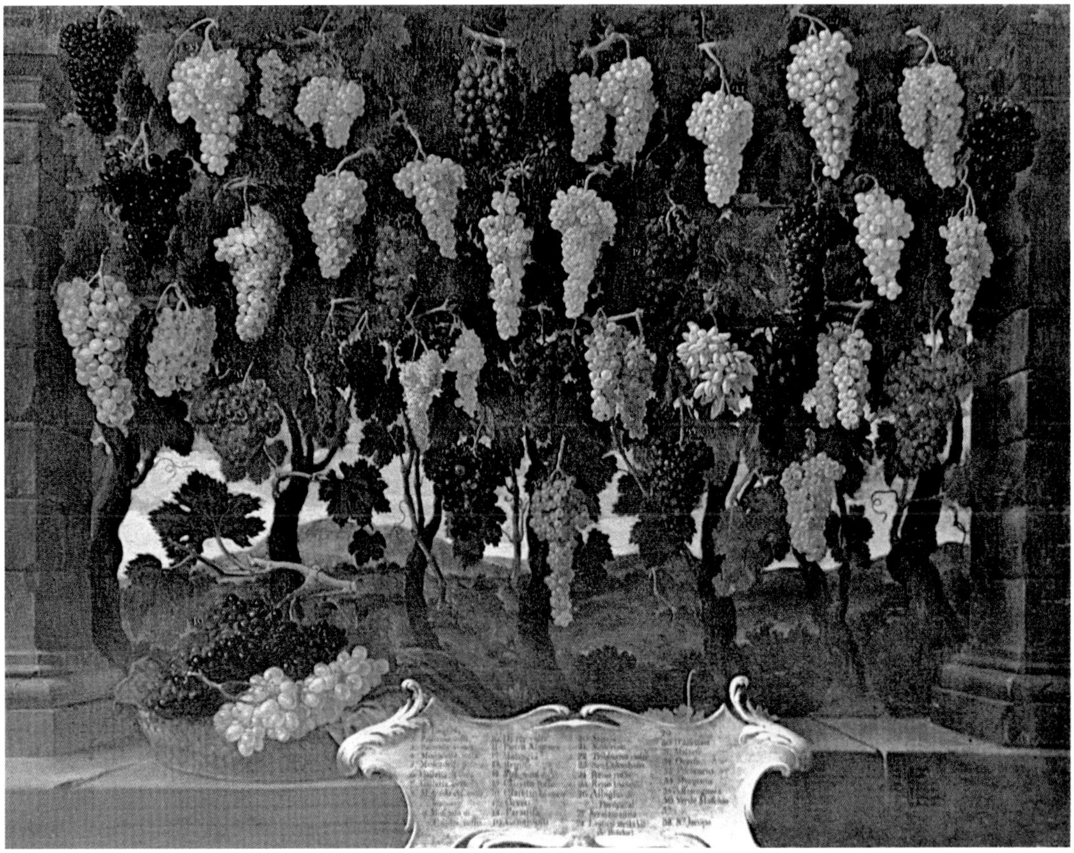

9. Bartolomeo Bimbi, *Grapevine*, 1700, oil
Villa Medicea, Poggio a Caiano

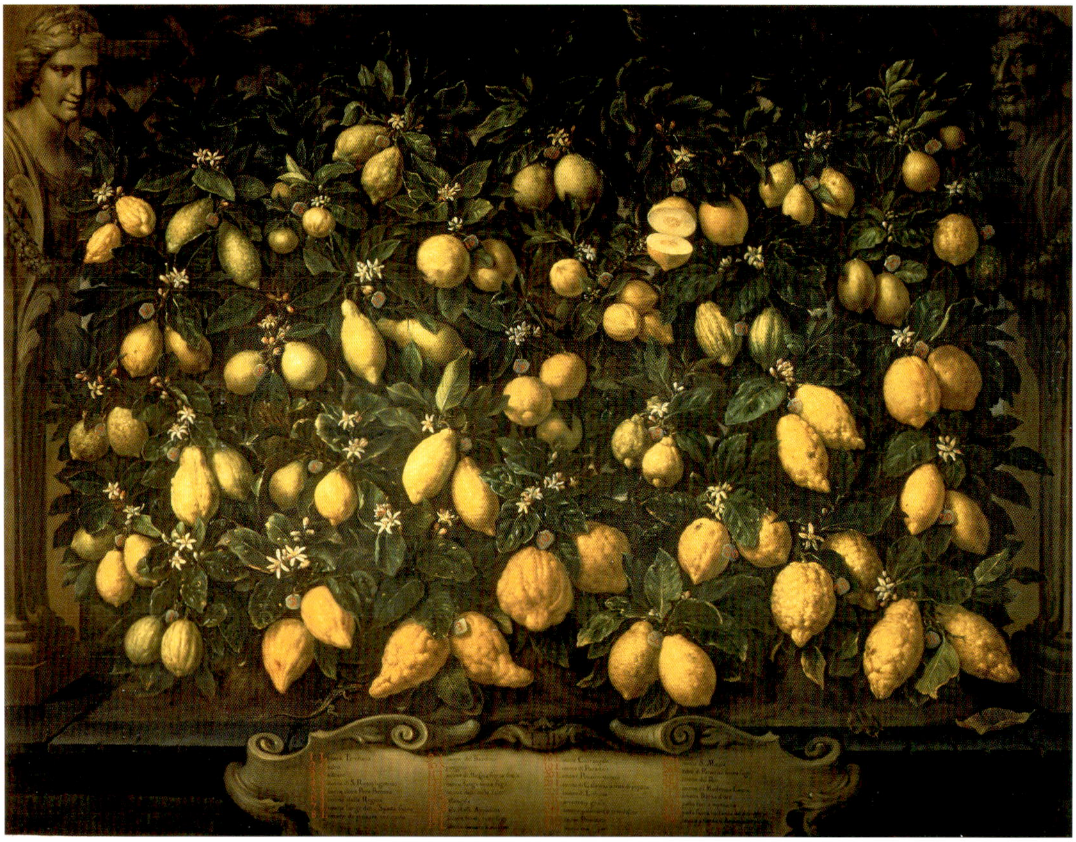

10. Bartolomeo Bimbi, *Citrus Fruit*, panel 1, 1715, oil
Villa Medicea, Poggio a Caiano; photograph Studio Fotografico Tosi, Florence

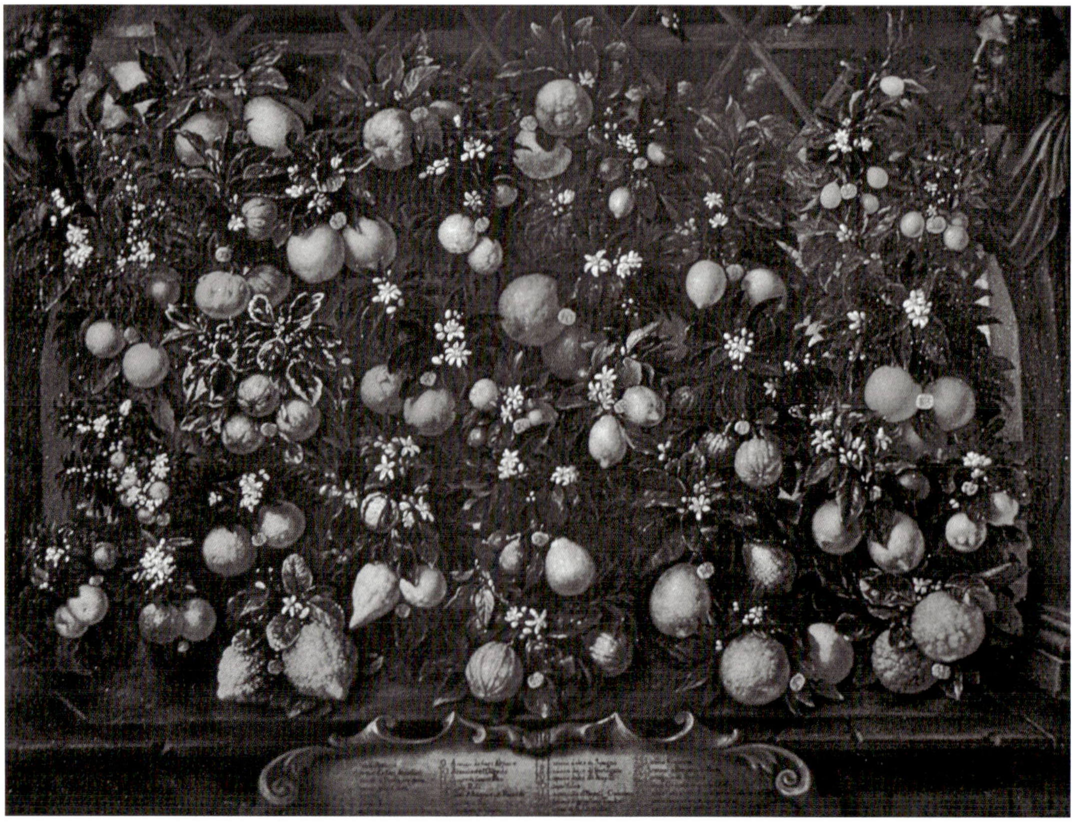

11. Bartolomeo Bimbi, *Citrus Fruit*, panel 4, 1715, oil
Villa Medicea, Poggio a Caiano

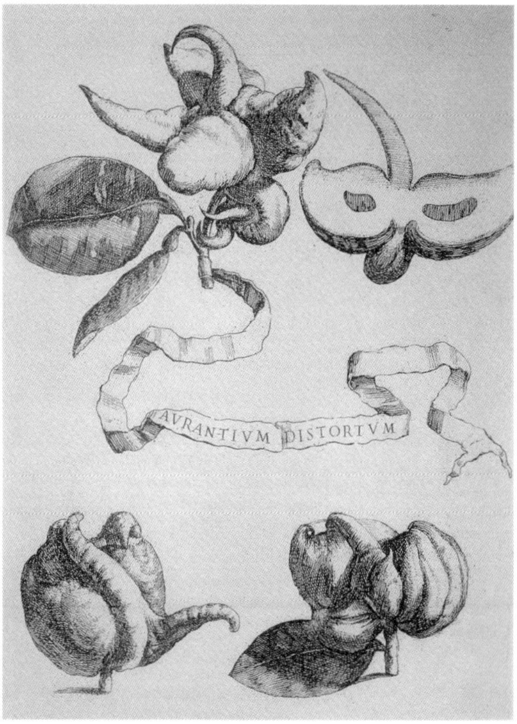

12. *Aurantium distortum,* from Giovanni Battista Ferrari, *Hesperides sive de malorum aureorum cultura et usu libri quatuor* (Rome, 1646), engraving
Private collection; author photograph

compendium that Bimbi could easily have used for his depictions. It is the treatise *Hesperides,* written by Giovanni Battista Ferrari (1584–1655), published in 1646 in Rome, and richly illustrated with eighty-seven engravings—most of them by Cornelius Bloemart (1603–1684). These plates are both scientifically accurate and of artistically high quality (fig. 12).[16] But, so far as we know, Bimbi did not copy from these engravings, nor did he copy from any other printed visual source. On the contrary, Bimbi's paintings appear to be based directly on the empirical and classifying work of the grand ducal botanists and not on older books and illustrations.

So we must recognize these pictures as genuine products of scientific research at the Cosmian court. Nonetheless, they stand in a long tradition of allegorical interpretation. Ferrari had distinguished—as did the Florentine botanist Pietro Nati after him—three types of citrus fruit, which are the *Malum citreum* (citron), the *Malum limonium* (lemon), and the *Malum aurantium* (orange), to which he gave the names of the nymphs of Hesperia: Aegle, Arethusa, and Hesperthusa.[17] Ferrari's equation of citrus fruit with the fruits that Hercules stole from the gardens of Hesperia had an important precursor in Florence, a lunette in the main hall of the Villa Poggio a Caiano. There, in 1582, Alessandro Allori (1535–1607) completed his representation of Hercules and Fortune in the garden of the Hesperides as a complex allegory of the happy and prosperous Medicean reign in Tuscany (fig. 13). The Hesperidical *Mala aurea* (golden apples) were deliberately interpreted as *Mala medica* (lemons) and as *Mala citrea* (citrons). Allori consequently depicted them as Medicean citrus fruit. It was possible in this way to identify the Tuscany of the Medici with the mythical garden of the golden fruit, particularly since Hercules himself was taken up by the Medici as a mythological identity figure.[18]

It would have been obvious for Bimbi to try to make a similar allegorical statement with his paintings of citrus fruit, but astonishingly there is no hint of allegory in any of his pictures. They seem to be a simple botanical inventory without any deeper significance, though further discussion will reveal that this appearance is deceptive.

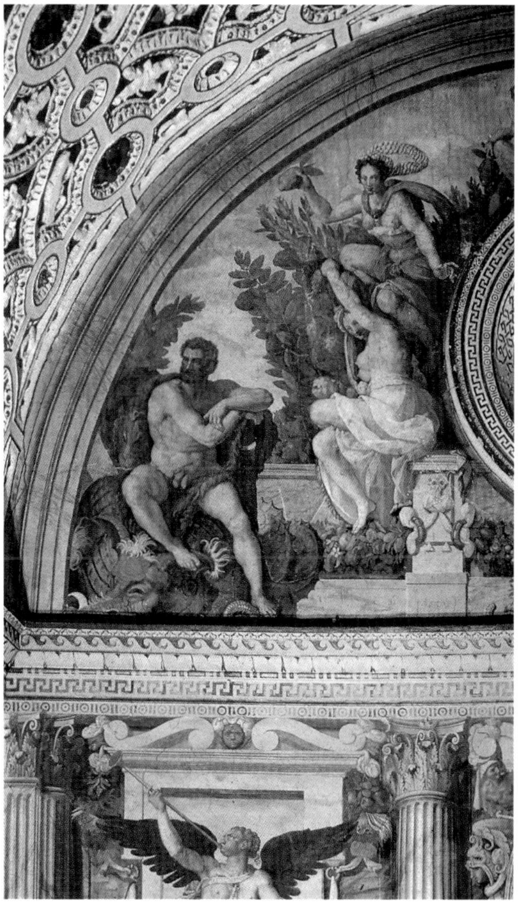

13. Alessandro Allori, *Hercules and Fortune in the Garden of the Hesperides* (detail), completed 1582, fresco
Villa Medicea, Poggio a Caiano

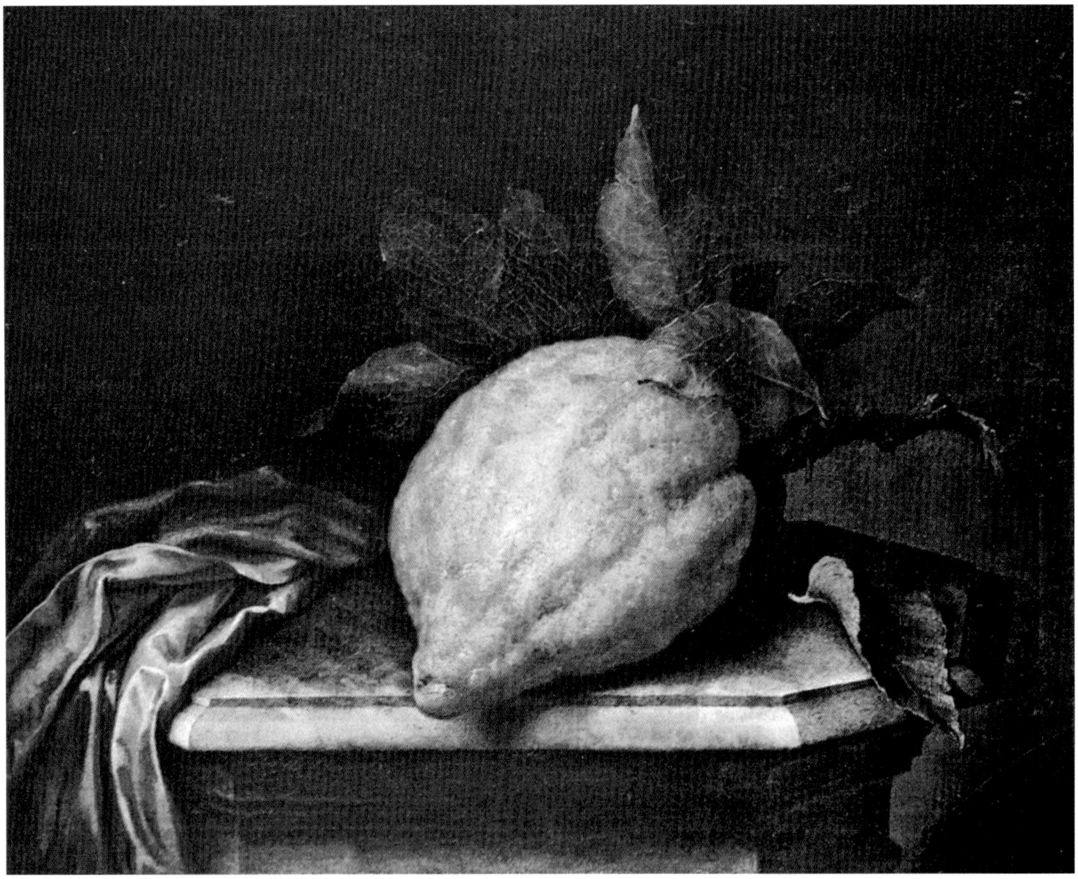

14. Bartolomeo Bimbi, *Citron*, c. 1715, oil
Palazzo Pitti, Florence

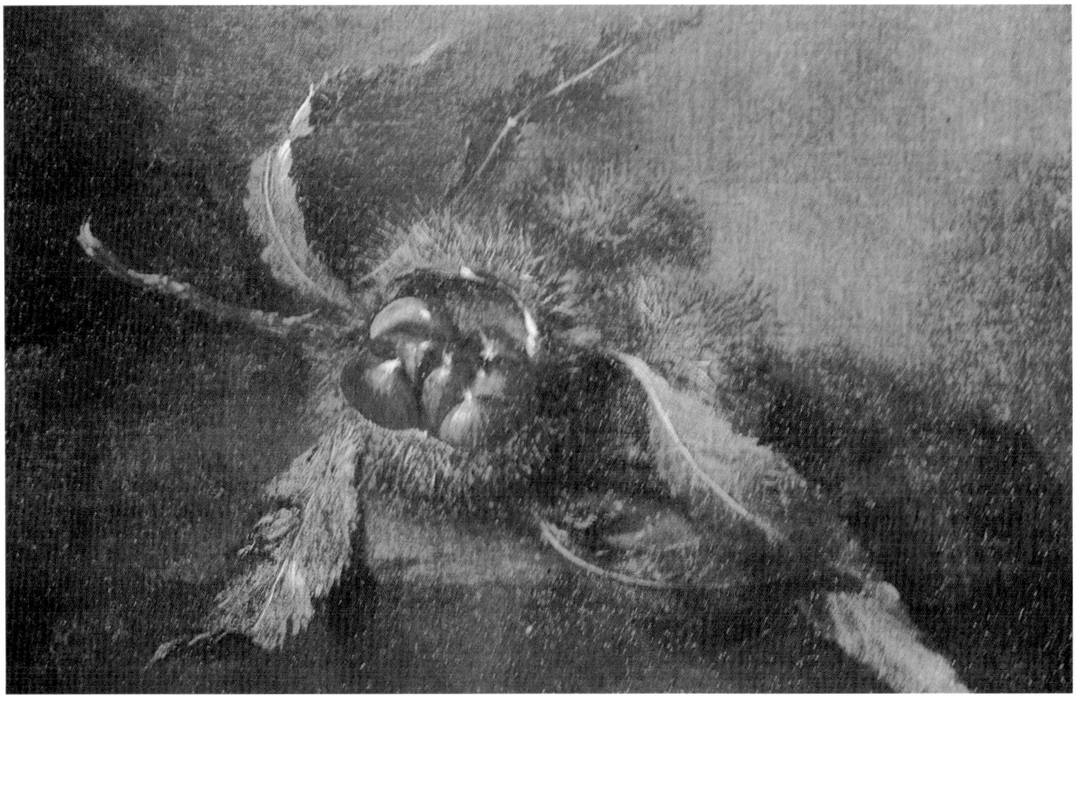

15. Bartolomeo Bimbi, *Chestnut*, 1717, oil
Palazzo Pitti, Florence

Let us return for a moment to the decoration of the Casino. Besides the twelve large paintings, there were at least thirty-six other medium and small pictures representing single fruits and plants (figs. 14 and 15). These single representations can be divided according to their subjects into three groups:

1. Subjects corresponding to one of the large fruit panels, such as the representation of a single lemon;

2. Paintings with subjects *not* corresponding to one of the large fruit panels, but thematically related because there was more than one painting with the same subject, for example the different representations of chestnuts; and

3. Paintings with subjects which were unique and consequently not groupable, for example, the canvas of a clump of grain.

Very probably, the paintings of the first group had hung in the same room as the large canvases to which they were thematically related. The paintings of the second group were probably hung together and the rest distributed purely according to aesthetic criteria. All these small paintings represented specific *meraviglie naturae,* or marvels of nature. In the seventeenth century, the notion of *meraviglia natura* meant products of nature that did not correspond to the norm or which were rare and therefore hard to find. That is why people thought they were worth collecting. Particular deviations from the norm, which were considered a kind of enhancement, were called *bizzarie,* oddities. In extreme cases, a given specimen was denoted a *mostruosità,* a monstrosity. *Meraviglie* and *bizzarie,* natural wonders and oddities, were classified positively, while monstrosities, in the public opinion, had negative connotations.

Bimbi's painting of a single lemon (fig. 14) was considered to be a natural wonder because of its enormous size; his painting of a chestnut (fig. 15) was considered to represent a marvel because it showed a single shell containing seven chestnuts. It is worth noting that these natural wonders were not inventions of artistic fantasy but depictions of real, existing *vegetabilia.* To emphasize the accuracy and legitimacy of the subjects depicted, Bimbi often included extensive information about the plants by means of labels. The large clump of grain growing on a dusty field, for example (fig. 16), was represented as a *meraviglia della natura,* as the inscription informs us, because in the year 1713, the 180 shafts of wheat depicted on this canvas sprang from one single wheat seed. This unusual incident occurred in the garden of Francesco Magnelli, a hereditary tenant of the hospital of Santa Maria Nuova, in the town of Legnaia, in the parish of San Pietro in Monticelli, in the form shown on this canvas.[19] Apparently Bimbi portrayed a specific plant, making a *ritratto al naturale* (portrait from nature) or *ritratto al vivo* (portrait from life), as his contemporaries would have put it. But this unusual specimen should not mislead us, because the motif of a clump of grain, as well as other of Bimbi's subjects, had a long iconic tradition. In 1563 Thiebold Berger printed a leaflet with the similar motif of a clump of wheat, in which all seventy-two shafts had grown from a single rootstock (fig. 17). As the inscription at the top says, the plant, grown in Klepsaw on the Rhine, was visited and admired by many senators and citizens of Strasburg, while the long song-text at the bottom praises the event as a divine wonder and as a call for conversion to the service of God.[20]

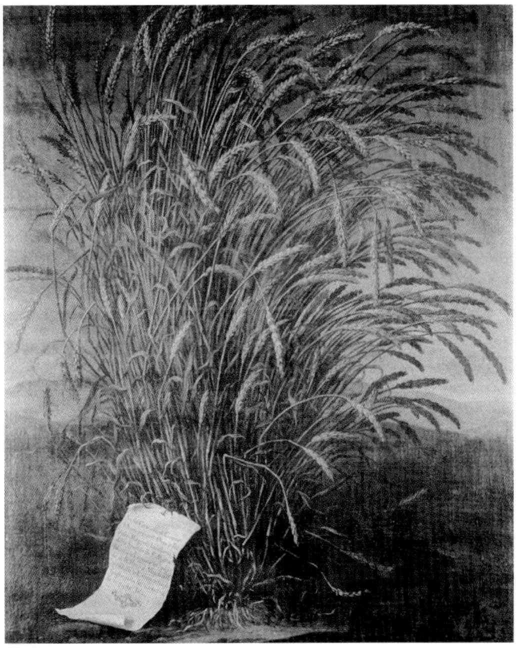

16. Bartolomeo Bimbi, *Clump of Grain,* c. 1715, oil
Palazzo Pitti, Florence

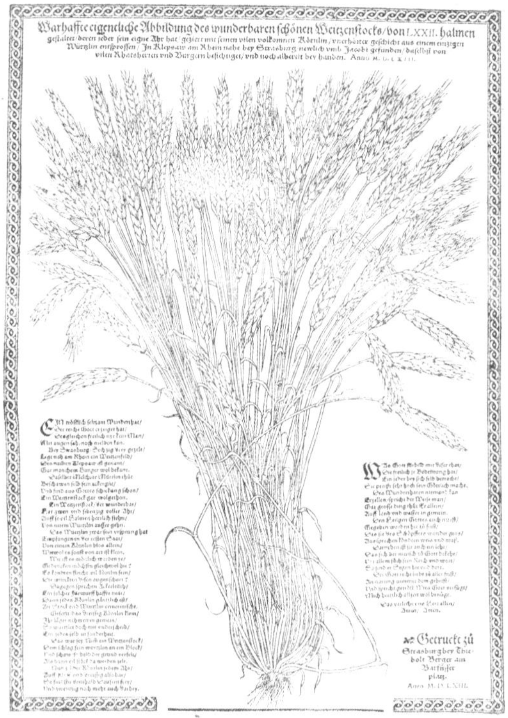

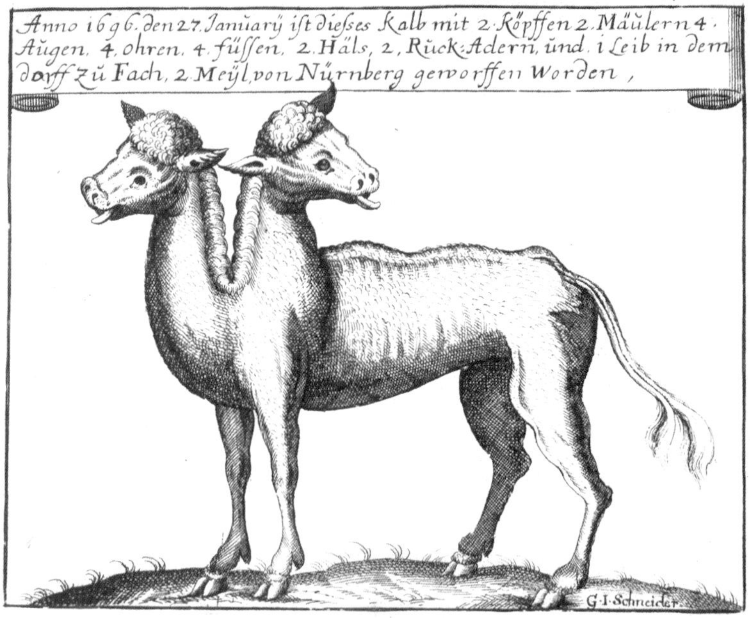

17. Thiebold Berger (printer), *Sheaf of Grain*, 1563, woodcut
Universitätsbibliothek, Erlangen

18. Georg Jacob Schneider (attributed to), *Two-headed Calf*, 1696, engraving
Universitätsbibliothek, Erlangen

It is very improbable that the Berger woodcut from more than 130 years earlier was known to anyone in Florence during the time of Bartolomeo Bimbi, any more than the engraving attributed to Georg Jacob Schneider of a dicephalous calf born in Nuremberg in 1696 (fig. 18).[21] Nevertheless, these prints are important testimonies to the general interest that such curiosities as natural wonders, oddities, and animal or human monstrosities provoked all over Europe. But it is not until the early seventeenth century that one finds examples in which such marvels of the plant kingdom were depicted in oil on canvas or panel, and thereby assumed the rank of accurate pictures intended to be framed and hung in collectors' cabinets or galleries of curiosities. Some very beautiful examples of such paintings were executed at the Florentine court, and perhaps it is not just by chance that one of them represents a giant citron.[22]

But the sheer quantity and great variety of gallery paintings which Bimbi produced for Cosimo de' Medici are completely unique. Apart from the bizarre clump of grain, Bimbi depicted other curiosities and rare subjects, including unusually large mushrooms, enormous sweet melons, bizarre grapes, big watermelons, giant carnations and other flowers, a giant truffle, a beetroot, a beanstalk, a plate with dates, and so on.

All these paintings once hung in the Casino della Topaia, and in that context it is helpful here to examine the structure of the building and the arrangement of the rooms (fig. 19). This austere building is in the shape of a long rectangle with a central loggia and an open terrace above, on the garden side. In the front is a lawn, and below—on a lower level—an extensive orchard subdivided into six fields, in which different kinds of dwarf fruit trees were once cultivated.[23] The main entrance was the garden loggia in the form of a Serliana with rusticated Tuscan columns on the central axis (see fig. 1). Behind this is a hall that divides the Casino into two asymmetrical parts of nearly equal size (plan at right in figure 19). Originally on the right (east) side of the central hall were the ducal apartments, consisting of a small chapel with sacristy (spaces numbered 7 and 8), a bedroom with a fireplace (6), and two other large vaulted rooms (4, 5).[24]

In front of the loggia, at the end of the sloping orchard, is a small exedra with a statue of Saint Fiacre as a focal point (fig. 20).

19. Casino della Topaia, plan of the ground floor (right) and upper floor, probably before 1787, pen and ink, with legend identifying, on the ground floor, (1) entrance, (2) hall, (3) stair, (4, 5) rooms, (6) room with fireplace, (7) sacristy, (8) chapel, (9) side entrance, (10) small room, (11) passage, (12) interior room behind kitchen, (13) kitchen with oven, (14) larder, and (15) stair; on the upper floor, (1) stair, (2) small kitchen, (3–9) rooms, and (10) six garrets

State Archive, Prague, Cabreo, 3, No. 54, Inv. No. VII, 2, (A 3)

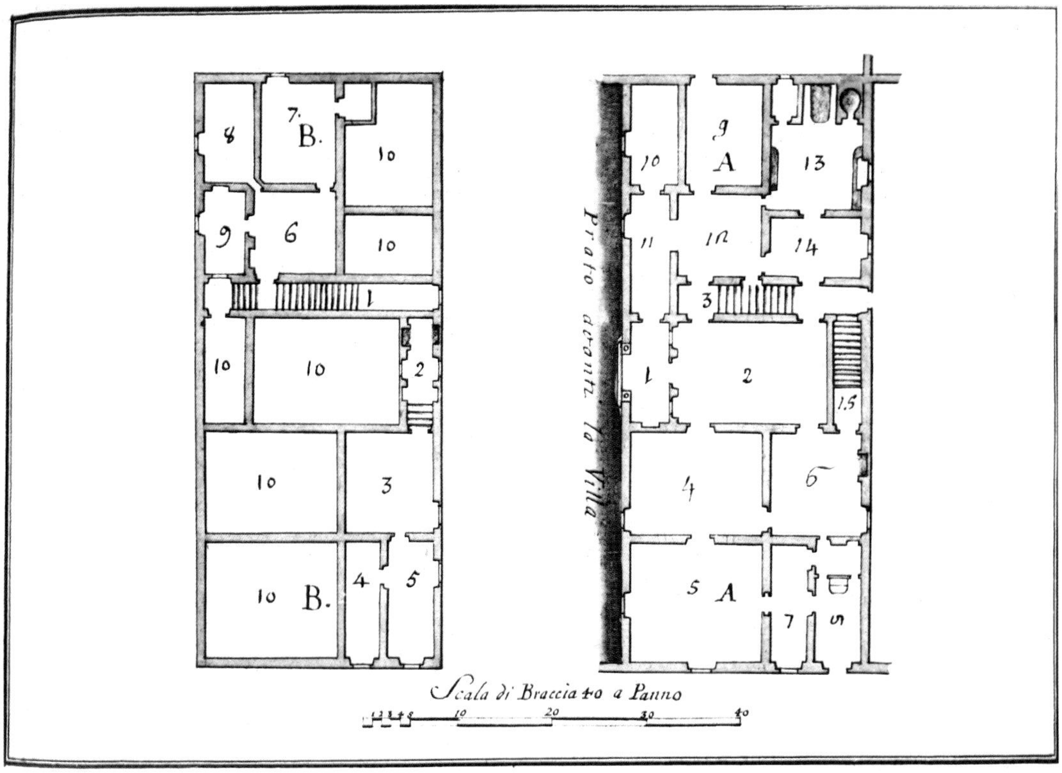

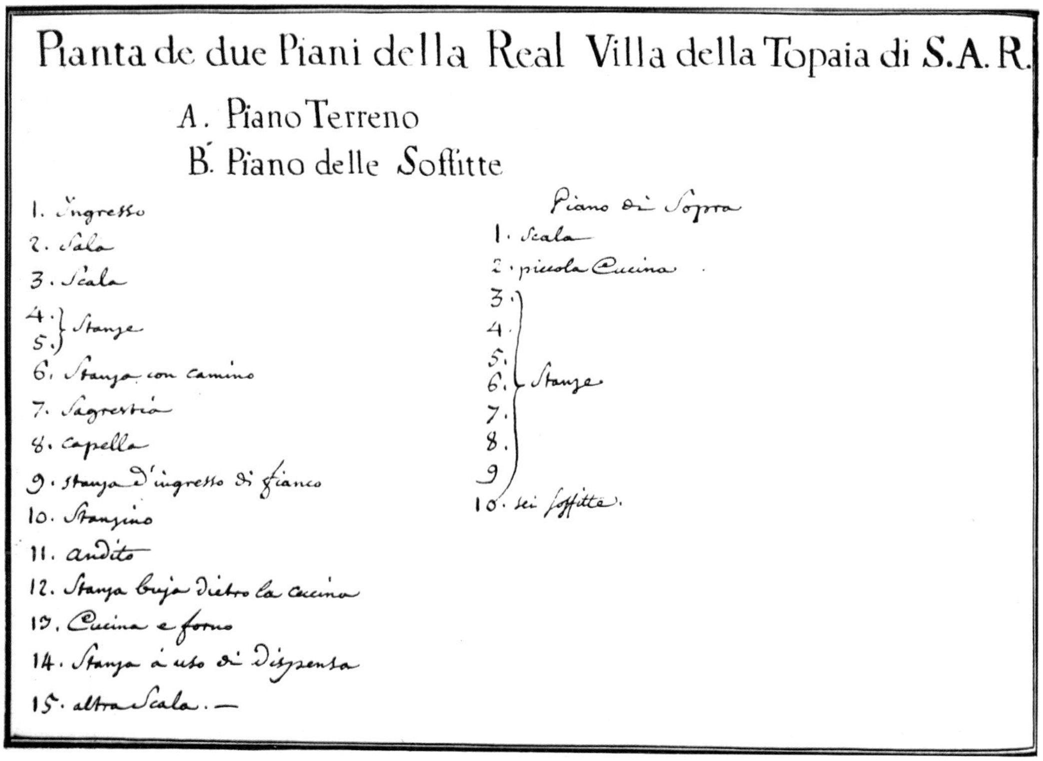

Pianta de due Piani della Real Villa della Topaia di S.A.R.

A. Piano Terreno

B. Piano delle Soffitte

1. Ingresso
2. Sala
3. Scala
4. } Stanze
5. }
6. Stanza con camino
7. Sagrestia
8. Capella
9. Stanza d'ingresso di fianco
10. Stanzino
11. andito
12. Stanza buja dietro la cucina
13. Cucina e forno
14. Stanza a uso di Dispensa
15. altra Scala. —

Piano di Sopra

1. Scala
2. piccola Cucina
3.–9. Stanze
10. sei soffitte.

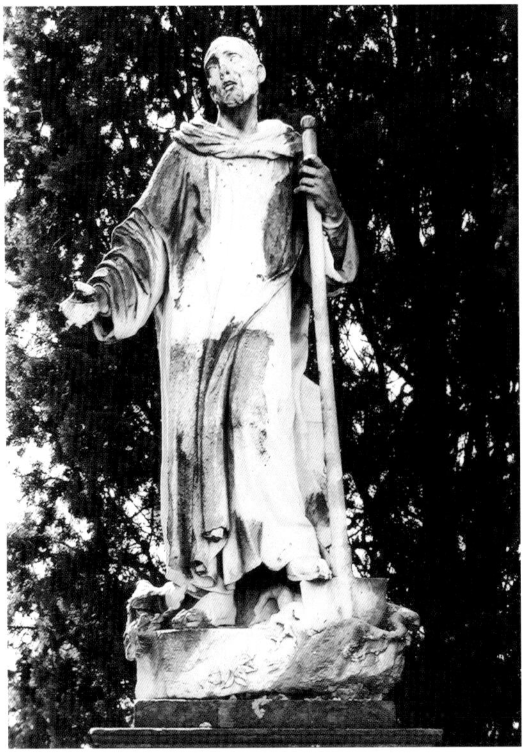

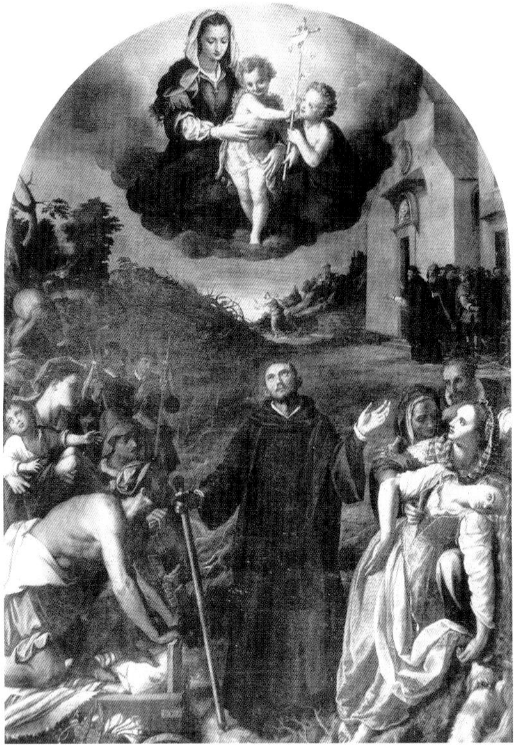

20. Giuseppe Piamontini (attributed to), *Saint Fiacre*, 1696, marble
Casino della Topaia, Florence; author photograph

21. Alessandro Allori, *Saint Fiacre Heals the Sick*, 1596, oil
Sacristy of Santo Spirito, Florence

This larger-than-life-size figure was probably made by the Florentine sculptor Giuseppe Piamontini (1664–1742) and can be dated by its inscription to 1696. Because it represents a very uncommon saint yet occupies a prominent position in the garden, it deserves explanation.[25] Saint Fiacre was a hermit of Irish origin who lived in the seventh century.[26] According to a tradition recorded by Hector Boethius (c. 1465–1536) in his book of Scottish history—*Scotorum historiae a prima gentis origine, cum aliarum et rerum et gentium illustratione non vulgari*—originally published in Paris in 1526, Fiacre was the son of a Scottish king.[27] The inscription on the pedestal clearly refers to this later tradition.[28] But, as it explains, Fiacre refused temporal power and instead of living the privileged life of a royal prince, he began a pilgrimage that brought him to Meaux, in northern France. There, Bishop Faro gave him a piece of land in a forest where Fiacre established his hermitage.[29] The legend tells us that Saint Fiacre, instead of digging, simply put his rod into the soil, which opened up, sprouting plants and fruit. Thus Saint Fiacre became the patron saint of gardeners, and that is why Cosimo III chose him as the custodian, patron, and benefactor of this *locus amoenus*.

Saint Fiacre was venerated mostly in northern countries such as France and Ireland. Although almost completely unknown in Italy, there is one important exception: Florence. It was the French princess Christine of Lorraine (1565–1636), wife of Grand Duke Ferdinand I and great-grandmother of Cosimo III, who introduced the cult of the venerated saint, as an herbalist and a healer, to the city on the Arno. She apparently did so both for her own consolation and for her people. Christine commissioned Alessandro Allori to create a large painting for the altar in the sacristy of the church of Santo Spirito (fig. 21). The painting was placed directly in front of the entrance, where it was installed in 1596, as an inscription testifies.[30] Perhaps Cosimo III intentionally erected the statue of Saint Fiacre in his garden exactly one hundred years later. Regardless of the motive, Fiacre was the ideal saint to express Cosimo's particular religious devotion to nature. But he did more than erect a garden statue of Saint Fiacre. In December 1695 he also asked Jacques Bénigne Bossuet, bishop of Meaux, for relics

22. Giovanni Battista Foggini (attributed to), stucco decoration with fruit motifs, c. 1696
Casino della Topaia, Florence; author photograph

of the saint to put in his private chapel in the Casino, thereby reviving the cult of Saint Fiacre at the Topaia.[31]

Unfortunately, there is no way to determine the precise original arrangement of Bimbi's paintings in the villa, but there is no doubt that at the end of the decoration campaign there was hardly a space left without a painting. Very probably the sets of large pendants—which were conceived to be seen together—hung in the same room, and probably the earlier painted pairs hung in the more intimate areas of the apartment—that is, the *camera,* or bedroom, and the two *stanze,* or living rooms. Those painted last—the four citrus fruit canvases—were probably in the entrance hall, which means that they were in a location analogous to that of the Allori fresco in the main hall of the Villa Poggio a Caiano (fig. 13). The stucco decoration in the vaults of the Casino, probably designed by Foggini, has motifs of fruit and foliage similar to those in Bimbi's paintings and their frames, an important clue that the building was conceived together with the plan for its decoration (fig. 22).

Two inscriptions in the hall substantiate this uniformity of concept, because they allude to the decoration and to the statue of Saint Fiacre in the garden. In a cartouche on the left side we read *LAETITIAE COSMUS DATOR,* which has a very interesting double sense and may be read either as "The cosmos is the giver of happiness" or "Cosimo is the giver of happiness." The inscription on the right-hand side, *REGIA MANV SUNT SATA,* directly alludes to the inscription on the pedestal of the garden statue of Saint Fiacre, *REGIA MANU SATAS.* This translates as "By the royal hand they were sown," of course referring to the grand ducal vineyards and orchards around the Casino, but referring also to all the fruits depicted by Bimbi in its rooms. The meaning of those *ritratti al naturale,* however, goes beyond pure, naturalistically faithful representation and the aesthetic delight that Cosimo wanted to provide for those who saw them. Several years ago, the art historian Ilaria della Monica studied the writings of the learned Jesuit Paolo Segneri (1624–1694) with regard to his concept of nature.[32] Segneri was a central figure at the Medicean court and, as father confessor to Cosimo III, exerted a great influence on him. In books dedicated to the members of the grand ducal family, Segneri expressed his religiously determined concept of nature. For Segneri, the multiplicity and variety of the products of nature gave visible evidence of the manifold creation of God.

In 1690 Segneri published a book entitled *L'incredulo senza scusa,* in which he formulated his criticism against the philosophical doctrine of atomism, which he considered too mechanistic and potentially agnostic.[33] Instead of studying nature exclusively with reason and intellect, Segneri invited his readers to open their hearts to perceive and recognize the beauty, the variety, and the wonders of nature. For Segneri, the richness of nature was the expression of the magnificence of God, of divine goodness and grace.[34] According to him this was also true of *meraviglie* and *bizzarie,* the wonderful and bizarre products of nature, some of which Cosimo had had painted for his Casino. But Segneri also saw the divine will of creation in the obviously horrible deformities of nature, like the monstrous

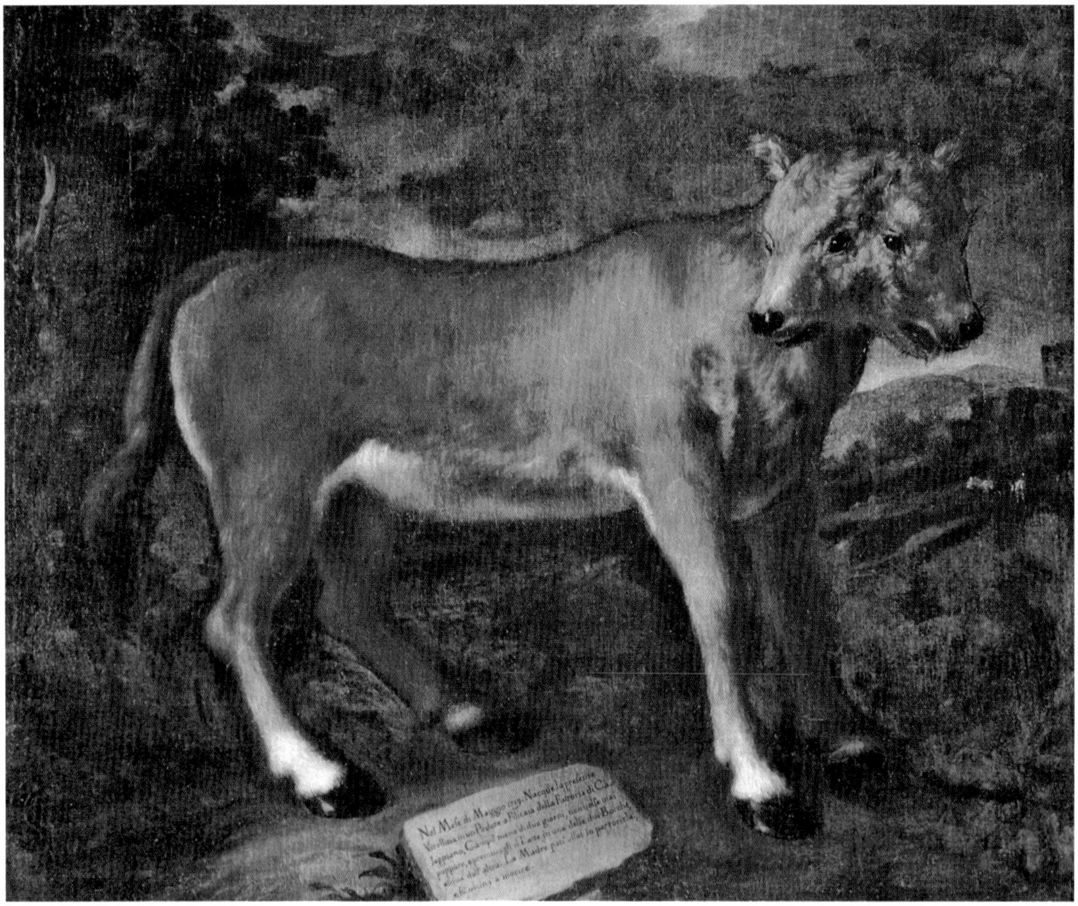

23. Bartolomeo Bimbi, *Two-headed Calf*, 1719, oil
Palazzo Pitti, Florence

two-headed calf and the two-headed lamb, born alive on Tuscan farms, which Bimbi portrayed at Cosimo's direction for the Villa Ambrogiana (fig. 23). This idea was not new, because Saint Augustine had already affirmed in his writings that even human monstrosities were created by a good and just God; that they, like all humans, were affected by original sin; and that they could enter into paradise.[35] This concept obtained broader acceptance and validity through the Council of Trent (1545–1563), and continued to play an important role in the eighteenth century, in the discussion undertaken by the Académie des sciences in Paris about the theory of biological preformation.[36]

With the discovery of the ovary in mammals around the mid-seventeenth century, the idea of the origin of human and animal life changed profoundly. The idea that a future being was already completely laid out in the ovule and, in order to develop, it had only to grow to full size, soon became accepted as the "preformation doctrine." It is not known how far and how deeply the naturalists at the Medicean court reflected upon this theory, but in Segneri's opinion, monstrous animals were visible evidence of the divine origin of creation, because the mechanistic interpretation of nature offered by the atomists apparently provided no convincing explanation for the biological folly of such creatures. The natural philosophical background of Bimbi's paintings was, of course, influenced by those religious ideas, and that is why, with his *ritratti al vivo,* Cosimo III glorified not only the variety of nature but also the richness of divine creation itself. Regarding the philosophical debate about nature between the Aristotelians and the atomists, the position of the grand duke was unequivocal: he terminated the teaching of atomism at the University of Pisa.[37]

The horizon of religious interpretation outlined very briefly here is but one aspect in better understanding this group of Bimbi's paintings. Equally relevant is Cosimo's self-definition as "Christian prince," and the specialized collection of paintings in the Casino della Topaia is, in some ways, a result of this notion. It was the Jesuit Roberto Bellarmine (1542–1621) who characterized the ideal Christian ruler and his tasks and duties in his book *De officii principi christiani.*[38] According to Bellarmine, the prince resembles God himself in his virtue of magnificence, and, as God the creator in his wisdom conferred extreme richness to nature, Bellarmine declared the prince to be the caretaker of nature with the important task of protecting and cultivating God's creation. This idea was developed further and became commonplace in the literature of the *Fürstenspiegel* during the seventeenth and eighteenth centuries.[39] No doubt Cosimo III was perfectly aware of the concept of the relation between prince and nature, and Cosimo was particularly able to identify himself with the role and the duties of a curator of nature, as the inscriptions in the hall of the Casino della Topaia testify. In the context of this essay, however, it is noteworthy that Cosimo was not generally interested in products of nature, but that he asked Bimbi to paint subjects of Tuscan flora and fauna exclusively. This means that the decorative concept of the Casino had no universal encyclopedic character, but rather that Cosimo restricted the program of decoration to fruit that flourished in his own domain.

In view of the universal endeavors of botanists at that time, this approach may seem somewhat unusual and archaic. But museum theory of the period expressly formulated that princely collections should represent the princely dominion.[40] In the context of the generally twofold division of princely collections into naturalia and artificialia, Bimbi's paintings represent naturalia, although they themselves are artifacts. The complexity of naturalia was again subdivided according to museum theory into the hierarchical order of animalia, vegetabilia, and mineralia. Traditionally, any objects that could not be collected because of their perishable or transitory nature were represented in a collection by pictorial depictions. And in fact, Cosimo did not collect or display any real plants in the Casino but exclusively used the gilt-framed gallery painting to re-create visually a particular natural phenomenon. In so doing, he obviously followed Horatio's principle of *delectare et prodesse,* to delight and to teach. If the paintings were supposed to offer delight for the eyes of the Prince and his guests, the labels and inscriptions were instead meant to instruct them as to the enormous variety of Tuscan nature. In fact, the terms "delight" and "usefulness" constitute a famous literary theme found frequently in the panegyric and botanical literature of the seventeenth century. Francesco Redi used it often, and he always distinguished clearly between the "delight" addressed to the prince, as benefactor of the collection, and the "usefulness" addressed to the naturalists.

Although in the Casino della Topaia inscription *LAETITIAE COSMUS DATOR* this distinction is somewhat diluted, Bimbi's botanically labeled gallery paintings still owe their specific form to this double function. On the whole, they glorify God's work of creation in Tuscany and they testify that this country thrives under the most excellent reign of its Christian prince and caretaker of nature. Cosimo's biographer Fra Domenico Sandrini (1673–1750) therefore wrote correctly around 1725: "It was to his glory that almost the entire state became the most fertile of all Italy under his care."[41] Bimbi's paintings in the Casino were also conceived as a symbolic expression of the way in which the Medici dominion's peacefulness and prosperity were pleasing to God.

NOTES

I acknowledge the kindness of the present owners of the Casino della Topaia, who made it possible for me to examine the structure of the building. I also express my gratitude to Riccardo Spinelli, Eike Schmidt, Wolfger Bulst, and Birgit Laschke, who discussed aspects of this work with me.

1. For this process in general, see Wolf Lepenies, *Das Ende der Naturgeschichte: Wandel kultureller Selbstverständlichkeiten in den Wissenschaften des 18. und 19. Jahrhunderts* (Munich, 1976), and Thomas L. Hankins, *Science and the Enlightenment*, Cambridge History of Science (Cambridge, 1985).

2. For more detail regarding scientific advances and collecting, see Paula Findlen, *Possessing Nature: Museums, Collecting, and Scientific Culture in Early Modern Italy* (Berkeley and Los Angeles, 1996), 217–220. On Redi, see Enrica Micheli Pellegrini, *Francesco Redi, letterato e poeta* (Florence, 1911); Dino Prandi, *Bibliografia delle opere di Francesco Redi; segue la riproduzione della lettera intorno all'invenzione degli occhiali scritta a Paolo Falconieri, dall'edizione delle "Lettere," 1779–1795* (Reggio Emilia, 1941); Luigi Belloni, *Francesco Redi, biologo* (Pisa, 1958); Lorella Mangani and Giuseppe Martini, eds., *Francesco Redi Aretino: Atti del Convegno di Studi, Arezzo, febbraio 1998* (Arezzo, 1999); *Francesco Redi, un protagonista della scienza moderna: Documenti, esperimenti, immagini*, ed. Walter Bernardi and Luigi Guerrini, Biblioteca di Nuncius, studi e testi 33 (Florence, 1999).

3. Riguccio Galuzzi, *Istoria del Granducato di Toscana sotto il Governo della Casa Medici*, 5 vols. (Florence, 1781), 409. The Gabinetto d'Istoria Naturale was then in the Villa Ambrogiana and would later be directed by Lorenzo Magalotti.

4. Francesco Redi, *Scritti di botanica, zoologia e medicina* (1685), ed. Piero Polito, Classici della società italiana 4 (Milan, 1975); Redi's text has been edited and annotated by Peter K. Knoefel, *Franceso Redi on Vipers* (Leiden and New York, 1988). For the metaphor of the "book of nature," see Hans Blumenberg, *Die Lesbarkeit der Welt*, 4th ed. (Frankfurt am Main, 1999).

5. For the Museum Kircherianum, see Giorgio de Sepi, SJ, *Romani Collegii Societatis Jesu Musaeum celeberrimum* (Amsterdam, 1678); Filippo Bonanni, SJ, *Musaeum Kircherianum sive Musaeum a P. Athanasio Kirchero in Collegio Romano Societatis Jesu iampridem incoeptum nuper restitutum, auctum, descriptum & Iconibus illustratum* (Rome, 1709); David Murray, *Museums, Their History and Their Use*, 3 vols. (Glasgow, 1904), 3:133–134; P. Conor Reilly, SJ, *Athanasius Kircher: Master of a Hundred Arts* (Wiesbaden and Rome, 1974); Joscelyn Godwin, *Athanasius Kircher: A Renaissance Man and the Quest for Lost Knowledge* (London, 1979); Thomas Leinkauf, *Mundus combinatus: Studien zur Struktur der barocken Universalwissenschaft am Beispiel Athanasius Kirchers SJ, 1602–1680* (Berlin, 1993); Thomas Leinkauf, " 'Mundus combinatus' und 'ars combinatoria' als geistesgeschichtlicher Hintergrund des Museum Kircherianum in Rom," *Macrocosmos in Microcosmo: Die Welt in der Stube, zur Geschichte des Sammelns 1450–1800*, ed. Andreas Grote (Opladen, 1994), 535–553.

6. For some aspects of mutual influences between art and science during this period, see Giuseppe Olmi, "Natura morta e illustrazione scientifica," in Francesco Porzio, Federico Zeri, and Carlo Pirovano, eds., *La natura morta in Italia*, 2 vols. (Milan, 1989), 1:69–91, and Giuseppe Olmi, "Ordine e fama: Il museo naturalistico in Italia nei secoli XVI e XVII," *Annali dell'Istituto storico italo-germanico in Trento* 8 (1982): 225–274.

7. The most important literature on Bimbi can be found in Maria Letizia Strocchi, "Bartolomeo Bimbi pittore naturalista alla corte di Cosimo III dei Medici," in *Agrumi, frutta e uve nella Firenze di Bartolomeo Bimbi pittore mediceo* (Florence, 1982), 7–16, with bibliography. See also Marco Chiarini, ed., *Il giardino del Granduca: Natura morta nelle collezioni medicee* (Turin, 1997); Silvia Meloni Trkulja and Lucia Tongiorgi Tomasi, eds., *Bartolomeo Bimbi: Un pittore di piante e animali alla corte dei Medici* (Florence, 1998); Daniela Savoia and Maria Letizia Strocchi, eds., *Le belle forme della natura: La pittura di Bartolomeo Bimbi (1648–1730) tra scienza e "maraviglia"* [exh. cat., Biblioteca Malatestiana, Cesena] (Bologna, 2001); Lucia Tongiorgi Tomasi and Gretchen A. Hirschauer, *The Flowering of Florence: Botanical Art for the Medici* [exh. cat., National Gallery of Art] (Washington, 2002).

8. The Dutch intarsia-worker, wood-carver, and ivory-carver Vittorio Crosten is documented in Florence from 1673. In the summer of 1688, he worked on the decoration for two coaches which were used on 8 January 1689 for the nuptial procession of Prince Ferdinando de' Medici and the Wittelsbach princess Violante Beatrix von Bayern. Between 1694 and c. 1715, Crosten made the frames for Bimbi's paintings, which were intended for the Casino della Topaia. Moreover, he carved ivory reliefs for a cupboard designed by Giovan Battista Foggini for the Appartamenti Monumentali of the Palazzo Pitti (completed in 1716) and made the ivory frame for the lens of Galileo Galilei's telescope (Museo di Storia della Scienza). Crosten carved a great wooden chandelier and furniture for the Palazzo Pitti. See Eike D. Schmidt, "Balthasar Permosers Schnitzwerk für den Brautwagen der Prinzessin Violante Beatrix von Bayern," *Opere e giorni: Studi su mille anni di arte europea dedicati a Max Seidel*, ed. Klaus Bergdolt and Giorgio Bonsanti (Venice, 2001), 673–686, with bibliography.

9. Most important in this sense are the studies published in Strocchi 1982.

10. See Chiarini 1997, 360. See also Trkulja and Tongiorgi Tomasi 1998 for provenances in individual catalogue entries.

11. A coat of arms of alliance on the exterior corners of the older (west) parts of the building shows the Medici balls and fragments of the Toledo arms. From this we can deduce a dating before 1562 (death of Cosimo's first wife, Eleonora of Toledo, 1519–1562).

12. Giovanni Fanelli, *Firenze: Architettura e città* (Florence, 1973; reprint 2003), 273–274.

13. Several elements of the Casino della Topaia correspond to Foggini's documented works in Florence. The very simple decoration is similar to that of Foggini's public granary in the Piazza di Cestello (from 1694 on). The cartouches with inscriptions in the main hall are nearly identical to the stucco cartouches around the oval ceiling fresco of the Palazzo Viviani in the Via Tornabuoni. In addition, the heads of the angels in the chapel of the villa are very similar to the heads of the cherubs in the church of Santa Maria di Candeli in Via dei Pilastri, all in Florence. For a more detailed discussion of the attribution, see Riccardo Spinelli, *Giovan Battista Foggini: "Architetto primario della casa Serenissima" dei Medici, 1562–1725* (Florence, 2003).

14. These measurements were taken by the author; not all of the measurements in the Bimbi monograph of 1998 are reliable.

15. The four canvases are numbered with Roman numerals, probably corresponding to the chronological order in which they were made. In the first painting, Bimbi depicted only twenty varieties of fruit, allowing him to arrange them more loosely than in the later pictures, where he had to place thirty different varieties in the same space.

16. Giovanni Battista Ferrari, *Hesperides sive de malorum aureorum cultura et usu libri quatuor* (Rome, 1646). Ferrari's book emerged from the circle of the diplomat, scientist, and collector Cassiano dal Pozzo (1588–1657) and is considered to be a milestone in the history of citrological studies because it was based mainly on empirical research. See David Freedberg and Enrico Baldini, *Citrus Fruit* (London, 1997). Ferrari's book greatly advanced the international literature on citrology, among which was the treatise of the Florentine court botanist Pietro Nati, *Florentina phytologica observatio de malolimonia citrata-aurantia, vulgo bizzaria* (Florence, 1674); Franciscus van Sterbeeck, *Citricultura oft regeringhe der uythemsche Boomen te weten Oranien, Citroenen, Limoenen, Granaten, Laurieren en andere* (Antwerp, 1682); Ulysses Aldrovandi, *Dendrologiae naturalis scilicet arborum historiae, libri duo, Sylva glandaria, acinosumo, pomarium ubi eruditiones omnium generum una cum botanicis doctrinis ingenia quaecumque non parum iuvant, et oblectant, Ovidius Montalvanus* (Bologna, 1668); Jan Commelyn, *Nederlantze Hesperides, dat is Oeffening en Gebruik van de Limonen- en Oranje-Bloemen; Gestelt na den Aardt, en Climaat der Nederlanden* (Amsterdam, 1676); Johann Christoph Volckamer, *Nürnbergische Hesperides, oder Gründliche Beschreibung der Edlen Citronat, Citronen und Pomarantzen Früchte* (Nuremberg, Frankfurt, and Leipzig, 1708–1714); also published in Latin as *Hesperidum Norimbergensium, sive, De malorum citreorum limonum aurantiorumque: Cultura et usu libri IIII bene multis iconibus in aes elegantissime, incisis ornati . . .* (Nuremberg, 1713).

17. Ada Segre, "Alla ricerca dell'orto delle esperidi: Un mito per i giardini d'agrumi," in *Il Giardino delle Esperidi: Gli agrumi nella storia, nella letteratura e nell'arte,* ed. Alessandro Tagliolini and Margherita Azzi Visentini (Florence, 1996), 19–40; David Freedberg, "From Hebrew and Gardens to Oranges and Lemons: Giovanni Battista Ferrari and Cassiano dal Pozzo," in *Cassiano dal Pozzo: Atti del Seminario Internazionale di Studi,* ed. Franceso Solinas (Rome, 1989), 36–72.

18. Erkinger von Schwarzenberg and Beatrice Paolozzi Strozzi, "Norzia o la costante Fortuna, la lunetta di Alessandro Allori a Poggio a Caiano," *Die Kunst des Cinquecento in der Toskana* (Munich, 1992), 197–206. The famous Florentine naturalist Francesco Redi referred to this model of identification when he wrote about Cosimo III in 1680: "And just as the transfer of the citrons from the African gardens of the Hesperides to Greece belongs to the glorious deeds of Hercules, so the sending of all possible foreign plants to the botanical gardens of Florence and Pisa belongs to the fame of our Prince Cosimo." MS Redi, Filza 26, c. 29r, Biblioteca Marucelliana, Florence.

19. As inscribed on the label shown in the painting, "Un sol granello di grano l'anno 1713 nell'orto di Francesco Magnelli, Livellario dello Spedalo di S. Maria Nuova à Legnaia nel popolo di S. Piero a Monticelli produsse le Cento ottanta spighe nella forma expressa in questa Tela."

20. Christina Hofmann-Randall, *Monster, Wunder und Kometen: Sensationsberichte auf Flugblättern des 16. bis 18. Jahrhunderts* [exh. cat., Universitätsbibliothek Erlangen] (Erlangen, 1999), 64–65.

21. Erlangen 1999, 54–55.

22. This is Filippo Napolitano's canvas in the Museo Botanico, Università degli studi di Firenze. See Mina Gregori, "La natura morta in Toscana," in Porzio et al. 1989, 2:529. Another early example (dated 1626), by an anonymous Netherlandish artist and labeled like Bimbi's paintings, is the representation of a giant radish and a carnation in the Rijksmuseum in Amsterdam. See Lucia Tongiorgi Tomasi, "Bartolomeo Bimbi 'pittore eccellente nei fiori, nelle frutte e negli animali' e la tradizione Toscana dell'immagine naturalistica," in Meloni Trkulja and Tongiorgi Tomasi 1998, 17–37, fig. 1.

23. This garden plan corresponds to the ideal garden as described in contemporary garden theory; see, for example, Jean de la Quintinye, *Il giardino francese overo trattato del tagliare gl'alberi da frutto, con la maniera di ben allevarli* (Venice, 1686). The circular fountain in the middle of the orchard is not original.

24. The west side of the building is the older part, dating to the sixteenth century. Although the interior partitioning was slightly changed by Grand Duke Cosimo III, there were no vaulted rooms in this area, and consequently there was room for servants on the upper floor (plan at left in figure 19). The west side of the ground floor, with its kitchen and larder, was used for housekeeping purposes.

25. The attribution is based on stylistic evidence. The general composition of the figure, its posture, and the style of the drapery are similar to Piamontini's statue *Saint John the Baptist* (1688) in the Florentine Baptistery and to his *Saint Bernard of Clairvaux Triumphing over the Demon* (1702) in the court of the former Cistercian cloister of San Frediano in Cestello (today the Seminario Maggiore) in Florence. See Giovanni Pratesi, ed., *Repertorio della scultura fiorentina del seicento e settecento,* 3 vols. (Turin, 1993), 1:55–56 and 3: figs. 407, 414, 415.

26. Literature on Saint Fiacre includes *Acta Sanctorum Augusti* (1743), 6:604 and following; *Catholic Encyclopedia* 7:67; James F. Kenney, *The Sources for the Early History of Ireland* (New York, 1929), 1:493; Louis Gougaud, "Le Culte de Saint Fiacre ermite," in *Revue Mabillon* 26 (1936): 197–201; Louis Gougaud, *Les Saints irlandais hors d'Irlande* (Louvain, 1936), 86–92; *Catholicisme* 4:1244 and following; Gaetano Moroni, *Dizionario di erudizione storico-ecclesiastica da S. Pietro sino ai nostri giorni* (Venice, 1844) 24:233; *Biblioteca sanctorum* (Rome, 1964) 5:668–672; Michael Buchberger, *Lexikon für Theologie und Kirche* (Freiburg, 1930–1938), 4:113.

27. Hector Boethius (Boece), *Scotorum historiae a prima gentis origine, cum aliarum et rerum et gentium illustratione non vulgari, libri XIX . . . continuatio per Joannem Ferrerium Pedemontanum recens & ipsa scripta & edita. Quae in hoc opere praecipua continentur sequens pagina indicabit* (Paris, 1574), 9:173. This seems to be the first time that the saint is declared to be not an Irishman but the son of a Scottish king.

28. The inscription reads *D. FIACRIO EVGENII IV. SCOTORVM/REGIS FILIO. CVI SOLITVDO PARADISVS/ET AD PARADISVM VIA. QVI AD REGNVM/DEI TOTO PECTORE ASPIRANS REGNVM/HOMINVM INGENTI ANIMO CONTEMPSIT/AC RESPVIT. IN AGRO MELDENSI. A S./PHARONE LOCI EPISCOPO SYLVA. ET/HORTVLO EI OB PIETATEM ASSIGNATO/ANIMAE SVAE SPONSVM QVEM DILIGEBAT/FLORES INTER ARBORESQ. REGIA MANV/SATAS QVAESIVIT. INVENIT NEC DIMISIT/COS. III. MAG. DVX ETRVRIAE/HARUM AMOENITATVM CVSTODI/PATRONO ET BENEFACTORI SVO/A. S. M D C LXXXXVI.*

29. This site eventually became the town of Saint-Fiacre-en-Brie.

30. Simona Lecchini Giovannoni, *Alessandro Allori* (Turin, 1991), 285, figs. 337–342. The first design (Uffizi 10259 F) showed the traditional iconography of the saint as patron of gardeners, but in the execution the theme was changed to a more spectacular scene of his miraculous healings together with a vision of the Madonna and Child with St. John the Baptist depicted in a cartouche in the heavens. Probably the new subject was expected to call forth greater popular devotion.

31. For more on this topic, see Charles Urbain and Eugene Levesque, eds., *Correspondance de Bossuet: Nouvelle édition augmentée de lettres inédites et publiés avec des notes et des appendices sous le patronage de l'académie française,* Les Grands Écrivains de la France, vol. 7 [January 1695–June 1696] (Paris, 1913): 257, no. 1308 bis, 287–288, 525–528. The relics were sent to Florence in January 1696; the year 1696 is also inscribed on the pedestal of the garden statue of the saint, which was obviously conceived together with the design of the villa. See also Johann Evangelist Stadler, *Vollständiges Heiligen-Lexikon* (Augsburg, 1861; reprint, Hildesheim and New York, 1975), 2:202.

32. Ilaria della Monica, "Bartolomeo Bimbi, Paolo Segneri e Cosimo III," *Artista, critica d'arte in Toscana* (1992): 36–65.

33. Paolo Segneri, *L'incredulo senza scusa, quale sia la vera religione* (Florence, 1690).

34. This notion was still current in the eighteenth century; see, for example, Irmgard Müsch, "Der Blick auf die Natur als Gottesbeweis: Johann Jakob Scheuchzers Kupfer-Bibel (1731–35)" in *Wahrnehmung der Natur—Natur der Wahrnehmung: Studien zur Geschichte visueller Kultur um 1800,* ed. Gabriele Dürbeck, et al. (Dresden, 2001), 87–116.

35. Augustine, *De civitate Dei,* 16:8, and *Contra Pelagos,* 1:16.

36. Bettina Sandritter, "Die Auseinandersetzung zwischen Louis Lémery und Jacques Benigne Winslow über die Entstehung von Missbildungen 1706–1743" (Ph.D. diss., Tübingen University, 1991), 95–101. The first large scientific compendium on anomalies appeared only in the nineteenth century under the authorship of Isidore Geoffroy Saint-Hilaire, son of Étienne Geoffroy Saint-Hilaire, the founder of modern teratology, with the title *Histoire générale et particulière des anomalies . . . ,* 4 vols. (Paris, 1832–1837). For the history of interpretations of natural anomalies, see Corrado Bologna, "Mostro," *Enciclopedia Einaudi* (Turin, 1980), 9:556–580.

37. For atomism and Aristotelianism in the seventeenth and eighteenth centuries, see Marie Cariou, *L'Atomisme, trois essais: Gassendi, Leibniz, Bergson et Lucrèce* (Paris, 1978); Konrad Moll, *Der junge Leibniz,* vol. 2, *Der Übergang vom Atomismus zu einem mechanistischen Aristotelismus: Der revidierte Anschluss an Pierre Gassendi* (Stuttgart, 1982); Lynn Sumida Joy, *Gassendi the Atomist: Advocate of History in an Age of Science* (Cambridge, 1987).

38. Roberto Bellarmine, SJ, *De officii principi christiani, libri tres* (Rome and Antwerp, 1619). The book is dedicated to Wladislaus, son of Sigismund III, King of Poland.

39. See, for example, Johann Jacob von Weingarten, *Fürsten-Spiegel oder Monarchia des höchlöblichen Ertz Hauses Österreich* (Prague, 1673), 660 and 670.

40. This concept was probably first expressed in Samuel Quiccheberg, *Inscriptiones vel tituli theatri amplissimi* (Munich, 1565). See Harriet Roth, *Der Anfang der Museumslehre in Deutschland: Das Traktat "Inscriptiones vel tituli theatri amplissimi" von Samuel Quiccheberg, lateinisch-deutsch* (Berlin, 2000).

41. In full: "Ma perché la sua propensione maggiore fu sempre all'agricoltura, e fin dai primi anni ebbe in essa moltissimo genio, egli ha avuto questa gloria, che quasi tutto il suo stato sia devenuto in coltura il giardino fertilissimo dell'Italia, poiché da esso la Nobiltà, e in Firenze, e in tutte le altre città della Toscana tutta ha appreso a metter molto l'animo nel coltivare si è veduto quasi da per tutto levare quanto vi era di salvatico, e rendere tutto domestico, e in quel modo, che è proprio della campagna di vicino a Firenze, che è quanto a dire, con modo e industria, che non vedesi altrove." Fra Domenico Sandrini, *Vita di Cosimo III Gran Duca di Toscana,* Biblioteca Nazionale Centrale di Firenze, Conventi soppressi, f. 9.978; vol. 2; II c. 110 v.

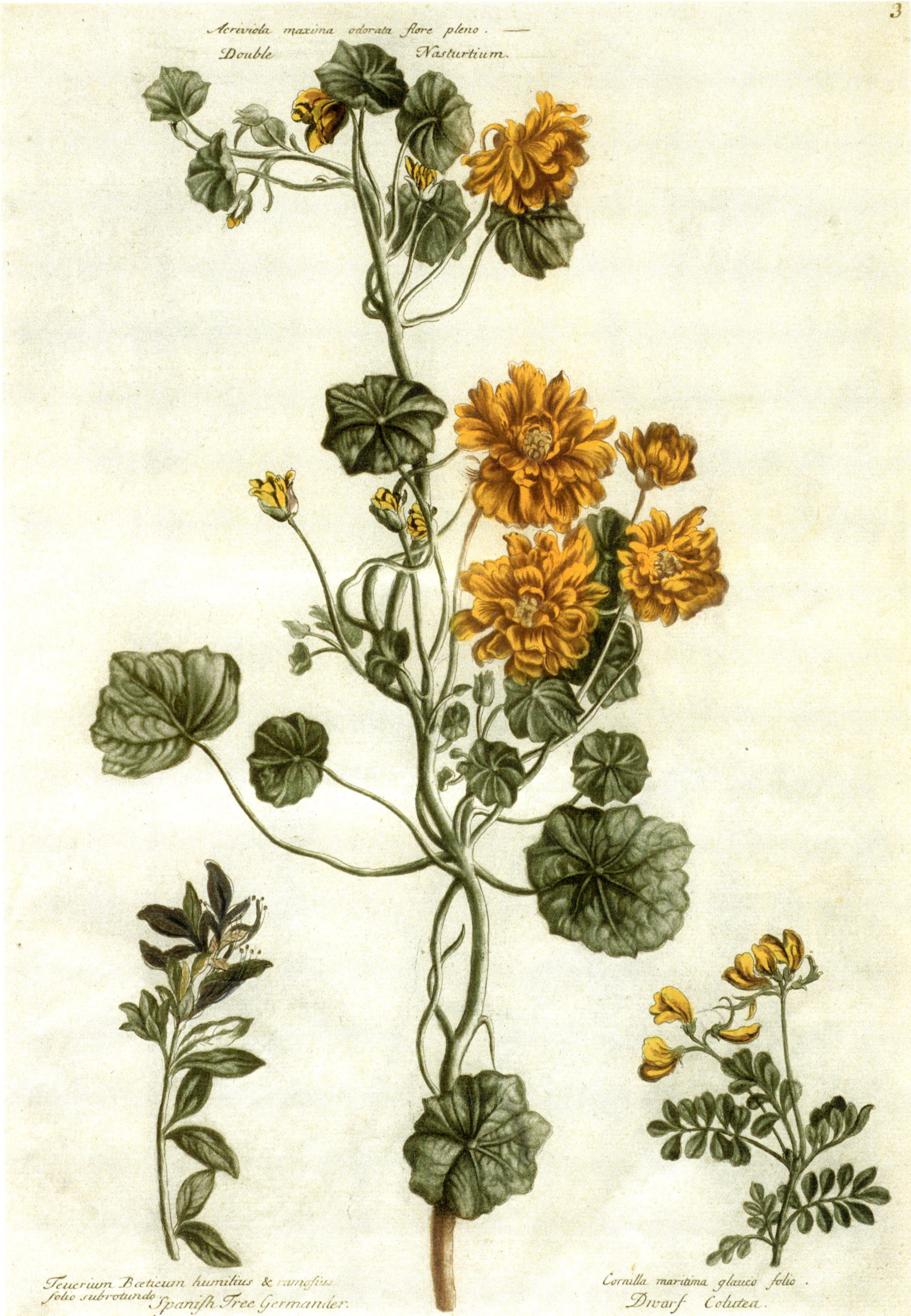
3
Acriviola maxima odorata flore pleno. —
Double Nasturtium.
Teucrium Bœticum humilius & ramosius folio subrotundo.
Spanish Tree Germander.
Cornilla maritima glauco folio.
Dwarf Colutea.

MARK LAIRD
Harvard University

The Congenial Climate of Coffeehouse Horticulture: The Historia plantarum rariorum *and the* Catalogus plantarum

In February 1722 the English naturalist-collector-artist Mark Catesby (1683–1749) embarked for Charleston, South Carolina. There, he would encounter a violent hurricane that caused the colony to suffer from inland flooding along the Savannah River. Catesby wrote home to his sponsor, the botanist William Sherard (1658/1659–1728), "The Deer were found frequently lodged on high trees." His later account, in *The Natural History of Carolina, Florida, and the Bahama Islands,* elaborated on this diluvial vision: "Panthers, Bears, and Deer, were drowned, and found lodg'd on the Limbs of Trees. The smaller Animals suffered also in this Calamity; even Reptiles and Insects were dislodged from their Holes, and violently hurried away, and mixing with harder Substances were beat in Pieces, and their Fragments (after the Waters fell) were seen in many Places to cover the Ground."[1]

It was also, coincidentally, in 1722 that the nurseryman Thomas Fairchild (1667–1729) published *The City Gardener* in London. The volume was intended to address another environmental issue. This was not a natural disorder, but what John Evelyn had called an "Avernus" of human making: the polluted air.[2] Fairchild explained how to grow plants that were tolerant of city smoke; he kept a personal record of what bloomed in his Hoxton nursery, month by month, from April 1722 to March 1723.[3]

When Mark Catesby returned to London in 1726, he took up residence with Thomas Fairchild in Hoxton. Perhaps they exchanged anecdotes about weather and other portentous events: the great storm of 26–27 November 1703[4] or the summer of 1725 (the coldest in the entire thermometer record)[5]—a summer that Catesby conveniently missed while studying fish, crustaceans, and plants in the Bahama Islands.

Gardeners have always had to contend with the vicissitudes of weather. But the 1720s and 1730s were exceptional in England as decades of climatic and horticultural change: a time of occasional extreme weather, yet a generally favorable environment for growing a palette of new plants. As the "Little Ice Age" of the sixteenth and seventeenth centuries began to wane, gardeners in England still encountered the "wide variability" of the previous hundred years.[6] While Fairchild and Catesby mused on how to grow a batch of North American plants in English soil, they found themselves blessed by an improving trend. This change meant that—overall at least—Carolina shrubs such as *Amorpha fruticosa* L., having survived the dismal summer of 1725, enjoyed warm winters and summers in their initial years in London. Apparently, too, the London smoke was not a hindrance to cultivating the plant known as bastard indigo.[7]

It is within this environmental matrix that one should review several new forms

Elisha Kirkall after Jacobus van Huysum, *Double Nasturtium [Tropaeolum majus* L. *flore pleno] with Spanish Tree Germander and Dwarf Colutea,* from Society of Gardeners, *Catalogus plantarum* (London, 1730), line engraving with mezzotint printed in color and hand-colored in watercolor
Typ 705.30.262, Department of Printing and Graphic Arts, Houghton Library, Harvard College Library

of botanical and horticultural literature that burgeoned at the end of the 1720s. For example, Robert Furber's "Borders of Cut Work," in *A Catalogue of English and Foreign Trees,* was printed in 1727; his *Twelve Months of Flowers* came out in 1730. Batty Langley's *New Principles of Gardening* came out by 1728, and the first twenty plates of Mark Catesby's *Natural History* appeared in 1729. Following the lead of Thomas Fairchild in *The City Gardener,* Langley's *New Principles of Gardening* promoted a new type of graduated planting. His account of how to embellish the front of a wilderness with ascending ranks of roses, lilac, laburnum, snowball tree, and tulip tree would prove the first inkling of a novel idea—the shrubbery of the picturesque pleasure ground (fig. 1).[8]

It might be misleading to imply by this that Catesby, Fairchild, and their contemporaries gauged the implications of the environmental moment. Yet, in engaging issues of classification and graduation, standardization and consumption, and, not least, the fraught question of acclimatization, these horticultural and botanical authors effectively contended with an unsettling interplay of human and biological forces. The plants of the American colonies were coming into a new home; they were placed in new planting structures, traded in a new marketplace, and subjected to all kinds of environmental stresses as well as the loving hand of personal cosseting.

From Watercolor to Mezzotint: The Production of the *Historia plantarum rariorum* and the *Catalogus plantarum*

Two volumes of this period are especially illuminating. One is *Historia plantarum rariorum,* by John Martyn (1699–1768). The initial plates of the *Historia* were printed in 1728, and the series was continued in five "decades," or parts, up to 1737. The *Historia* showcased rare plants from the Cape of Good Hope to the islands of the Caribbean. Many were introduced from the Carolinas, the Bahamas, Mexico, and the West Indies, by Mark Catesby and William Houston (1695 or c. 1704–1733). The other volume, the Society of Gardeners' *Catalogus plantarum,* published in 1730, illustrated exotic and

1. Reconstruction (1997), as elevation, of the narrative account "Manner of Disposing and Planting Flowering Shrubs," in Batty Langley, *New Principles of Gardening* (London, 1728), watercolor
Author's reconstruction

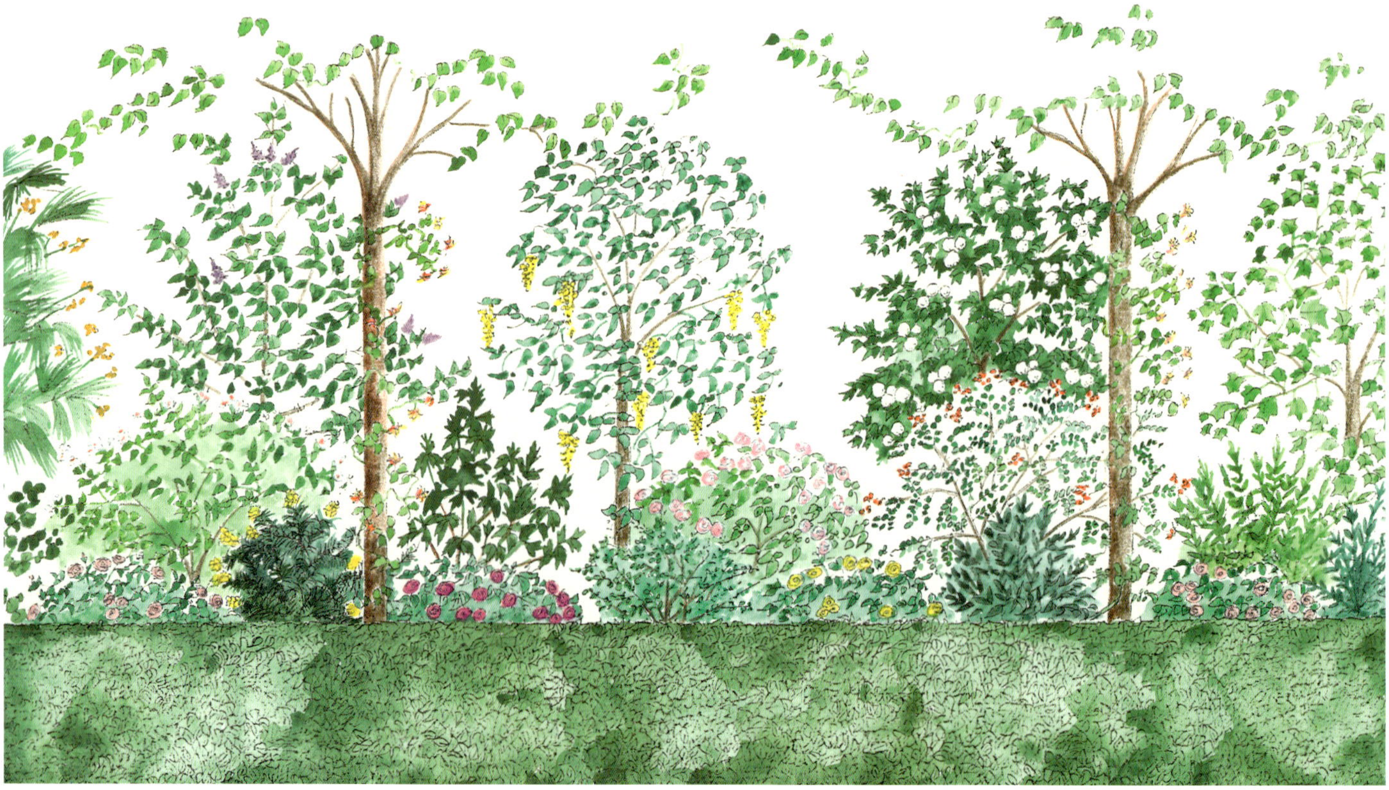

domestic plants "propagated for Sale, In the Gardens near London. Divided, according to their different Degrees of Hardiness."[9]

The plates of the *Historia* were engraved by Elisha Kirkall (or Kirkhall, c. 1682–1742) in what art historian Wilfrid Blunt called "a half-hearted imitation of true mezzotint." Blunt was referring to the fact that Kirkall had probably used an engraving tool known as a roulette wheel to create tone on certain areas of the printing plate, rather than a rocker, the traditional tool used to incise an overall dot pattern on the surface of the plate.[10] The plates of the *Catalogus* were etched or engraved in mezzotint by Henry Fletcher and Elisha Kirkall. The results were, in Blunt's rather blunt words, "no better and no worse than those in Martyn's *Historia*."[11] The Dutch artist Jacobus van Huysum (c. 1687–1746) painted most of the original plant portraits for both works. He was a younger brother of the great Dutch flower painter, Jan van Huysum. Jacobus, born in Amsterdam, had come to London in the early 1720s and, according to Horace Walpole, found employment with Walpole's father in Chelsea until his dismissal for "drunken dissolute conduct."[12] Little else seems to be known about him. After his work on the folios and some oil paintings, he disappears from the historical record. According to Blanche Henrey, he died in obscurity in 1746.

Some background to the *Historia plantarum rariorum* is provided in Henrey's *British Botanical and Horticultural Literature before 1800.* She writes of the *Historia:*

> It bears the imprimatur of the Royal Society, with the name of Sir Hans Sloane as President, and is dedicated to the President, Council, and Fellows. Shortly before the appearance of the first decade, a notice, dated 10 November 1727, was issued giving the conditions on which it was proposed to publish the work. One plate was to be dedicated to each subscriber, and the subscribers were asked to send their names with titles and arms to Elisha Kirkall, at Wine-office Court in Fleet Street, for engraving. The work is of particular interest as it contains some of the earliest examples of colour-printing from a single metal plate. These plates were executed by Kirkall in a mixture of line-engraving—for the titles and coats of arms of the dedicatees—etching, and mezzotinting. Many of the prints in the *Historia* are printed in green but some are printed in two or more colours from a single plate at one impression. All were afterwards touched by hand.[13]

Jacobus van Huysum's original folio of plant portraits is kept in the Royal Society Library. Hence it is still possible to compare the watercolor painting to the mezzotint reproduction: for example, the exquisite "Amaranthus spica" (*Celosia argentea* L.), its crimson-pinkish hues contrasting with the more muted version in the *Historia* (figs. 2 and 3).

Rather confusingly, the Royal Society folio also contains other van Huysum watercolors. These are duplicate or variant versions of plant portraits appearing in two British Museum folios that were the intended source of images for the *Catalogus plantarum.* Only some of the British Museum originals were actually used for published mezzotints, as will become clear below. A good example for comparing watercolor to mezzotint is the double nasturtium (*Tropaeolum majus* L. *flore pleno*), a plant that was subsequently lost to cultivation and reintroduced to gardening only in the twentieth century.[14] The better (or better-preserved) version is in the Royal Society folio, and it may be compared to its reverse depiction in the *Catalogus,* which is combined in the book's third plate with a Spanish tree germander and a dwarf colutea (figs. 4 and 5).

The manuscript note pasted on the flyleaf of the first volume in the British Museum helps explain the disjunction between the hundreds of watercolors and the mere twenty-one mezzotint plates. It reads: "The original paintings of plants by Jacobus van Huysum, some of which were publish'd in 21 plates by a Society of gardeners, anno 1730, & the rest design'd for a continuation of that work."[15] The continuation never materialized, probably for the financial reasons that the *Historia* circumvented by way of subscription plates. According to Blanche Henrey, these two albums were once in the library of Sir Hans Sloane, and the descriptive names on some of the paintings appear to be in the hand of the gardener Philip Miller (1691–1771). Of the twenty-one plates published in the *Catalogus plantarum,* seven were created by Elisha

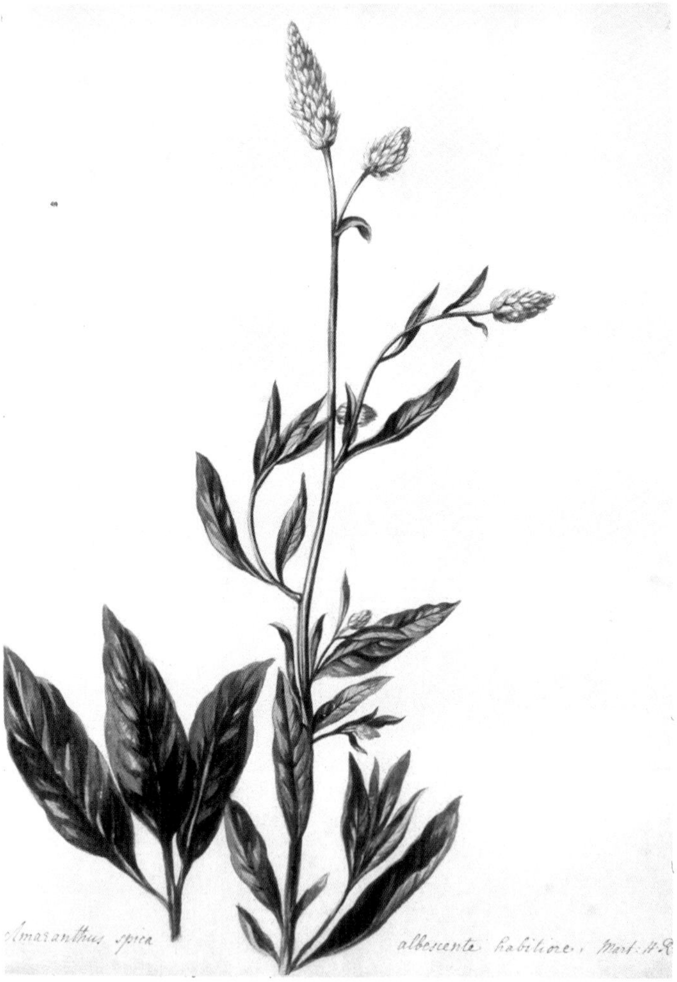

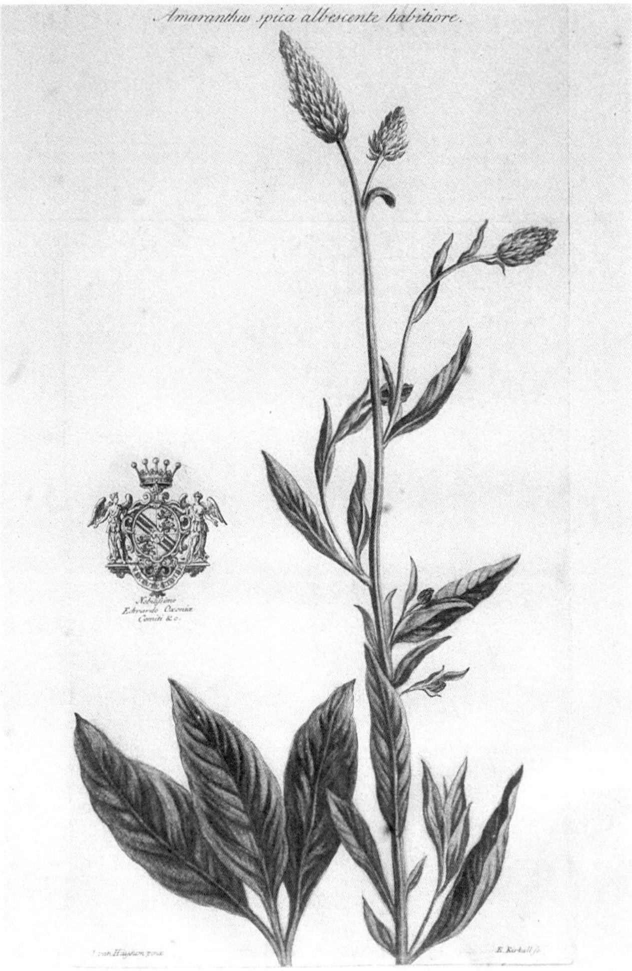

2. Jacobus van Huysum, *Amaranthus spica* (*Celosia argentea* L.), drawing for John Martyn, *Historia plantarum rariorum* (London, 1728–1737), watercolor
© The Royal Society, London

3. Elisha Kirkall after Jacobus van Huysum, *Amaranthus spica* (*Celosia argentea* L.), from John Martyn, *Historia plantarum rariorum* (London, 1728–1737), line engraving and mezzotint printed in color and hand-colored in watercolor
Courtesy of the Library of the Arnold Arboretum, Harvard University

Kirkall following the color printing technique he had used in the *Historia,* and fourteen were prepared for printing in black ink by Henry Fletcher. All were then colored by hand.

The organization of the Society of Gardeners, its decision to publish the *Catalogus,* and its methods of book production all stand in marked contrast to the production of the *Historia.* As a tangential offspring of Martyn's Botanical Society, and with backing from the Royal Society, the *Historia* was a "professional" work, whereas the *Catalogus* was a "trade" publication. Blanche Henrey provides a good summary of the genesis of the Society of Gardeners, founded in 1724, and the origins of the publication:

> It seems that Philip Miller as Secretary, or "Clerk", of the Society, had much to do with its preparation. It is inscribed to Thomas Herbert, eighth Earl of Pembroke (c. 1656–1733), a patron of gardening, and the owner of fine gardens at Wilton in Wiltshire. The preface is signed by twenty gardeners and nurserymen. . . . According to the preface, much trouble had been caused to the trade and to the public by the same plant being sold under several different names. Thus, in an attempt to establish a nomenclature the above Society, consisting of a number of the most eminent gardeners and nurserymen situated near London, agreed to hold monthly meetings. At these meetings members were required to bring exhibits of their flowers and fruits for examination and comparison. A Register was kept recording names and descriptions. After a period of five or six years it was decided to have the various plants drawn and painted "by an able hand."[16]

That able hand was, of course, Jacobus van Huysum. Not all the watercolors, it should be added, fulfilled the promise. However,

4. Jacobus van Huysum, *Acriviola maxima* (*Tropaeolum majus* L. *flore pleno*; double nasturtium), drawing for Society of Gardeners, *Catalogus plantarum* (London, 1730), watercolor

5. Elisha Kirkall after Jacobus van Huysum, *Double Nasturtium [Tropaeolum majus* L. *flore pleno] with Spanish Tree Germander and Dwarf Colutea*, from Society of Gardeners, *Catalogus plantarum* (London, 1730), line engraving with mezzotint printed in color and hand-colored in watercolor
Typ 705.30.262, Department of Printing and Graphic Arts, Houghton Library, Harvard College Library

some of the best—for example, the double spread of *Campsis radicans* (L.) Seem., or trumpet creeper—confirm his flair for composition and coloration (fig. 6). Compare this to Martyn's *Historia* plate *Geranium Africanum (Pelargonium inquinans* [L.] L'Hérit. ex Aiton; fig. 7), whose expansive rendering is further enhanced by the device of the coat of arms. In contrast to the exclusive circulation of the *Historia*, the *Catalogus* was produced for a wide audience. Indeed, according to Henrey, the Society of Gardeners achieved popular renown within a year or two of publication. As a token of this, two obscene skits on its writings appeared in London in 1732. The first was J. Wilkinson's *The Natural History of the Arbor Vitae: Or, the Tree of Life*, "published by a Society of Gardeners." The second was *The Natural History of the Frutex vulvaria or Flowering Shrub . . . By Philogynes Clitorides, Botanist, and One of the Missionaries for Propagating Knowledge in Foreign Parts*, by W. Jones.[17] Thomas Fairchild's work on hybridization—the production of what he called the "Mule Pink"—had by then made tangible the questions of sexual generation discussed by botanists of his time. His links to the Society of Gardeners may help explain the thrust of these satires.

Just how the selection of the plants from the van Huysum originals was made for inclusion in the *Catalogus* is a matter for further consideration. Why, for example, did the trumpet creeper miss selection? Why did particular groupings—for instance, the double nasturtium, Spanish tree germander, and dwarf colutea—end up assembled together on a particular plate? Similarly, careful analysis should be given to the relationship of the duplicates or variants of the Royal Society folio to the full British

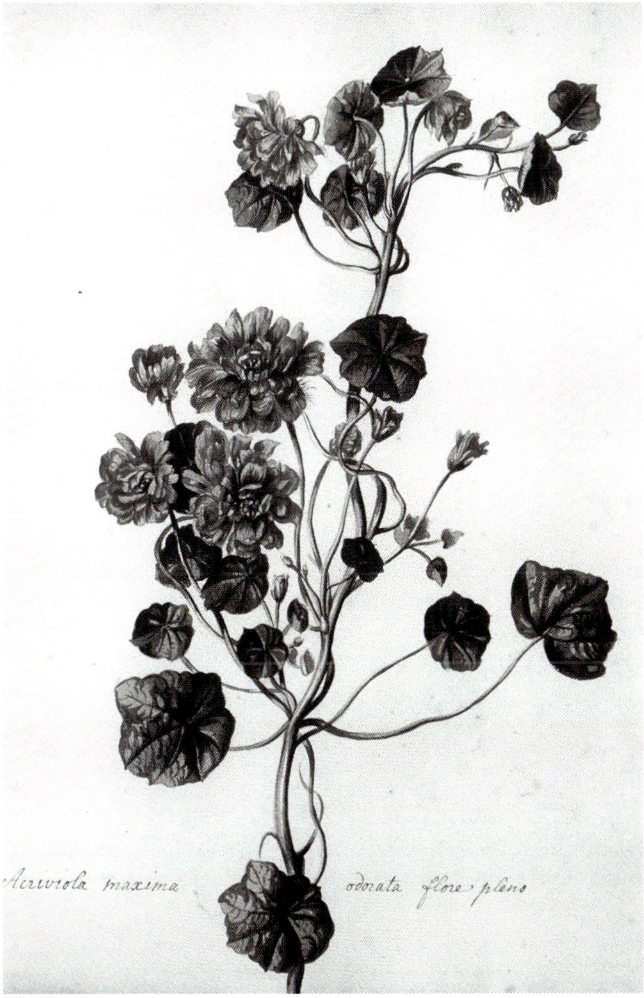

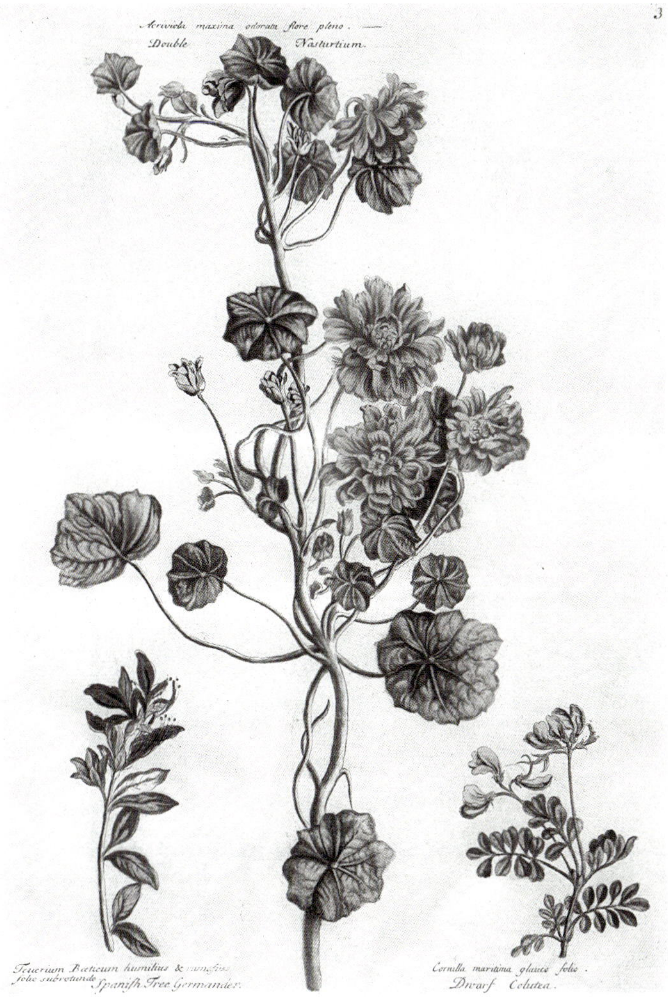

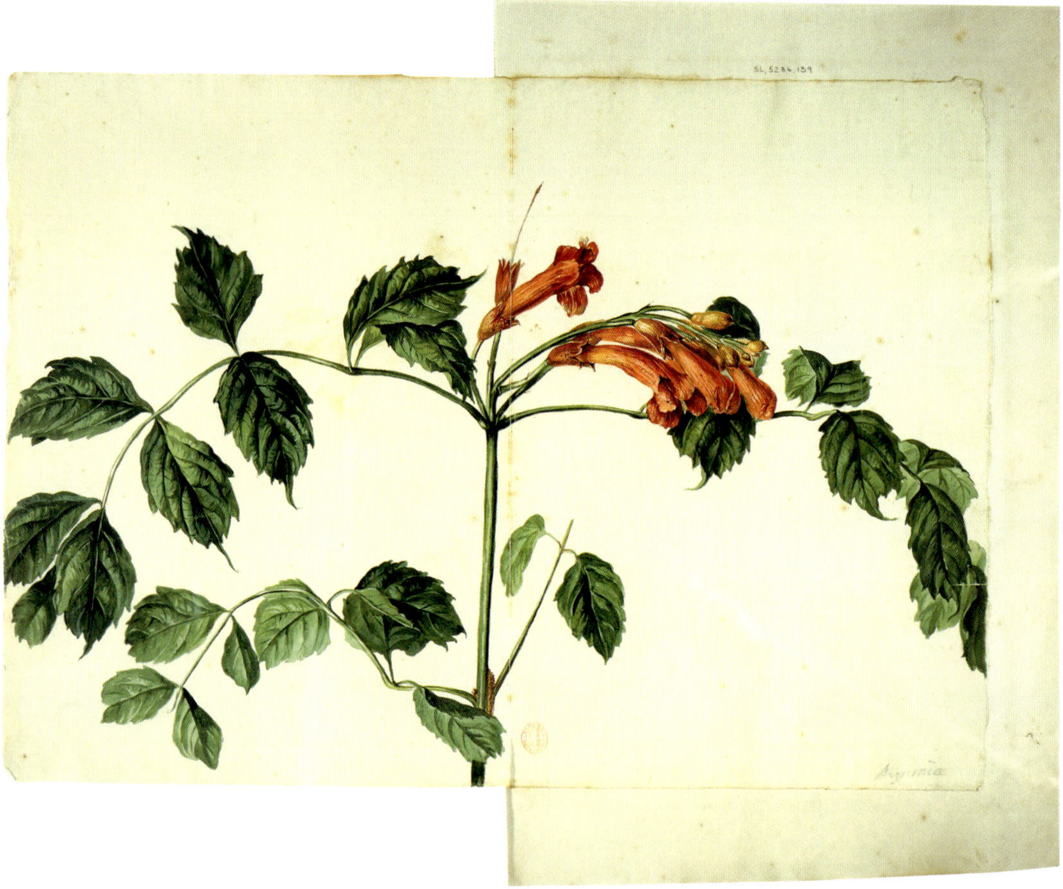

6. Jacobus van Huysum, *Campsis radicans* (L.) Seem. (trumpet creeper), drawing for Society of Gardeners, *Catalogus plantarum* (London, 1730), watercolor
© The British Museum, London, Department of Prints and Drawings, Sloane MS 5284, folio 139

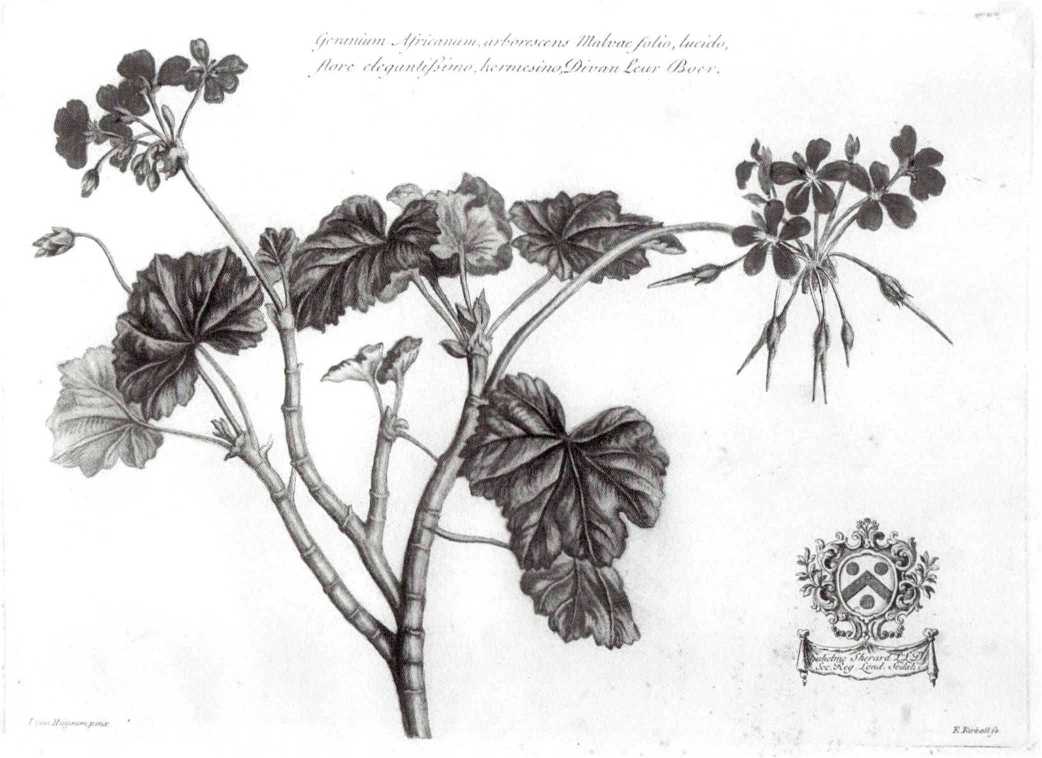

7. Elisha Kirkall after Jacobus van Huysum, *Geranium Africanum* (*Pelargonium inquinans* [L.] L'Hérit. ex Aiton), from John Martyn, *Historia plantarum rarorium* (London, 1728–1737), line engraving and mezzotint printed in color and hand-colored in watercolor
Courtesy of the Library of the Arnold Arboretum, Harvard University

Museum folios. For the purposes of this essay, however, the focus is on the horticultural context—the planting legacy of the two publications.

Coffeehouse Clubs and Horticultural Culture

Each publication ostensibly served a distinctive purpose. The *Catalogus* was designed for a wide audience in English and was primarily concerned with acclimatization and taxonomic standardization. By contrast, the *Historia* served an exclusive group of medicobotanical readers who were conversant with Latin. It covered tender plants as much as hardy ones or acclimatized exotics. To understand fully their respective places in the horticultural burgeoning of the late 1720s, we must look back to the previous generation: to Henry Compton, bishop of London (1632–1713), for example, who was already cultivating American plants such as *Physocarpus opulifolius* (L.) Maxim., or common ninebark.[18] It appears that, in the decades before 1730, city clubs—often located in coffeehouses—helped create preconditions for the publication of the *Historia* and the *Catalogus*. The activities of the club at the Temple Coffee House—an offshoot of the Royal Society—were at the epicenter of a seismic shift in botanical knowledge. Sir Hans Sloane (1660–1753) and James Petiver (c. 1663–1718) were representative of its medicobotanical roots.

The arrival of coffee in Europe in the seventeenth century would, by the 1680s, lead to a flourishing of London coffeehouses. Scholars have written much about coffeehouses as key places in creating new culture among new cultural communities.[19] As snug centers of conversation and conviviality, they functioned as "penny universities," or as proto–post offices, libraries, and galleries. They helped generate the modern stock market and modern insurance industries. Here was a milieu for exchanging ideas and sponsoring art and science. As science became a public matter, the coffeehouse took on the trappings of a laboratory. Cultural leaders of all kinds recognized in the congenial climate of coffeehouse exchange the potential for swapping more than just gossip and rumors. Groups—grand or intimate—gathered to trade ideas: the Scriblerians at the Smyrna, or Addison and Steele at Button's. Virtuosi swapped plant specimens while florists shared tips on cultivating auriculas. Natural philosophers and ordinary horticulturists thus found coffeehouses an ideal venue for regular but loosely structured gatherings. A gathering could become a "club" or a "society."

In his book *The Pleasures of the Imagination*, John Brewer wrote, "Coffee-house clubs and tavern associations were involved in all the processes by which culture was shaped: the creation of works of art and the imagination, their communication, reception and consumption."[20] With regard to the club at the Temple Coffee House, Brewer's assertion begs the following question: What part did London coffeehouse society play in the generation of "horticultural culture"? Three instances of urban conviviality are key to the answer. First, there is the elusive evidence of activities at the Temple Coffee House from 1689 to c. 1706. Second, we have the better documented but ultimately abortive proceedings of the Botanical Society, which met at the Rainbow Coffee House in Watling Street from 1721; and third, the productive gatherings of the Society of Gardeners at Newall's Coffee House in Chelsea in the second half of the 1720s that led to the *Catalogus*. A full century before the founding of the Horticultural Society in 1804, these informal groups were shaping botanical science, plant collecting, and the nursery trade.

Unlike the fragmentary records of the club at the Temple Coffee House (discussed by Margaret Riley in the appendix to this essay), documentation of John Martyn's Botanical Society is rather complete. In an article published in 1967, David E. Allen pointed out that this was an aspiring collection of surprisingly young men, largely middle class, and with medicobotanical interests.[21] It was not the intellectual equal of the loosely constituted Temple Coffee House group—a coterie that Allen called "an unofficial appendage of the Royal Society"[22] (and that Riley has now classified as a significant gathering of virtuosi, not confined to the medical fraternity, and interested in dried specimens as much as living ones).

However, under the tutelage of Isaac Rand (d. 1743), and in the context of the "herbarizing" excursions from Chelsea Physic Garden, where Philip Miller had become gardener in 1722, the Botanical Society emerged as an influential entity. In founding the society in 1721, the twenty-two-year-old John Martyn—acting as secretary—turned over the presidency to Johann Jakob Dillenius (1684–1747). Dillenius had just arrived from Giessen under the sponsorship of William Sherard. Lord Petre would dismiss the cantankerous Dr. Dillenius as nothing but "an arrant old woman,"[23] and his irascibility may have helped break up the society in 1726. But, while it lasted, every member took turns to bring in a collection of plants. Each became a "Demonstrator of Plants" for the evening. Allen was able to reconstruct the biographies of some twenty-three members: from the significant George Charles Deering (c. 1695–1749) to the lesser apothecary Joseph Forsitt (act. c. 1702–1763). All but six were under age twenty-five in 1721, and Miller himself was just thirty. One-third of the members were apothecaries, one-third physicians, and most of the rest were either surgeons or surgeon-apothecaries.

Alongside the medicobotanists, the nurserymen and gardeners constituted a second significant grouping. One of the members of the Botanical Society, Philip Miller, had sufficient social or intellectual standing to bridge the gap between these professional and trade worlds. As has been noted, he was secretary or "clerk" of the Society of Gardeners, a guild founded in 1724 to represent the interests of practicing gardeners and nurserymen. Members included Thomas Fairchild, Robert Furber (c. 1674–1756), and Christopher Gray (1693/1694–1764), all of whom stocked the recently imported exotics of the New World. As we have already seen, part of the impetus for the *Catalogus* was to demonstrate which plants were hardy; the parallel incentive was to eliminate nomenclatural confusion in pre-Linnaean taxonomy. Elite clients branded the nurseryman as "either a Knave or a Blockhead" (as the preface claimed) for appearing to sell a product under a false name. Membership in the society was defined by "trade," but the publication achieved social and professional status by its dedication to the Earl of Pembroke. The preface listed distinguished "Patriots of Horticulture" who had pioneered exotic acclimatization. This list ranged from Bishop Compton to Sir Charles Wager (1666–1743), Sir Henry Goodrick (1677–1738), James Sherard (1666–1738), Peter Collinson (1694–1768), and Charles du Bois (1658–1740).[24] Du Bois, an amateur naturalist, was linked to the club at the Temple Coffee House, while du Bois, Goodrick, and James Sherard's brother—William Sherard—sponsored Mark Catesby's collecting endeavors in North America.

The Milieu of Patronage: Book Sponsorship and Subscriptions

David Brigham has reconstructed the milieu of patronage surrounding Mark Catesby's voyage and publication.[25] From this basis, it is possible to look at the intersection of coffeehouse activities and book sponsorship. Patronage in publishing was as much in flux as patronage of plant collecting. As John Brewer puts it, the role of patron was changing from "commissioner and controller of literary work to its promoter, consumer and distributor."[26] He argues that the transformation is illustrated by the "practice of book subscription which increased as the nature of patronage changed. The number of subscription volumes rose rapidly between the late seventeenth century and the 1730s."[27] Subscription publication brought together the interests of author, patron, and bookseller. Nevertheless, in some cases an author, receiving payment only on delivery, could lose out if publication were protracted. Catesby's *Natural History of Carolina, Florida, and the Bahama Islands,* which appeared from 1729 to 1747, is an example of subscription publishing in which the author took a major risk. Approximately one-third of his 155 subscribers died before his volume was completed, and only an interest-free loan from Peter Collinson (a London woolen-draper, plant collector, horticulturist, and patron of science) allowed Catesby to proceed.

Robert Furber's *Twelve Months of Flowers* represented a rather different subscription base (fig. 8). Around one-third of the sponsorship was attached to women of varying rank, from duchess and countess down to plain

8. Henry Fletcher after Pieter Castels, frontispiece showing list of subscribers, from Robert Furber, *Twelve Months of Flowers* (London, 1730), engraving, hand-colored in watercolor
Dumbarton Oaks, Library, Rare Books Collection, Washington, D.C.

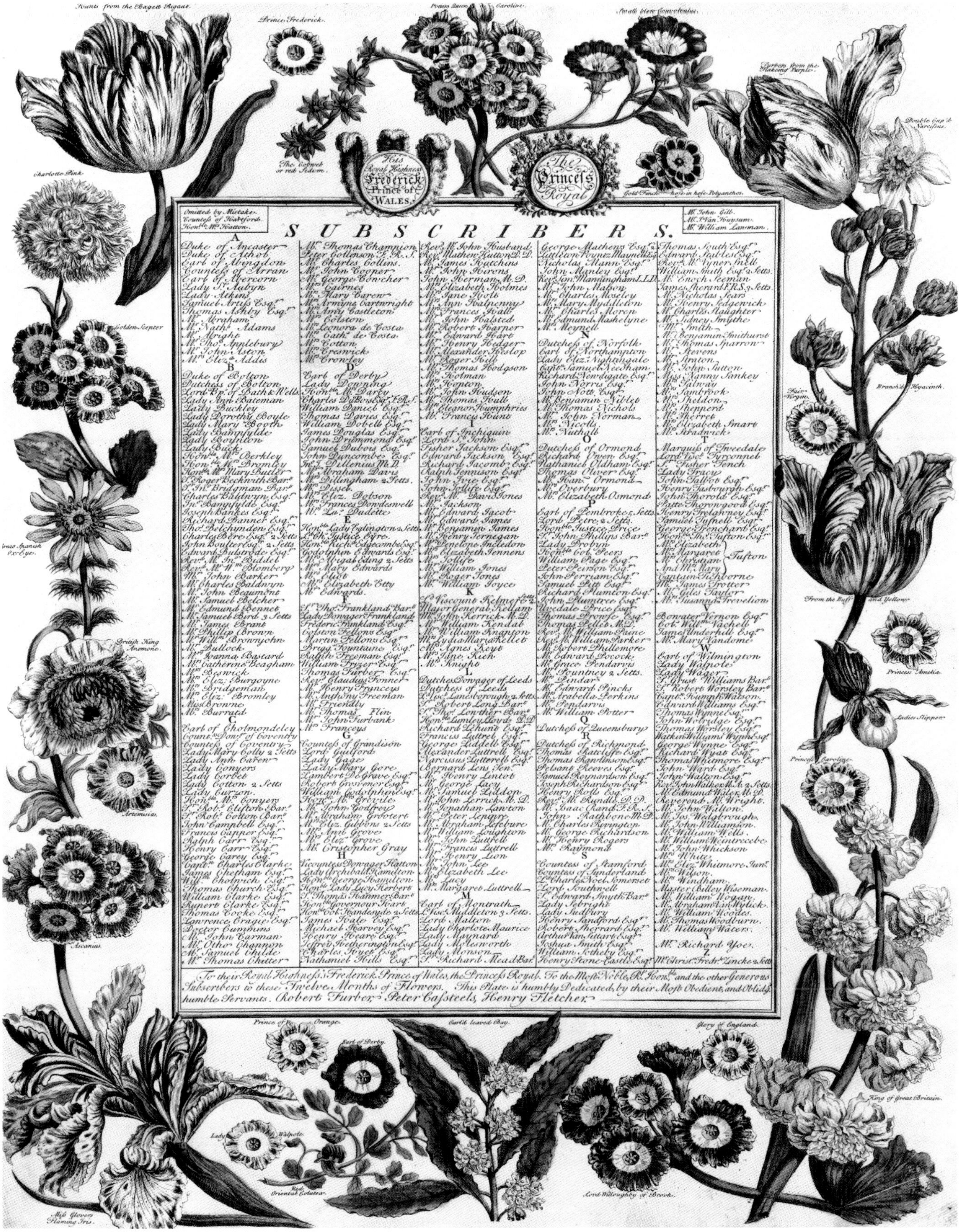

Omitted by Mistake.
Countess of Hartford.
Honble Mrs Hatton.

Mr John Gill.
Mr J. Van Huysum.
Mr William Lanman.

SUBSCRIBERS.

A
Duke of Ancaster
Duke of Athol
Earl of Abingdon
Countess of Arran
Earl of Abercorn
Lady St Aubyn
Lady Atkins
Samuel Artiss Esqr
Thomas Ashby Esqr
Mr Abraham
Mr Nathl Adams
Mr Aright
Mr Thos Applebury
Mr John Aston
Mrs Eliza Addis

B
Duke of Bolton
Dutchess of Bolton
Lord Bp. of Bath & Wells
Lady Ann Bateman
Lady Buckley
Lady Dorothy Boyle
Lady Mary Booth
Lady Bampfylde
Lady Boynton
Lady Buck
Honble Mrs Berkley
Honble Mrs Bromley
Honble Mrs Mary Butler
Sr Roger Beckwith Bart
Sr Jno Bridgman Bart
Charles Baldwyn Esqr
Jno Bampfylde Esqr
Joseph Bankes Esqr
Richard Banner Esqr
Thos Belchynden Esqr
Charles Bere Esqr 2 Setts
John Boulter Esqr 2 Setts
Edward Bulstrode Esqr
Revd Mr Jno Buddel
Revd Mr Wm Blomberg
Mr John Barker
Mr Charles Baldwyn
Mr John Beaumont
Mr Samuel Belcher
Mr Edmund Bennet
Mr Samuel Bird 3 Setts
Mr James Brant
Mr Phillip Brown
Mr Willm Brownjohn
Mr Bullock
Mrs Joanna Bastard
Mrs Catherine Beagham
Mrs Beswick
Mrs Eliz: Burgoyne
Mrs Bridgeman
Mrs Eliz: Bromley
Miss Browne
Mrs Burratt

C
Earl of Cholmondeley
Countss Dowr of Coventry
Countess of Coventry
Lady Mary Colly 2 Setts
Lady Ann Caren
Lady Conyers
Lady Corbet
Lady Cotton 2 Setts
Lady Curzon.
Honble Mrs Conyers
Sr Robt Clifton Bart
Sr Robt Cotton Bart
John Campbell Esqr
Francis Capper Esqr
Ralph Carr Esqr
Henry Carr Esqr
George Carey Esqr
Captn Charles Clarke
James Chetham Esqr
Willm Cholwich Esqr
Thomas Church Esqr
William Clarke Esqr
Rupert Clarke Esqr
Thomas Cooke Esqr
Lawrence Cragie Esqr
Doctor Cummins
Mr John Carman
Mr Otho Channon
Mr Samuel Childe
Mr Thomas Chitter
Mr Thomas Champion
Peter Collenson F.R.S.
Mr Charles Collins.
Mr John Cooper
Mr George Cowcher
Mrs Cairnes
Mrs Mary Caren
Mrs Armyne Cartwright
Mrs Amy Castleton
Mrs Colston
Mrs Leonora de Costa
Mrs Cathn de Costa
Mrs Cotton
Mrs Creswick
Mrs Crowley

D
Earl of Derby
Lady Downing
Honble Mrs Darcy
Charles DuBois Esqr F.R.S.
William Daniel Esqr
Thomas Davies Esqr
William Dobell Esqr
James Douglas Esqr
John Drummond Esqr
Samuel Dubois Esqr
John Duncombe Esqr
Jno J. Dillenius M.D.
Mr Abraham Davis
Mrs Dillingham 2 Setts.
Mrs Dissel.
Mrs Eliz. Dobson
Mrs Frances Dowdeswell
Mrs La: Dudette

E
Honble Lady Eglington 2 Setts
Ld Ch. Justice Eyre.
Honble Richd Edgecombe Esqr
Godolphen Edwards Esqr
Mrs Abigal Eding 2 Setts
Mrs Mary Edwards
Mrs Eliot
Mrs Elizabeth Etty
Mr Edwards.

F
Sr Thos Frankland Bart
Lady Dowager Frankland
Frederic Frankland Esqr
Colston Fellows Esqr
Martin Fellows Esqr
Brigg Fountaine Esqr
Ralph Freeman Esqr
William Frizer Esqr
Thomas Furber Esqr
Revd Claudius Fonner
Mr Henry Franceys
Mr Anthony Freeman
Mr Friendly
Mr Thomas Flin
Mr John Furbank
Mrs Franceys

G
Countess of Grandison
Lord Guilford
Lady Gage
Lady Mary Gore
Lambert De Grave Esqr
Robert Grosvenor Esqr
William Godolphin Esqr
Honble Mrs Grevile
Mr John Godfrey
Mr Abraham Grootert
Mrs Eliz. Gibbons 2 Setts
Mrs Ann Grove
Mrs Eliz: Grove
Mr Cristopher Gray

H
Vicountess Dowager Hatton
Lady Archibald Hamilton
Honble George Hamilton
Honble Lady Lucy Herbert
Sr Thomas Hanmer Bart
Honble Governour Hart
Honble Col. Handesyde 2 Setts
James Hale Esqr
Michael Harvey Esqr
Henry Heare Esqr
Jeffrey Hetherington Esqr
Charles Huet Esqr
Nathaniel Hills Esqr
Revd Mr John Husband
Revd Mathew Hutton D.D.
Mr James Hutchins
Mr John Hirons
John Herman M.D.
Mrs Elizabeth Holmes
Mrs Jane Holt
Mrs Ann Halpenny
Mrs Frances Hall
Mr John Halsted
Mr Robert Harper
Mr Edward Hart
Mr Henry Hedger
Mr Alexander Heslop
Mr Roger Hill
Mr Thomas Hodgson
Mrs Hotman
Mrs Hopton
Mr John Hudson
Mr Thomas Hull
Mrs Eleanor Humphries
Mr Francis Hunt

I
Earl of Inchiquin
Lord St John
Fisher Jackson Esqr
Edward Jackson Esqr
Richard Jacomb Esqr
Ralph Jennison Esqr
John Ivie Esqr
John Jollife Esqr
Revd Mr David Jones
Mr Jackson
Mr Edward Jacob
Mr Edward James
Mr Benyamin James
Mr Henry Jernegan
Mrs Penelope Incledon
Mrs Elizabeth Jennens
Mrs Jollife
Mr William Jones
Mr Roger Jones
Mr William Joyce

K
Ld Viscount Kelmore 2 Setts
Major General Kellam
Mr John Kerrick M.D.
Mr William Kendal
Mr William Knapton
Mrs Lydia Maria Kellet
Mrs Agnes Keyt
Mrs Jane Kich
Mrs Knight

L
Dutchess Dowager of Leeds
Dutchess of Leeds
Ld Visct Lanesborough 2 Setts
Sr Robert Long Bart
Sr Thos Lowther Bart
Honble Lumley Lloyd D.D.
Richard Lehunt Esqr
Francis Luttrel Esqr
George Liddell Esqr
Alexander Luttrell Esqr
Narcissus Lutterell Esqr
Bernard Lens Senr
Mr Henry Lintot
Mr George Lacy
Mr Samuel Liddon
Mr John Lerrick M.D.
Mr Jonathan Lawton
Mr Peter Lepipre
Mr Abraham Lefebure
Mr William Loughton
Mr John Luttrell
Mr Francis Luttrell
Mr Henry Lion
Mr John Lee
Mrs Elizabeth Lee
Mrs Lucy
Mrs Margaret Luttrell

M
Earl of Montrath
Ld Visct Middleton 3 Setts.
Lord Malton
Lady Charlote Maurice
Lady Maynard
Lady Molesworth
Lady Monson
Sr Richard Mead Bart
George Mathews Esqr 2
Littleton Poynz Maynell Esqr
Nicholas Mann Esqr
John Manley Esqr
Revd Sim. Manningham L.L.D.
Mr John Malson
Mr Charles Moseley
Mrs Mary Myddleton
Mr Charles Moren
Mr Edmund Maskelyne
Mrs Meynell

N
Dutchess of Norfolk
Earl of Northampton
Lady Eliz: Nightingale
Captn Samuel Needham
Richard Newdigate Esqr
John Norris Esqr
John Nott Esqr
Mr Benjamin Niblet
Mr Thomas Nichols
Mr John Norman
Mrs Nicolls
Mrs Nuthall

O
Dutchess of Ormond
Richard Owen Esqr
Nathaniel Oldham Esqr
Thomas Oliver Esqr
Mrs Han: Ormond
Mrs Overbury
Mrs Elizabeth Osmond

P
Earl of Pembroke 5 Setts
Lord Petre 2 Setts.
Honble Justice Price
Sr John Phillips Bart
Lady Probyn
Honble Col. Peers
William Page Esqr
Peter Peirson Esqr
John Perriam Esqr
Samuel Pitt Esqr
Richard Plumton Esqr
John Plumtree Esqr
Uvedale Price Esqr
Thomas Prowse Esqr
Thomas Pellet M.D.
Revd Mr William Paine
Revd Mr William Parker
Mr Robert Phillemore
Mr Edward Pocock
Mrs Grace Pendarvis
Mr Pountney 2 Setts.
Mrs Pindar
Mr Edward Pincks
Mrs Arabella Perkins
Mrs Pendarvis
Mr William Potter

Q
Dutchess of Queensbury

R
Dutchess of Richmond
Thomas Ratcliffe Esqr
Thomas Rawlinson Esqr
Pelsant Reeves Esqr
Samuel Reynardson Esqr
Joseph Richardson Esqr
Henry Rolls Esqr
Revd Mr Rundle D.D.
Mr Isaac Rand F.R.S.
John Rathbone M.D.
Mr Charles Rivington
Mr George Richardson
Mr Henry Rogers
Mrs Raymond

S
Countess of Stamford
Countess of Sunderland
Ld Charles Noel Somerset
Lord Southwell
Sr Edward Smyth Bart
Lady Sebright
Lady Sudbury
Henry Sandford Esqr
Robert Sherrard Esqr
Arthur Van Sittart Esqr
Joshua Smith Esqr
William Sotheby Esqr
Henry Stone-Castle Esqr
Thomas South Esqr
Edward Stables Esqr
Revd Mr Vyner Snell
William Smith Esqr 2 Setts.
Mr Enoch Seeman
James Sherard F.R.S. 3 Setts.
Mr Nicholas Sears
Mr Henry Sedgewick
Mr Charles Slaughter
Mr Sidney Smythe
Mr Smith
Mr Benjamin Smithurst
Mr Thomas Sparrow
Mr Stevens
Mr Straton
Mr John Sutton
Miss Nanny Sankey
Mrs Salway
Mrs Sambrook
Mrs Sheldon
Mrs Shepperd
Mrs Skerret
Mrs Elizabeth Smart
Mrs Stradwick

T
Marquiss of Tweedale
Lord Visct Tyrconnel
Sr Fisher Tench
Lady Tracy
John Talbot Esqr
Henry Tasbourgh Esqr
John Thorold Esqr
Patte Thorowgood Esqr
Henry Trelawney Esqr
Samuel Tufnell Esqr
George Trenchard Esqr
Honble Thos Tufton Esqr
Mrs Elizabeth, Mrs Margaret, Mrs Christian, And Mrs Mary — Tufton
Captain Tichborne
Mr James Trotter
Mr Giles Taylor
Mrs Susanna Trevelion

V
Bowater Vernon Esqr
Col. Willm Vachell
James Underhill Esqr
Mrs Mary Vandome

W
Earl of Wilmington
Lady Walpole
Lady Wager
Sr Grust Williams Bart
Sr Robert Worsley Bart
Captn Humph. Watson.
Edward Williams Esqr
Thomas Wynne Esqr
John Wobridge Esqr
Thomas Worsley Esqr
Watkin Williams Wynn Esqr
George Wynne Esqr
Richard Wyat Esqr
Thomas Whitmore Esqr
John Ward Esqr
John Walton Esqr
Revd John Walker M.A. 2 Setts.
Mr Edmund Waler M.D.
Reverend Mr Wright.
Mr John Wilton.
Mr Jos. Wedgbrough.
Mr John Williamson.
Mr William Wells.
Mr William Weintrecebe
Mr John Whickson
Mrs White
Mrs Eliz: Whitmore Junr
Mrs Wilson.
Mrs Windham.
Master Billey Wiseman.
Mr William Wogan.
Mr Abraham Van Wylick.
Mr William Woodes.
Mr Thomas Woodburn.
Mr William Waters.

Y
Mr Richard Yoe.

Z
Mr Christn Fredr. Zincke 2 Setts

To their Royal Highness's Frederick Prince of Wales, the Princess Royal. To the Most Noble, Rt. Honble and the other Generous Subscribers to these Twelve Months of Flowers. This Plate is humbly Dedicated, by their Most Obedient, and Oblidg humble Servants. Robert Furber, Peter Casteels, Henry Fletcher.

Mrs. or Miss. It is worth speculating here on the gender implications.[28] For example, in contrast to *Twelve Months*, only some 6 to 7 percent of subscribers to Catesby's *Natural History* were women. Was the latter, in other words, a more "masculine" alliance associated with colonial enterprise, science, medicine, and exotic collecting, as opposed to the "feminine" realm of decorative floral display? Or was there a shift in book production and consumption?

I suspect that the gender associations are complex, as the Duchess of Beaufort's dedication to "science" clearly implies.[29] D. E. Allen, one of the first to equate female patronage with "decorative" display, also argued a central place for women in scientific, exotic collecting: from the Duchess of Beaufort to the Duchess of Portland. In 1976, in his social history *The Naturalist in Britain*, he proposed:

> It was presumably thanks to this appeal of the exquisite that entomology was able to count among its followers at this time a remarkably high proportion of women. Mrs Eleanor Glanville (of the Glanville Fritillary) is the best known and was certainly one of the first of these, receiving much encouragement from Petiver. Another was Mary Somerset, the first Duchess of Beaufort, the chief patron of Albin and inspirer of his earliest book. It was she who first told Richard Bradley that every species of butterfly and moth has its own special food-plant: "I believe [she] has bred a greater variety of English Insects, than were ever rightly observ'd by any one Person in Europe," he wrote admiringly in 1721. A volume of drawings of English lepidoptera is still preserved in what was once her library at Badminton. Rather later, in the 1740s, we find the Hon. Mrs Walters mentioned by Wilkes as a breeder of rare moths. Indeed, out of over a hundred persons listed by Wilkes in 1750 as subscribers to or 'encouragers' of his principal work, women comprise no less than a quarter.[30]

A review of the list of subscribers to Eleazar Albin's *A Natural History of English Insects* (London, 1720) elicits a comparatively high engagement of women in sponsorship. The volume, dedicated to "Her Royal Highness the Princess of Wales," contains the names of a total of 171 subscribers (that is, if one puts the booksellers "Messieurs William and John Innys," who ordered fourteen copies, on an equal footing with all others including the Princess of Wales). Of these, some 63—or around 37 percent—were women. Among these women, many—such as the Duchess of Grafton or the Countess of Northampton—were aristocratic subscribers. On the other hand, as with Furber's *Twelve Months*, the subscription list is notable for the assortment of women without title: Mrs. Bovey, Mrs. Ermin Cartwright, Mrs. Mary Goopye, Mrs. Howland, Mrs. Lambert, and Mrs. Weld of Lulworth Castle in Dorset. Ordinary readers were contributing as consumers of natural history.

Some indication of the impetus to solicit subscribers is provided in the preface to the volume. Here Eleazar Albin writes how Joseph Dandridge introduced him to a "Mrs. *How*, Widow of the late famous Physician of that Name." He continues:

> For this curious Lady I painted a great Number of both *Caterpillars* and *Flies*, and likewise several Things relating to Natural History for Sir *Hans Sloane*: After this I was introduc'd to Her Grace *Mary* the late Dutchess Dowager of *Beaufort*, who imployed me in the same manner; this excellent Lady first persuaded me to undertake the following Work, and encouraged me by procuring me Subscriptions from several Persons of the first Quality: While this good Lady lived it went on apace, and I am persuaded had been finished long since if it had pleased God to have spared her; but after the loss of my Patroness, Subscriptions coming in slowly, and my Circumstances (having a great Family to provide for) not being able to carry it on without, retarded it. These Considerations I hope will dispose those honourable and worthy Persons who subscribed early to pardon me that they have expected it so long.[31]

If the sponsorship began with a widow of a physician (Mrs. How) and was promoted by another physician (Sir Hans Sloane), its momentum depended on an aristocratic "patroness" (the Duchess of Beaufort), a lady who had family and friends to commandeer. To what extent the presence of women in the list indicates active involvement in natural history is unclear. For example, was Mrs. Etheldred Hovel—wife to the archbishop of Canterbury—persuaded to subscribe for reasons of status or connection, or did she have an independent interest in natural history as opposed to delight in the decorative or "exquisite"? No doubt, passive and active engagement fused in the process.

Medicobotanists are just as evident as women in the sponsorship of Albin's *Natural History.* The physician, apothecary, surgeon, and fellow of the Royal Society are denoted by the professional qualifications at the bottom of each plate: for example, "Sir David Hamilton, Physician to ye Princess, & F.R.S." Perhaps, in this sense, we may classify Albin's "exquisite" entomological work as a step away from the exclusively male sponsorship of John Martyn's *Historia* plates. Martyn's subscribers were still linked to the "masculine" realms of science, medicine, and pharmacy. Substantial numbers were fellows of the Royal Society; many were physicians or apothecaries. Cambridge University played its role, too. They were part of the overlapping circles of book and collecting patronage—an alliance of medicobotanists, sponsors, and collectors. For example, we find the names of Edward, Earl of Oxford (1689–1741), and William Sherard on the list of twelve sponsors of Catesby's voyage of 1722 1726. They then reappear as the respective sponsors of the plates in the *Historia* depicting "Amaranthus spica" and "Geranium Africanum" (see figs. 3 and 7). The Catholic Robert James, the eighth Lord Petre (1713–1742), who adopted the passion flower (*Passiflora serratifolia* L.; fig. 9) for his sponsorship of the *Historia,* was among the subscribers to Catesby's *Natural History.*

The circles are, however, by no means uniform or consistent. There is no obvious clue to the wider botanical connections of "Johanni Carparo Scheuthzer M.D. Soc. Reg. Lond. Sodali" as the patron of the "Lychnidea Caroliniana" (*Phlox carolina* L.; fig. 10) or to "Conrado Sprengell [Sir Conrad Joachim Sprengwell] . . . Medico Lond. & Soc. Reg. Sodali" as the sponsor of "Bidens Caroliniana" *(Coreopsis lanceolata* L.; fig. 11).[32] Both were members of the Royal Society, yet neither is among the twenty-three members whom David Allen identified in John Martyn's Botanical Society; nor do they feature in a more extensive milieu—for example as subscribers to Catesby's *Natural History.*

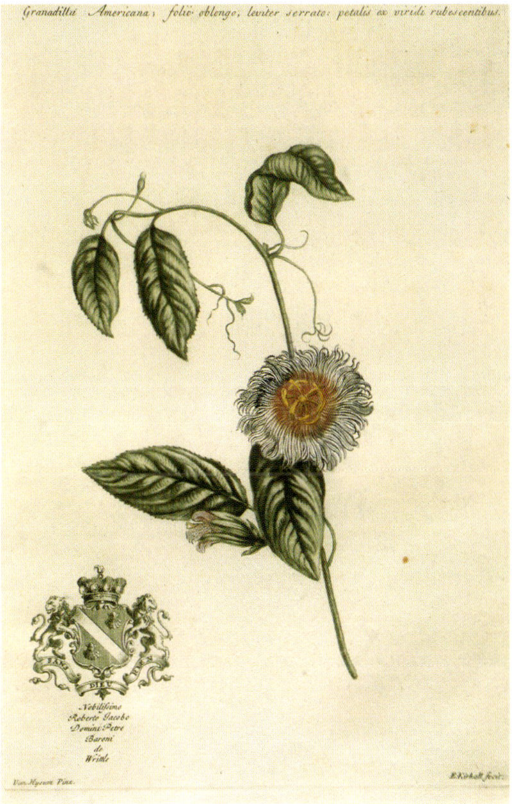

9. Elisha Kirkall after Jacobus van Huysum, *Grandilla Americana* (*Passiflora serratifolia* L.; passion flower), from John Martyn, *Historia plantarum rariorum* (London, 1728–1737), line engraving and mezzotint printed in color and hand-colored in watercolor
Courtesy of the Library of the Arnold Arboretum, Harvard University

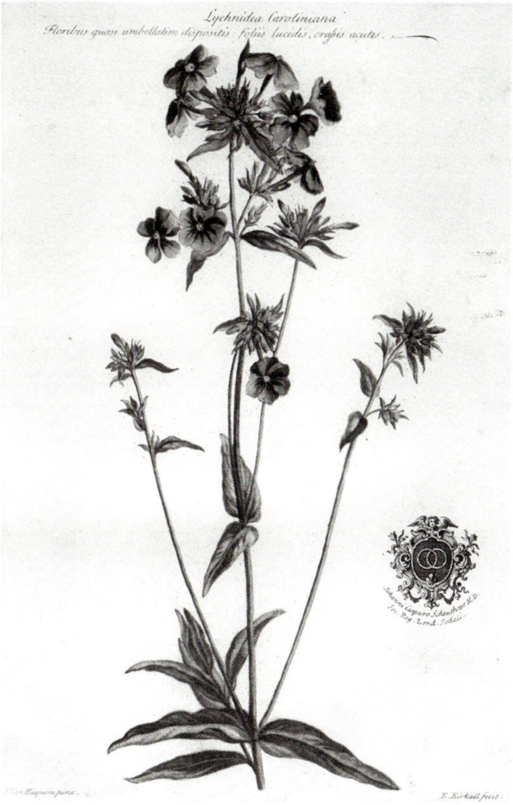

10. Elisha Kirkall after Jacobus van Huysum, *Lychnidea Caroliniana* (*Phlox carolina* L.), from John Martyn, *Historia plantarum rariorum* (London, 1728–1737), line engraving and mezzotint printed in color and hand-colored in watercolor
Courtesy of the Library of the Arnold Arboretum, Harvard University

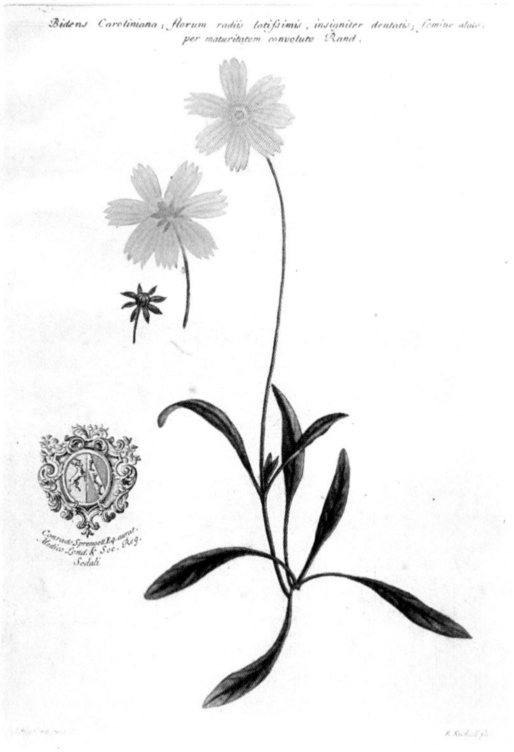

11. Elisha Kirkall after Jacobus van Huysum, *Bidens Caroliniana* (*Coreopsis lanceolata* L.), from John Martyn, *Historia plantarum rariorum* (London, 1728–1737), line engraving and mezzotint printed in color and hand-colored in watercolor
Courtesy of the Library of the Arnold Arboretum, Harvard University

For all that, the professional profiles of Botanical Society members and *Historia* subscribers do appear consistent. Like the Royal Society, this was an exclusively male preserve. Of the fifty plates, over a dozen are linked to physicians like Richard Myddleton (or Middleton) Massey (c. 1678–1743), who sent letters to Sloane at the Temple Coffee House in 1706, 1707, and 1708.[33] If surgeons, apothecaries, physicians, and professors of anatomy are counted together, they amount to nearly half the subscribers. Alexander Stewart, M.D., who died in 1742, is a good example. He shows up as sponsor to the *Historia*; he was also a subscriber to the *Natural History*.[34] Meanwhile, John Wilmer represents the apothecaries connected to Chelsea Physic Garden, where he was appointed to the office of demonstrator in 1748. He is a vital link between Martyn's Botanical Society and Martyn's *Historia*, for which he was a sponsor of one "golden rod" plate. According to Allen, it was he who introduced the nineteen-year-old Martyn to botany in 1718.

Beyond the distinctions of gender (and thereby audience, as much as patronage), Eleazar Albin's *A Natural History of English Insects* shares with John Martyn's *Historia* the device of direct patronage of an individual plate. Yet there is a fundamental difference in the form of representation. In Albin's *Natural History*, the subscriber is represented as a dedicatee: for example, "To the Most Noble Mary Capell Dutchess Dow.^r of Beaufort this Plate is humbly Dedicated by Eleazar Albin." In contrast, the plates of the *Historia* include not merely the name of the sponsor but also his coat of arms, or some other emblem if the patron were without title. Though it does appear unusual, this practice was not unique; nor was it John Martyn's own innovation. Historians Blanche Henrey and Michael Hunter both independently point to Robert Morison's *Plantarum umbelliferarum distributio nova* of 1672 as significant. Each of the twelve copper plates was the gift of a donor and bore his name, office, and heraldic arms. Similarly, Henrey points out that the "Pars tertia" of Robert Morison's *Plantarum historia universalis oxoniensis*, published in 1699, included 69 of the total 166 plates that bear the names of donors and their heraldic arms.[35] In this lineage of medicobotanical publications, family pedigree counted as much as did male professional status.

New Alliances beyond Patronage: The Nurserymen and "Patriots of Horticulture"

The *Catalogus*, by contrast to the *Historia*, was not a subscription publication. This seems strange given that the preface, dedicated to the Earl of Pembroke, included lists of "Procurers of Plants," "Encouragers of Gardening," and "Patriots of Horticulture." The preface highlights four figures among those who had shaped the early history of exotic collecting at the turn of the century: Henry Compton, bishop of London, at Fulham; Samuel Reynardson of Hillingdon (d. 1721); the Reverend Robert Uvedale of Enfield (1642–1722); and Mary Somerset, Duchess of Beaufort, at Badminton (1630–1715). The duchess, it was claimed, had specialized in "tender Exotick Plants." What separated the Earl of Pembroke from these four predecessors was his collection of "acclimatized trees." This was indicative of another shift in the horticultural environment. While Dr. Uvedale had evidently planted some of his exotics "in full ground," Reynardson was more cautious, and perhaps more typical. The account states, "As he kept them for the most Part confin'd to Pots and Tubs, preserving them in Greenhouses in Winter, never attempting to naturalize them to our Climate, so soon after his Death, that valuable Collection was so dispersed, as at present to be hardly known what he was possessed of."[36] Dispersal after death, or posthumous destruction, affected many of these early collections, including Bishop Compton's at Fulham. Hence the very sporadic strides made in the cultivation of exotics outdoors before the 1720s.[37]

The new generation, following Pembroke's success with "Tryals," included some distinguished names: the Earl of Islay, Lord Wilmington, Sir Charles Wager, Sir Henry Goodrick in Yorkshire, Lumley Lloyd of Cheam in Surrey, Henry Trelawny at Buttshead near Plymouth, Peter Collinson, and many others, including Charles du Bois of Mitcham, to whom special praise was extended.

Sir Henry Goodrick and Charles du Bois were, as noted above, among the twelve sponsors of Mark Catesby's collecting trip of 1722–1726. And they—along with the Earl of Islay (later Duke of Argyll, 1682–1761), the Earl of Wilmington (c. 1673–1743), Sir Charles Wager, Lumley Lloyd (possibly Rev. Robert Lumley, died 1730), Henry Trelawny (act. 1700s), and Peter Collinson—subscribed to Catesby's *Natural History.* Here, in other words, in the overlapping circles of establishment patronage, a new group was emerging. The emphasis was shifting from the alliance of "science" and "medicine" (what David Allen portrays in his first chapter of *The Naturalist in Britain* as the respective activities of the Society of Apothecaries, the Royal Society, and the club at the Temple Coffee House).[38] These new, and rather loose, alliances focused instead on the practice of acclimatizing plants for decorative as well as commercial purposes. The collecting syndicate that Peter Collinson would establish with the help of the nurseryman John Bartram in Philadelphia illustrates the new momentum. The collective enterprises of the Botanical Society and the Society of Gardeners were giving way to the individual collecting pursuits of aristocrat and nurseryman alike.[39] Periodicals, magazines, and the Horticultural Society itself would eventually come to bind individuals to a new collectivity. But the flowering of the English landscape garden gained initial impetus as clubs dissolved into looser syndicates.

Just why the society left the patrician "patriots of horticulture" unsolicited for subscriptions remains unclear. Instead, the *Catalogus* was paid for by the members of the society at "their own Expense."[40] The result was the release of plates in several parts. Doubtless, then, lack of resources explains the ultimate failure to translate all the British Museum watercolors into the proposed three volumes of engravings. In one sense, the inconsequential career of Jacobus van Huysum seems mirrored in the abortive venture of the *Catalogus plantarum.* On the other hand, through the many editions of *The Gardeners Dictionary* (1731–1768), Philip Miller—dispensing with the costly production of images—effectively sustained the main objectives of the Society of Gardeners: the standardization of names and the registration of plants that could or could not be acclimatized.

In short, Miller found a way to manage the lack of class connection that had undermined the *Catalogus.* Without an active patron in the aristocracy—as Eleazar Albin stated—subscriptions were hard to come by. After all, the members of the Society of Gardeners were primarily nurserymen of fairly humble origins. They are listed on the title page with the location of their business (or sometimes the nearest inn): "Robert Furber, at Kensington. John Alston [act. 1730s–1740s], near Chelsea College. Philip Miller, at the Physick Garden in Chelsea. Obadiah Lowe [act. 1720s], at Battersea. John Thompson [act. 1720s–c. 1758], at the Rose in Chelsea. Christopher Gray [1693/1694–1764], at Fulham. Francis Hunt [1691–1763], at Putney. Moses James [act. 1719–1731], at Lambeth. George Singleton [act. 1730–d. 1735], at the Neat Houses. Wm. Hood [act. 1730–1733], at the Wheatsheaf near Hyde-Park Corner. Benjamin Whitmill [act. 1722–1730], at Hoxton. Richard Cole [act. 1730s], at Battersea. Samuel Hunt [act. 1730–d. 1763], at Putney: And Stephen Bacon [1709–1734], at Hoxton."

Interestingly, six additional names appear at the end of the preface. These are Thomas Fairchild, Samuel Driver (act. 1717–1730), Thomas Bickerstaff (act. 1730), William Welstead (act. 1730), John James (possibly the architect, c. 1672–1746), and William Spencer (act. 1730). John Harvey speculated that they "were apparently not concerned in the production and may have been only associates."[41]

Although some of these figures seem insignificant nurserymen,[42] the remaining members proved—at least individually rather than collectively—to have influential careers. In contrast to John Martyn (who, having started brilliantly, "produced not a single pupil who later adorned the subject—with the conspicuous exception of his son"[43]), the group that met at Newhall's Coffee House in Chelsea had a considerable impact on the next generation. Indeed, they largely defined the world of horticulture and botany in the mid-eighteenth century.

Of these influential individuals, Philip Miller stands out as the primary figure. By 1730—when he was in his late thirties—he was already an authority, and through the various editions of *The Gardeners Dictionary* (1731–1768) he continued the society's work.[44] The lives of Robert Furber, Christopher Gray, and Samuel Driver are well documented in John Harvey's *Early Nurserymen*;[45] and the impact of Furber and Gray on garden design is considered in *The Flowering of the Landscape Garden*.[46] Above all, the milieu of the nurserymen in Hoxton—Thomas Fairchild, Stephen Bacon, and Benjamin Whitmill—has been imaginatively, if only partly, reconstructed by Michael Leapman in his recent study, *The Ingenious Mr Fairchild*.

Thomas Fairchild, whose early life as an apprentice clothworker led to his later career as a nurseryman, demonstrates one unconventional path of an artisan. His career contrasts with the generally more predictable vita of the medicobotanist. It appears that in 1704 Fairchild, by then thirty-seven, accepted his freedom from the Clothworkers' Company as a first step to paying his redemption fee to the Worshipful Company of Gardeners.[47] Regulating the qualifications of a guild gardener to counteract the activities of itinerant and unskilled laborers appears akin to the efforts of the Society of Gardeners to regulate the nursery trade to standardize names and commodities.[48] By 1720, at age fifty-two, Thomas Fairchild had presented his hybrid Mule Pink—his 1717 cross between a carnation and a sweet william—to the Royal Society. Because Fairchild himself lacked professional status, his friend Patrick Blair (c. 1666–1728)—a doctor, an amateur botanist, and a fellow of the Royal Society—had to represent him. On his death in 1729, Fairchild left modest sums to relatives and one guinea for a ring to his friend and witness, Mark Catesby. Yet for all that, his legacy in horticultural terms was considerable.

At the time, little was made of Fairchild's early experiment in hybridization. Yet his success as an artisan florist points to a significant trend in the history of "horticultural culture." Just as a shift had occurred from the influence of the medicobotanists to the marketing and commercial pull of the nurseries, so too there was a change in the consumer marketplace. The technical skills of the florists and nurserymen were competing on equal terms with the investment capital and social connections of the wealthy collectors. By the time Dr. John Hope visited English gardens in 1766, the nurserymen could rival Kew, Whitton, and Painshill in the scope of their exotic collections.[49]

The Shift from Collecting Exotics to the Acclimatization of Exotics

In short, the long-term influence of the Society of Gardeners proved more sustained than the inconsequential activities of John Martyn's Botanical Society. Similarly, for the planting of the English landscape garden, the hardy plants featured in the *Catalogus* turned out to be more useful than the mixed tender and hardy exotics of the *Historia*. Nevertheless, certain plants featured in the *Historia*—broadly illustrating the shifting linkages among collecting, patronage, publishing, and gardening—stand out as significant enough. Mark Catesby is central to all such spheres, and acclimatization is the central issue. It was Catesby, after all, who, by supplying seeds to the nursery trade, helped turn the rare exotics of his *Natural History* into the affordable commodities of the nurseryman's catalogue. Take the notable *Catalpa bignonioides* Walt., for example. When the Prince of Wales procured a single thirteen-foot specimen of the catalpa in 1734, it came from Furber's nursery and cost the huge sum of three guineas. Furber had presumably obtained the plant from Catesby. Hence the prince was gaining first access to Catesby's direct importation of 1726, but via a nurseryman. By 1755 the price for an average-sized tree had come down to 1s. 6d.[50]

The *Amorpha fruticosa* L., or "bastard indigo" of the *Catalogus*, was another success of acclimatization. Jacobus van Huysum captures its bottle-brush inflorescence: purple and orange (fig. 12). Although first introduced to Bishop Compton's Fulham garden before 1713, it probably died out until being reintroduced as seed by Catesby in 1724.[51] Within six years of reintroduction, the *Catalogus* could record it as hardy and growing to around eight to ten feet.

I have tracked its dispersal through the gardening literature: how it comes down in price from 2s. or 1s. 6d. to 9d. toward the end of the century; how it was growing at Burton Constable, Adlestrop, Hartwell House, Whitton, and Orford Hall—in other words, throughout England; and how its vulnerability to frost, especially in the Midlands and the North, might have kept it in a permanently juvenile condition. The Carolina kidney bean tree (*Wisteria frutescens* [L.] Poir.; fig. 13) was also sent as seed in 1724 (although earlier shipments may have reached Thomas Fairchild before 1722).[52] It flowered for the first time in Robert Furber's nursery. We can picture it trained up a stake or entwined around a tree trunk—what was later called "garnishing." It proved useful in the midcentury shrubbery. Thus, what began as Langley's wilderness embellishment of 1728 would, by the 1770s, become the mature shrubbery. The tulip tree was still at the back, but the "bastard indigo" now took center stage (compare figs. 1 and 14). The plants first introduced under the auspices of the medicobotanists had left behind their initial patronage, becoming free agents in the commercial marketplace.

Of course, not all of Catesby's introductions were suitable for outdoor cultivation. Martyn's *Historia* included the tender "Cassia Bahamensis" (*Cassia ligustrina* L., figs. 15 and 16), which was sent from the Bahama Islands in 1726. It first flowered at Chelsea Physic Garden and Sir Charles Wager's private residence, but acclimatization was never an option. Nevertheless, two hardy herbaceous plants featured in the *Historia* had an enormous influence on flower gardening as they became acclimatized to English conditions. The first is "Bidens Caroliniana" (*Coreopsis lanceolata* L.; see fig. 11), which Catesby introduced in 1724. One specimen, eventually ending up in the Duchess of Portland's Bulstrode, was listed in a James Gordon nursery bill of 1765 as a "new plant."[53] This ostensibly false label may indicate a slow rate of dispersal after 1724, or indeed a fresh reintroduction after the initial import was lost. Whatever the reason, it points to haphazard dissemination (or to a nursery trade, liberated from the ignorance of the "blockhead," and manipulating the market in some fashion).

The collector of the second hardy herbaceous flower—the "Lychnidea Caroliniana" (fig. 17 and fig. 10)—is unknown. However, I have grounds for suspecting that, when *Phlox carolina* L. came to John Cowell's nursery in Hoxton before 1728, the shipment was from Catesby's voyage of 1722–1726. A number of early unidentified specimens in the Sherard Herbarium point to Catesby: one is indeed labeled "Mr Catesby Pale Red," another "Mr. Catesby: Carolina 1722"; and two specimens identified by James Reveal in 1982 as *Phlox amoena* Sims (hairy phlox) are associated with Mark Catesby in 1724. More conclusively, we know the species *P. carolina* was growing at Chelsea Physic Garden in 1733, when it was submitted as a specimen to the Royal Society.[54] By 1774 Mrs. Delany would create a collage of *Phlox carolina*, and its cultivation can be traced throughout the middle years of the eighteenth century.[55] On the other hand, according to *Botanical Magazine*, at some point subsequently *Phlox carolina* was lost to cultivation, only to be reintroduced by Fraser around 1810 (fig. 18). Was the loss something to do with interrupted supplies of seed from North America, or with

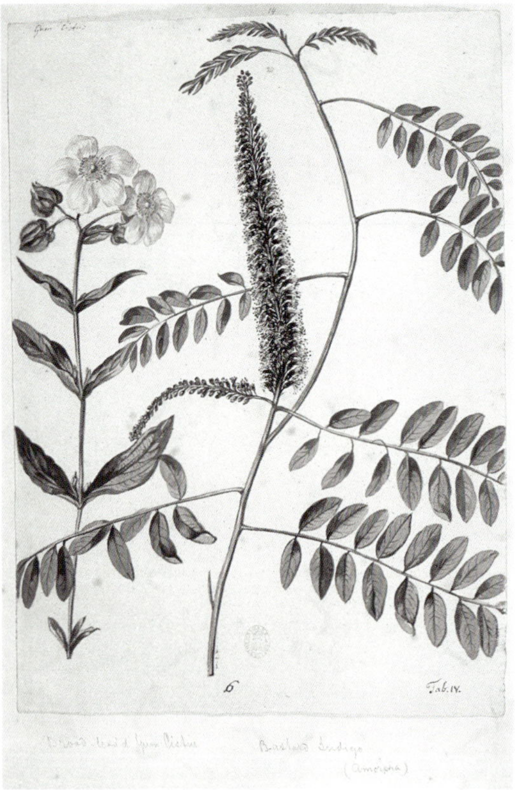

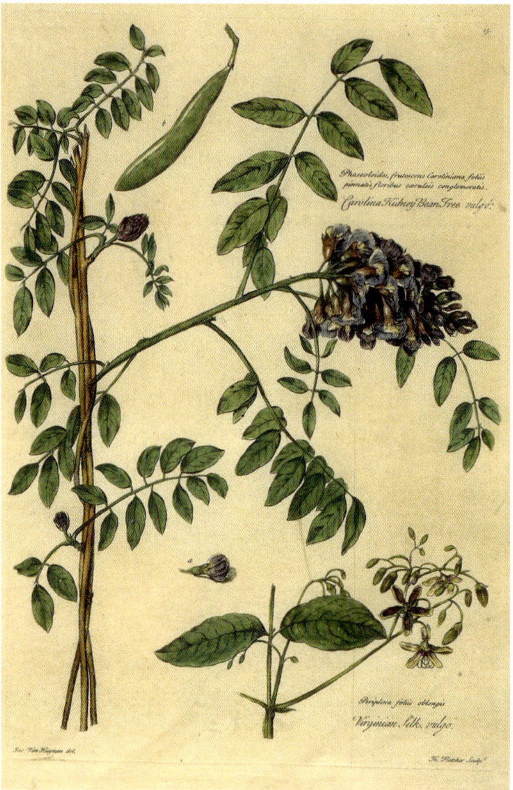

12. Jacobus van Huysum, *Bastard Indigo [Amorpha fruticosa* L.*] with Gum Cistus*, watercolor drawing for the Society of Gardeners, *Catalogus plantarum* (London, 1730)
© Copyright The British Museum, London, Department of Prints and Drawings, Sloane MS 5284, folio 14

13. Henry Fletcher after Jacobus van Huysum, *Carolina Kidney Bean Tree [Wisteria frutescens* L. Poir.*] with Virginian Silk*, from the Society of Gardeners, *Catalogus plantarum* (London, 1730), line engraving and mezzotint, hand-colored in watercolor
Typ 705.30.262, Department of Printing and Graphic Arts, Houghton Library, Harvard College Library

weather-related factors in England in the final quarter of the eighteenth century, or simply due to plain happenstance? Here the shifting environment of exotic acclimatization in the late eighteenth century takes us beyond the congenial climate of coffeehouse horticulture.

The Shifting Natural and Cultural Environment: Ramifications for Horticulture

In the 1730s, Mark Catesby had been confident of the ability of the English gardener to acclimatize the best of North American plants to English conditions. Without a true understanding of the Gulf Stream, he based his vision of "domesticated exoticism" on a simple observation: that the mild winters of London were more conducive to growth than those of the southern colonies in the New World.[56] As Joyce Chaplin has pointed out, the *Magnolia grandiflora* L. was a test case. In his *Natural History,* Catesby explained how, in America, it could not grow north of South Carolina. Yet in England, which lay far to the north, it was already "naturalized." By 1737 Christopher Gray would choose to place Catesby's etching of the *Magnolia grandiflora* at the center of his advertisement for his nursery at Fulham. It was the most expensive exotic, the most beautiful in every respect, and—much to the delight of wealthy collectors—it grew in England's balmy northern latitudes.

Thomas Fairchild, listing the plants in bloom in his Hoxton nursery in 1722/1723, recorded that "Mr. Catesby's new Virginian Starwort" (*Aster grandiflorus* L., introduced in 1720 and appearing in the *Historia*)[57] was still flowering in October, along with the "Chrisanthemum Tree from Carolina" (which John Harvey postulated was *Helenium autumnale* L., common sneezeweed). "Mr. Catesby's fine blue Startwort" (presumably again *Aster grandiflorus*) was still blooming in December 1722.[58] These examples are indicative of several things. First, they imply that the introduction of North American flowers was helping to lengthen the flowering season. By contrast, some twenty years earlier, Henry Wise possessed only the North American "Red and Blue Cardinal Flower" (*Lobelia cardinalis*

L. and *L. syphilitica* L.). These he included in his September entry. Apparently there was insufficient bloom to itemize flowers from October to December.[59] Second, it suggests that despite the fumes of sea coal, Fairchild—"a truly great plantsman and one of the greatest florists"—was sometimes able to overcome the effects of pollution that had troubled John Evelyn at Sayes Court.[60] And third, it shows that, although the *Magnolia grandiflora* was yet to arrive (it was not introduced until c. 1734), the mild seasons of the 1720s were already favoring new exotics. The "Scarlet flower'd Horse Chestnut of Virginia" (*Aesculus pavia* L., introduced in 1711) figured among the new North American woody exotics in flower at Hoxton in April 1722.[61]

The winter of 1723/1724 and eight of the winters of the 1730s proved among the mildest on record.[62] The average temperature in December to February 1733/1734, for example, was 6.1°C (43°F), making it the sixth warmest on record (for the years 1659–1979).[63] Similarly, the summers of the late 1720s and 1730s were warm (close behind recent hot summers in the era of "climate change"). From 1730 to 1739 there was a sequence of ten good harvests; the hot summer of 1734 permitted large wheat exports.[64] These must have lulled the gardener into a false confidence, which was rudely broken by the bitter winter of 1739/1740, when the *Magnolia grandiflora* and other American exotics like the *Callicarpa americana* L. were decimated.[65] Indeed, 1740 proved the coldest calendar year in English temperature records (kept from 1659 onward), averaging only 6.8°C (44°F) in central England. Yet, even before the shock of that winter (which averaged –0.4°C (31°F) for December, January, and February, making it the coldest winter since 1683/1684 and the second coldest on record for the years 1659–1979), there had been one severe warning. The summer of 1725 was the coldest on record, averaging only 13.1°C (56°F) for the months of June, July, and August. July 1725 was described as "more like winter than summer." Catesby missed its impact because of his sojourn in the Bahamas, but Fairchild doubtless kept record of stunted flowering to put alongside

14. Reconstruction (1997), as elevation, of one section of the diagram "Disposition of Deciduous Trees and Shrubs for a Plantation," in James Meader, *The Planter's Guide; or, Pleasure Gardener's Companion* (London, 1779), showing a tulip tree (top left) and a "bastard indigo" (center left), watercolor
Author's reconstruction

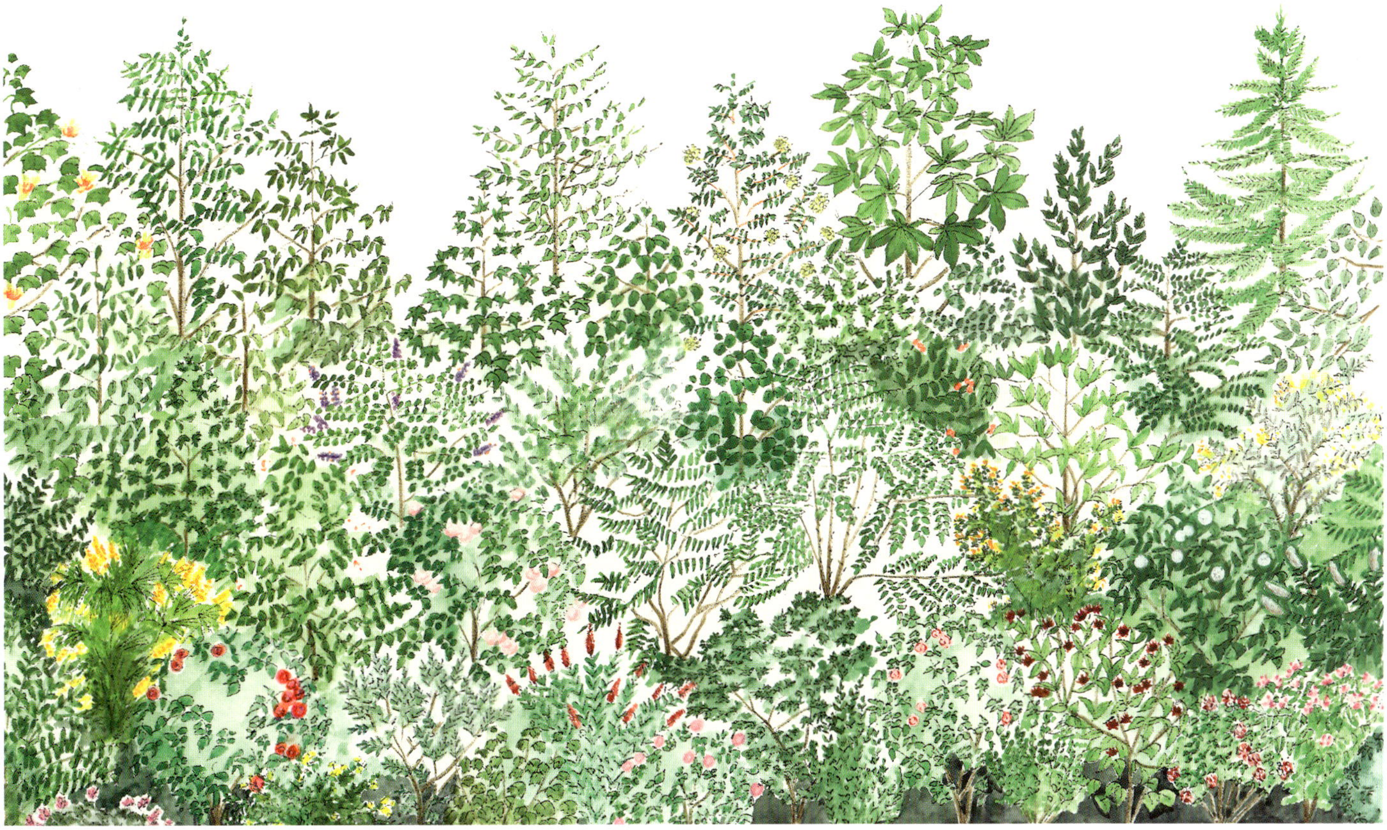

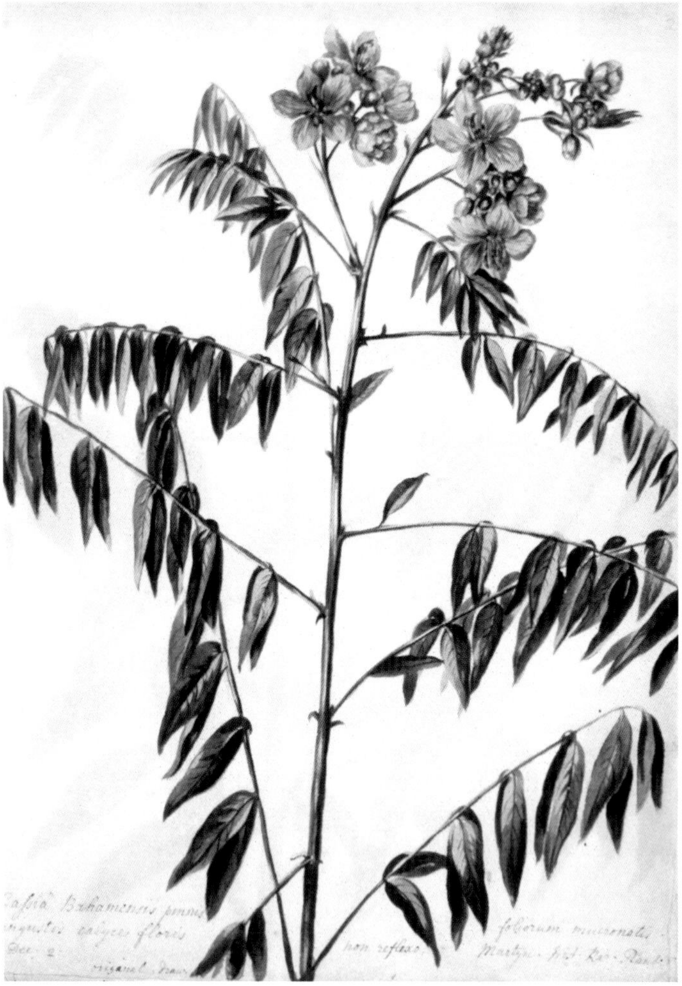

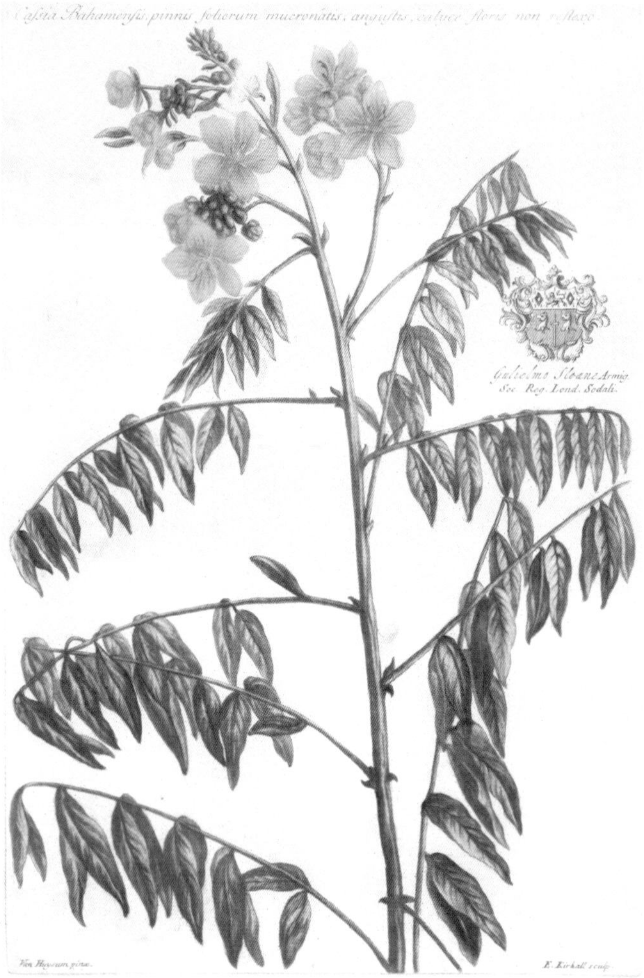

15. Jacobus van Huysum, *Cassia Bahamensis* (*Cassia ligustrina* L.), drawing for John Martyn, *Historia plantarum rariorum* (London, 1728–1737), watercolor
© The Royal Society, London

16. Elisha Kirkall after Jacobus van Huysum, *Cassia Bahamensis* (*Cassia ligustrina* L.), from John Martyn, *Historia plantarum rariorum* (London, 1728–1737), line engraving and mezzotint printed in color and hand-colored in watercolor
Courtesy of the Library of the Arnold Arboretum, Harvard University

the favorable year 1722/1723. If we add the London smoke as an inhibiting factor, it is clear that the environment was by no means invariably conducive to growing the newly acclimatized exotics.

This may seem a fittingly truncated conclusion to the story of John Martyn's *Historia* and the Society of Gardeners' *Catalogus*. After all, Martyn's Botanical Society had promised so much. The *Catalogus plantarum* had been conceived as an ideal four-volume compendium: to present each exotic with its exact name, the plants "Divided, according to their different Degrees of Hardiness." Yet in practice, John Martyn, abandoning the expensive *Historia* for lack of funds after five "decades," declined with his generation—sinking into the "lurid depths of Grub Street journalism."[66] Jacobus van Huysum died in obscurity. The intended four volumes of the *Catalogus plantarum* never materialized. And the hard work of acclimatization over twenty years was reversed by a single bitter winter.

However, this temporary and very typical setback in the cruel months of 1739/1740 was balanced by repeated runs of predominantly favorable seasons until the end of the century.[67] Moreover, the publishing spurt of the late 1720s had an enduring legacy. Just as Langley's 1728 model of planting would help define the shrubbery for the rest of the century (see fig. 1), so too Robert Furber's "Borders of Cut-Work" of 1727 set the characteristic pattern for the flower garden of the 1770s: a baroque model still serving the picturesque Nuneham Courtenay. At Nuneham and Hartwell House, the second Earl Harcourt and his sister Lady Elizabeth Lee would go on to plant an improved form of *Phlox maculata* L.—or possibly James Edward Smith's *Phlox pyramidalis* (fig. 19).[68] It graced the "theatrical" beds of their flower gardens (fig. 20). While the introduction of *Phlox maculata* in 1740 probably goes back to Christopher Witt or John Bartram, Mark Catesby had already encountered the species in the Carolinas in the 1720s (as his specimens in the Sherard Herbarium in Oxford clearly demonstrate).[69] In 1799, in other words, on the eve of the founding of the Horticultural Society in 1804, the work of the Botanical Society and the Society of Gardeners, promoted by the art of Jacobus van Huysum, would yield results through an unbroken lineage in acclimatization that stretches from Catesby to Bartram and from Bartram to Smith.

The Legacy

What began, in essence, in the physic gardens of the seventeenth century as an effort to re-create the Garden of Eden[70] had effectively shifted by the end of the eighteenth century to an Edenic (as well as imperial) endeavor in the pleasure grounds of the English landscape garden. The change was under way by the 1720s. Mark Catesby, witness to the violence of the natural world, returned to gentle (but polluted) Hoxton in 1726 and, with Thomas Fairchild's help, set about reconstructing paradise on an island: "an American garden that was the opposite of an anarchic mangrove swamp."[71] Despite the city smoke, the dramatic fluctuations of the weather, and the resistance of many eastern North American plants to adapting to English conditions, and despite an unheeded violence to transported fauna,[72] a felicitous harmony resulted: the "theatrical" shrubbery and flower beds of the English pleasure ground.

The publications of the late 1720s promoted a re-creative vision that banished the mayhem of the deluge. The *Historia plantarum rariorum*—written in Latin for a medicobotanical audience, with subscriptions from apothecaries, physicians, and fellows of the Royal Society, and featuring tender as well as hardy plants—appears more closely allied to the coffeehouse culture from which Catesby's sponsorship grew. The *Catalogus plantarum*—marketed in English by practical nurserymen for the use of gentry and

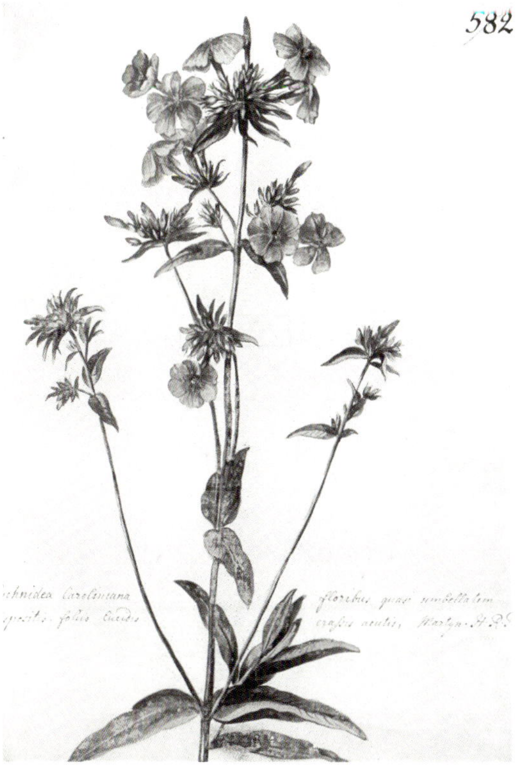

17. Jacobus van Huysum, *Lychnidea Caroliniana* (*Phlox carolina* L.), drawing for John Martyn, *Historia plantarum rariorum* (London, 1728–1737), watercolor
© The Royal Society, London

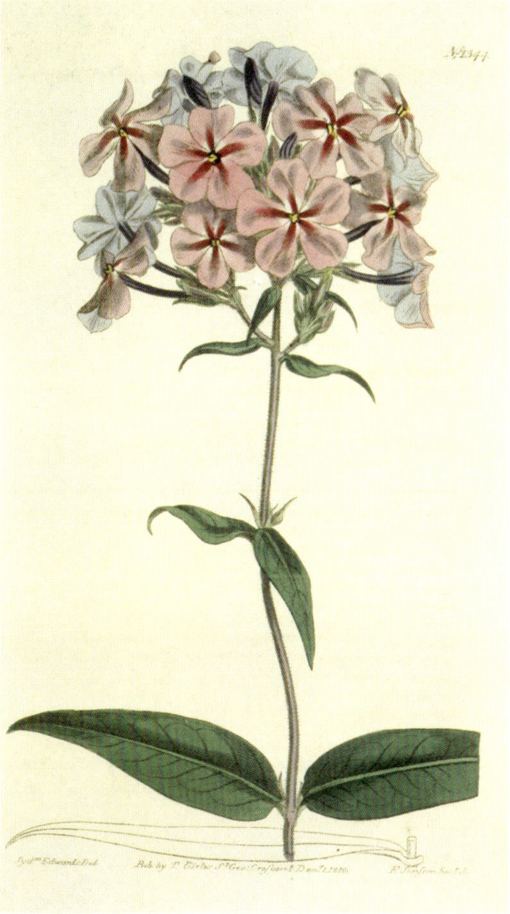

18. Francis Sansom after Sydenham Edwards, *Phlox carolina* (*Phlox carolina* L.), originally from *Botanical Magazine* (1810), plate 1344, engraving with watercolor
Dumbarton Oaks, Library. Rare Books Collection, Washington, D.C.

middle classes alike, and featuring "acclimatized" plants to grace both town and country—suggests, in contrast, a commercialized direction for the future. Of course, the scientific collections of Philip Miller's Chelsea Physic Garden and the Princess of Wales' Kew would continue. Yet, as early as the mid-1730s, the Prince of Wales would turn to the nurserymen Robert Furber and Richard Butt to supply him with the mostly exclusive, ostentatious, and costly exotics for his garden at Carlton House.[73]

The long-term consequences of the activities that engaged Mark Catesby and Thomas Fairchild in the 1720s are, of course, with us today. As a by-product of that early consumer revolution, plant collecting would lead to the invasive species of global exchange; hybridization has been replaced by the genetic modification of plants and the patenting of nature in the name of science, medicine, and human well-being. The direct link to climate change of the very violent hurricanes of 2005 has been hotly debated in the journals *Nature* and *Science.*[74] Catesby and Fairchild had no inkling of such matters. Yet, as John Harvey pointed out, Fairchild "is especially worthy of honour in that, unlike so many scientists, he had serious moral scruples about what he was doing. The production of new forms, even by the artificial mixing of those created, smacked to him of interference with the plan of the Creator."[75] In Fairchild's last will and testament, apparently trying to "discharge his soul of impiety," he ensured a personal legacy that was later perpetuated by subscriptions from Sir Hans Sloane, Lady Walpole, Shoreditch vinegar-maker Cornelius Wittenoom, and eventually Dr. Denne and the Royal Society.[76]

Fairchild's will stated that a sum of twenty-five pounds be invested toward the payment of twenty shillings in interest each year for the preaching of a sermon. Saint Leonard Shoreditch was the chosen venue; the afternoon of Tuesday in every Whitsun week was the appointed time. The subject was specified to be the "wonderfull works" of God and the certainty of resurrection in relationship to the animal and vegetable parts of "the Creation."[77]

Thomas Fairchild also took care in the will to leave money for his servants and

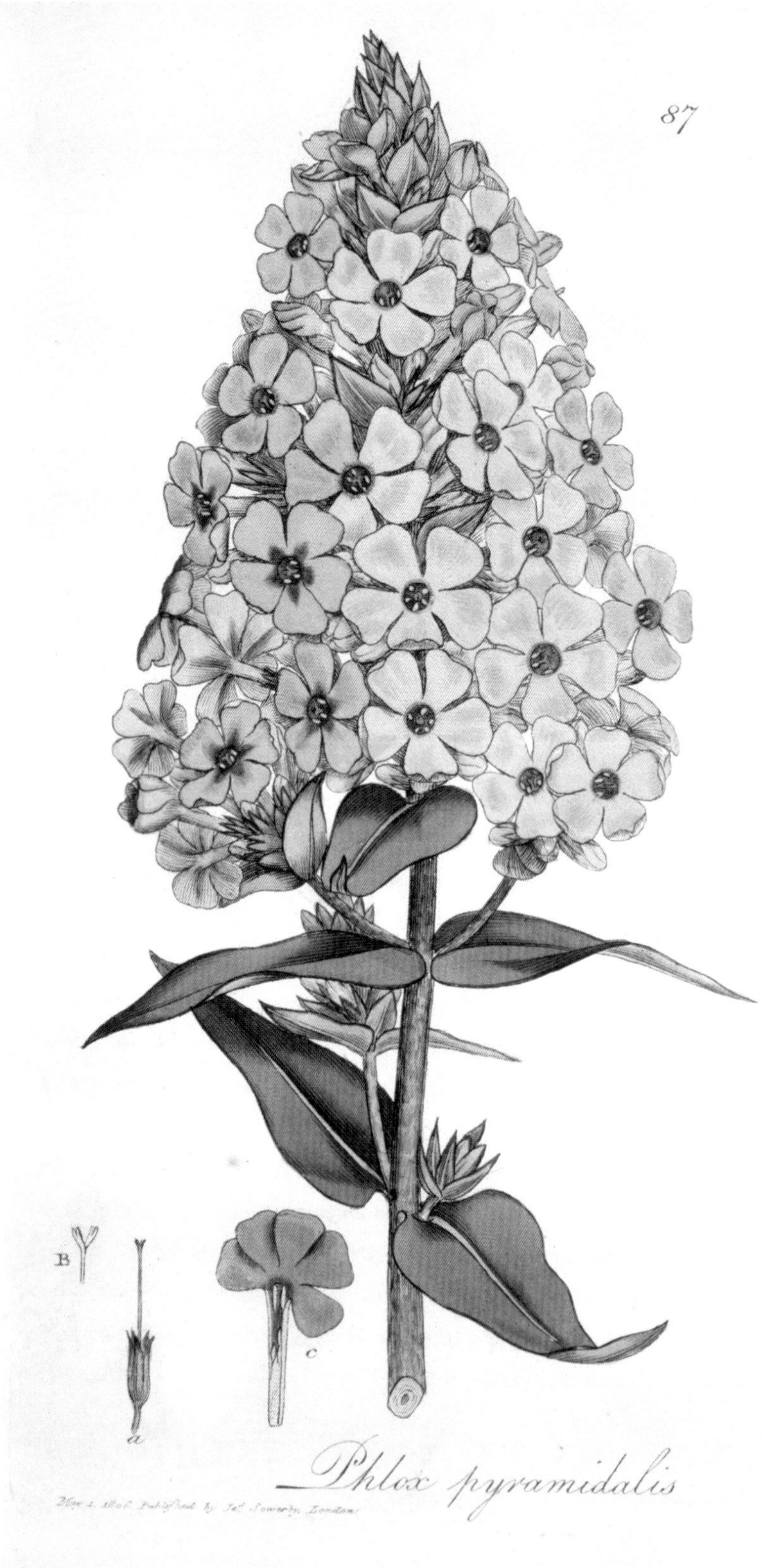

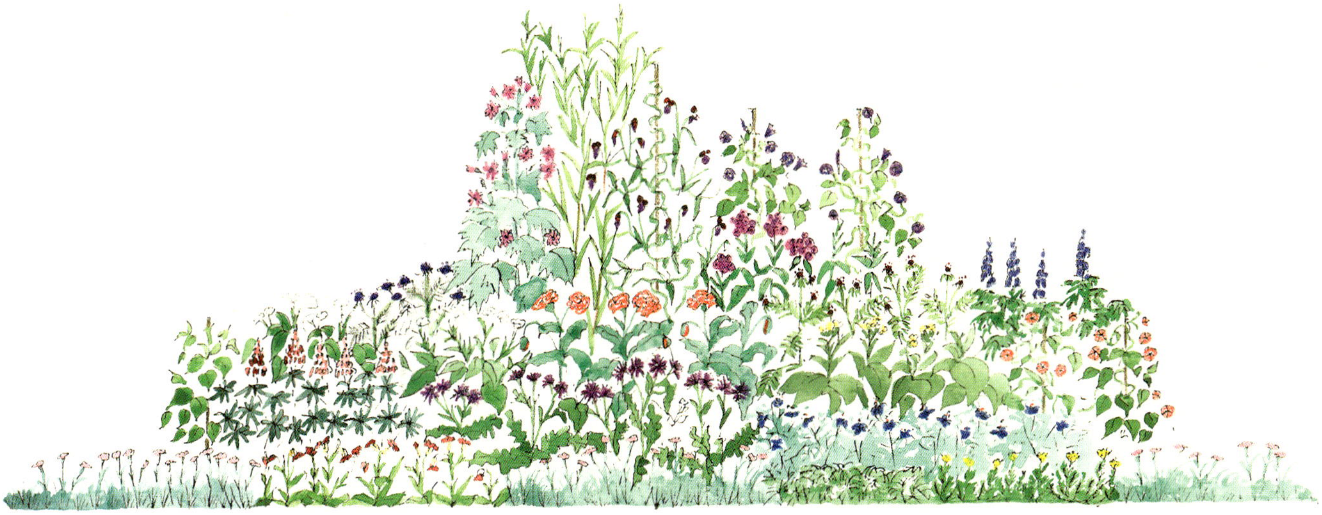

19. J. Sowerby, *Phlox pyramidalis* (*Phlox maculata pyramidalis* [Smith] Wherry), from James Edward Smith, *Exotic Botany* (London, 1804–1805), volume 2, plate 87, engraving with watercolor
Dumbarton Oaks, Library. Rare Books Collection, Washington, D.C.

20. Reconstruction (1989), as elevation, of planting plan 36 for "Round Bed" at Lady Elizabeth Lee's Hartwell House, Buckinghamshire (1799), watercolor
Author's reconstruction; original planting plan, Bodleian Library, Oxford University, MS Top. Gen. b.55, folio 36r

his relatives. The residue—including his lands, goods, garden stock, and personal estate—went to his nephew Stephen Bacon, nurseryman of Hoxton and signatory to the preface of the *Catalogus plantarum.* Given that Stephen was already down for his copy of the *Catalogus,* there was one important exclusion: "my right and title to a subscription of a Booke belonging to the Society of Gardners subscribed thereto which I gave to my nephew John Bacon."[78] The following year, 1730, the book came out (and hopefully went to John); and on 19 May the initial Fairchild Lecture—the first of what were later called the "Vegetable Sermons"—was delivered at Saint Leonard Shoreditch. The summer would prove hot and dry; a good harvest resulted. But we are left wondering: what was the weather like that Tuesday afternoon before Whitsun as the preacher, the Reverend Dr. John Denne, climbed up to the high pulpit?

NOTES

I am grateful to Margaret Riley for her help with aspects of this essay, and for agreeing to append her research on the club at the Temple Coffee House to this essay. I am also indebted to John Edmondson for tracking down an article that contains modern identifications of the plants featured in the *Historia plantarum rariorum:* I have followed the currently accepted names as determined by Hess in Ian McPhail and William J. Hess, "The Date of the Linnaean Index in John Martyn's *Historia plantarum rariorum*," *Huntia* 3, no. 2 (1979): 73–81. Phyllis Andersen and Lisa Pearson are to be thanked for sending me a copy of the *Huntia* article from the Arnold Arboretum library. Thanks also are due to Monica S. Rumsey for her supportive editing of this essay. Her guidance on mezzotint was especially appreciated.

1. Mark Catesby, *The Natural History of Carolina, Florida, and the Bahama Islands* (London, 1731–1743 [1729–1747]), 2:45. Quoted from Joyce E. Chaplin, "Mark Catesby, a Skeptical Newtonian in America," in *Empire's Nature: Mark Catesby's New World Vision*, ed. Amy R. W. Meyers and Margaret Beck Pritchard (Chapel Hill and London, 1998), 74.

2. See Mark Laird, "Sayes Court Revisted," in *John Evelyn and his Milieu*, ed. Frances Harris and Michael Hunter (London, 2003), 115–144. Avernus, a lake in the Campagna region of Italy, was thought in the ancient world to give off an effluvium that killed overflying birds.

3. This list, in the form of three letters sent to Richard Bradley's *Monthly Register*, is reproduced in transcription in John Harvey, *Early Nurserymen* (London and Chichester, 1974), appendix 4, 150–159.

4. For John Evelyn's account of this "Hurecan," see Esmond S. de Beer, *The Diary of John Evelyn*, 6 vols. (Oxford, 1955), 5:550.

5. See Hubert H. Lamb, *Climate, History and the Modern World*, 2nd ed. (London and New York, 1995), 219, for discussion of the 1703 storm, and 244 for the summer of 1725.

6. Lamb 1995, 243–244.

7. Philip Miller's account in *The Gardeners Dictionary* of 1768 supports this view (entry "Amorpha Fruticosa"): "The seeds of this plant were sent to England from Carolina, by Mr. Mark Catesby, F.R.S. in 1724, from which many plants were raised in the gardens near London; these were of quick growth, and many of the plants produced flowers in three years." It is interesting to note that, by 1768, Miller associated *Amorpha* uniquely with "shrubbery," a term he otherwise avoided in favor of the old term "wilderness."

8. For a full account of Langley's role in the development of shrubbery, see Mark Laird, *The Flowering of the Landscape Garden: English Pleasure Grounds, 1720–1800* (Philadelphia, 1999).

9. The title is *Catalogus plantarum, tum exoticarum tum domesticarum, quae in hortis haud procul a Londino sitis in venditionem propagantur. A Catalogue of Trees, Shrubs, Plants, and Flowers, both Exotic and Domestic, Which are propagated for Sale, In the Gardens near London. Divided, according to their different Degrees of Hardiness, into particular Books, or Parts . . .* (London, 1730).

10. Wilfrid Blunt and William T. Stearn, *The Art of Botanical Illustration* (1950; new edition, Woodbridge, Suffolk, U.K., 1994), 147. So-called true mezzotint is an engraving technique that produces tones, rather than lines, in the final print. The tones are achieved by first creating an overall texture on a metal printing plate by using a rocker and then by alternately scraping and burnishing that texture to create darker and lighter tones on the plate. The textured areas hold more ink and will appear as continuous tone in the resulting print. The roulette wheel, an alternate tool for creating tone in an incised print, is more often used in the process known as stipple engraving. See Paul Goldman, *Looking at Prints, Drawings and Watercolours: A Guide to Technical Terms* (London and Malibu, Calif., 1988), 42, 53.

11. Blunt and Stearn 1994, 150.

12. Quoted from Blunt and Stearn 1994, 147. Van Huysum's oil paintings of flowers by season, reminiscent of Furber's *Twelve Months of Flowers*, are in the Fitzwilliam Museum, Cambridge. See Sam Segal, *Flowers and Nature: Netherlandish Flower Painting of Four Centuries* (The Hague, 1990). See also Sam Segal with Mariël Ellens and Joris Dik, *The Temptations of Flora: Jan van Huysum, 1682–1749* (Zwolle, 2007), 23–24, 321–325. Segal states that the twelve paintings of flowers by season are dated between 1731 and 1733. He gives van Huysum's dates of birth and death as 1688–1740.

13. Blanche Henrey, *British Botanical and Horticultural Literature before 1800*, 3 vols. (London, 1975), 2:52–53. Some explanation of the color printing process may be helpful here: In the early eighteenth century, some innovative print designers for illustrated scientific treatises sought ways to increase the efficiency of print production and to enhance the accuracy of the images themselves. The traditional method of producing colored illustrations for such books involved first creating the images in watercolor, then copying the design onto a metal plate, either by etching, line engraving, or mezzotint engraving. Black ink would then be rubbed into the incised lines and tonal areas of the plate, and a "sandwich" comprising the inked plate and a damp sheet of paper would be passed through the rollers of the printing press. Finally, teams of trained colorists would painstakingly hand-color each print with watercolors to match the original watercolor design, often supervised by the artist who had created it. The new, still rudimentary, method of "color printing" was aimed at eliminating—at least partially—the need for hand-coloring the prints after the fact. This

process allowed color to be added to the plate before it was put through the printing press. Once the plate had been incised by etching, engraving, or mezzotinting, the printer would mix printing inks in one or more colors, taking care to simulate the exact hues in the original watercolor. The colored ink—such as green—would then be rubbed into the incised metal plate. Additional colors, such as brown or other tones, could be applied to other specific areas of the plate. The result would be a "color print" that could either stand on its own or be further enhanced with watercolors applied by hand, producing a print described as "hand colored." For more information on early color printing processes in America, see James N. Green, "Hand-Coloring vs. Color Printing in Early-Nineteenth-Century Natural History Color-Plate Books," in *The Culture of Nature: Art and Science in Philadelphia, 1740–1840*, ed. Amy R. W. Meyers (New Haven and London, 2008).

14. Blunt and Stearn 1994, 150. Robert Furber described it in 1732: "This Plant we lately received from *Holland*, but it was first raised in *Italy*, and many Contrivances were used before it could be brought to *Holland*; it first bore a great Price, and was esteem'd as a great Rarity, and by planting it of Cuttings it is now become pretty plentiful. . . ."

15. Quoted from Henrey 1975, 2:213.

16. Henrey 1975, 2:212.

17. See Henrey 1975, 2:213.

18. The links between John Banister and Bishop Compton and the growing of American plants such as *Physocarpus opulifolius* are discussed in Mark Laird, "Exotics and Botanical Illustration," in *Sir John Vanbrugh and Landscape Architecture in Baroque England, 1690–1730*, ed. Christopher Ridgway and Robert Williams (Stroud, Gloucestershire, 2000), 95–97 and fig. 74.

19. For background, see the standard work by Bryant Lillywhite, *London Coffee Houses: A Reference Book of Coffee Houses of the Seventeenth, Eighteenth, and Nineteenth Centuries* (London, 1963). John Brewer gives a good overview of the coffeehouse club in *The Pleasures of the Imagination: English Culture in the Eighteenth Century* (London, 1997), 34–40. For a recent survey of the literary culture of the eighteenth-century coffeehouse, see Markman Ellis, ed., *Eighteenth-Century Coffee-House Culture*, 4 vols. (London, 2006). I am grateful to Monica Rumsey for this latter reference.

20. Brewer 1997, 50.

21. David Elliston Allen, "John Martyn's Botanical Society: A Biographical Analysis of the Membership," *Proceedings of the Botanical Society of the British Isles* 6 (May 1967): 305–324.

22. Allen 1967, 306.

23. Allen 1967, 308.

24. I am grateful to Margaret Riley for confirming that, according to his baptismal record, Charles du Bois was born in 1658 rather than c. 1656 as previously given (on the basis of his tombstone inscription).

25. David R. Brigham, "Mark Catesby and the Patronage of Natural History in the First Half of the Eighteenth Century," in *Empire's Nature: Mark Catesby's New World Vision*, ed. Amy R. W. Meyers and Margaret Beck Pritchard (Chapel Hill and London, 1998), 91–146.

26. Brewer 1997, 164.

27. Brewer 1997, 164.

28. Mark Laird, "The Culture of Horticulture: Class, Consumption, and Gender in the English Landscape Garden," in *Bourgeois and Aristocratic Cultural Encounters in Garden Art, 1550–1850*, ed. Michel Conan (Washington, D.C., 2002), 247–248. A calculation of percentages depends, of course, on whether equal weighting is given to those who ordered more than one copy. Yet, even with a factor allowed for two or three "setts," the proportion of women remains around one-third. It is interesting to note that a "Mr. J. Van Huysum" is among the subscribers whose names—initially omitted—are listed at the top of the page.

29. See Douglas Chambers, "'Storys of Plants': The Assembling of Mary Capel Somerset's Botanical Collection at Badminton," *Journal of the History of Collections* 9, no. 1 (1997): 47–58.

30. David Elliston Allen, *The Naturalist in Britain: A Social History* (London, 1976), 28.

31. Eleazar Albin, *A Natural History of English Insects* (London, 1720; here in the 1724 edition), first page of the preface.

32. Johanni Carparo Scheuthzer seems likely to be Jean Gaspard Scheuchzer (1702–1729), the son of a Swiss scientist, Johan Jakob Scheuchzer, correspondent of Sherard, etc. Sloane gave the son a job looking after his library, and he died in Sloane's house. According to the Royal Society Web site, Sprengell, or Sprengwell (d. 1740), was elected a fellow of the society in 1721 and was physician-in-ordinary to kings William III and George I. I am indebted to Margaret Riley for pointing this out to me.

33. Richard Myddleton Massey is important as a key figure in relationship to the Temple Coffee House, but not necessarily in relationship to the club. On 23 February 1703/1704, he wrote to Edward Lhwyd and asked that his next letter be sent to the Coffee House (Bodleian: Ashmole MSS 1816 folio 346). On 4 September 1706, and again in October, he addressed letters to Sloane at the coffeehouse (Sloane MSS 4040, folios 214–215 and folio 238). There is no mention of the club. On 27 October 1707, he used the address again (Sloane MSS 4041, folio 48). However, he was unsure whether an earlier letter addressed to the coffeehouse had reached Sloane, and his final letters

addressed to the Temple Coffee House are following 24 March 1707/1708 (Sloane MSS 4041, folio 121); after 5 May 1708 (Sloane MSS 4041, folio 134), he used Sloane's home address. This, along with Sherard's letter of 1706, is circumstantial evidence that Sloane was using the coffeehouse as headquarters for the club until 1706. All information was kindly supplied to me by Margaret Riley through personal communication. Plate 29 of the third decade of the *Historia* was drawn by Massey.

34. It may be reasonable to assume that "Alexander Stewart" is the same as Alexander Stuart (c. 1673–1742), a ship's surgeon who collected plants in Asia for Sloane and du Bois and was familiar with the Temple Coffee House. See the reference to Stuart in Margaret Riley's appendix to this essay.

35. Henrey 1975, 1:122 and 124–126. I am very grateful to Margaret Riley for the latter reference. See also Michael Hunter, *Science and Society in Restoration England* (Cambridge, 1981; Aldershot and Brookfield, Vermont, 1992), 143. Hunter (69) refers to *Historia Piscium* of 1686 by Francis Willughby (1636–1672) as an example of a publication with plates paid for by virtuosi.

36. All quotations in this paragraph are from the preface to the *Catalogus plantarum.*

37. See Laird 2000, 96, where I discuss the fact that, by 1723, only around 12 percent of John Banister's introductions to Bishop Compton's garden in Fulham had entered the nursery trade.

38. Allen 1976, 6–7: "The study of plants had obvious practical application for the infant science of medicine. . . . The agency that we have to thank for this is the Society of Apothecaries, one of the old-established livery companies of the City of London." Michael Leapman's account of the correspondence between Richard Bradley (c. 1688–1732) and James Petiver points to an amusing aspect of plant knowledge in early medicine. In 1714, while in the Netherlands to collect plants and specimens, Bradley masqueraded as a doctor to pay off debts. The apothecary Petiver helped Bradley's ruse by supplying advice and recipes—for example, oil of elder for swollen testicles. See Michael Leapman, *The Ingenious Mr Fairchild: The Forgotten Father of the Flower Garden* (London, 2000), chapter 3.

39. See Laird 1999, chapter 2, and Mark Laird, "From Callicarpa to Catalpa: The Impact of Mark Catesby's Plant Introductions on English Gardens of the Eighteenth Century," in Meyers and Pritchard 1998, 184–227.

40. *Catalogus* 1730, preface.

41. Harvey 1974, 77–80. Thomas Fairchild was, of course, dead by 1730, even though his name appears first in the signatories' list.

42. See, for example, Harvey 1974, 80.

43. Allen 1976, 15.

44. Philip Miller, *The Gardeners Dictionary: Containing the Methods of Cultivating and Improving the Kitchen, Fruit and Flower Garden* (London, 1731–1768).

45. Harvey 1974, 78–79 and 80.

46. Laird 1999.

47. Leapman 2000, 87–88.

48. For a good summary of the role of the Worshipful Company of Gardeners in regulating the trade, see Todd Longstaffe-Gowan, *The London Town Garden, 1740–1840* (New Haven and London, 2001), 154–159.

49. See John H. Harvey, "A Scottish Botanist in London in 1766," *Garden History* 9, no. 1 (Spring 1981): 40–75. James Lee of the Vineyard Nursery, Hammersmith, and James Gordon of Mile End Nursery stand out. Harvey commented in his introduction (43): "From Hope's observations, supported by perusal of catalogues of the time, it seems that the nurseries of Gordon and Lee and some gardens such as that of Collinson at times had plants in advance of Kew." For further discussion of Hope, see Mark Laird, "This Other Eden: The American Connection in Georgian Pleasure Grounds: From Shrubbery and Menagerie to Aviary and Flower Garden," in Meyers 2008.

50. For further discussion of the catalpa, see Laird 1998, 219–224; and Laird 2002, 244.

51. For a complete history of *Amorpha fruticosa* in Georgian gardens, see Laird 1998, 190–202.

52. See John H. Harvey, "The English Nursery Flora, 1677–1723," *Garden History* 26, no. 1 (Summer 1998): 63–64. Its cultivation is discussed in Laird 1998, 206.

53. Laird 1999, 222.

54. A later submission from Chelsea Physic Garden (RS no. 3189, 1785) implies sustained cultivation of this species. For further information on dates of submission of Royal Society specimens, see William T. Stearn, "Philip Miller and the Plants from the Chelsea Physic Garden Presented to the Royal Society 1722–1796," *Transactions of the Botanical Society of Edinburgh* 41 (1972): 293–307; and Ruth Stungo, "The Royal Society Specimens from the Chelsea Physic Garden 1722–1799," *Notes and Records of the Royal Society* 42, no. 2 (1993): 213–224.

55. See Laird 1999, 221, which suggests that *Phlox carolina* was cultivated at Bulstrode in the mid-eighteenth century.

56. See Chaplin 1998, 79.

57. *Aster grandiflorus* appears in the *Historia* as plate 17, "Aster virginiana."

58. Harvey 1974, 150–159; and Harvey 1998, 60.

59. Harvey 1974, 150.

60. Harvey 1974, 150. Fairchild's list for April 1722, for example, includes "Anemonies of various Kinds, the best Sorts." Evelyn had struggled to cultivate anemone in Deptford, as I discuss in Laird 2003, 116.

61. It should be pointed out, on the other hand, that—apart from *Robinia pseudoacacia*—these woody American exotics do not feature prominently in Fairchild's *The City Gardener* (1722). A few American flowers mentioned in the book include the perennial sunflower, the evening primrose, and the Michaelmas daisy.

62. Lamb 1995, 244. All subsequent statistics in this paragraph are drawn from Lamb 1995, 230 and 244.

63. Lamb 1995, 230.

64. Eric L. Jones, *Seasons and Prices: The Role of the Weather in English Agricultural History* (London, 1964), 137.

65. See Laird 1998, 190.

66. Allen 1976, 15.

67. According to Lamb 1995, 230 and 246, the middle decades of the eighteenth century were notable for wet summers, and the summers of 1762 and 1781 rank among the fourteen hottest on record in central England: "The latter part of the eighteenth century could be described as a time of mainly warm summers and cold winters." But, while the improving trend may have helped in the cultivation of exotics, extreme phenomena still offset the favorable years: notably, the frosts of January 1768 and January 1776, and the drought summers of 1781 and 1783 (in the latter case, heat combined with the occluding effects of volcanic dust). These were discussed by Gilbert White in his letters 61–65, in *The Natural History of Selborne* (1789; this edition, Oxford, 1993).

68. See Mark Laird, "'Our Equally Favorite Hobby Horse': The Flower Gardens of Lady Elizabeth Lee at Hartwell and the 2nd Earl Harcourt at Nuneham Courtenay," *Garden History* 18, no. 2 (Autumn 1990): especially John H. Harvey's appendix, 143.

69. The Sherard Herbarium specimen in Oxford (304), identified by James Reveal in 1982 as *Phlox maculata* L., is labeled in Catesby's hand: "These flowers are red grows in Sedgy Moyst places blows in May."

70. See John Prest, *The Garden of Eden: The Botanic Garden and the Re-creation of Paradise* (New Haven and London, 1981).

71. Chaplin 1998, 77.

72. See Laird in Meyers 2008.

73. Laird 2002, 248; see also David Coombs, "The Garden at Carlton House of Frederick, Prince of Wales, and Augusta, Princess Dowager of Wales: Bills in Their Household Accounts, 1728–1772," *Garden History* 25, no. 2 (Winter 1997): 153–177.

74. See, for example, *Nature* 437 (22 September 2005): 461. The impact on flora and fauna may return as part of that debate. An issue of *Nature* in November 2005 reported the findings of a team of researchers at Britain's National Oceanography Centre who have detected a considerably decreased flow in the Gulf Stream since 1957. Although associated with global warming, this trend may make the English climate substantially colder. Aside from the devastating socioeconomic effects of such a climate change, it would no longer be possible to grow acclimatized plants like *Magnolia grandiflora.*

75. Harvey 1974, 76.

76. Leapman 2000, 215.

77. See Harvey 1974, 77; Leapman provides a full transcription of the will, but it should be noted that his "wonderful world" does not match Harvey's "wonderfull works."

78. Quoted from Harvey 1974, 77.

The Coffee Tree.

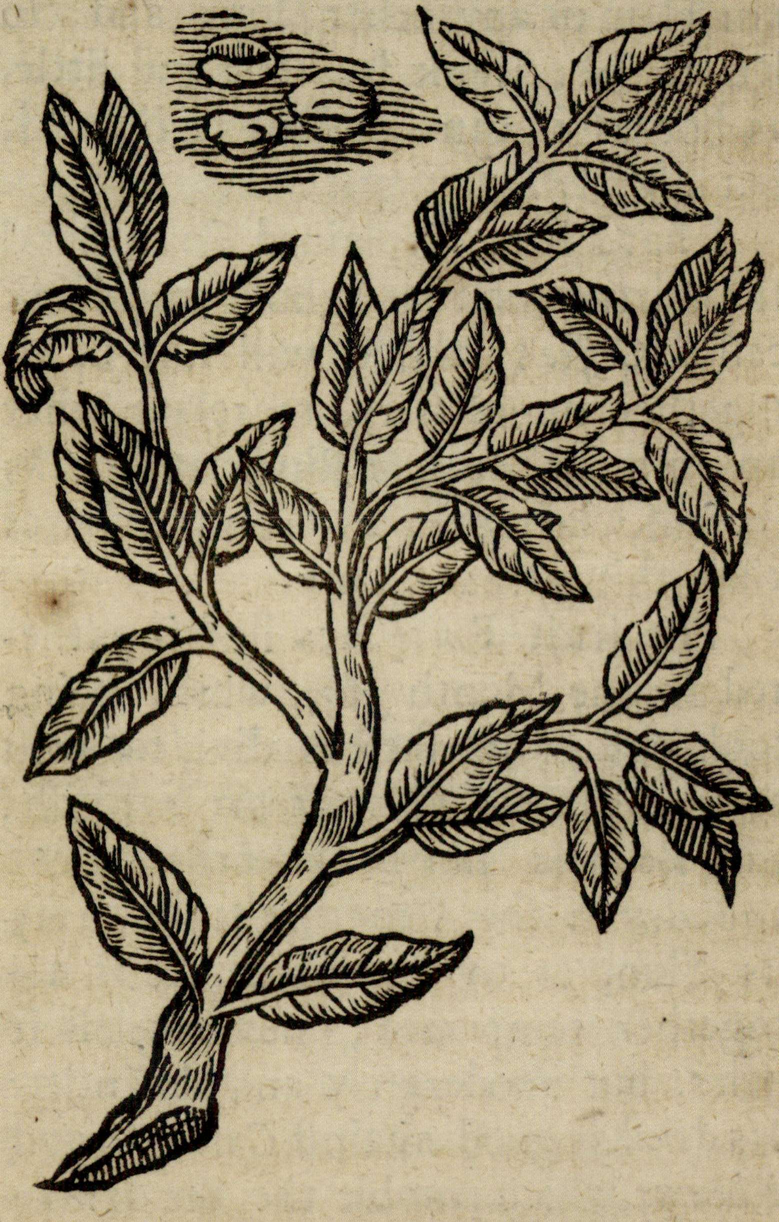

The Inſtrument.

APPENDIX

The Club at the Temple Coffee House

MARGARET RILEY
University of Buckingham

The Scottish journalist John Macky (d. 1726), writing of his travels around England in the early eighteenth century, described London as having "an infinity of clubs or societies for the improvement of learning and keeping up good humour and mirth."[1] Between 1689 and about 1706, naturalists from across the country and abroad clubbed together in the convivial surroundings of the Temple Coffee House near Fleet Street, to pore over plant specimens and discuss the latest botanical discoveries from the New World and beyond. Two individuals associated with the club throughout its existence were Hans Sloane (1660–1753) and the apothecary James Petiver (c. 1663–1718). Upon completing his medical studies in France, Sloane rapidly established his reputation in scientific circles and was elected to the Royal Society in 1685. He was appointed as its secretary in 1693.[2] Petiver concluded his apprenticeship in the same year as Sloane's election, setting up his own shop in the City soon afterward. Although hampered by a lack of "Academicall Learning," he was an extraordinarily dedicated naturalist and collector, specializing in botany and entomology; through Sloane's influence, he was made a fellow of the Royal Society in 1695.[3]

A club was defined by Samuel Johnson as "An assembly of good fellows, meeting under certain conditions,"[4] but the sketchy records of the club at the Temple Coffee House leave a great deal of ambiguity about its participants and their activities. Past interpretations of this material give the impression that more is known about it than the facts support, and a reappraisal is called for. This essay sets out to discuss the archival sources, and to show that additional evidence found in an early eighteenth-century satire substantiates the existing documentation, while shedding further light on the club's activities.

Evidence in The Transactioneer

The Transactioneer, a satirical tract, was published anonymously in 1700 but was later attributed to the lawyer and humorist William King (1663–1712), an associate of the writer Jonathan Swift (1667–1745). King set out to poke fun at the Royal Society, mocking the quality of the society's *Philosophical Transactions* and attacking its editor, who was then Hans Sloane, and the contributions of his friends. Sloane's relationship with the poorly educated Petiver, ridiculed for his lack of erudition, was especially singled out. Cast as "The Transactioneer" in dialogue with a "Gentleman," Sloane was made to exclaim:

> I can never be to seek where to begin, as long as there is such a Personage as Mr. J—Pet—r in the Philosophical World. He is a F. of the R.S. indeed! I made him so. 'Tis my way of Rewarding my Friends and Benefactors.[5]

In the lines that follow, the apothecary was presented as the "Darling of the *Temple-Coffee-House Club,*" and "The Transactioneer" went on to berate his unenlightened interrogator:

> Oh, lay! Why don't you know? Where can you have lived? Why you must be an utter Stranger to Philosophy, and all *pretty Things*! Never heard of the *Temple* Club? Oh for shame, lets see you there a *Friday* night. I'm President there, and I'll assure you there are many *odd Things.* And Mr. Pet—r is, Gad he's every thing. He is the very Muffti, the Oracle of our Club.[6]

The mocking words of King's parody may belittle the group's purpose, but they provide a valuable printed record of the club's existence and tell us how it was run.

George Pasti discovered the club at the Temple Coffee House in 1950 when he was researching a biography of the distinguished botanist William Sherard (1658/1659–1728); Raymond Stearns supplemented these findings in his paper on Petiver. King's satirical observations are consistent with Stearns' conclusion that meetings were held on "Friday evenings" to "mix social intercourse with the exchange of botanical information, specimens, and communications from correspondents at home and abroad." The satire also implies, however, that the club was sufficiently ordered by 1700 to require officers rather than being "unofficial, probably without formal organization," as Stearns thought.[7]

According to Peter Clark's survey, professional men were particularly well placed to found societies in the seventeenth and eighteenth centuries, operating as they did "on the flank of the upper and middle-classes, able to move and mediate between different social groups."[8] The energetic Sloane—with his Royal Society connections and a web of contacts within virtuosi circles—was an obvious president of a club that brought together not only amateur collectors like Sherard, but

The Coffee Tree, The Instrument, from Philippe Sylvestre Dufour, *The Manner of Making of Coffee, Tea, and Chocolate* (London, 1685), woodcut
Library of Congress

also apothecaries who depended on plant knowledge for their trade. Sloane's own reputation as a naturalist was boosted after he spent fifteen months in the West Indies, where he was personal physician to the Duke of Albemarle, governor of Jamaica (1687–1690). During this time he collected a great many plants, drying "as fair Samples of them as I could" to bring home. His return in 1689 with a rich haul of eight hundred specimens, "most whereof were New," to show off to "all lovers of such Curiosities," perhaps provided the incentive for the founding of the club.[9]

Petiver's role was not precisely specified, but at the most basic level of organization a society had just one or two stewards to set up the meetings.[10] He was conceivably in charge of arranging the evening's proceedings. Fanatical in his devotion to the study of natural history and to collecting, he was exceptionally well placed to provide novelties for members to discuss. John Ray (1627–1705) believed he had "the greatest correspondence both in East and West Indies of any man in Europe."[11] Sloane found Petiver indispensable, a fact not lost on the pair's humorous tormentor, as the following extract shows:

> Sir, he and I are all one. You must know we club Notions, laying them up in a kind of Joynt Stock, and have all things in common: Sometimes he draws, and sometimes I, as we have occasion. But he pays in most plenteously. By my good-will, I would never be without him. I call him the Philosophick *Sancho*, and he me *Don*.[12]

Petiver's shop, at the "sign of the White Cross in Aldersgate Street," was familiar to "shipmasters, merchants, planters, physicians, surgeons, ministers of the gospel," or anyone going abroad, whom the apothecary pressed into sending back "hundreds of consignments of seeds, dried plants, insects, serpents, birds, fishes, and small animals."[13] All "*Curious Persons* and *Lovers of Natural History*" not only overseas, but also at home, were asked to add to his collections.[14] It is clear that sometimes club members were entertained by travelers such as the Cambridge fellow William Vernon (c. 1667–c. 1715) when he returned from Maryland in 1698 with a beautiful collection of dried plants to discuss.[15] Or Sherard, who wrote on the way back from Italy that he had "something to show them at my return tho not so much as they can me, or as I had y[e] last voyage."[16] The stay-at-home Petiver, however, relied on his many contacts for a continual supply of curiosities to interest his scientist friends.

Interpreting the Source Material

Described by David E. Allen as "the earliest natural history society in Britain and probably in the world" and "an unofficial outgrowth of the Royal Society," the club at the Temple Coffee House was a significant body, existing for about fifteen years.[17] And yet no set of rules, list of members, or any formal papers have ever come to light. Lack of such evidence led Allen to conclude that it was purely an informal gathering of friends. Clark found, however, that the survival of official records from clubs and societies of the early modern period is rare. Data are more likely to come from a "pot-pourri of external sources: diaries and correspondence, sermons, ephemera like poems, tickets for the feast day . . . and newspapers," while club life was a favorite subject for caricature, whether in cartoons or literature, fact or fiction.[18] It is consistent with these findings that references to the club were first discovered among archives of correspondence and then corroborated by a satire.

Pasti came across references to the group scattered through the letters of late seventeenth- and early eighteenth-century naturalists among the Sloane manuscripts at the British Library in London. He found further evidence in the Sherard papers at the Royal Society, and in the Ashmole and Lister manuscripts at the Bodleian Library in Oxford. Stearns later added to these discoveries. Subsequent authors paraphrased their accounts, most notably Blanche Henrey and Allen, whose own works are also frequently quoted by others.[19] Serious doubts about their conclusions were raised by museum curator Les Jessop when he was investigating the life of Charles du Bois (1658–1740),[20] a possible member.[21]

The Name

The name by which the club was known remains uncertain, and perhaps this was never officially determined. Pasti referred to "The Temple Coffee House Club," and Stearns, to "The Temple Coffee House Botany Club," but as Jessop has pointed out, neither author provided supporting evidence for his particular choice. Pasti's title, however, does appear in *The Transactioneer*, which gives it some authenticity.[22] Manuscript sources often mention the "Club" ("Clubb," or "Clubbe"), but occasionally the "Botanic (or "Botanick") Club" and even the "Botanic Societie." Stearns' interpretation is therefore also plausible.[23]

The Location

Even the identity of the Temple Coffee House, Jessop suggested, was not entirely clear. In his book *London Coffee Houses*, Bryant Lillywhite mentioned several similarly named establishments existing in the Fleet Street area in the early eighteenth century but was unable to discover whether there was any connection among them. There was also the Temple Exchange Coffee House near Temple Bar.[24] The chances of confusion are confirmed in a letter to Petiver from the Scottish botanist George Preston (1665/1666–1749) who, missing an opportunity to see him one evening, protested in exasperation: "I came to the Temple Coffee-house before 7 o'clock last night and there stayed til it was past 9 and then I found out my mistake that it was not the Coffee house within the Temple where it seems you was but it was not my fault for I thought that there had not been any other."[25] It is more than likely, however, that the venue was in Devereux Court, near the Fleet Street end of the Strand, because Sloane used this address for general delivery, as did Dr. Richard Myddleton Massey (c. 1678–1743) when he visited London in 1704.[26] The coffeehouse must have been close to the famous Grecian, a favorite haunt of Royal Society members.[27]

The Membership

An incorrect claim that there were forty members of the club in 1691 resulted, as Jessop discovered, from a misreading of a manuscript. This document described "the examination of forty plants by an unspecified number of people" rather than "the examination of plants by forty botanists."[28] It referred to a set of illustrations of South African species that were shown at a club meeting. These were given to Henry Compton (1632–1713), bishop of London, during his visit to the Netherlands earlier that year.[29] Jessop also found that Pasti and Stearns did not strictly separate activities carried out by the club from those "undertaken by the group of naturalists that existed in the London area in the late seventeenth and early eighteenth century."[30] According to Stearns, "On Sundays and occasional holidays during the summer months members of the Botany club often embarked upon botanizing expeditions."[31] But one of the manuscripts he cited, Petiver's instructions to Joseph Dandridge, Joseph Miller, Mr. Tindsley, and Mr. Boucher, made no mention of the club: "We meet this Evening between 6 & 7 at Waltons Coffe house in Warwick lane the corner of Newgate markett, in order to goe from thence to Morrow morning at 4 of ye clock to Chistlehurst bogg, a herborizing."[32]

The archival material is particularly ambiguous when it is used to identify the membership. This is clearly apparent when David Krieg (1669–1710), a German surgeon and naturalist who traveled to Maryland at the same time as William Vernon, asked Petiver to pass on "My humble Service to all Friends especially to Dr Sloan, Dr Lister, Dr Haar, Dr Woodward, & the noble club att the Temple Coffe house. to Mr Watts, Mr Ayrey, Mr Doody, Mr Stapphes [Staphorst?] at the Apothecary hall, Mr Dare, Mr Adare: etc. etc."[33] His greetings say something about London botanical circles, but although he mentioned the club, it is impossible to say which of those named were connected with it. Dr. John Woodward (1665–1728), for example, was unlikely to be a member as he and Sloane were enemies, openly feuding over Royal Society matters.[34] Woodward also vindictively blocked William Vernon's application to Archbishop Tenison (1636–1715) for a post as a physician and naturalist in the West Indies because the Cambridge fellow "kept company" with Sloane, Robinson, and Lister, and "that would not doe." Vernon refused to "take leave of my Club friends on Fryday night."[35]

Apart from Sloane and Petiver, whose connection with the group is evident from their correspondence and confirmed by William King's satire, documentation about the regular members and the extent of their commitment to the club remains sparse and inconclusive. Vernon's grievances indicate that Dr. Martin Lister (1639–1712), a fellow of the Royal Society and an amateur zoologist, was probably a member. Sloane's close friends, Dr. Tancred Robinson (c. 1657–1748), another Royal Society fellow, and the collector and dealer William Courten (also known as "Charleton," 1642–1702) were clearly involved from the early days; sometimes meetings were held in Courten's house. Petiver's good friend and fellow apothecary, Samuel Doody (1656–1706), who succeeded John Watts (act. 1670s–1701) in 1692 as keeper of the Apothecaries' Garden at Chelsea, was also included. Doody was elected to the Royal Society at the same time as Petiver. There is evidence that Captain Charles Hatton (1635–c. 1705) took part; a keen horticulturist, he was the younger son of Christopher, first Baron Hatton (c. 1605–1670) of Kirby, Northamptonshire, and, like Sloane, Robinson, and Lister, he was a patron of John Ray. Sir John Hoskins (1634–1705), "a most learned virtuoso" and a vice president of the Royal Society, is also mentioned in connection with the club.[36] It has already been noted that Sherard and Vernon were both active supporters when in London.

A degree of flexibility toward membership in the club is apparent, as visitors were also welcomed to evenings at the Temple Coffee House to enjoy "pleasant and virtuous conversation," which was considered "the soul of good living."[37] Meetings were valuable opportunities to make one's mark with like-minded men, and a number of individuals wrote later asking to be remembered to "ye Gentlemen of ye Clubb."[38] Their backgrounds were diverse. Among them was Dr. Charles Preston (1660–1711) from Edinburgh; Thomas Tanner (1674–1735), a fellow of All Souls, Oxford, who became bishop of St. Asaph in 1732; and Gedeon Bonnivert (act. 1673–1703), a trooper in King William's army in Ireland. Bonnivert was a keen botanist and found time when not soldiering to correspond with Sloane and send observations to the Royal Society.[39] The Scottish surgeon and naturalist James Cuninghame (act. 1698–1709), one of Petiver's correspondents, wrote from China in 1700, also recalling the club.[40] And Krieg, newly arrived in Riga, was "so unhappy" at leaving London that he could not bring himself to say goodbye to Sloane and missed the company of his friends at the "Temple Coffe House."[41]

Further material is more conjectural but nevertheless worth considering. Archival and printed sources indicate that meetings were held on Friday evenings, which could help identify others who were connected with the club.[42] Stearns thought Nehemiah Grew (1641–1712) was a member.[43] He may have based this conclusion on Grew's request to Sloane: "if you intend to be at our Club this afternoon, pray, as you go will you call on me?" But the letter concerned was dated 17 March 1691/1692, a Thursday, and the botanist was perhaps referring to a different meeting place.[44] Petiver, on the other hand, could have been describing a Temple Coffee House gathering to the Danzig merchant Jacob Breyne (1637–1697) when he wrote on Saturday, 24 December 1692, of a "worthy meeting of Botanick friends" that took place the previous night.[45] Sloane, Robinson, and Doody—all actively involved—were present, while others attending included Dr. Leonard Plukenet (1642–1706), who was appointed Queen's Botanist in 1689; the apothecary and physician Samuel Dale (1659–1739), John Ray's neighbor in Essex; London botanist, Dr. James Newton (1639–1718); and the city merchant Charles du Bois, a collector of exotic rarities in his garden at Mitcham. Four years earlier, Plukenet, du Bois, Petiver, and Doody each paid Tancred Robinson two shillings and sixpence toward subscriptions for a new edition of Gabriel Grisley's *Viridarium lusitanum*, a work on plant species found in Portugal that was first published in 1661. The four men's receipt was dated 20 December 1689, a Friday, and it is tempting to wonder whether the occasion of this exchange was an early meeting of the club at the Temple Coffee House.[46]

Activities

The slightly later society that the botanist John Martyn (1699–1768) organized from 1721 until about 1726 drew up formal rules. These specified that members take turns to demonstrate plants, naming them and explaining their uses, an activity that reflected the men's largely medical concerns.[47] By contrast, evidence that the aims of the club at the Temple Coffee House were ever stipulated is slight, but its members were avid collectors of naturalia, both for their cabinets of curiosities and for scientific classification. Given the participants' wide range of interests and the sparseness of surviving data, the possibility that discussion turned to zoology, entomology, or even mineralogy cannot be ruled out. Manuscripts indicating the club's purpose, however, mention plants and botany.[48] The pursuit of collecting dried plant specimens for *horti sicci* united many of the Temple Coffee House fraternity, and most of the botanists connected with the group are listed as contributors in Dandy's account of the Sloane herbarium. In his introduction to the work, Spencer Savage, librarian to the Linnaean Society, summarized the motives of the seventeenth-century collector as obtaining "new plants somewhat in the same way as the ardent stamp-collector obtains new issues of postage stamps." The incomplete and sometimes chaotic arrangement of certain volumes in Sloane's vast collection perhaps led Savage to conclude somewhat sweepingly that often "no serious attempt" was made "to determine taxonomically the plants obtained—it sufficed that he had the rarest plants."[49]

In the early days of the club, members discussed the possibility of sponsoring their own collector. Edward Lhwyd (1660–1709), who longed to escape his tedious job as keeper of the Ashmolean Museum in Oxford, learned with excitement in the summer of 1690 that "the Botanic Club in London have entertain'd some thoughts of sending me to the Canarie Islands to make what Discoveries I can in Plants."[50] But England was at war with France, and the enemy fleet lay menacingly in the Channel, "masters at sea," as Evelyn vividly recounted, "with that power as gave terror to the whole nation, in daily expectation of a descent."[51] A plant-collecting expedition was out of the question. Lhwyd discovered that the club's plans amounted to no more than "wishes and talk, without any real & fix'd intention," until the return of "quiet and settled times."[52]

No indication has been found that any similar undertakings were ever carried out, although members continued to support plant-collecting endeavors abroad throughout the 1690s. The Temple Coffee House evenings provided valuable opportunities for overseas travelers to meet sponsors as well as to show off their discoveries on their homecoming. Before leaving Maryland in 1698, Vernon wrote to inform Sloane that he proposed to "bring every Fryday night a collection of plants to be discussed by . . . ye Honourable Club" on his return.[53] Vernon's specimens and those brought back by David Krieg were eventually divided between Sloane and several other botanists.[54] Hugh Jones (c. 1671–1702), who was sent as a minister to Maryland in 1696, was also familiar with the club. In order to prepare him for the post, Jones, Lhwyd's young deputy at the Oxford museum, was given extremely perfunctory training as both a clergyman and a naturalist. During this time, he was befriended by Petiver and other members of the London botanist circle, and perhaps attended Friday night meetings in the weeks before he left.[55] Jones certainly knew about the club, as a letter to him from Petiver concerning Vernon's impending return from the colony indicates: "Mr Vernon . . . says he will . . . bring over as many Plants, Shells, Insects, Fossils, Serpents &c. as will take our Botanick Club and Royal Society a Twelve month the lookin' over."[56] James Cuninghame, already mentioned in connection with the club, sent plant specimens from China to English botanists, including Sloane, Petiver, Plukenet, and du Bois.[57] Alexander Stuart (c. 1673–1742) collected for Sloane and du Bois when he was employed as a ship's surgeon in Asia; he, too, was familiar with the Temple Coffee House.[58]

The Demise of the Club

The club probably broke up in about 1706: after that date, references to it disappear from the naturalists' correspondence. Because Petiver had become more involved with the running of the Apothecaries' Garden in Chelsea by the end of that year, it is conceivable that he no longer had time to devote time to club matters.[59] He was still taking part in club activities in 1704, however, when Thomas Greenhill, a surgeon, wanted proposals distributed for a book he intended to publish and told him "if ye please to think of it at y^{r} Club, I believe several will subscribe."[60] Sherard, who left England in 1703 to take up a consularship with the Levant Company in Smyrna, sent good wishes to "friends at ye club" when writing to Sloane in 1706. But as he was no longer in close touch with events back in London, his evidence is unreliable.[61] Myddleton Massey continued to address letters for Sloane to the coffeehouse until 1708, but did not mention the club.[62]

Many of those closely associated with the group either died or left London during the first decade of the eighteenth century. Courten died in 1702 and Sir John Hoskins in 1705; Charles Hatton's death also occurred around that time. Plukenet and Doody followed them in 1706. Vernon returned to Cambridge in 1702, but it is possible that within a few years he, too, was dead.[63] Lister moved to Epsom in 1702 and Sherard remained abroad until late 1717.[64] As support for the meetings became ever more depleted, it seems likely that the club simply ceased to carry on.

Conclusion

Les Jessop's examination of works discussing the club at the Temple Coffee House has revealed that certain errors about its history were being perpetuated. He discovered, in particular, that activities were attributed to it without foundation and that the idea that there were forty members in 1691 was the result of a mistake in transcription. When the archival sources were reexamined for this appendix, only ten or eleven participants were definitely identified, together with various visitors. Circumstantial evidence, however, points to the possible involvement of a number of others from among the naturalists' circles.

No official papers relating to the club's business have been found, but it is apparent from the pages of *The Transactioneer* that proceedings were not as informal as previously thought. By 1700, at least, there were two officers, including a president, at the Friday evening meetings. The satirical comments indicate, furthermore, that the club was sufficiently well known to justify a mention in a popular work. They also show that Hans Sloane and James Petiver, leading figures in late seventeenth- and early eighteenth-century scientific circles, were crucial to its organization.

Details of subscribers to the club at the Temple Coffee House are sketchy, but unlike John Martyn's later botanical society, it was clearly a gathering of virtuosi, and not largely confined to the medical fraternity. Gentlemen, academics, and a variety of professional men were associated with it, and many were not exclusively interested in the study of plants; some members also belonged to the Royal Society.

William King referred simply to "The Temple-Coffee-House Club," but it was described in contemporary correspondence on several occasions as a "Botanic" club. The amount of evidence about the club's activities is small, given that it existed for about fifteen years, but plants were plainly a leading topic for discussion. It flourished, however, during a period when there was enormous enthusiasm for collecting all kinds of natural objects, and the possibility that other areas of natural history were also discussed cannot be ruled out. Even though no documents have thus far been found to indicate that the club ever succeeded in financing a botanist or a naturalist to go abroad, the Friday-night gatherings did provide a congenial venue for collectors travelling overseas to meet their supporters and their clientele.

In light of the scattered and fragmentary nature of the documentation, it remains wise to exercise caution in reaching conclusions about the activities of the club at the Temple Coffee House. The gradual accumulation of material about its meetings nevertheless confirms that the club was both a significant focal point for scientific virtuosi and a fertile environment for the proliferation of botanical knowledge in the late seventeenth and early eighteenth centuries.

NOTES TO APPENDIX

This material was originally published in *Archives of Natural History* 33, part 1 (April 2006). It is republished here in slightly reedited form by permission of the Society for the History of Natural History.

I thank Mark Laird for his generous help in the preparation of this paper, and for bringing Les Jessop's article about the club to my attention. Thanks also to Dr. Jane Ridley of the University of Buckingham for her valuable advice.

1. Quoted from Peter Clark, *British Clubs and Societies, 1580–1800: The Origins of an Associational World* (Oxford, 2000), 1, citing John Macky, *A Journey through England*, 2 vols. (1714; 4th ed., London, 1724), 1:269.

2. These details are taken from Arthur MacGregor, ed., "The Life, Character and Career of Sir Hans Sloane," in *Sir Hans Sloane: Collector, Scientist, Antiquary* (London, 1994), 11–35.

3. Petiver's biographical details are taken from Raymond Phineas Stearns, "James Petiver: Promoter of Natural Science, c. 1663–1718," in *Proceedings of the American Antiquarian Society* 62 (1952): 243–365.

4. Samuel Johnson, *Dictionary of the English Language*, 2 vols. (London, 1755), cited in Clark 2000, 10.

5. Quoted from *The Transactioneer, with some of his Philosophical Fancies* (London, 1700), 33; reprinted in *The Original Works of William King, LL.D., with Historical Notes, and Memoirs of the Author*, ed. John Nichols, 3 vols. (London, 1776).

6. Quoted from *The Transactioneer* 1700, 34.

7. George Pasti, Jr., "Consul Sherard: Amateur Botanist and Patron of Learning, 1659–1728" (Ph.D. diss., University of Illinois, 1950), 48–55; quoted in Stearns 1952, 253.

8. Quoted from Clark 2000, 252.

9. Pasti 1950, 49; quoted from Hans Sloane, *A Voyage to the Islands Madera, Barbados, Nieves, S. Christophers and Jamaica*, 2 vols. (London 1707–1725), 1: preface.

10. Clark 2000, 251.

11. Quoted from Walter H. G. Armytage, "The Royal Society and the Apothecaries, 1660–1722," in *Royal Society Notes and Records* 11 (1954–1955): 28, citing Robert W. T. Gunther, ed., *Further Correspondence of John Ray* (London, 1928), 279.

12. Quoted from *The Transactioneer* 1700, 34.

13. Quoted from Stearns 1952, 243.

14. Quoted from Stearns 1952, 260, citing James Petiver, *Musei Petiveriani. Centuria (-decima), rariora naturae continens* (London, 1695–1703).

15. Vernon to Sloane, 24 July 1698, British Library, Sloane MSS 4037, ff. 102–103. See James E. Dandy, ed., *The Sloane Herbarium* (London, 1958), 226.

16. Sherard to Sloane, Paris, 7 October 1699, Sloane MSS 4037, ff. 340–341.

17. David Elliston Allen, *The Naturalist in Britain: A Social History* (London, 1976), 10.

18. Clark 2000, 9.

19. Blanche Henrey, *British Botanical and Horticultural Literature before 1800*, 3 vols. (London, 1975), 2:78; Allen 1976, 10–11.

20. Du Bois' date of birth is usually given as c. 1656 (Ray Desmond and Christine Ellwood, *Dictionary of British and Irish Botanists and Horticulturists* [London, 1994]), calculated from the inscription on his tombstone at Mitcham, Surrey, which was engraved "20 Oct. A.D. 1740, aetat. 83." The revised date of 1658 was taken from the baptismal records of the parish of St. Mary the Virgin Aldermanbury, 1538–1722.

21. Les Jessop, "The Club at the Temple Coffee House: Facts and Supposition," in *Archives of Natural History* 16 (1989): 263–274. I am grateful to Mark Laird for this reference. Laird discusses the club in "Exotics and Botanical Illustration," in *Sir John Vanburgh and Landscape Architecture in Baroque England, 1690–1730*, ed. Christopher Ridgway and Robert Williams (Stroud, 2000), 93–113.

22. Jessop 1989, 271; *The Transactioneer* 1700, 34.

23. Sherard to Sloane, 11 June 1692, Sloane MSS 4036, f. 124; George Preston to Sloane, 17 March [?] 1698/1699, Sloane MSS 4037, f. 126; and Gedeon Bonnivert to Sloane, 5 August 1703, Sloane MSS 4039, f. 167, for example, are among many instances in which the club is only mentioned. On the other hand, Edward Lhwyd to Martin Lister, 1 July 1690, Lister MSS 36, f. 11, Bodleian Library, Oxford, referred to "the Botanic Club of London"; James Cuninghame to James Petiver, 20 December 1700, Sloane MSS 3321, f. 65, "the Botanic Societie"; and Petiver to Hugh Jones, 6 October 1698, Sloane MSS 3333, ff. 170–171, wrote of the "Botanick Club."

24. Jessop 1989, 274. Bryant Lillywhite, *London Coffee Houses* (London, 1963), 568; 1323, Temple Coffee House in Clifford's Inn, Fleet Street; 1324, Temple Coffee House "near the Temple Fleet Street" Temple Bar; 1325, Temple Coffee House, Devreux [i.e., Devereux] Court; 1326, Temple Exchange Coffee House.

25. George Preston to Petiver, n.d., Sloane MSS 4067, f. 114.

26. Vernon to Sloane, 15 February 1698/1699, Sloane MSS 4037, ff. 209–210; Dr. Richard Myddleton Massey to Sloane, 4 September 1706, Sloane MSS 4040, f. 215, addressed the letter "to be left at the Temple Coffee-house in Devreux Court, High Temple Bar, London." Massey used the address frequently when writing to Sloane, and for his own correspondence when in London, for example, on 23 February 1703/1704, he told Lhwyd: "When ye write next pray to direct to be left at the Temple Coffee-house in Devreux Court for yr obliged servant." Bodleian, Ashmole MSS 1816, f. 346.

27. Aytoun Ellis, *The Penny Universities: A History of the Coffee-Houses* (London, 1956), 82.

28. Quoted from Jessop 1989, 268, citing a draft letter from W. C. [William Courten or Charleton] to Mr. Sherwood [Sherard], 11 May 1691, Sloane MSS 3961, f. 41, "I wisht (wish we could have had [crossed out]) yow at our club ye other night for an Hower or 2 to have had a sight of a curious book of plants (of ye Cape of good Hope) in miniature presented by ye States to ye Bishop of London. There were to ye number of 40—& as (ye learned said not [crossed out]) our chief Botanists said most of ym not described." Pasti misread the letter to mean that "in the spring of 1691 . . . forty members were present at one of their regular Friday meetings." This claim was innocently repeated by Stearns (1952), 254; Henrey (1975) 2:78; and Allen (1976), 10. Allen drew attention to Jessop's findings in *Naturalists and Society: The Culture of Natural History in Britain, 1700–1900* (Aldershot, 2001).

29. Sloane MSS 5286 includes copies of these illustrations.

30. Quoted from Jessop 1989, 273.

31. Quoted from Stearns 1952, 253.

32. Quoted from Jessop 1989, 269–270, citing Sloane MSS 3336, f. 11, 7 July 1707.

33. Quoted from Jessop 1989, 267, citing Sloane MSS 4067, f. 7, n.d.

34. MacGregor 1994, 19.

35. George F. Frick, James L. Reveal, C. Rose Broome, and Melvin L. Brown, "Botanical explorations and discoveries in colonial Maryland, 1688 to 1753," in *Huntia* 7 (1987): 24; Vernon to Sloane, 28 December [1698], Sloane MSS 4061, ff. 233–234.

36. Pasti 1950, 48, citing Sherard to Sloane, 11 June 1692, Sloane MSS 4036, f. 126, "My humble service to all at our clubb particularly to Mr Charleton Dr Robinson and Mr Doody"; and 12 July 1699, Sloane MSS 4037, f. 299, "all Gentlemen that use to come on Fryday night to the Temple Coffe-house and to Mr Charleton address." Pasti 1950, 64, also cites Sherard to Sloane, 25 October 1694, Sloane MSS 4036, f. 186; and 75 cites Sherard to Sloane, 9 May 1698, Sloane MSS 4037, ff. 64–65. Esmond Samuel de Beer, ed., *The Diary of John Evelyn*, 6 vols. (1951; reprint Oxford, 2000) 4:296 makes this reference to Hoskins' qualities.

37. Krieg to Petiver, n.d., Sloane MSS 4067, f. 7.

38. Michael Hunter, *Science and Society in Restoration England* (1981; reprint Aldershot, 1992), 47.

39. Charles Preston to Sloane, 11 September 1697, Sloane MSS 4036, f. 349; Tanner to Martin Lister, 13 November 1699; Bodleian, Lister MSS 36, f. 237; Bonnivert to Sloane, 4 September 1702, Sloane MSS 4039, f. 21. Royal Society, Journal Book Copy 9:3, 28 October 1696; Journal Book Copy 10:17, 27 January 1702/1703.

40. Cuninghame to Petiver, Chusan, 20 December 1700, Sloane MSS 3321, f. 65.

41. Krieg to Sloane, 12 July 1699, Sloane MSS 4066, f. 299.

42. Friday is, of course, mentioned in *The Transactioneer*, but also in Vernon to Sloane, 24 July 1698, Sloane MSS 4037, ff. 102–103; Sherard to Sloane, 20 September 1698, Sloane MSS 4037, f. 123; and Krieg to Sloane, 12 July 1699, Sloane MSS 4037, f. 299.

43. Stearns 1952, 254.

44. Grew to Sloane, 17 March 1691/1692, Sloane MSS 4036, f. 113. For the day of the week for this date, see John J. Bond, *Handy Book of Rules and Tables for Verifying Dates of Historical Events and of Public and Private Documents* (London, 1866).

45. Petiver to Jacob Breyne, Christmas Eve, 1692, Sloane MSS 4067, f. 81. Jessop, 1989, 271.

46. Sloane MSS 4019, f. 164. Another edition of Grisley's work was not published until c. 1714.

47. Allen 1976, 12–13. Analysis of the minutes reveals that one-third of the twenty-three members listed became apothecaries, one-third physicians, and most of the remainder either surgeons or surgeon-apothecaries.

48. Jessop 1989, 273.

49. Quoted from Dandy 1958, 10.

50. Lhwyd to Lister, 1 July 1690, Bodleian, Lister MSS 36, f. 11.

51. De Beer 1951, 5:29, 17 July 1690.

52. Lister to Lhwyd, 10 July 1690, Bodleian, Ashmole MSS 1816, f. 81.

53. Vernon to Sloane, 24 July 1698, Sloane MSS 4037, ff. 102–103.

54. Dandy 1958, 151 and 226; Sloane, Plukenet, Petiver, and Bobart are mentioned. Hermia Newman Clokie, *An Account of the Herbaria of the Department of Botany in the University of Oxford* (Oxford, 1964), 79 and 86; Clokie notes that there are a few of Krieg's and Vernon's specimens in the Sherardian Herbarium, and a great many in Charles du Bois' collection.

55. Frick et al. 1987, 11.

56. Quoted from Frick et al. 1987, 20, citing Sloane MSS 3333, ff. 170–171.

57. Dandy 1958, 117–122; *Oxford Dictionary of National Biography* (Oxford, 2004; hereafter *ODNB*), 14:653.

58. Alexander Stuart to Sloane, n.d., Sloane MSS 4061, f. 144.

59. Penelope Hunting, *A History of the Society of Apothecaries* (London, 1998), 124 and 277, n. 39.

60. Thomas Greenhill to Petiver, 8 June 1704, Sloane MSS 3321, f. 142.

61. Sherard to Sloane, 29 June 1706, Sloane MSS 4040, f. 187.

62. Sloane MSS 4040, ff. 214–215, 4 September 1706; MSS 4041, f. 134, 5 May 1708.

63. John Ray, who was for years such a brilliant inspiration to natural historians and to Sloane and his friends in particular, died in 1705. Vernon's death is usually given as c. 1715 (Desmond and Ellwood 1994), but Petiver wrote to Joseph Pitton de Tournefort on 29 June 1706, "Dr Plukenett is lately dead, and is Mr Ray and Mr Vernon some time since," which suggests his death may have occurred earlier. See Sloane MSS 3335, f. 22.

64. *ODNB*, 33:986–987; Pasti 1950, 188.

Contributors

Lina Bolzoni is professor of Italian literature at the Scuola Normale Superiore in Pisa, where she directs the Centro per l'Elaborazione Informatica di Testi e Immagini nella Tradizione Letteraria. She is the author of many books and articles on Italian Renaissance literature and the art of memory and the editor of volumes of collected works by Giulio Camillo and Tommaso Campanella. Her recent publications in English include *The Web of Images: Vernacular Preaching from Its Origins to St. Bernardino da Siena* (2003) and *The Gallery of Memory: Literary and Iconographic Models in the Age of Printing* (2001).

Jesús Carrillo Castillo is lecturer in art history at the Universidad Autónoma of Madrid. His publications on the representation of the natural world and technology in early modern Spain include *Naturaleza e imperio: la representación del mundo natural en la "Historia general y natural de las Indias" de Gonzalo Fernández de Oviedo* (2004) and *Tecnología e imperio: Turriano, Lastanosa, Herrera, Ayanz; ingenios y leyendas del siglo de Oro* (with Nicolás García Tapia; 2003). His areas of research also include the relationship among technology, art, and power in contemporary culture, on which he wrote *Arte en la red* (2003), and the role of art institutions in contemporary Spain. He is the editor of *Desacuerdos: sobre arte, políticas y esfera pública en el estado español,* volumes 1–3 (2003–2006).

Erik A. de Jong teaches history and theory of landscape architecture at Wageningen University and Research Center, the Netherlands. He has lectured and published widely in Europe and the United States. His publications include *Nature and Art: Dutch Garden and Landscape Architecture, 1650–1740* (2001) and *Aardse Paradijzen: de tuin in de Nederlandse Kunst* (two volumes, 1996, 1999). His most recent book (edited with Erika Schmidt and Brigitt Sigel) is *Der Garten: Ein Ort des Wandels; Perspektiven für die Denkmalpflege* (2006). He is working on a book project entitled "Landscape in Translation: The Dutch and Their Designed Landscapes." He is a consultant on landscape history and restoration projects and project leader for the building of a new natural history museum in Amsterdam.

Hans W. Hubert is professor of art history and director of the Kunstgeschichtliches Institut of Albert-Ludwigs-Universität Freiburg. His main fields of research are Italian architecture and art from medieval to early modern times and their relationship to art theory. His work focuses especially on the arts and nature and the arts and the four elements. He spent many years at the Kunsthistorisches Institut in Florence, where he developed his study on the Casino della Topaia and the panel paintings of Bartolomeo Bimbi.

Mark Laird is senior lecturer in the history of landscape architecture in the Graduate School of Design, Harvard University. He is the author of *The Flowering of the Landscape Garden: English Pleasure Grounds, 1720–1800* (1999). He is a consultant in historic planting preservation with extensive projects in Europe and North America. For his replanting work at Painshill Park, Surrey, England, he was joint recipient in 1998 of a Europa Nostra medal for exemplary restoration. He lives in Toronto, where he has taught landscape history at the University of Toronto.

Amy R. W. Meyers is director of the Yale Center for British Art and formerly curator of American art at the Huntington Library, Art Collections, and Botanical Gardens. Her numerous publications on the history of natural history illustration include *The Culture of Nature: Art and Science in Philadelphia, 1740–1840* (2008), *Empire's Nature: Mark Catesby's New World Vision* (coedited with Margaret Beck Pritchard; 1998), *Art and Science in America: Issues of Representation* (1998), and *Mark Catesby's Natural History of America: The Watercolours from the Royal Library, Windsor Castle* (with Henrietta McBurney; 1997).

Janice Neri is assistant professor of art history and visual culture at Boise State University. Her research and publications address the ways in which concepts of truth and accuracy in visual images were developed in media ranging from scientific illustration to the decorative arts during the period 1500–1800. She is currently working on a book entitled "The Insect and the Image: Visualizing Nature in Early Modern Europe, 1500–1700."

Therese O'Malley is associate dean of the Center for Advanced Studies in the Visual Arts, National Gallery of Art, where she oversees the Center's programs of special meetings and publications. She has published and lectured widely on the history of landscape architecture and garden design, primarily in the eighteenth and nineteenth centuries, concentrating on the transatlantic exchange of ideas, plants, and people. Her recent publications concern the professionalization of landscape architecture, the design of botanical gardens, and the history of glasshouse architecture. Her current projects include a reference work entitled *Keywords in American Landscape Design,* in press with Yale University Press, and a book on American gardens, to be published in the Norton/Library of Congress Visual Sourcebooks in Architecture, Design, and Engineering.

Margaret Riley has a postgraduate diploma in the conservation of historic landscapes, parks, and gardens from the Architectural Association School of Architecture in London and is currently working on a Ph.D. at the University of Buckingham. Her thesis compares the scientific lives of the late seventeenth- and early eighteenth-century English botanists Charles du Bois and the brothers William and James Sherard.

Pamela H. Smith is professor of history at Columbia University and the author of *The Body of the Artisan: Art and Experience in the Scientific Revolution* (Chicago, 2004) and *The Business of Alchemy: Science and Culture in the Holy Roman Empire* (Princeton, 1994). She has published numerous articles on artisanal knowledge and culture in early modern Europe, and in current research she is attempting to reconstruct the vernacular knowledge of early modern European metalworkers from a variety of disciplinary perspectives.

Claudia Swan is associate professor of art history at Northwestern University, where she teaches early modern northern European art. Her research focuses on the relationship between art and science in early modern Holland in particular. She is the author of *Art, Science, and Witchcraft in Early Modern Holland: Jacques de Gheyn II (1565–1625)* (2005) and *The Clutius Botanical Watercolors: Plants and Flowers of the Renaissance* (1998) and is coeditor, with Londa

Schiebinger, of *Colonial Botany: Science, Commerce, and Politics in the Early Modern World* (2004).

Lucia Tongiorgi Tomasi is vice rector of the Università di Pisa and professor of the history of art at the same university. She has held appointments as a member of the committee of senior fellows in landscape architecture at Dumbarton Oaks, Trustees of Harvard University, Washington, D.C., and as Isaiah Berlin Visiting Professor in the department of history of art, University of Oxford. Her principal field of research is relationships between art and natural sciences from the sixteenth through the eighteenth century, with particular attention to naturalistic illustration, still life, and garden history. She has lectured at universities in Europe, the United States, and Brazil and presented papers at international symposia. She has curated exhibitions at the Gabinetto Disegni e Stampe, Galleria degli Uffizi, and was cocurator, with Gretchen A. Hirschauer, of *The Flowering of Florence: Botanical Art for the Medici* (National Gallery of Art, Washington, 2002).

Alessandro Tosi is professor of art history at the Università di Pisa. His main fields of research are the history of the visual arts and the relationship between art and science from the sixteenth to the twentieth century. Among his numerous publications in these areas, he is the author of *Inventare la realtà: Giuseppe Zocchi e la Toscana del Settecento* (1997) and editor of *Ulisse Aldrovandi e la Toscana: carteggio e testimonianze documentarie* (1989). He is also coeditor, with Marco Beretta, of *Linnaeus in Italy: The Spread of a Revolution in Botany* (2007).

Alain Touwaide is historian of sciences in the department of botany of the National Museum of Natural History, Smithsonian Institution. He has extensively researched the history of botany, medicine, and pharmacology in the Mediterranean world from antiquity to the Renaissance and has published widely on these topics. In the field of botanical illustration, he has recently coedited, with Jean Givens and Karen Reeds, *Visualizing Medieval Medicine and Natural History, 1200–1500,* and written, in collaboration with Arsenio Ferraces and Jose Manuel Canas, a volume of commentary on the *Herbolarium* contained in the manuscript Lucca 296 (Biblioteca Governativa, Lucca). Currently he directs a research program on medicinal plants of antiquity and is preparing a book entitled "A Flora of Classical Antiquity."

Index

Italicized page references indicate illustrations.

A

B

C

D

Studies in the History of Art

Published by the National Gallery of Art, Washington

This series includes Studies in the History of Art, collected papers on objects in the Gallery's collections and other art-historical studies (formerly Report and Studies in the History of Art); Monograph Series I, a catalogue of stained glass in the United States; Monograph Series II, on conservation topics; and Symposium Papers (formerly Symposium Series), the proceedings of symposia sponsored by the Center for Advanced Study in the Visual Arts.

[1] *Report and Studies in the History of Art,* 1967

[2] *Report and Studies in the History of Art,* 1968

[3] *Report and Studies in the History of Art,* 1969 [In 1970 the National Gallery of Art's annual report became a separate publication.]

[4] *Studies in the History of Art,* 1972

[5] *Studies in the History of Art,* 1973

[The first five volumes are unnumbered.]

6 *Studies in the History of Art,* 1974

7 *Studies in the History of Art,* 1975

8 *Studies in the History of Art,* 1978

9 *Studies in the History of Art,* 1980

10 *Macedonia and Greece in Late Classical and Early Hellenistic Times,* edited by Beryl Barr-Sharrar and Eugene N. Borza. Symposium Series I, 1982

11 *Figures of Thought: El Greco as Interpreter of History, Tradition, and Ideas,* edited by Jonathan Brown, 1982

12 *Studies in the History of Art,* 1982

13 *El Greco: Italy and Spain,* edited by Jonathan Brown and José Manuel Pita Andrade. Symposium Series II, 1984

14 *Claude Lorrain, 1600–1682: A Symposium,* edited by Pamela Askew. Symposium Series III, 1984

15 *Stained Glass before 1700 in American Collections: New England and New York (Corpus Vitrearum Checklist I),* compiled by Madeline H. Caviness et al. Monograph Series I, 1985

16 *Pictorial Narrative in Antiquity and the Middle Ages,* edited by Herbert L. Kessler and Marianna Shreve Simpson. Symposium Series IV, 1985

17 *Raphael before Rome,* edited by James Beck. Symposium Series V, 1986

18 *Studies in the History of Art,* 1985

19 *James McNeill Whistler: A Reexamination,* edited by Ruth E. Fine. Symposium Papers VI, 1987

20 *Retaining the Original: Multiple Originals, Copies, and Reproductions.* Symposium Papers VII, 1989

21 *Italian Medals,* edited by J. Graham Pollard. Symposium Papers VIII, 1987

22 *Italian Plaquettes,* edited by Alison Luchs. Symposium Papers IX, 1989

23 *Stained Glass before 1700 in American Collections: Mid-Atlantic and Southeastern Seaboard States (Corpus Vitrearum Checklist II),* compiled by Madeline H. Caviness et al. Monograph Series I, 1987

24 *Studies in the History of Art,* 1990

25 *The Fashioning and Functioning of the British Country House,* edited by Gervase Jackson-Stops et al. Symposium Papers X, 1989

26 *Winslow Homer,* edited by Nicolai Cikovsky Jr. Symposium Papers XI, 1990

27 *Cultural Differentiation and Cultural Identity in the Visual Arts,* edited by Susan J. Barnes and Walter S. Melion. Symposium Papers XII, 1989

28 *Stained Glass before 1700 in American Collections: Midwestern and Western States (Corpus Vitrearum Checklist III),* compiled by Madeline H. Caviness et al. Monograph Series I, 1989

29 *Nationalism in the Visual Arts,* edited by Richard A. Etlin. Symposium Papers XIII, 1991

30 *The Mall in Washington, 1791–1991,* edited by Richard Longstreth. Symposium Papers XIV, 1991, 2002

31 *Urban Form and Meaning in South Asia: The Shaping of Cities from Prehistoric to Precolonial Times,* edited by Howard Spodek and Doris Meth Srinivasan. Symposium Papers XV, 1993

32 *New Perspectives in Early Greek Art,* edited by Diana Buitron-Oliver. Symposium Papers XVI, 1991

33 *Michelangelo Drawings,* edited by Craig Hugh Smyth. Symposium Papers XVII, 1992

34 *Art and Power in Seventeenth-Century Sweden,* edited by Michael Conforti and Michael Metcalf. Symposium Papers XVIII (withdrawn)

35 *The Architectural Historian in America,* edited by Elisabeth Blair MacDougall. Symposium Papers XIX, 1990

36 *The Pastoral Landscape,* edited by John Dixon Hunt. Symposium Papers XX, 1992

37 *American Art around 1900,* edited by Doreen Bolger and Nicolai Cikovsky Jr. Symposium Papers XXI, 1990

38 *The Artist's Workshop,* edited by Peter M. Lukehart. Symposium Papers XXII, 1993

39 *Stained Glass before 1700 in American Collections: Silver-Stained Roundels and Unipartite Panels (Corpus Vitrearum Checklist IV),* compiled by Timothy B. Husband. Monograph Series I, 1991

40 *The Feast of the Gods: Conservation, Examination, and Interpretation,* by David Bull and Joyce Plesters. Monograph Series II, 1990

41 *Conservation Research.* Monograph Series II, 1993

42 *Conservation Research: Studies of Fifteenth- to Nineteenth-Century Tapestry,* edited by Lotus Stack. Monograph Series II, 1993

43 Eius Virtutis Studiosi: *Classical and Postclassical Studies in Memory of Frank Edward Brown,* edited by Russell T. Scott and Ann Reynolds Scott. Symposium Papers XXIII, 1993

44 *Intellectual Life at the Court of Frederick II Hohenstaufen,* edited by William Tronzo. Symposium Papers XXIV, 1994

45 *Titian 500,* edited by Joseph Manca. Symposium Papers XXV, 1994

46 *Van Dyck 350,* edited by Susan J. Barnes and Arthur K. Wheelock Jr. Symposium Papers XXVI, 1994

47 *The Formation of National Collections of Art and Archaeology,* edited by Gwendolyn Wright. Symposium Papers XXVII, 1996

48 *Piero della Francesca and His Legacy,* edited by Marilyn Aronberg Lavin. Symposium Papers XXVIII, 1995

49 *The Interpretation of Architectural Sculpture in Greece and Rome,* edited by Diana Buitron-Oliver. Symposium Papers XXIX, 1997

50 *Federal Buildings in Context: The Role of Design Review,* edited by J. Carter Brown. Symposium Papers XXX, 1995

51 *Conservation Research 1995.* Monograph Series II, 1995

52 *Saint-Porchaire Ceramics,* edited by Daphne Barbour and Shelley Sturman. Monograph Series II, 1996

53 *Imagining Modern German Culture, 1889–1910,* edited by Françoise Forster-Hahn. Symposium Papers XXXI, 1996

54 *Engraved Gems: Survivals and Revivals,* edited by Clifford Malcolm Brown. Symposium Papers XXXII, 1997

55 *Vermeer Studies,* edited by Ivan Gaskell and Michiel Jonker. Symposium Papers XXXIII, 1998

56 *The Art of Ancient Spectacle,* edited by Bettina Bergmann and Christine Kondoleon. Symposium Papers XXXIV, 1999

57 *Conservation Research 1996/1997.* Monograph Series II, 1997

58 *Olmec Art and Archaeology in Mesoamerica,* edited by John E. Clark and Mary E. Pye. Symposium Papers XXXV, 2000, 2006

59 *The Treatise on Perspective: Published and Unpublished,* edited by Lyle Massey. Symposium Papers XXXVI, 2003

60 *Hans Holbein: Paintings, Prints, and Reception,* edited by Mark Roskill and John Oliver Hand. Symposium Papers XXXVII, 2001

61 *Italian Panel Painting of the Duecento and Trecento,* edited by Victor M. Schmidt. Symposium Papers XXXVIII, 2002

62 *Small Bronzes in the Renaissance,* edited by Debra Pincus. Symposium Papers XXXIX, 2001

63 *Moche Art and Archaeology in Ancient Peru,* edited by Joanne Pillsbury. Symposium Papers XL, 2001, 2005

64 *Large Bronzes in the Renaissance,* edited by Peta Motture. Symposium Papers XLI, 2003

65 *Tilman Riemenschneider, c. 1460–1531,* edited by Julien Chapuis. Symposium Papers XLII, 2004

66 *Circa 1700: Baroque Architecture in Europe and the Americas,* edited by Henry A. Millon. Symposium Papers XLIII, 2005

68 *Nationalism and French Visual Culture, 1870–1914,* edited by June Hargrove and Neil McWilliam. Symposium Papers XLV, 2005

69 *The Art of Natural History: Illustrated Treatises and Botanical Paintings, 1400–1850,* edited by Therese O'Malley and Amy R. W. Meyers. Symposium Papers XLVI, 2008

70 *Collecting Sculpture in Early Modern Europe,* edited by Nicholas Penny and Eike D. Schmidt. Symposium Papers XLVII, 2008

72 *French Genre Painting in the Eighteenth Century,* edited by Philip Conisbee. Symposium Papers XLIX, 2007

Forthcoming

71 *Romare Bearden, American Modernist,* edited by Ruth Fine and Jacqueline Francis. Symposium Papers XLVIII

73 *The East Building in Perspective,* edited by Anthony Alofsin. Symposium Papers L

74 *Dialogues in Art History,* edited by Elizabeth Cropper. Symposium Papers LI

75 *The Woodcut in Fifteenth-Century Europe,* edited by Peter Parshall. Symposium Papers LII

76 *Orsanmichele and the History and Preservation of the Civic Monument,* edited by Carl Brandon Strehlke. Symposium Papers LIII

77 *Art and the Early Photographic Album,* edited by Stephen Bann. Symposium Papers LIV